FUNDAMENTALS OF MICROWAVE PHOTONICS

WILEY SERIES IN MICROWAVE AND OPTICAL ENGINEERING

KAI CHANG, Editor
Texas A&M University

A complete list of the titles in this series appears at the end of this volume.

FUNDAMENTALS OF MICROWAVE PHOTONICS

VINCENT J. URICK Jr.
JASON D. McKINNEY
KEITH J. WILLIAMS

WILEY

Published by John Wiley & Sons, Inc., Hoboken, New Jersey
Published simultaneously in Canada

Library of Congress Cataloging-in-Publication Data:

Urick, Vincent J., Jr. (Vincent Jude), 1979-
Fundamentals of microwave photonics / Vincent J. Urick Jr., Jason D. McKinney, Keith J. Williams.
pages cm – (Wiley series in microwave and optical engineering)
Includes bibliographical references and index.
ISBN 978-1-118-29320-1 (cloth)
1. Microwave communication systems. 2. Photonics. I. McKinney, Jason D. (Jason Dwight), 1975- II. Williams, Keith J. (Keith Jake), 1964- III. Title.
TK5103.4833.U75 2015
621.36′5–dc23

2014040933

Cover Image courtesy of iStockphoto © jmcreation.

Typeset in 11/13pt TimesTenLTStd by Laserwords Private Limited, Chennai, India

10 9 8 7 6 5 4 3 2 1

1 2015

For Cindy, Amanda and Vicki

CONTENTS

PREFACE

This volume provides what we believe to be a thorough treatment of the microwave photonics field, sometimes referred to as RF or analog photonics. The intended audience ranges from an advanced undergraduate student in engineering or physics to experts in the field. The treatment is fundamental in nature and could be used in an advanced undergraduate or graduate-level course to introduce students to microwave photonics. Although a problem set is not included, there are instances throughout where an inventive instructor could devise assignments. It is our hope that seasoned veterans of the field will find this book most useful for a variety of reasons. We have tried to provide as much of the basic underlying physics as is possible in a work of this size. Sometimes, this information can be lost in a field as applied as microwave photonics. Plots that give bounds on performance for a variety of scenarios abound. A thorough list of references is provided for each chapter, including original sources where applicable. Design equations in easy-to-use forms are provided throughout and are intended for quick reference. Indeed, we plan to keep this volume readily accessible in the laboratory, in the field, or during design meetings.

We intended for this book to flow continuously from the first to the last page and believe we have succeeded in this endeavor. Beginners in the field are encouraged to read continuously, as the later chapters build on foundations laid in the earlier ones. Those more experienced in the field should find that navigation of the individual chapters is readily achievable. Chapter 1 gives an introduction to microwave photonics and stands on its own, pointing to later chapters where more detail is provided. Chapter 2 describes the radio-frequency metrics that are most important to quantifying performance of microwave photonics systems and is largely divorced from optics. Chapters 3 through 5 provide fundamental treatments of noise, distortion, propagation, and fiber nonlinearities as they pertain to microwave photonics. These three chapters do not concentrate on any single modulation mechanism but rather are intended to provide a generalized treatment. Specific modulation and corresponding demodulation techniques are covered in Chapters 6 through 8, using the material in the previous four chapters. In Chapter 6, intensity modulation with direct detection employing an external Mach–Zehnder modulator is detailed. This technique is arguably the most prevalent today and therefore receives the most thorough treatment. Phase modulation is covered in Chapter 7 but with slightly less detail. Complete but relatively brief analyses of numerous other modulation formats are conducted in Chapter 8. Chapter 9 is concentrated on high power photodetectors. System and subsystem applications are covered in Chapter 10, which also describes some of the present trends in the field.

We ourselves acquired a more complete knowledge of many topics while writing this book and the work inspired many new concepts. We sincerely hope the same is true for all who pick up this volume.

Vincent Urick
Jason McKinney
Keith Williams
Washington, DC, April 2014

ACKNOWLEDGMENTS

This book was written as a private work, and as such, the opinions expressed in this book are those of the authors and do not reflect the official position of the US Naval Research Laboratory (NRL), the US Navy, or the US Government. That being said, this work would not have been possible without the support of NRL throughout our careers. The work environment provided at NRL has made it possible to make steady progress in developing a thorough understanding, both experimental and theoretical, of microwave photonics technology. This would not have happened without the support of the management at NRL, specifically the Superintendents and Branch Heads who were instrumental in supporting our ability to make progress in this important technology area. Those individuals include Dr. Francis Klemm, Dr. Thomas Giallorenzi, Dr. John Montgomery, Dr. Joseph Weller, Dr. Ronald Esman, Mr. Michael Monsma, and Dr. Don Northam. We would also like to acknowledge those staff at NRL, both past and present, who have contributed to the development of microwave photonics.

We are indebted to the countless colleagues and collaborators that we have had the pleasure to work with over the years. The citations in the text name many of those who have inspired us but some are deserving of special mention. Firstly, Dr. Frank Bucholtz at NRL has provided significant insight into the analysis and understanding of analog optical links. His work is cited where applicable but his contributions to our progress go well beyond those instances. Professor Nicholas Frigo of the US Naval Academy Physics Department assisted with the development of sections pertaining to polarization effects. Mr. Carl Villarruel, NRL (retired), has spent countless hours discussing the technical fine points of microwave photonics with us, particularly in areas concerning optical fiber effects. Dr. Preetpaul Devgan of the US Air Force Research Laboratory stimulated important concepts pertaining to modulation formats. Dr. Andrew Kowalevicz from Raytheon Company inspired useful viewpoints on optical fields in various media. Dr. Marcel Pruessner at the NRL provided valuable feedback on silicon integration for microwave photonics applications. Dr. Olukayode Okusaga, US Army Research Laboratory, gave insight into the subtitles of optoelectronic oscillators. Mr. Bill Jacobs, US Space and Naval Warfare Systems Command, provided alternative views on applications of microwave photonics and also assisted with professional responsibilities while this book was being written. We acknowledge Dr. Thomas Clark Jr. at Johns Hopkins Applied Physics Laboratory for discussing aspects of multioctave millimeter-wave photonics and signal processing. Finally, we wish to thank all the ambitious students we have instructed and those we have mentored for allowing us to pass on what we have learned. It is in those instances when one realizes that you don't truly understand something until you can teach it to someone else, a concept that was reinforced tenfold while writing this book.

Beyond the mainly professional acknowledgements mentioned previously, there are numerous individuals who have influenced us in profound ways. This work would have never come to be if it weren't for our parents and families. Our wives and children were supportive during the writing process, making numerous concessions along the way. We are forever grateful to our parents, Vincent Urick Sr., Susanne Urick, Dwight McKinney, Deborah McKinney, and Gertrude Williams. They nurtured intellectual curiosity in us and instilled a work ethic that was required to complete this book. Paul Urick, an old-time farmer from Pennsylvania, and Norman Zlotorzynski, a kind man who survived Omaha Beach in 1944, always provided inspiration when it was needed most. They both passed away while this book was being written and would like to have seen the completed work.

CHAPTER 1

INTRODUCTION

Microwave photonics is a multidisciplinary field that encompasses optical, microwave, and electrical engineering. The microwave photonics field must therefore span frequencies of below 1 kHz in the radio-frequency (RF) domain to frequencies of hundreds of terahertz associated with the optical domain. The field originated from the need to solve increasingly complex engineering problems when radio engineers ventured outside their discipline to the optical domain in search of new capabilities. Generally, the field is applied in nature stemming from its roots and driven by present-day system needs. However, many basic research areas are associated with the underlying component technologies.

Although the field of microwave photonics was not formalized internationally until the late 1980s and the early 1990s (Berceli and Herczfeld, 2010), its history spans more than a few decades. The use of RF for telegraph communications in the early to mid-1800s gave birth to the need for radio engineers. However, it was not until the expanded development of radar during World War II (Page 1962) to search for aircraft electronically did the need for those with analog or radio engineering skills increase dramatically. Nearly as quickly as

Fundamentals of Microwave Photonics, First Edition.
Vincent J. Urick Jr, Jason D. McKinney, and Keith J. Williams.

radar was established as a useful tool to aid in detection, radar countermeasures were developed to confuse and deny the radar operators effective use of their new tools. Countermeasures necessitate radar redesign in order to render countermeasures ineffective. This iterative countermeasure/counter-countermeasure battle continues today and will so long into the future as the radar designer is constantly trying to "see and not be seen" (Fuller 1990). The use of higher frequencies and the desire to delay those frequencies created a need for low loss delay lines. The early promise of microwave photonics technologies for low loss long delay lines is closely linked to this radar and electronic countermeasure battle.

Today's society makes abundant use of the electromagnetic spectrum for communication. Radio and television broadcasts, cell phones, satellite communications, push-to-talk radios, and many other techniques have been developed to facilitate communication between two or more parties. These systems make use of RF signal transmission and processing within the devices. Due to the expansion of microelectronic circuits and their size/power/speed advantages, many of these systems have moved from strictly analog systems to mixed-signal implementations. In these systems, analog signals are digitized, processed, and/or transported in digital form before being converted back to continuous waveforms for use in the analog world. Although modern RF systems increasingly use digital signal processing (DSP), analog fiber optic links offer the radio engineer significant and useful tools in the design of these systems. The ability to process a signal in the analog domain can simplify overall system design, especially in wide bandwidth systems, where bandwidth demands are difficult to achieve with DSP. However, the analog system engineer should use the best analog tools along with the features that DSP can provide to make the most efficient and powerful system possible.

In its most basic form, an analog photonic link is a delay line containing an electrical-to-optical (E/O) converter to transform the RF signal into the optical domain, an optical transmission medium, and an optical-to-electrical (O/E) converter. Figure 1.1 illustrates a functional block diagram for a multichannel fiber optic link. One or more RF inputs are converted into the optical domain by E/O converters. Once the RF signal has been transformed into the optical domain, it can be delayed in time with optical fiber, processed, and delivered to one or more O/E converters where the optical signals are demodulated back into electrical RF signals. The processing elements can take many forms, including switching, routing, filtering, frequency translation, and

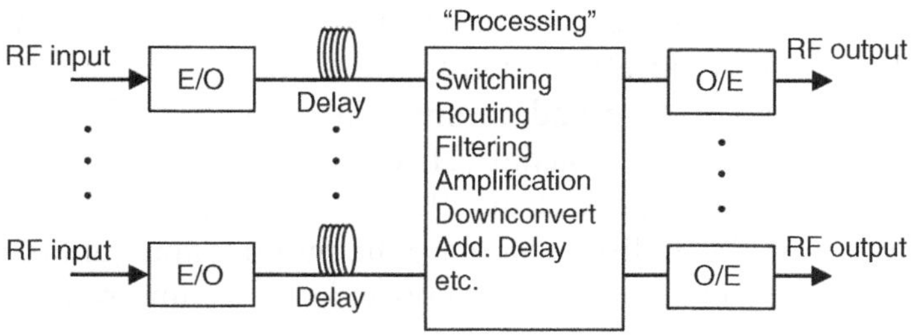

Figure 1.1. Basic block diagram of an array of fiber optic links.

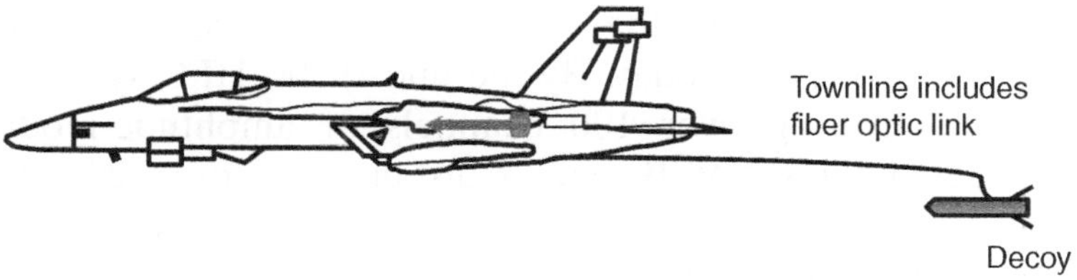

Figure 1.2. A depiction of an RF towed decoy from an F/A 18.

amplification, to name a few. The performance of various forms of such analog photonic links will be treated throughout the middle chapters.

Fiber optic links have proved to be advantageous over their electronic (coaxial cable) counterparts for a number of applications. One of the early military applications was the use of a fiber optic link in an airborne towed decoy as shown in Figure 1.2, the ALE-55. The concept of a towed RF transmitter from an aircraft to distract an RF-guided missile away from its intended target has existed since, at least, the 1960s (Norman and Meullen, 1964), with fiber optic versions appearing later (Toman 1989). In early designs, a receiving antenna on the decoy detected a threat, amplified, and then re-transmitted a higher power return signal. However, due to the size limitations necessitated by aerodynamics of the decoy, only a limited amount of signal processing can be performed on the decoy itself. The use of a fiber optic cable to connect the airplane and the decoy makes it possible to use sophisticated signal processors onboard the aircraft, remoting processed signals to the decoy where amplification and transmission occur. This allows the decoy to be used in a multiphase approach for defeating a threat missile including suppressing the radar's ability to track the aircraft, deceiving the radar with jamming techniques, and seducing the missile away from the aircraft by presenting a more attractive target. Fiber optics minimizes the size of the decoy and reduces the tension on the decoy towline, allowing it to be useful on a wider variety of aircraft.

One of the first widespread commercial uses of analog fiber optic links was in hybrid fiber-coaxial (HFC) systems for cable television signal distribution (Chiddix et al. 1990). HFC solutions offered cable system operators the ability to increase the number and quality of video signals delivered to the home and to provide upstream broadband data services at low cost with high reliability. HFC systems transformed the role of the cable industry from being strictly a provider of video to a viable competitor in the local access market, traditionally served by the telephone system. Combined with the expansion of the Internet, this has helped to shape the competitive broadband information infrastructure as it exists today. By the mid-1990s, HFC systems were capable of delivering over 100 channels of amplitude-modulated vestigial-sideband (AM-VSB) video distances of over 20 km with a variety of optical link designs. The key to this success was the ability to deliver video signals optically having high carrier-to-noise ratios (CNR) and low composite second-order (CSO) and composite triple-beat (CTB) distortion levels. Significant early work on improving the linearity of analog optical links was performed for this application, including work on linearizing external modulation (Nazarathy et al. 1993) and study of crosstalk due to optical fiber nonlinearities (Phillips and Ott 1999). A significant portion of this book is devoted to sources of nonlinearity in analog optical links. Almost as quickly as HFC changed the cable and telephone industry, conversion from AM-VSB video distribution to compressed digital video (CDV) began. Although the conversion was slow due to the cost of replacing an entrenched and expensive infrastructure, CDV has now displaced much of the AM-VSB video distribution technologies. As with the legacy AM-VSB signals, fiber optic links remain the transmission medium of choice for such modern telecommunication systems.

In radio astronomy, large antennas are used to detect RF emissions from space. Microwave engineering plays a crucial role in radio astronomy, with analog fiber optic links being used in modern systems (Webber and Pospieszalski 2002). The Greenbank Telescope (GBT), located in West Virginia and operating from 0.1 to 115 GHz, is the world's largest fully steerable single antenna (Lockman 1998, Prestage et al. 2009). The 100-m-diameter parabolic antenna is used to enhance scientific understanding in areas such as the detection of gravitational waves (through precision pulsar timing); the formation of stars, galaxies, and galaxy clusters; and the composition of planets. The antenna is used for the detection of atomic and molecular emission lines spanning from high red-shift situations (emissions near black holes) to those where the measurement of weak, spatially extended spectral lines can be used to detect new organic molecules in space. The GBT uses an analog fiber optic link for remoting signals to a processing laboratory (White 2000). For higher spatial resolution, smaller dish antennas can be used in a phased-array configuration to take advantage of long baselines to measure small phase changes. Such an array was inaugurated in 2013 in the mountains of Chile (Testi and Walsh 2013), a portion of which is shown in Figure 1.3. Fiber optic links to remote the millimeter wave signals have shown potential utility in large radio astronomy antenna arrays (Payne and Shillue 2002). Because the RF signals originate at astronomical distances and are thus very low power, large antenna systems with very low noise figures are assembled and operated as large phased arrays. Such systems take advantage of the array gain from a large effective aperture and the phase sensitivity of a long baseline. In some systems, as many as 64 12-m dish antennas operating over a 16-km baseline must have their RF signals coherently summed at a central location. Since the frequencies of interest may reach hundreds of gigahertz, relative path differences must be precisely accounted for. This is very challenging, even for fiber optics (Thacker and Shillue 2011), as temperature variations, polarization drift, and chromatic dispersion all lead to length errors requiring active compensation. Additional details on the fiber links for the GBT and ALMA are provided in Chapter 10.

Figure 1.3. The Atacama large millimeter/sub-millimeter array (ALMA) interferometer in Chile. (*Credit:* ALMA (ESO-NAOJ-NRAO), J. Guarda.)

The aforementioned applications are just a few of the many within RF, microwave, or millimeter-wave systems where fiber optic links have proven useful. Microwave photonics provides utility in areas spanning the military, industrial, and academic sectors. Other applications include radio-over-fiber for wireless communications, delivering power to and from antenna feeds for antenna and array calibration, signal routing and true time delay beamforming in arrays, optical signal processing, filtering, waveform synthesis, optoelectronic oscillators for the precision generation of RF signals, optical clocks for precision timing, and RF downconverters and upconverters. The underlying technology and components contained within analog optical links are the subjects of this book, including more detail on applications of the technology in Chapter 10.

1.1 ENABLING TECHNOLOGICAL ADVANCES AND BENEFITS OF FIBER OPTIC LINKS

The frequency range of interest to the field of microwave photonics depends to a large degree on Mother Nature. Figure 1.4 (Liebe 1983) shows the atmospheric attenuation of RF radiation at sea level under different atmospheric conditions. As can be seen, there are

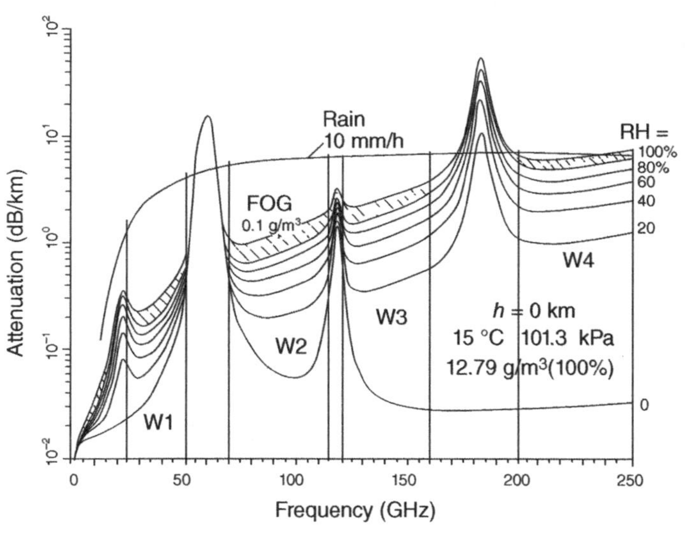

Figure 1.4. Specific atmospheric attenuation at sea level for various levels of relative humidity (RH), including fog and rain. Transmission windows are designated W1–W4 (Liebe 1983).

strong absorption bands near 23, 60, 119, and 182 GHz. Between these frequencies are transmission "windows" with comparatively less loss. Systems using frequencies below 20 GHz have proliferated for ground-based or sea-level applications, with a few systems operating in the second and third windows, centered around 35 and 94 GHz, respectively. Inspection of Figure 1.4 implies that these systems are functioning with an atmospheric attenuation of 0.3 dB/km or less at sea level. At altitudes above 9.2 km, the attenuation decreases in these atmospheric transmission windows to below 0.05 dB/km at frequencies up to 300 GHz (Wiltse 1997). Given that 0.3 dB/km is an acceptable level of attenuation at sea level, it then becomes plausible to consider the use of frequencies up to 300 GHz at high altitudes such as in air-to-air applications. In terms of fractional bandwidth, 300 GHz is only 0.16% of the bandwidth of an optical carrier at 1550 nm (193 THz). This small fractional bandwidth allows many applications to be realized in photonics, including RF signal multiplexing. In addition, many photonic device technologies have been shown to be feasible in the 100–300 GHz range, making the technology suitable throughout this entire frequency range (see Section 10.5). The field of microwave photonics evolved largely due to such application needs. However, before the technology could prosper, several significant breakthroughs were needed, including low loss optical fibers and efficient high bandwidth transducers (E/O and O/E).

Figure 1.5 shows a typical cross-section and index profile for a step index optical fiber. A high index glass core having index of refraction n_1 and diameter d_1 is surrounded by a slightly lower index glass cladding

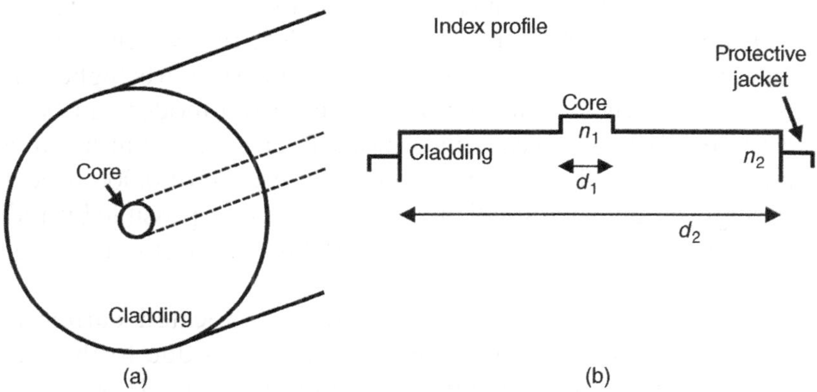

Figure 1.5. (a) Depiction of single mode fiber core and cladding regions with index profile (b) for a step index waveguide design.

having index n_2 and diameter d_2. The cladding is sufficiently thick such that the evanescent electric field of the propagating mode(s) exponentially decays in this region. The cladding glass is usually coated with a lower index polymer for environmental protection. Typical core and cladding diameters are from 8 to 50 μm and from 60 to 125 μm, respectively. The core–cladding index difference and the diameter of the core determine how many propagating modes the fiber waveguide can support for a particular wavelength.

Maxwell's equations describe the propagation of waves within the dielectric waveguide of an optical fiber. From a solution to the wave equations, a normalized frequency or V-number for the fiber can be defined as

$$V = \frac{\pi d_1}{\lambda}\sqrt{(n_1^2 - n_2^2)}, \tag{1.1}$$

where λ is the wavelength. For typical optical fibers, the normalized index difference, $\Delta = (n_1 - n_2)/n_1$, is usually $\ll 1$, and Equation (1.1) reduces to

$$V = \frac{\pi d_1}{\lambda} n_1 \sqrt{2\Delta} = \frac{\pi d_1}{\lambda}\mathrm{NA}, \tag{1.2}$$

where NA is the numerical aperture of the fiber. In ray optics, $\mathrm{NA} = n_0 \sin(\theta)$, where θ is the acceptance half-angle, and n_0 is the index of the material in front of the fiber interface ($n_0 = 1$ for air). The NA is a measure of the light-gathering capacity of a fiber whereby light impinging on the fiber at an angle greater than θ relative to the propagation axis does not excite a guided mode. One can show that for all values of V up to the first zero of the Bessel function J_0 such that $J_0(V) = 0$ (see Appendix VI) that the waveguide can only support the lowest order hybrid mode, HE11 (Ramo et al. 1994). Thus, for $V < 2.405$, the waveguide is single mode. When V exceeds 2.405, the waveguide supports higher order modes, and for large V, the number of supported modes can be estimated to be $V^2/2$. A typical single mode fiber at 1550 nm has a core diameter of 10 μm, allowing for an index difference of 0.006 or less to remain single mode. Such small index differences are possible by adding dopant materials such as GeO_2, P_2O_5, or B_2O_3 to pure fused silica glass (SiO_2).

Multimode fibers with larger cores were fabricated earlier than single-mode fiber and typically achieved lower loss due to the higher tolerances to waveguide dimensional imperfections. However, RF photonic links at high frequencies use single-mode fibers almost exclusively to avoid power fading experienced in multimode fibers due to

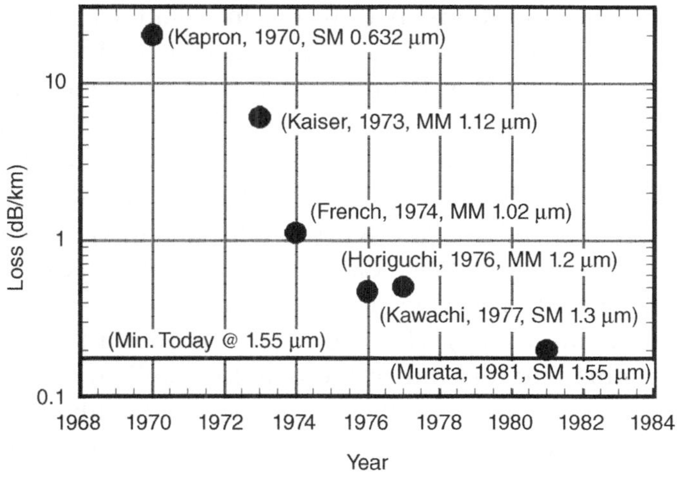

Figure 1.6. Reported losses in optical fiber over time for single-mode (SM) and multimode (MM) fibers at various wavelengths (French et al., 1974;Horiguchi, 1976; Kaiser, 1973; Kapron, 1970; Kawachi, 1977; and Murata and Inagaki, 1981).

modal dispersion. Figure 1.6 shows the progress over time of the optical losses of multimode and single-mode fibers in terms of propagation loss. Fundamentally, the loss is limited by Rayleigh scattering in the fiber, which amounts to a loss of 0.175 dB/km at 1550 nm. As can be seen from Figure 1.6, fiber loss decreased to below 1 dB/km by 1974 and was within 10% of the Rayleigh scattering limit by 1981. It will be demonstrated in later chapters that for many link modulation formats, the RF loss in a microwave photonic link is twice that (in decibels) of the optical loss. Therefore, by 1981, RF delay line propagation loss would have been as low as 0.4 dB/km at 1550-nm wavelength. Since the wavelength dependence of the loss is minimal over a few nanometers bandwidth (hundreds of gigahertz bandwidth at 1550 nm), the RF propagation loss is practically frequency independent.

Low optical fiber loss offered the promise of substantial performance advantages in RF delay lines if the subsequent transducers from E/O and O/E could be developed in the frequency ranges of interest. Initially, the most important frequency range of interest was the region below the first atmospheric absorption feature including frequencies up to 20 GHz (Figure 1.4) where a substantial number of deployed RF systems existed. On the E/O side, the semiconductor laser was an early choice due to the sub-ns photon lifetimes in GaAs (wavelengths up to 860 nm) and InGaAsP (wavelengths up to 1600 nm). Direct modulation of the pump current for these lasers provides a straightforward E/O mechanism. Demonstrations up to 10 GHz modulation bandwidth were prevalent by the mid-1980s (Su and Lanzisera 1986). The first demonstration of a semiconductor laser to surpass 20 GHz bandwidth was at 1.3 μm, using a buried heterostructure in a bulk material (Olshansky et al. 1987). Research continued in this area to improve differential efficiency (leading to higher E/O conversion efficiency) and to increase bandwidth. It was widely expected that multiple quantum well laser designs would help to improve differential efficiency because of their carrier confinement properties and low carrier densities required for inversion (Okamoto 1987). However, it was not until the high speed carrier transport into and out of the quantum wells was studied and understood (Nagarajan et al. 1992) that the bandwidths of quantum well lasers exceeded those made without quantum confinement. Distributed feedback (DFB) laser designs quickly followed, allowing for single-longitudinal-mode operation. While 20 GHz bandwidth lasers satisfy a large number of RF system applications, semiconductor laser intensity noise near the modulation bandwidth limit peaks, leading to lower signal-to-noise ratios (SNR). This intensity noise (or relative intensity noise—RIN) peak can be mitigated by increasing the modulation bandwidth; DFB lasers achieving 25 GHz bandwidth at 1550 nm (Morton et al. 1992) and over 40 GHz bandwidth (Weisser et al. 1996) have been reported.

On the back end of the link, an O/E converter is required to convert RF modulation impressed on the optical carrier back into an RF signal. The most significant device for this is the p–n junction photodiode incorporating a depleted intrinsic region to reduce capacitance, referred to as a p–i–n photodiode. Early work on high speed photodiodes yielded substantially higher bandwidths than their high speed laser counterparts (Bowers et al. 1985), and photodiodes were generally not the bandwidth-limiting device within the first links. There are design trades for these photodiodes when implemented in bulk surface-illuminated structures (Bowers and Burrus 1987); increasing the depletion region thickness lowers capacitance (increases bandwidth) and improves absorption efficiency but causes carrier transit times to increase (decreasing bandwidth). This tradeoff can be avoided by using waveguide or distributed traveling wave designs that improve both efficiency and bandwidth at the expense of device and packaging complexity.

In addition to low propagation loss, the information bandwidth available and the frequency independence of the loss in fiber are just as

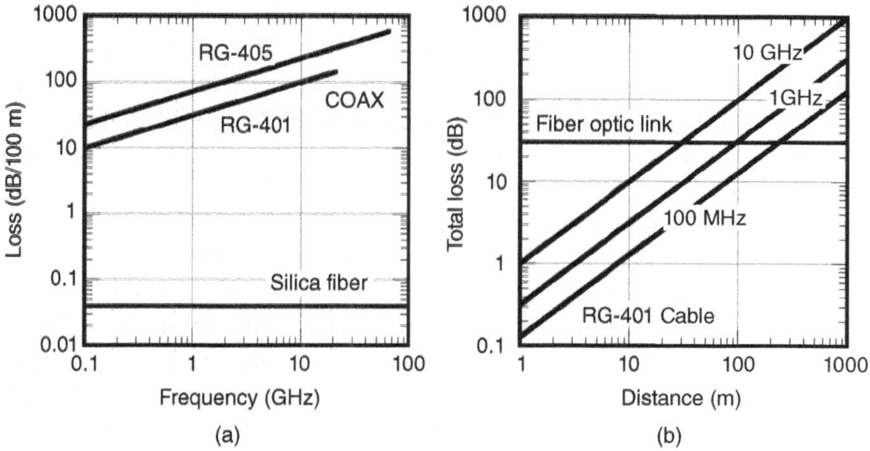

Figure 1.7. Loss as a function of (a) frequency including only propagation loss in the cable for RG-401, RG-405 and silica fiber and (b) propagation distance for RG-401 at three frequencies. In (b), the fiber optic loss includes a 30 dB fixed loss due to E/O and O/E conversion.

important for RF fiber optic links. This is in stark contrast to propagation loss in an RF coaxial cable that tends to have a square root dependency with frequency. As an example, consider Figure 1.7(a) where the propagation losses in two coaxial cables, RG-401 and RG-405, are plotted versus frequency along with the propagation losses of optical fiber. In general, larger diameter cables such as RG-401 tend to have lower loss but also have a lower cutoff frequency for the waveguide to remain single mode. Note how the coaxial cable loss increases by one decade for every two decades in frequency, characteristic of losses that have a square root dependency with frequency. Note also that the propagation losses in coaxial cable are two or three orders of magnitude higher than those of optical fiber. This reason by itself has led the push for the further development of microwave photonics technology through the present day.

When E/O and O/E transducer losses are included with the propagation loss in the comparison between coaxial cable and fiber, the differences are not quite as pronounced as Figure 1.7(a) might suggest. The total loss in a fiber optic link and the propagation loss in RG-401 at three different frequencies are plotted in Figure 1.7(b) as a function of distance. Included in the fiber optic link loss is a 30-dB transducer loss due to the E/O and O/E conversion losses. Because of the exceptionally low propagation loss, there will always be a length for which the fiber optic link will outperform coaxial cable from a loss perspective. This crossover distance tends to be higher at lower frequencies, but distances between tens of meters to a few hundred meters are typical. If loss were the only factor, long distance links would always use fiber; however, factors other than loss also contribute to the decision matrix. Cost, noise performance, phase stability, size, immunity to electromagnetic interference (EMI), and other factors can all play a role. These additional considerations can tip the scales toward fiber optics even for very short links. For example, the relative phase change after propagating an optical fiber is compared to that for a coaxial cable using normalized units of parts per million (ppm) in Figure 1.8. Coaxial cable comprises many different materials including solid and stranded metals, different metal types, and various dielectric materials, all having their own coefficients of thermal expansion. This causes the group velocity of coaxial cable to be a complicated function of temperature. In contrast, optical fiber is primarily made from fused silica. Changes in the propagation delays with temperature are due to the temperature dependencies in both the physical waveguide length and in the index of refraction (also see Section 5.3). Uncoated fiber, if it is not mechanically attached to another material with a large thermal expansion coefficient, has an 8 ppm change in delay per unit length per degree of temperature (Hartog et al. 1979). This includes both the material and waveguide dimensional temperature dependencies. The length fluctuation is both very low and very predictable over a wide temperature range, so long as the temperature dependencies associated with fiber coating or cabling

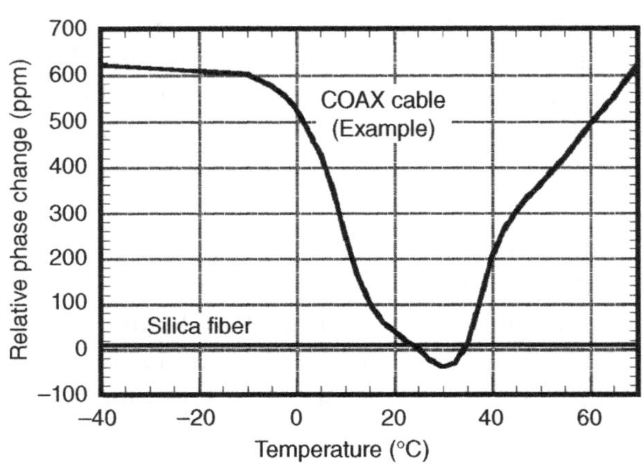

Figure 1.8. Relative phase change versus temperature for a coaxial cable and for optical fiber.

techniques are minimized. This property can be very advantageous in systems where phase stability or phase predictability in the link is a requirement.

Other often-cited advantages associated with fiber optic links include (i) the available bandwidth of over 10,000 GHz, (ii) the reduced size of cable, where sub-millimeter diameters of optical fibers compare to 3–10 mm or larger diameter coaxial cables, (iii) the associated reduction in weight if one can minimize the protective materials needed for cabling, (iv) nonconductive or nonmetallic elements, making the fiber useful in cases where electrical isolation between transmitter and receiver is needed, (v) environmental advantages such as being submersible in fluids, liquid nitrogen, and so on, and (vi) being impervious to corrosion. Analog fiber optic links afford additional less-obvious advantages that are difficult or impossible to achieve electrically. These features include the ability to achieve variable true time delay or RF signal multiplexing. For the latter, the advantages of bundling small fibers into close proximity within a single cable allows for a reduction in the temperature dependence between fiber links (Roman et al. 1998a). This allows for better phase tracking among multiple fiber links, which may be used in phased array applications. As an alternative to multiple fibers, the exceptionally wide bandwidth in the fiber can be used to multiplex numerous RF signals onto one fiber link using different optical carriers. Such multiplexed links and the associated nonlinearities were first studied as a means to distribute cable television channels (Phillips and Ott 1999) and later for higher frequency microwave signals from antenna arrays (Campillo et al. 2003). Many of these advantages and their impact on link performance are discussed throughout this text.

1.2 ANALOG VERSUS DIGITAL FIBER OPTIC LINKS

The RF photonics technology that exists today would not be possible if it were not for the use of fiber optics in digital communication systems. The use of optical fiber to transport digital bits of information across the globe has fundamentally changed the way the world communicates. The Internet and an associated thirst for bandwidth have necessitated the rapid development and deployment of multichannel fiber optic data links to squeeze every last bit of information capacity from a single strand of fiber. An additional benefit of the widespread use of optical fiber for telecommunications is the availability of a vast array of components, many of which can be leveraged for microwave photonics.

Economies of scale and the commoditization of many of these devices have reduced the cost of analog links, except in those cases where specialized components are needed that have no dual use in digital systems.

The differences between analog and digital optical communication links can be substantial. In the digital domain, ones and zeroes can be encoded into optical links as groups of photons (an optical pulse) or the absence of photons. Whether the one or the zero is associated to the actual pulse is not relevant. Noise and timing uncertainty can corrupt the signal during modulation, propagation, and/or detection. So long as the noise and timing uncertainty are small, an integrator can accurately distinguish a pulse from the absence of a pulse using a threshold-like decision in a given time window. In early optical communication links, electrical regenerators periodically removed the noise and timing uncertainty and regenerated the information, thus allowing for propagation over very long distances. In contrast, analog systems must account for the presence of or minimize the effects of this noise and timing uncertainty. In many digital systems today, electronic regenerators are minimized or avoided altogether due to cost implications. Therefore, many long-haul digital communication links are essentially analog, in the sense that the quantization occurs at the link output after transmission.

To expand on this point, Figure 1.9 shows a block diagram of a typical long-haul digital communications link. A digital signal (sequence of ones and zeroes) is input to an E/O converter. Since the attenuation over the entire length of propagation would not allow for detection with a low error rate, the signal must be amplified periodically by several optical amplifiers, typically erbium-doped fiber amplifiers (EDFAs). At the end of the link, O/E conversion returns the waveform to the electrical domain for processing with electronics. The input digital waveform (shown on the middle left) is a series of ones and zeroes denoted by two voltage states. This is simply a baseband RF waveform and can be represented by its Fourier transform or equivalently its spectral content as shown in the lower left plot. A periodic pseudorandom non-return-to-zero (NRZ) waveform has a spectral content of individual lines having an amplitude envelope of a $\mathrm{sin\,c}^2(f)$ function with frequency spacing that is the inverse of the pattern length (Redd and Lyon 2004). Also shown in Figure 1.9 are noise levels. At the output of the link, noise is added due to the amplification stages. In this illustration, the fundamental clock frequency associated with the bit rate has been enhanced as might occur when a small level of chromatic dispersion in the link causes pulse broadening. Such a "digital" link

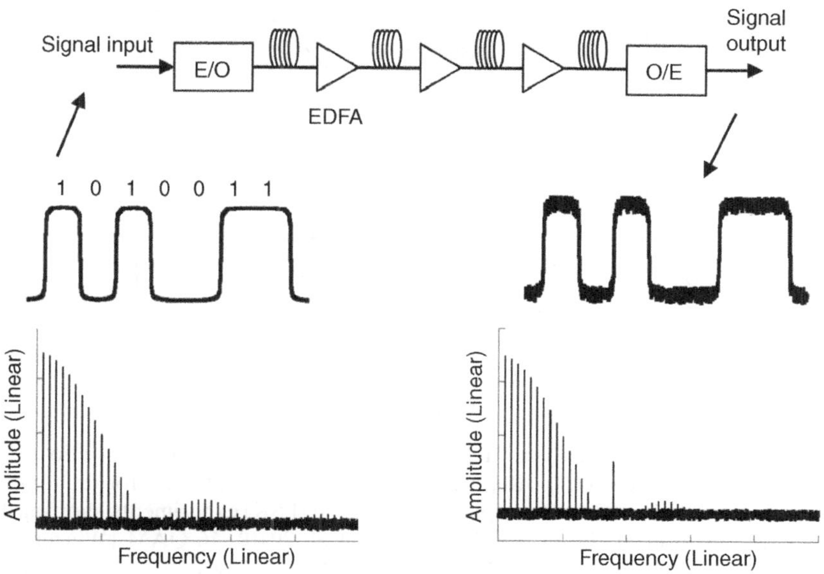

Figure 1.9. Depiction of a long-haul fiber optic link intended to transport digital waveforms. The input digital waveforms with their characteristic baseband RF spectrums are shown both at the input and at the output. The output spectrum depicts a higher noise level due to periodic amplification and distortion leading to the rise of the clock frequency component.

can be viewed as an analog optical system that transports a multitude of RF frequencies from one end to the other. This is only true so long as there are no decision elements involved in the transport as would occur if there were an in-line digital regenerator that detects, retimes, and reshapes the waveform. A current emphasis in digital fiber optic communication systems is to extend the total transmission distance between these electrical regenerations (or repeaters) or even eliminate them altogether to reduce cost and complexity.

Because of the seemingly ever-growing need for bandwidth, optical communications systems have pushed the limits of spectral efficiency within a single optical fiber (Essiambre et al. 2010). Shannon (1949) published a seminal article on the SNR needed to transport a certain number of bits per second in a given unit of bandwidth for a linear information channel, whether that channel is fiber optic, coaxial cable, or otherwise. Shannon's limit can be expressed as

$$C = B \log_2 (\mathrm{SNR} + 1), \tag{1.3}$$

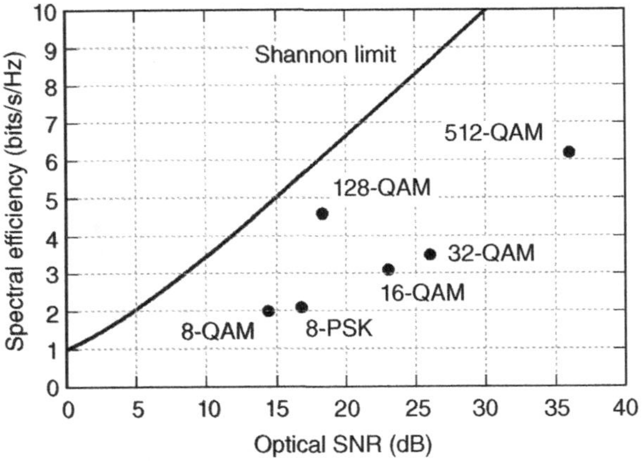

Figure 1.10. Channel capacity as a function of optical signal-to-noise ratio (SNR) according to Equation (1.3). Reported results are also plotted for quadrature-amplitude-modulation (QAM) and phase-shift-keyed (PSK) experiments, showing data for 8-PSK (Zhou et al. 2008), 8-QAM (Zhou et al. 2010), 16-QAM (Winzer et al. 2010), 32-QAM (Takahashi et al. 2010), 128-QAM (Nakazawa 2010) and 512-QAM (Okamoto et al. 2010).

where C is the bit rate in a channel with bandwidth B, and SNR is the output signal-to-noise ratio (linear form, not expressed in dB). The ratio C/B is the channel capacity per unit bandwidth and is denoted as the spectral efficiency in units of bits per second per hertz. Figure 1.10 plots the spectral efficiency given by Equation (1.3) as a function of SNR. Also plotted in Figure 1.10 are several results for various modulation formats. The limiting factor for increasing the information capacity of optical fiber is SNR; however, SNR cannot continue to increase without bound. Since noise cannot be completely eliminated, the signal level must be increased to raise the SNR. At high SNR, fiber nonlinearities limit the attainable spectral efficiency. The nonlinear Shannon limit (Mecozzi and Essiambre 2012, Essiambre et al. 2013) must be employed in this regime, which results in smaller spectral efficiency than predicted by Equation (1.3). It is also important to maximize the output SNR in analog RF systems, as this improves the analog noise figure. Thus, an analog optical RF engineer can leverage work in the optical communications field. In fact, the distinction between lightwave links transporting digital and analog information is becoming blurred. It is not surprising that the performance of many components in the more advanced multilevel digital systems has more stringent performance metrics such as frequency response ripple, laser linewidth, and photodetector amplitude balance, parameters that are typically important in analog links.

With the large and expanding capacity of digital transport and processing systems, it may appear that all analog signals can be digitized immediately and processed in digital form. This is not always possible as discussed in the following section. According to Nyquist (1928), an analog signal can be converted to digital form without deformation as long as the signal is sampled at regular intervals with a minimum of two samples per period of the highest frequency present. To represent the samples in digital form after sampling, the amplitude of the samples must be quantized into discrete levels and to each assigned a digital number. Assigning the amplitude of the sample to one of these discrete levels leads to a quantization noise. This quantization noise is additive to other noise in the signal itself. Let the ratio of the squares of maximum signal voltage to the minimum discernible signal voltage be denoted as a SNR,

$$\mathrm{SNR[dB]} = 20\log\left(V_{\mathrm{max,\,rms}}/V_{\mathrm{min,\,rms}}\right). \tag{1.4}$$

If the difference between the quantization levels is uniform (not necessarily a requirement) and comprise 2^n discrete levels, then the SNR reduces to (Walden 1999),

$$\mathrm{SNR[dB]} = 20\log\left(\frac{\sqrt{3}}{\sqrt{2}}2^n\right) = (6.02)n + 1.76. \tag{1.5}$$

Thus, it is often specified that an n-bit analog-to-digital converter (ADC) has a dynamic range of 6.02 dB per effective bit (neglecting the 1.76 dB offset).

Consider a phased-array radar consisting of 1000 receiving elements operating with 10 GHz instantaneous bandwidth. Radars often require 90 dB or more SNR (Roman et al. 1998b), requiring an ADC with 15 bits. Sampling at the Nyquist frequency requires 20 Giga samples per second (Gsps), with each sample being represented by a 15-bit quantization word. The result in this case is an aggregate minimum bit rate of 300 Gb/s per antenna. Assume that an ADC with these specifications is feasible, although this is beyond current technology capabilities (Walden 2008, Khilo et al. 2012). Now, because this is a phased array, the output of all 1000 elements must be processed coherently, and thus the total data rate that must be processed is 300 Tb/s. Although feasible, this is an overwhelming amount of data to manage and process. Analog signal processing or beamforming may always be necessary, as the system requirements for larger dynamic range, instantaneous bandwidth, and array size may increase at a faster rate than the improvements made in digital processing.

1.3 BASIC FIBER OPTIC COMPONENTS

A thorough understanding of analog photonic links and systems cannot be obtained without first understanding the characteristics and performance of the underlying component technologies. A few basic components are introduced in this chapter, so the reader can relate the metrics in the next few chapters back to the basic link configuration block diagram of Figure 1.1. First and foremost, devices that convert electrical signals into the optical domain are required. In the infancy of fiber optics, the diode laser provided this function using direct modulation of the injection current. Figure 1.11 shows a conceptual transfer characteristic for a directly modulated laser. The output power from a laser diode increases rapidly and approximately linearly with injection current above a certain threshold. The threshold current is necessary to cause carrier inversion and to overcome losses in the laser cavity. In either analog or digital modulation, the laser is biased above threshold. An input modulation current causes the intensity of the laser output to vary. In the digital scenario, the laser is nearly turned off for the digital "zero" and to some higher value for the digital "one." This results in nearly 100% modulation depth of the laser diode. For analog links, the

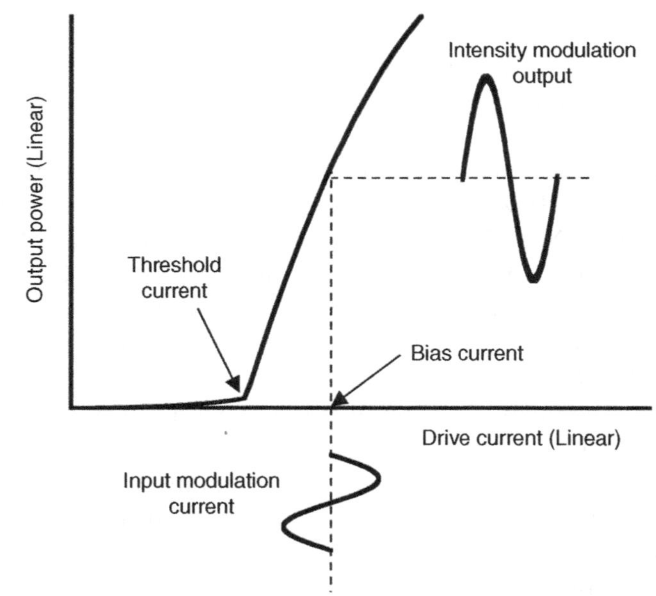

Figure 1.11. Conceptual transfer characteristics of a directly modulated laser diode. The diode is biased with a fixed DC current, and the current is modulated by an RF input signal about this point.

modulation depth varies as the input modulation current amplitude depends on the RF signal level input. To support high dynamic range (a wide range of input RF signal levels), a high bias current is employed to prevent large signals from clipping as the negative swings in signal amplitude push the current down towards the threshold current. A second and often competing issue is the linearity of the current–power curve. As the drive current increases, many diode lasers exhibit lower differential quantum efficiency where an incremental change in current (at high drive current) yields a smaller change in power relative to the response at lower drive currents. This appears as curvature in the current–power curve as shown in Figure 1.11, which can distort the input waveform.

Another important characteristic of directly modulated diode lasers is the rate at which the current can be changed and the corresponding modulation bandwidth. The maximum direct modulation frequency is related to the relaxation-oscillation frequency of the laser cavity. The relaxation-oscillation frequency can be derived from the laser rate equations and is a characteristic frequency that is determined by the interrelationship between the oscillating field in the laser resonator and the atomic inversion. The relaxation-oscillation frequency for a semiconductor laser was given by Lau and Yariv (1985) as

$$f_{\mathrm{r}} = \frac{1}{2\pi}\sqrt{\frac{Ap_0}{\tau_{\mathrm{p}}}}, \tag{1.6}$$

where A is a gain coefficient, p_0 is the intracavity photon density, and τ_{p} is the photon lifetime. Since p_0 is proportional to the drive current above threshold, the modulation bandwidth of the laser diode increases as the drive current is increased (also see Section 8.1). A limitation of direct modulation laser diodes is the inherent noise properties of the source. The relaxation oscillation peak [Equation (1.6)] must be on the same order or higher as the maximum operating frequency. Therefore, any noise present in the drive current below this frequency can manifest as laser RIN in the output light and then eventually in the RF signal output. Thus, there is a tight coupling between the operational bandwidth of a direct-modulation fiber optic link and laser noise. External modulation offers the option to separate these two metrics (laser noise and modulation bandwidth) and thus to optimize each separately.

In external modulation, a continuous wave (CW) laser is modulated by a separate modulator. These devices can take many forms including intensity, phase, or polarization modulators. As mentioned previously, external modulation allows the laser noise to be optimized independent of the operational bandwidth of the link. A common technique to impress the microwave signal onto a CW laser (E/O conversion) makes use of the electro-optic effect (Kaminow and Turner 1966). The electro-optic effect is a change in the index of refraction of a material due to a change in the applied electric field. This effect is available within certain crystals and is dependent on crystal orientation as described by the crystal's electro-optic coefficients. One common material used in fiber optic links is lithium niobate ($LiNbO_3$), although gallium arsenide, lithium tantalate, and many different organic polymers have also been employed. Figure 1.12 depicts a cross-section of a $LiNbO_3$ modulator. A bulk $LiNbO_3$ substrate is oriented as z-cut, meaning that the z-axis of the crystal is aligned with the primary lines of the input electric field of the optical wave. $LiNbO_3$ has a strong electro-optic coefficient along the z-axis (r_{33}) and a slightly weaker coefficient along the x-axis (r_{23} or r_{13}, depending on the orientation of the coordinate system). Titanium can be diffused into the $LiNbO_3$ substrate to form optical waveguides. After depositing a buffer layer of SiO_2, metal electrodes can then be patterned to form an RF waveguide. Figure 1.12 depicts a coplanar waveguide (CPW) ground-signal-ground (GSG) microwave waveguide where two optical waveguides are diffused below the signal waveguide and one of the ground waveguides. This allows for the applied RF energy to interact (via the electric field lines between the signal-ground electrodes) with the propagating optical signal. The applied RF signal then modifies the index of refraction of the optical waveguide and therefore the phase of the light traveling down the waveguide. In this particular design,

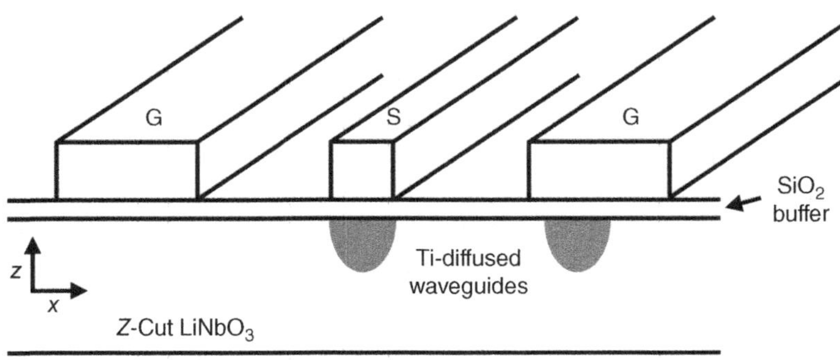

Figure 1.12. Cross-sectional depiction of a z-cut $LiNbO_3$ electro-optic modulator. Metallic ground-signal-ground (GSG) electrodes are patterned over two titanium-diffused optical waveguides.

the optical phase shift in the waveguide under the signal electrode is changed approximately twice as much and with the opposite sign as the phase shift under the ground electrode. Much research has been devoted to the exact construction of these modulators to improve the overlap between the RF fields and optical waveguides (Gopalakrishnan et al. 1992). It is particularly important to ensure that the optical and RF waves travel with the same group velocity and that the transfer of energy from the RF to optical fields is achieved with peak efficiency (Haga et al. 1986). Such "velocity matching" can be quite challenging because of the large difference between the RF and optical indices of refraction for $LiNbO_3$.

The previous discussion illustrates how a microwave signal can interact to change the phase of an optical field within an electro-optic crystal. This is sufficient for creating optical phase modulators, but many optical links use intensity modulation to enable simple demodulation with a photodiode. To convert optical phase modulation into intensity modulation, an optical interferometer is often employed. A common interferometer architecture is the Mach–Zehnder configuration. Figure 1.13(a) shows a basic functional diagram of a Mach–Zehnder modulator (MZM). An optical input is split evenly into two waveguides. The two optical fields are brought together to interfere after some length of propagation. In this example, the two interfering signals are coupled into outputs 1 and 2. Depending on the phase difference between the optical fields, $(\phi_1 - \phi_2)$, the light can constructively or destructively interfere. Figure 1.13(b) shows the intensity at the two optical outputs as a function of the phase difference $(\phi_1 - \phi_2)$. Note how the output intensities are complementary. An intensity modulator can be constructed from this interferometer by phase modulating the fields in either waveguide. Figure 1.13(a) shows an RF CPW over just one of the interferometer waveguides, but in practice, the two optical waveguides can be made very close together as in Figure 1.12. The second waveguide can be patterned directly under the inner ground electrode to increase the RF-to-optical conversion efficiency due to the push–pull nature of the phase shift between the two optical waveguides. Intensity modulation is therefore possible by adjusting the static phase bias $(\phi_1 - \phi_2)$ somewhere along this transfer function, applying an RF input voltage to phase modulate the light and allowing the interferometer to convert this phase modulation into an intensity modulation. Intensity modulation of output I_1 is shown in Figure 1.13(b), but a complementary (180° phase shift) intensity modulation occurs on output I_2. This is useful in the cancellation of

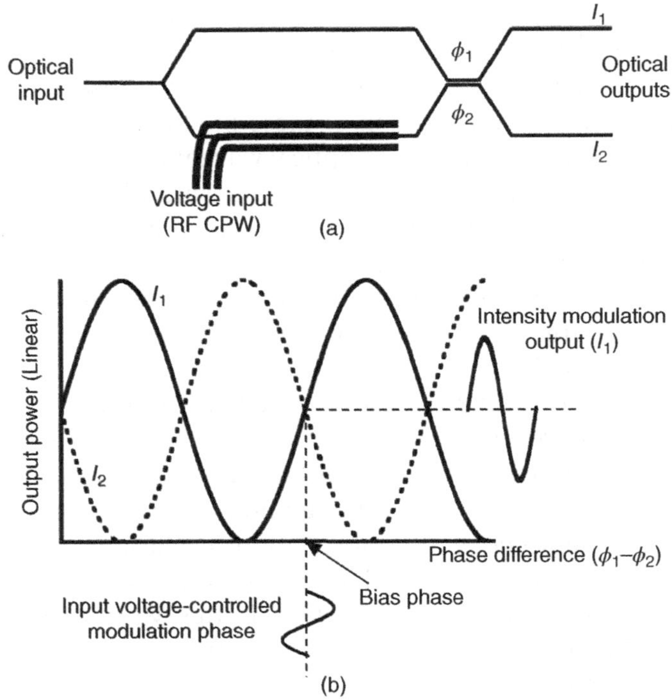

Figure 1.13. (a) Basic Mach-Zehnder modulator (MZM) structure. The RF coplanar waveguide overlaps with a single optical waveguide, but in practice, one ground electrode is patterned over the second optical waveguide as in Figure 1.12. (b) Typical transfer characteristics of an MZM with complementary optical outputs. The modulator is biased with a fixed DC voltage, and an RF signal modulates the lightwave.

some types of optical amplitude noise as demonstrated in Chapter 6. Many other external modulators including Franz–Keldysh semiconductor absorption modulators (Leeson et al. 1988), polarization modulators (Ranalli and Sonek 1991), and modulators based on the index change due to free carrier effects in silicon (Liu et al. 2005) have been demonstrated. Modulators will be discussed in more detail in Chapters 6–8.

Many passive optical components can be used between the E/O and O/E conversions shown as the "processing" block in Figure 1.1. Such components include optical splitters, directional couplers, filters, resonators, attenuators, isolators, circulators, polarizers, and switches. Kashima (1995) provides an extensive description of such optical components. A few devices are described in this chapter. The first is the fused-tapered fiber optic coupler illustrated in Figure 1.14. The fused-tapered fiber optic coupler was proposed by Villarruel and

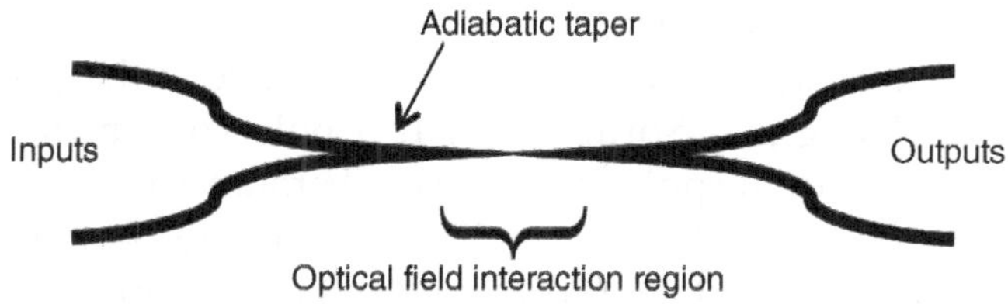

Figure 1.14. Depiction of a 2 × 2 fused-tapered fiber optic coupler.

Moeller (1981) and is a critical component for many systems. The device operates on the concept of electric field resonant coupling between two propagating waveguide modes where the optical evanescent field profiles of the two modes overlap in space. To accomplish this, the waveguide modes must be sufficiently close to one another, as the evanescent fields penetrate only a few micrometers into the cladding of the optical fiber. Therefore, the fibers must be adiabatically tapered to allow the fundamental modes of the fiber to expand into air-guided modes. As the fibers are reduced in diameter, the core can no longer support a guided mode, and the fundamental mode becomes guided by the surrounding air. This allows the electric field of each guided mode to interact, coupling energy from each waveguide into the other. The coupling ratio can be varied if the length over which this field interacts is controlled. This same electric field coupling effect can be used as the output coupler in the interferometric modulator of Figure 1.13(a).

Most passive devices are reciprocal, meaning that their transfer functions are the same in both directions. The fiber optic coupler shown in Figure 1.14 is a common example of a reciprocal device. However, optical isolators and circulators are the often-used nonreciprocal devices. An isolator is commonly employed in laser packages to prevent light backscatter into the laser cavity, which can lead to instabilities. Many lasers, such as DFB lasers, are sensitive to reflections re-entering the cavity (Tkach and Chraplyvy 1986). Most nonreciprocal devices rely on the Faraday Effect, where the polarization state of light is rotated an amount that is proportional to the magnetic field component in the direction of propagation. An optical isolator can be made by inserting a 45° Faraday polarization rotator between two linear polarizers rotated relative to one another by 45°. An incident linear polarization is passed by one polarizer, has its polarization rotated by 45° by the Faraday rotator, and is aligned to pass through the second polarizer. A signal entering from the reverse direction that passes through the output polarizer has its polarization rotated 45° in the opposite direction and impinges on the input polarizer at 90° relative to the transmission polarizer and is blocked. If two polarizing beamsplitters are combined to handle both linear polarization states, the device can be made to be polarization independent (Chang and Sorin 1990). If a polarizing beamsplitter is added between one of the polarizers and the Faraday rotator, backwards travelling light that would be normally be attenuated by the isolating polarizer could be redirected and coupled out into another fiber resulting in a three-port device that acts as a circulator. Such an optical circulator has the operation characteristics depicted in Figure 1.15. Three low loss transmission states occur as well as three high loss (isolation) paths. Circulators are often used to collect reflections from other devices. For example, they can be used to turn reflective-based optical filters into spectrum separating or combining devices such as wavelength division demultiplexers/multiplexers.

Active devices are also important and are often used in the "processing" block of Figure 1.1. One of the most widely used active components is an optical amplifier. Amplifiers are used to overcome signal attenuation such as from propagation loss, excess loss in components, and loss due to splitting or distribution. Optical amplifiers come in many forms including semiconductor (O'Mahony, 1988), Brillouin (Tkach and Chraplyvy 1989), Raman (Rottwitt and Stentz 2002), and rare-earth-doped fiber amplifiers (Poole et al. 1986). Each has its own advantages and disadvantages (see Sections 3.5 and 4.3); however, the rare-earth-doped fiber amplifier has been the most widely used. In such amplifiers, the core of an optical fiber is doped with a small number of rare-earth ions, Er^{3+} in the case of an EDFA. A basic EDFA design is shown in Figure 1.16. Wavelengths of 980 and 1480 nm have strong absorption cross-sections in Er^{3+} and can therefore be employed as pumps for erbium-doped fiber. This pump light can be combined with a 1550-nm signal into the same fiber using a wavelength division multiplexer (WDM). With the medium inverted, the 1550-nm input light can

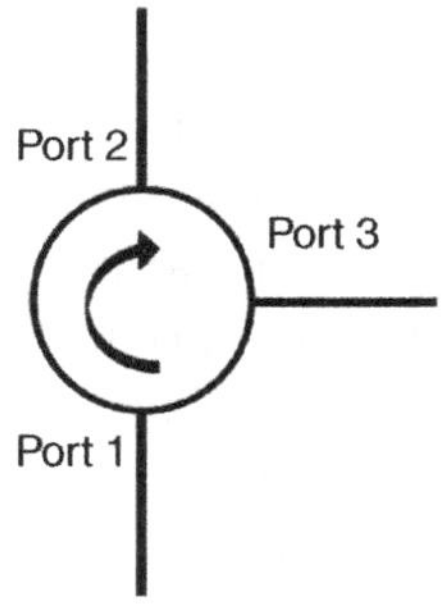

Low loss (Transmission)	High loss (Isolation)
Port 1→ Port 2	Port 1→ Port 3
Port 2→ Port 3	Port 2→ Port 1
Port 3→ Port 1	Port 3→ Port 2

Figure 1.15. Functional diagram of a three-port circulator.

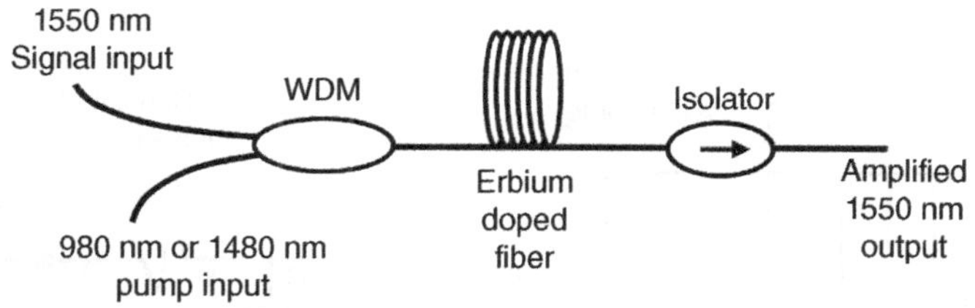

Figure 1.16. Basic components of an erbium-doped fiber amplifier (EDFA).

stimulate the transition of excited erbium ions to their ground states, leading to stimulated emission of photons and thus coherent optical gain. An optical isolator on the output can be employed to prevent reflected light from re-entering the active fiber, which can produce instabilities. Two of the most important properties of the EDFA that make it an ideal amplifier are the low spontaneous emission factor (low noise figure) and the long (milliseconds) upper-state lifetime. The millisecond lifetime does not permit rapid gain fluctuations, and therefore, the signal distortion on high frequency signals ($\gg$kilohertz) is minimal (see Chapter 4). Quantifying the noise performance of an EDFA within an analog link will be covered in Chapter 3.

At the distal end of the link, an O/E converter is required to return the optical signal back into an electric signal. This is most often accomplished with a photodetector. One particular and quite popular photodetector, the p–i–n photodiode, has been used in various forms. Basic architectures include surface-illuminated, waveguide, or travelling-wave designs. The basic structure of a p–i–n photodiode is shown in Figure 1.17, and its performance has been studied extensively. A good review of p–i–n photodiodes is provided by Bowers and Burrus (1987). An intrinsic (undoped) semiconductor is placed between doped p- and n-type materials. The particular materials can be the same (homojunction) or different (heterojunction) and in some special cases do not need to be latticed matched (Ejeckam et al. 1995). A bias voltage is applied to deplete the intrinsic region forming a small parallel-plate capacitor, with higher voltages resulting in a wider depletion region and lower device capacitance. Light can then be absorbed in the intrinsic region by illumination through the p- or n-type materials (surface-illuminated style) or from the edge (waveguide style) creating electron–hole pairs. The photogenerated carriers transit the depletion region and cause image currents to flow in the external circuit. Since the absorption of a photon results in the flow of current, the photodetector acts similarly to a square-law device where it responds only to

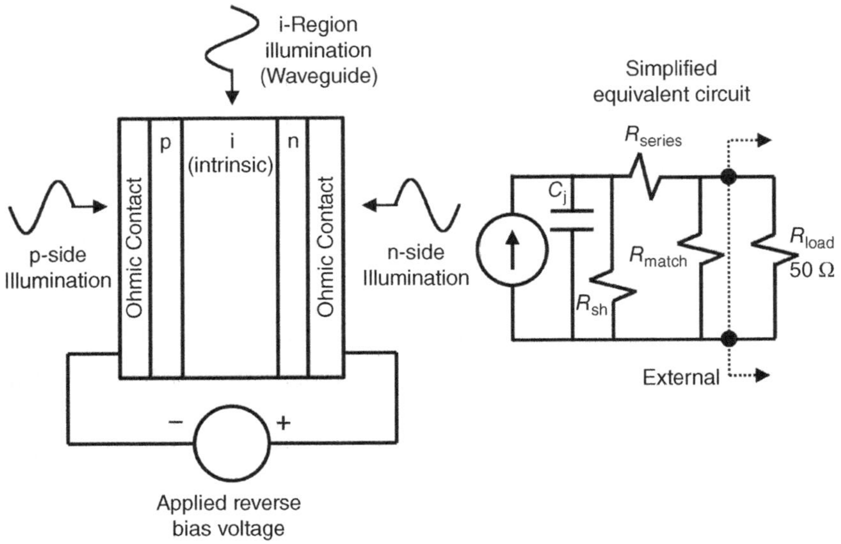

Figure 1.17. A p–i–n photodiode and an equivalent electrical circuit. The photodiode can be surface illuminated either through the n- or p-contacts or parallel to the intrinsic region as would be the case in waveguide architectures.

the optical intensity, thus emitting the optical intensity modulation envelope (the RF input signal) in the form of an output current. The response of these devices is specified in terms of their responsivity in units of Amperes per Watt, how many Amperes of current they produce for every Watt of incident power. Electrically, photodiodes have very simple equivalent circuits if packaging-related impedances are neglected. A photodiode can be viewed as a current source in parallel with a junction capacitance (C_j) and a shunt resistance (R_{sh}), connected to the output through a small series resistance (R_{series}). The junction capacitance is determined by the specific material properties and the geometry, such as the volume of the depletion region. The series resistance is related to the contact resistance and the resistance of the bulk p- and n-type semiconductors. The shunt resistance arises from current leakage and is usually quite high (1 kΩ or more) relative to the RF characteristic impedance of 50 Ω. Therefore, the shunt resistance is usually neglected in photodiode analysis. Since the shunt resistance is usually large, an external matching resistance (R_{match}) is often added to provide a good RF match to the load. This matching resistor has the deleterious effect of shunting half of the photogenerated current and will thus lower link gain by 6 dB. Further details of photodiodes, including other photodiode types, will be discussed in Chapters 4 and 9.

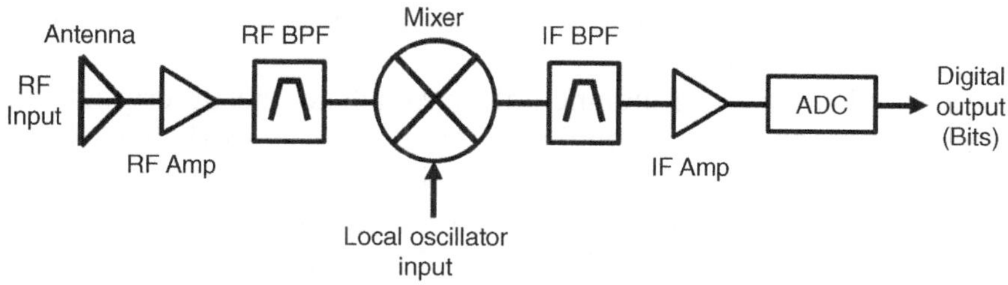

Figure 1.18. Block diagram and typical components contained within an RF receiver.

1.4 ANALOG LINKS WITHIN RF SYSTEMS

Consider the block diagram of a typical RF receiver as shown in Figure 1.18. The goal of such an architecture may be to digitize received RF signals in order to determine the RF situational environment. An antenna receives RF signals from the environment, and those signals are amplified and band limited with a bandpass filter. Since the ADC operates only over a specific bandwidth, a mixer is used to convert the center frequency of the RF signal to that of the ADC. After frequency conversion, the signal is filtered and amplified to a level within the dynamic range of the ADC, which has n bits of resolution at a given sampling rate.

There are numerous advantages of applying RF photonics to the architecture shown in Figure 1.18. For example, a fiber optic link can be inserted between the antenna and the input RF amplifier or between the input RF amplifier and the electronic mixer in order to remote the back end from the front end. In this case, a system trade space analysis considering loss (see Figure 1.7), noise figure, and dynamic range is appropriate. If a multichannel antenna system is required, as in direction-finding applications, phase stability might also be considered in the analysis. A fiber optic solution affords improved phase stability over all-electronic approaches (see Figure 1.8). If the noise figure performance of a photonic link is acceptable, the post-antenna amplifier can potentially be removed. Removing this amplifier can help to reduce distortion because low noise RF amplifiers can exhibit relatively small dynamic range. Photonic channelization with resolution approaching what is required in the electronic domain is becoming a realization; thus, the filter in Figure 1.18 might be replaced with an optical equivalent. Other real possibilities for the injection of photonics in such a system include optical down conversion and optical ADC. The possibilities are limited only by the ingenuity of the system designer and the available photonic technologies.

This book entails a fundamental approach to the description and analysis of microwave photonics technology. The work begins with basic principles as a foundation in Chapters 2 through 5. An analysis of numerous link architectures follows in Chapters 6 through 8. Chapter 9 deals with the design and performance of high current photodetectors. The last chapter focuses on specific applications and system-level considerations for microwave photonics. The text presented in this book provides a thorough treatment of the field—hopefully one that will provide a spark for insightful scientists and engineers to innovate new photonic solutions at the RF system level.

REFERENCES

Berceli, T. and Herczfeld, P. R., "Microwave photonics—a historical perspective," *IEEE Trans. on Microwave Theory and Techniques*, 58(11), 2992–3000 (2010).

Bowers, J. E., Burrus, C. A., and McCoy, R. J., "InGaAs PIN photodetectors with modulation response to millimetre wavelengths," *Electronics Letters*, 21(18), 812–814 (1985).

Bowers, J. E., and Burrus, C. A., "Ultrawide-band long-wavelength p-i-n photodetectors," *Journal of Lightwave Technology*, 5(10), 1339–1350 (1987).

Campillo, A., Tulchinsky, D., Funk, E., and Williams, K., "RF phase distortion due to crosstalk in an 8 channel wavelength division multiplexed analog delay line," *Optical Fiber Communications Conference* 2003 729–730 (2003).

Chang, K. W. and Sorin, W. V., "High-performance single-mode fiber polarization-independent isolators," *Optics Letters*, 15(8), 449–451 (1990).

Chiddix, J. A., Laor, H., Pangrac, D. M., Williamson, L. D., and Wolfe, R. W., "AM video on fiber in CATV systems: need and implementation," *IEEE Journal on Selected Areas in Communications*, 8(7), 1229–1239 (1990).

Ejeckam, F. E., Chua, C. L., Zhu, Z. H., Lo, Y. H., Hong, M., and Bhat, R., "High-performance InGaAs photodetectors on Si and GaAs substrates," *Applied Physics Letters*, 67(26), 3936–3938 (1995).

Essiambre, R. J., Kramer, G., Winzer, P. J., Foschini, G. J., and Goebel, B., "Capacity limits of optical fiber networks," *Journal of Lightwave Technology*, 28(4), 662–701 (2010).

Essiambre, R. J., Tkach, R. W., and Ryf, R., "Fiber nonlinearity and capacity: single-mode and multimode fibers," in *Optical Fiber Telecommunications VIB*, Academic (2013).

French, W. G., MacChesney, J. B., O'Connor, P. B., and Tasker, G. W., "Optical waveguides with very low losses," *Bell System Technology Journal*, 53(5), 951–954 (1974).

Fuller, K. L., "To see and not be seen," *IEE Proceedings F, Radar and Signal Processing*, 137(1), 1–10 (1990).

Gopalakrishnan, G. K., Bulmer, C. H., Burns, W. K., McElhanon, R. W., and Greenblatt, A. S., "40 GHz, low half-wave voltage Ti:LiNbO3 intensity modulator," *Electronics Letters*, 28(9), 826–827 (1992).

Haga, H., Izutsu, M., and Sueta, T., "$LiNbO_3$ traveling-wave light modulator/switch with an etched groove," *IEEE Journal of Quantum Electronics*, 22(6), 902–906 (1986).

Hartog, A. H., Conduit, A. J., and Payne, D. N., "Variation of pulse delay with stress and temperature in jacketed and unjacketed optical fibers," *Optical and Quantum Electronics*, 11(3), 265–273 (1979).

Horiguchi, M., and Osanai, H., "Spectral losses of low-OH content optical fibers," *Electronics Letters*, 12(12), 310–312 (1976).

Kaiser, P., Tynes, A. R., Astle, H. W., Pearson, A. D., French, W. G., Jaeger, R. E., and Cherin, A. H., "Spectral losses of unclad vitreous silica and soda-lime-silicate fibers," *Journal of the Optical Society of America*, 63(9), 1141 (1973).

Kaminow, I. P. and Turner, E. H., "Electrooptic light modulators," *Applied Optics*, 5(10), 1612–1628 (1966).

Kapron, F. P., "Radiation losses in glass optical waveguides," *Applied Physics Letters*, 17(10), 423–425 (1970).

Kashima, N., *Passive Optical Components for Optical Fiber Transmission*, Boston: Artech House, USA (1995).

Kawachi, M., Kawana, A., and Miyashita, T., "Low-Loss Single-Mode Fiber at Material Dispersion-Free Wavelength of 1.27 μm," *Electronics Letters*, 13(15), 442–443 (1977).

Khilo, A., et al., "Photonic ADC: overcoming the bottleneck of electronic jitter," *Optics Express*, 20(4), 4454–4469 (2012).

Lau, K. Y. and Yariv, A., "Ultra-high speed semiconductor lasers," *IEEE Journal of Quantum Electronics*, 21(2), 121–138 (1985).

Leeson, M. S., Payne, F. P., Mears, R. J., Carroll, J. E., Roberts, J. S., Pate, M. A., and Hill, G., "Design and fabrication of a planar, resonant Franz-Keldysh optical modulator," *IEE Colloquium on Optics in Computing*, 8 (1988).

Liebe, H. J., "Atmospheric EHF Window Transparencies near 35, 90, 140, and 220 GHz," *IEEE Transactions on Antennas and Propagation*, AP-31(1), 127–135 (1983).

Liu, A., Samara-Rubio, D., Liao, L., and Paniccia, M., "Scaling the modulation bandwidth and phase efficiency of a silicon optical modulator," *IEEE Journal of Selected Topics in Quantum Electronics*, 11(2), 367–372 (2005).

Lockman, F. J., "The Green Bank Telescope: an overview," *Proceedings of SPIE*, 3357, 656–665 (1998).

Mecozzi, A., and Essiambre, R., "Nonlinear Shannon Limit in Pseudolinear Coherent Systems," *Journal of Lightwave Technology*, 30(12), 2011–2024 (2012).

Morton, P. A., Logan, R. A., Tanbun-Ek, T., Sciortino Jr., P. F., Sergent, A. M., Montgomery, R. K., and Lee, B. T., "25 GHz bandwidth 1.55 μm GaInAsP p-doped strained multiquantum-well lasers," *Electronics Letters*, 28(23), 2156–2157 (1992).

Murata, H., and Inagaki, N., "Low-loss single-mode fiber development and splicing research in Japan," *IEEE Journal of Quantum Electronics*, 17(6), 835–849 (1981).

Nagarajan, R., Ishikawa, M., Fukushima, T., Geels, R. S., and Bowers, J. E., "High speed quantum-well lasers and carrier transport effects," *IEEE Journal of Quantum Electronics*, 28(10), 1990–2008 (1992).

Nakazawa, M., "Ultrahigh spectral density coherent optical transmission techniques," in *High Spectral Density Optical Communications Technologies*, M. Nakazawa, K. Kikuchi and T. Miyazaki (editors), pp. 51–80, Springer (2010).

Nazarathy, M., Berger, J., Ley, A. J., Levi, I. M., and Kagan, Y., "Progress in externally modulated AM CATV transmission systems," *Journal of Lightwave Technology*, 11(1), 82–105 (1993).

Norman, R. F. and Meullen, N. F., "Towed Target," US Patent 3,135,511, 2 1964.

Nyquist, H., "Certain Topics in Telegraph Transmission Theory," *Transactions of the American Institute of Electrical Engineers*, 47(2), 617–644 (1928).

Okamoto, H. "Semiconductor Quantum-Well Structures for Optoelectronics–Recent Advances and Future Prospects," *Japanese Journal of Applied Physics*, 26(Part 1, No. 3), 315–330 (1987).

Okamoto, S., Toyoda, K., Omiya, T., Kasai, K., Yoshida, M., and Nakazawa, M., "512 QAM (54 Gbit/s) coherent optical transmission over 150 km with and optical bandwidth of 4.1 GHz," in *European Conf. on Optical Communications*, 1–3 (2010).

Olshansky, R., Powazinik, W., Hill, P., and Lanzisera, V., "InGaAsP buried heterostructure laser with 22 GHz bandwidth and high modulation efficiency," *Electronics Letters*, 23(16), 839–841 (1987).

O'Mahony, M. J., "Semiconductor laser optical amplifiers for use in future fiber systems," *Journal of Lightwave Technology*, 6(4), 531–544 (1988).

Page, R. M., "The Early History of Radar," *Proceedings of the IRE*, 50(5), 1232–1236 (1962).

Payne, J. M. and Shillue, W. P., "Photonic techniques for local oscillator generation and distribution in millimeter-wave radio astronomy," *IEEE International Topical Meeting on Microwave Photonics*, MWP 2002, *Paper* W1-3, 9–12 (2002).

Phillips, M. R., and Ott, D. M., "Crosstalk due to optical fiber nonlinearities in WDM CATV lightwave systems," *Journal of Lightwave Technology*, 17(10), 1782–1792 (1999).

Poole, S., Payne, D., Mears, R., Fermann, M., and Laming, R., "Fabrication and characterization of low-loss optical fibers containing rare-earth ions," *Journal of Lightwave Technology*, 4(7), 870–876 (1986).

Prestage, R. M., Constantikes, K. T., Hunter, T. R., Kinig, L. J., Lacasse, R. J., Lockman, F. J., and Norrod, R. D., "The Green Bank Telescope," *Proceedings of the IEEE*, 97(8), 1382–1390 (2009).

Ramo, S., Whinnery, J. R., and Duzer, T. V., *Fields and Waves in Communications Electronics*, p. 413, Wiley, New York (1994).

Ranalli, E. R., and Sonek, G. J., "Narrow bandwidth electrooptic polarization modulator using GaAs quantum-well waveguides," *IEEE Photonics Technology Letters*, 3(4), 320–323 (1991).

Redd, J. and Lyon, C., "Spectral content of NRZ test patterns," *EDN Magazine*, 49(9), 10–14 (2004).

Roman, J., Frankel, M., Williams, K., and Esman, R., "Optical fiber cables for synchronous remoting of numerous transmitters/receivers," *IEEE Photonics Technology Letters*, 10(4), 591–593 (1998a).

Roman, J. E., Nichols, L. T., Williams, K. J., Esman, R. D., Tavik, G. C., Livingston, M., and Parent, M. G., "Fiber-optic remoting of an ultrahigh dynamic range radar," *IEEE Transactions on Microwave Theory and Techniques*, 46(12), 2317–2323 (1998b).

Rottwitt, K. and Stentz, A. J., "Raman amplification in lightwave communication systems," in *Optical Fiber Telecommunications IVA*, I. Kaminow and T. Li (editors), Academic (2002).

Shannon, C. E., "Communication in the presence of noise," *Proceedings of the IRE*, 37(1), 10–21, (1949).

Su, C., and Lanzisera, V., "Ultra-high-speed modulation of 1.3-μm InGaAsP diode lasers," *IEEE Journal of Quantum Electronics*, 22(9), 1568–1578 (1986).

Takahashi, H., Amin, A. A., Jansen, S. L., Morita, I., and Tanaka, H., "Highly spectrally efficient DWDM transmission at 7.0 b/s/Hz using 8x65.1-Gb/s coherent PDM-OFDM," *Journal of Lightwave Technology*, 28(4), 406–414 (2010).

Testi, L. and Walsh, J., "The inauguration of the Atacama large millimeter/submillimeter array," *The Messenger*, 152, 2–6 (2013).

Thacker, D. L., and Shillue, B., "Atacama large millimeter array local oscillator: how photonics is enabling millimeter-wave astronomy," in *2011 Optical Fiber Communication Conference*, paper OThJ1, (2011).

Tkach, R., and Chraplyvy, A., "Regimes of feedback effects in 1.5-μm distributed feedback lasers," *Journal of Lightwave Technology*, 4(11), 1655–1661 (1986).

Tkach, R. W., and Chraplyvy, A. R., "Fibre Brillouin amplifiers," *Optical and Quantum Electronics*, 21, S105–S112 (1989).

Toman, D. "Towed Decoy with Fiber Optic Link," US Patent 4,808,999, 28 Feb (1989).

Villarruel, C. A. and Moeller, R. P., "Fused single mode fibre access couplers," *Electronics Letters*, 17(6), 243–244 (1981).

Walden, R. H., "Analog-to-digital converter survey and analysis," *IEEE Journal on Selected Areas in Communications*, 17(4), 539–550 (1999).

Walden, R. H., "Analog-to-digital converters and associated IC technologies," in *IEEE Compound Semiconductor Integrated Circuits Symposium*, 1–2 (2008).

Webber, J. C. and Pospieszalski, M. W., "Microwave instrumentation for radio astronomy," *IEEE Transactions on Microwave Theory and Techniques*, 50(3), 986–995 (2002).

Weisser, S., Larkins, E. C., Czotscher, K., Benz, W., Daleiden, J., Esquivias, I., Fleissner, J, Ralston, J. D., Romero, B., J., Sah, R. E., Schonfelder, A., and Rosenzweig, J., "Damping-limited modulation bandwidths up to 40 GHz in undoped short-cavity In0.35Ga0.65As-GaAs multiple-quantum-well lasers," *IEEE Photonics Technology Letters*, 8(5), 608–610 (1996).

White, S. D., "Implementation of a photonic automatic gain control system for correcting gain variations in the Green Bank Telescope fiber optic system," *Review of Scientific Instruments*, 71(8), 3196–3199 (2000).

Wiltse, J. C., "Corrections to published curves for atmospheric attenuation in the 10 to 1000 GHz region," *IEEE Antennas and Propagation Society International Symposium*, 4, 2580–2583 (1997).

Winzer, P. J., Gnauck, A. H., Doerr, C. R., Magarini, M., and Buhl, L. L., "Spectrally Efficient Long-Haul Optical Networking Using 112-Gb/s Polarization-Multiplexed 16-QAM." *Journal of Lightwave Technology*, 28(4), 547–556 (2010).

Zhou, X., Yu, J., Qian, D., Wang, T., Zhang, G., and Magill, P. D., "8x114 Gb/s, 25-GHz-spaced, PolMux-RZ-8PSK transmission over 640 km of SSMF employing digital coherent detection and EFDA-only amplification," in *Optical Fiber Communication Conference*, Post-Deadline Paper PDP1 (2008).

Zhou, X., Yu, J., Huang, M. F., Shao, Y., Wang, T., Magill, P., Cvijetic, M., Nelson, L., Birk, M., Zhang, G., Ten, S., Matthew, H. B., and Mishra, S. K., "Transmission of 32-Tb/s capacity over 580 km using RZ-shaped PDM-8QAM modulation format and cascaded multimodulus blind equalization algorithm," *Journal of Lightwave Technology*, 28(4), 456–465 (2010).

CHAPTER 2

ANALOG PERFORMANCE METRICS

The optical aspects of microwave photonics are covered throughout this text, with the results more often than not cast in terms of radio-frequency (RF) performance metrics. Thus, the purpose of the present chapter is to define the most important performance parameters for analog systems. The metrics are defined for RF systems in general, without much discussion pertaining to optics, although some data from microwave photonic links are used for illustrative purposes.

The treatment begins with a simple analysis of linear two-port networks using a scattering matrix, from which the small-signal RF gain is defined. The standard metric for quantifying the performance of a noisy RF component or system, the noise figure, is then defined including a brief discussion of thermal noise. Both the RF gain and noise figure are defined for linear two-port networks, as they are very important metrics for describing the signal throughput and sensitivity of RF systems. In reality, physical RF systems will be nonlinear. A Taylor series analysis is conducted for a single sinusoidal drive and a two-tone sinusoidal signal in order to describe the nonlinearity of a RF component or system in a standard manner. A single-tone signal is used to define the compression dynamic range (CDR), whereas the analysis of a two-tone drive introduces the spurious-free dynamic range (SFDR). Both of these

Fundamentals of Microwave Photonics, First Edition.
Vincent J. Urick Jr, Jason D. McKinney, and Keith J. Williams.

dynamic ranges are used to describe the fidelity of an analog RF system. The chapter is concluded with a prescription on how to cascade the individual performance metrics in a chain of stages to facilitate system analysis.

2.1 THE SCATTERING MATRIX

An elementary description of the scattering matrix for an RF network is given with the purpose of defining the average power gain and the reflection coefficient for a two-port network. The interested reader should consult Pozar (2005) or Ramo et al. (1994) for a description of network matrices beyond the introduction given in this chapter. Consider a linear N-port network where V_n^{in} is the amplitude of the voltage wave into the nth port and V_n^{out} is the amplitude of the voltage wave out of the nth port. The scattering matrix for such a linear network is defined in the equation

$$\begin{bmatrix} V_1^{\mathrm{out}} \\ V_2^{\mathrm{out}} \\ \vdots \\ V_{\mathrm{N}}^{\mathrm{out}} \end{bmatrix} = \begin{bmatrix} S_{11} & S_{12} & \cdots & S_{1\mathrm{N}} \\ S_{21} & & & S_{2\mathrm{N}} \\ \vdots & & & \vdots \\ S_{\mathrm{N}1} & S_{\mathrm{N}2} & \cdots & S_{\mathrm{NN}} \end{bmatrix} \begin{bmatrix} V_1^{\mathrm{in}} \\ V_2^{\mathrm{in}} \\ \vdots \\ V_{\mathrm{N}}^{\mathrm{in}} \end{bmatrix}, \tag{2.1}$$

where the individual elements of the scattering matrix are known as scattering coefficients. The majority of the analysis in this book will be for two-port networks (Figure 2.1), and in this case, Equation (2.1) reduces to

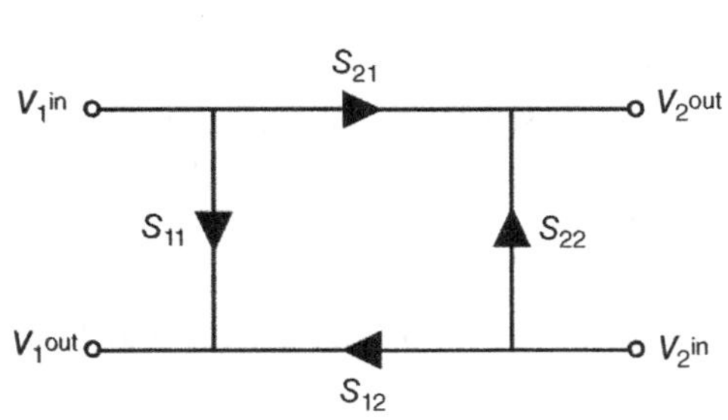

Figure 2.1. A pictorial description of a two-port network, including the four scattering coefficients and the input and output voltages on each port.

$$\begin{bmatrix} V_1^{\text{out}} \\ V_2^{\text{out}} \end{bmatrix} = \begin{bmatrix} S_{11} & S_{12} \\ S_{21} & S_{22} \end{bmatrix} \begin{bmatrix} V_1^{\text{in}} \\ V_2^{\text{in}} \end{bmatrix}. \tag{2.2}$$

A physical understanding of the scattering coefficients in Equation (2.2) is obtained by considering the case where a voltage source is applied to Port 1 and Port 2 is impedance matched to eliminate any reflections at Port 2. Under these conditions, $V_2^{\text{in}} = 0$, $V_1^{\text{out}} = S_{11}V_1^{\text{in}}$, and $V_2^{\text{out}} = S_{21}V_1^{\text{in}}$. Therefore, S_{11} is the reflection coefficient for Port 1 with Port 2 connected to a matched termination. Likewise, the S_{21} parameter is the transmission coefficient from Port 1 to Port 2 with Port 2 connected to a matched termination. A similar analysis can be applied in the reverse order to determine that S_{22} is the reflection coefficient for Port 2 and that S_{12} is the transmission coefficient from Port 2 to Port 1, both for a matched load on Port 1.

The RF power gain factor from Port 1 to Port 2 is related to the S_{21} parameter as

$$g \equiv \frac{P_{\text{out}}}{P_{\text{in}}} = |S_{21}|^2. \tag{2.3}$$

In Equation (2.3), g is defined as the ratio of the average output power to the average input power, $P_{\text{out}} = |V_2^{\text{out}}|^2/(2Z_2)$ and $P_{\text{in}} = |V_1^{\text{in}}|^2/(2Z_1)$, where Z_1 and Z_2 are the impedances at Port 1 and Port 2, respectively ($Z_1 = Z_2$ due to the impedance-matching conditions mentioned previously). Logarithmic scales (Appendix I) are often used in microwave engineering because of the range of values encountered; a commonly accepted unit is the decibel, abbreviated as dB, which is based on a power ratio of $10^{0.1}$ (Martin 1929). Therefore, in most instances, the RF gain, $G = 10\log(g)$, will be given in dB as opposed to g in linear units. Furthermore, G is typically specified for a small-signal drive, which is in agreement with the definition of the scattering matrix being for a linear network.

The S parameters for an RF component or system are typically measured as a function of frequency with a vector network analyzer (see e.g., Agilent 2006). The results of such a measurement on a fiber optic link using an Agilent N5247A vector network analyzer are shown in Figure 2.2. The fiber optic link comprised a laser-modulator-photodiode configuration. Of note is the strong frequency dependence in the data, stressing the fact that the performance of microwave systems must be specified as a function of frequency. The $|S_{21}|^2$ in Figure 2.2, plotted in decibels and equal to G, is negative across the measurement range

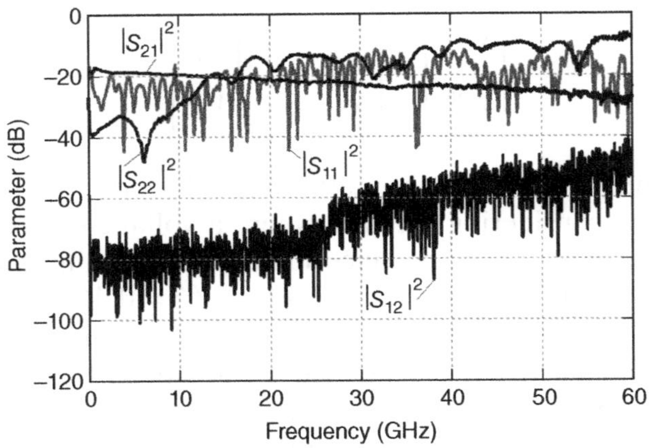

Figure 2.2. Measured magnitude-squared scattering coefficients for an analog fiber optic link as a function of frequency. Shown are the loss from the input to the output ($|S_{21}|^2$), the isolation from the output to the input ($|S_{12}|^2$), the magnitude-squared reflection coefficient for the modulator at the input ($|S_{11}|^2$), and the magnitude-squared reflection coefficient for the photodiode at the output ($|S_{22}|^2$). There are 1601 evenly spaced points for each curve.

($g < 1$), which indicates loss. The S_{11} and S_{22} exhibit periodic peaks and nulls that are to be expected with reflections (the network analyzer is calibrated to 50-Ω impedances on both ports) and quantify the reflection coefficients for the electro-optic modulator and photodiode, respectively. Finally, the S_{12} in Figure 2.2 is a measure of the network analyzer noise floor because the photodiode cannot transmit an RF signal back through the fiber optic link nor can the modulator demodulate an optical signal carrying RF information.

2.2 NOISE FIGURE

The typical metric for describing the noise performance of an RF component or system is the noise factor. According to a standard written in 1959 (Haus 1960), "the noise factor, at a specified input frequency, is defined as the ratio of 1) the total noise power per unit bandwidth at a corresponding output frequency available at the output port when the noise temperature of the input termination is *standard* (290 K) to 2) that portion of 1) engendered at the input frequency by the input termination." Let the numerator in this ratio be N_{out}, the total output noise power spectral density with units of power per unit bandwidth. The denominator is the noise power spectral density generated by the input termination multiplied by g. Two seminal articles from 1928 (Johnson and Nyquist) yield expressions for this "thermal" noise generated by the input termination. They determined that the root mean square (RMS) voltage across a resistor with resistance R at temperature T due to thermal agitation of the electrons in the conductor is given by[1]

$$V = \sqrt{4k_B TRB}, \tag{2.4}$$

where k_B is Boltzmann's constant, and B is the bandwidth over which the measurement of the voltage is made. The thermal noise power transferred from the input termination to the output port is determined by considering a circuit of two parallel resistances. As described by Nyquist (1928), thermal agitation of a resistor in a parallel circuit with a second matched resistor causes a current in the matched resistor of $I = V/(2R)$, where V is given by Equation (2.4). In other words, the current associated with the voltage across one resistor [Equation (2.4)] is divided equally between the two equivalent resistances in the parallel circuit. The average power exchanged between these two resistors in thermal equilibrium is therefore

$$P_{th} = I^2 R = \frac{V^2}{4R} = k_B TB. \tag{2.5}$$

Multiplying Equation (2.5) with g and then normalizing to a unit bandwidth according to the definition given previously gives the noise factor as

$$F \equiv \frac{N_{out}}{g k_B T_s}, \tag{2.6}$$

where $T_s = 290$ K = 17 °C = 62 °F is the standard noise temperature. The noise factor is often given in dB as the noise figure,

$$\text{NF[dB]} \equiv 10\log(F) = 174 + N_{out}\text{[dBm/Hz]} - G\text{[dB]}, \tag{2.7}$$

where the dBm/Hz unit is a dB-scale representation of power spectral density relative to 1 mW (Appendix I), and the thermal noise power spectral density is $10\log(k_B T_s) = -174$ dBm/Hz. With thermal noise as the minimum possible noise level, the allowable ranges on noise factor and noise figure as defined previously are $F \geq 1$ and NF ≥ 0 dB.

[1]This is actually an approximation that is valid at microwave frequencies and the standard noise temperature of 290 K. See Section 3.2 for further explanation.

An alternative but equivalent statement of the noise factor is the quotient of the signal-to-noise ratio (SNR) at the input to that at the output, taken for a thermal-noise-limited input at T_s. Mathematically, this can be written as

$$F = \frac{\mathrm{SNR}_{\mathrm{in,th}}}{\mathrm{SNR}_{\mathrm{out}}}, \tag{2.8}$$

where $\mathrm{SNR}_{\mathrm{in,th}} = P_{\mathrm{in}}/(k_B T_s B)$ and $\mathrm{SNR}_{\mathrm{out}} = P_{\mathrm{out}}/(N_{\mathrm{out}} B)$. Inserting these values into Equation (2.8) with the conditions mentioned previously and using Equation (2.3) reproduces Equation (2.6).

The prescription for measuring the noise figure is given by Equation (2.6). The input to the device under test is terminated in an impedance that is matched to the input at 290 K and N_{out} is measured. With the g also measured, F can then be calculated from Equation (2.6). Noise figure measurements can be conducted with a variety of test equipment including noise figure analyzers, spectrum analyzers, networks analyzers, noise parameter test sets, power meters, and true-RMS voltmeters (Agilent 2010). The results of such a measurement are shown in Figure 2.3. The data in Figure 2.3 show NF as a function of frequency for a state-of-the-art fiber optic link (Urick et al. 2011). These data were obtained with a noise figure analyzer (HP 8970B) and stress the frequency dependence of such measurements.

To exemplify the utility of the F metric, consider the minimum detectable signal (MDS) for an RF receiver. Sometimes referred to as

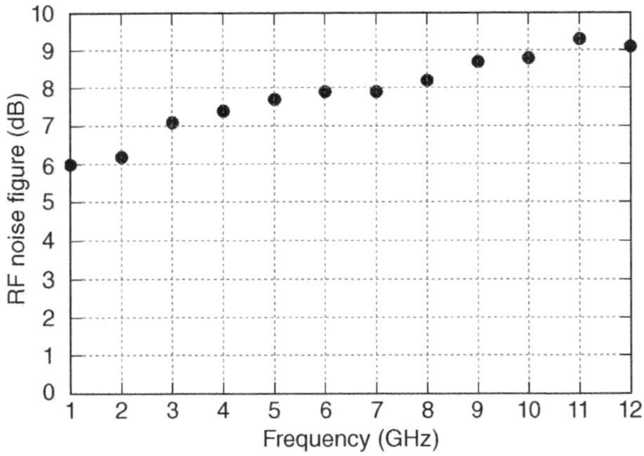

Figure 2.3. Measured RF noise figure as a function of frequency for an analog fiber optic link (Urick et al. 2011).

sensitivity, the MDS is typically used for a receiver but is sometimes applied to components preceding the receiver in a system chain. The MDS is

$$\mathrm{MDS} = F k_\mathrm{B} T_\mathrm{s} B(\mathrm{SNR}_\mathrm{min}), \tag{2.9}$$

where $\mathrm{SNR}_\mathrm{min}$ is the minimum SNR required of the receiver to detect a signal, sometimes referred to as the "margin." In practice, receivers can be cooled to $T \ll T_\mathrm{s}$, such as in radio astronomy (Webber and Pospieszalski 2002), in which case the operational MDS will be lower than that predicted by Equation (2.9).

2.3 DYNAMIC RANGE

The treatment until this point has been concerned with linear systems. No RF system, however, is perfectly linear. In the following section, a Taylor series is used to describe small nonlinearities in terms of compression, harmonic distortion, and intermodulation distortion (IMD). Two dynamic ranges are defined to describe the most important effects. The CDR is associated with gain compression and defined for a single-tone input signal. The SFDR deals with unwanted frequency components generated by nonlinearity and is defined in terms of a two-tone stimulus.

2.3.1 Compression Dynamic Range

A power series expansion is a useful tool for the analysis of nonlinear systems under small-signal excitations. A commonly used power series for describing nonlinear circuits is the Taylor series. The Taylor series (Feigenbaum 1985) for a function $f(x)$ is

$$f(x) = \sum_{m=0}^{\infty} \frac{(x-a)^m}{m!} \cdot \left.\frac{d^m f}{dx^m}\right|_{x=a}. \tag{2.10}$$

The point $x = a$ at which the derivatives in Equation (2.10) are evaluated is typically the bias point for the device when applied to circuit analysis. Application of the Taylor series restricts the analysis to small-signal excitations. Further restrictions include treating only nonlinear systems without memory, that is, systems where the output at some instant in time depends on the input only at that same instant in time.[2] Furthermore, a power series expansion is conventionally used to describe amplitude nonlinearity, although power series expansions have be employed to include phase nonlinearity as well (Heiter 1973). Only amplitude nonlinearities are considered in this chapter for simplicity.

The linearity of a general circuit can be analyzed by representing the output voltage ($V_{\rm out}$) as a function of the input voltage ($V_{\rm in}$) in a Taylor series:

$$V_{\rm out}(V_{\rm in}) = a_0 + a_1(V_{\rm in} - V_{\rm b}) + a_2(V_{\rm in} - V_{\rm b})^2 + a_3(V_{\rm in} - V_{\rm b})^3 + \cdots, \tag{2.11}$$

where $V_{\rm b}$ is the bias voltage. The coefficients in Equation (2.11) are defined by Equation (2.10) as

$$a_m = \frac{1}{m!} \cdot \left. \frac{d^m V_{\rm out}}{dV_{\rm in}^m} \right|_{V_{\rm in}=V_{\rm b}} \tag{2.12}$$

with $a_0 = V_{\rm out}(V_{\rm b})$. A relatively simple measure of the linearity of an analog component or system is its response to a single sinusoidal drive voltage. Such a single-tone test can be analyzed by inserting an input signal of the form $V_{\rm in}(t) = V_{\rm b} + V\sin(\Omega t)$ into Equation (2.11), where $f = \Omega/(2\pi)$ is the frequency. After some algebra and using standard trigonometric identities (Appendix IV), a resulting expression is

$$V_{\rm out} = \left(a_0 + \frac{a_2 V^2}{2}\right) + \left(a_1 V + \frac{3a_3 V^3}{4}\right)\sin(\Omega t) - \frac{a_2 V^2}{2}\cos(2\Omega t) - \frac{a_3 V^3}{4}\sin(3\Omega t) + \cdots. \tag{2.13}$$

Equation (2.13) has a nonoscillatory (DC) term, a term at the fundamental drive frequency, and its harmonics. For single-octave systems (where the highest frequency is less than twice the lowest frequency), harmonic distortion is not a major concern. However, a nonlinear response at the fundamental frequency will result for $a_3 \neq 0$, which will typically produce compression at the output when a_1 and a_3 have an opposite sign.

An important metric often determined by a single-tone test is the CDR. The x-dB CDR ($\mathrm{CDR}_{x\mathrm{dB}}$) is defined as the range of input powers

[2] A Volterra series is an alternate method capable of handling nonlinear systems with memory that was first applied to nonlinear circuits by Wiener (1942). Later seminal papers on the Volterra series for such applications include those by Bedrosian and Rice (1971) and Bussgang et al. (1974).

over which the signal is above the noise floor at the output, and the output power is compressed by x-dB or less relative to a linear response. Mathematically,

$$\mathrm{CDR}_{x\mathrm{dB}} = \frac{P_{x\mathrm{dB}} 10^{x/10}}{N_{\mathrm{out}} B}, \tag{2.14}$$

where $P_{x\mathrm{dB}}$ is the output power at x-dB compression. Typically, the 1-dB CDR (sometimes referred to as the linear dynamic range) is specified.

Determining the $\mathrm{CDR}_{1\mathrm{dB}}$ entails measurement of the fundamental response of the device under test and the output noise. The fundamental response can be obtained with a network analyzer or with a signal source and electrical spectrum analyzer. The output noise can be measured in a variety of ways, such as those listed previously in the context of NF. Shown in Figure 2.4 is a graphical representation of $\mathrm{CDR}_{1\mathrm{dB}}$ employing normalized measured data from a fiber optic link for illustrative purposes. The link response at the fundamental frequency is compared to a linear response; the output power where

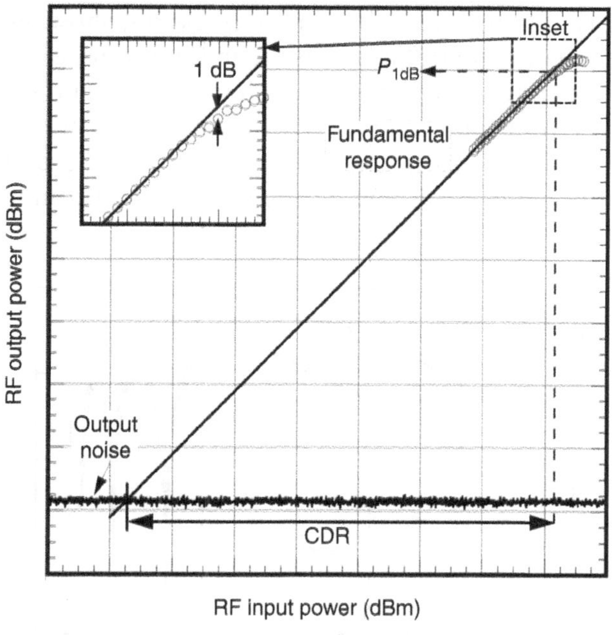

Figure 2.4. Graphical definition of compression dynamic range (CDR) in terms of the output power at 1-dB compression ($P_{1\mathrm{dB}}$) and the output noise power. The circles are normalized measured data from a photonic link, and the solid line has a slope of one. The inset is an expansion near the 1-dB compression point showing the 1-dB deviation between a linear response and the measured data.

the measured response is 1-dB less than the linear response is P_{1dB}. When plotted using a dBm scale as in Figure 2.4, the CDR_{1dB} can be written in dB form using Equation (2.14): $CDR_{1dB}[dB] = P_{1dB}[dBm] + 1 - N_{out}[dBm/Hz] - 10\log(B[Hz])$. The assumptions in this expression are that the total noise power is not a function of frequency over B and that the response is linear from the noise floor up to the specified P_{1dB} level for all B. Given these assumptions, it is often convenient to normalize the CDR_{1dB} to $B = 1$ Hz, specified in units of $[s^{-1}]$ or $[dB \cdot Hz]$, from which the unitless CDR_{1dB} can be obtained by dividing by the bandwidth of interest.

Shown in Figure 2.5 are the measured normalized single-tone responses for an electronic RF amplifier and a microwave photonic link. The behaviors of the two beyond their linear regimes and into hard compression are quite different. The RF amplifier compresses and then saturates at a maximum output power. Once saturated, increasing power at the fundamental frequency results in hardly any change of the output power at that frequency. The added RF power in this case manifests as new frequency components (harmonics) at the output of the amplifier due to its nonlinearity. The microwave photonic link exhibits a very different response in compression, showing peaks and nulls at the fundamental frequency as the input power is increased. One would rarely operate an RF system very hard into compression because of poor power transfer efficiency. However, the curves in Figure 2.5 do stress that compression characteristics can vary greatly between different components or systems, particularly in hard compression.

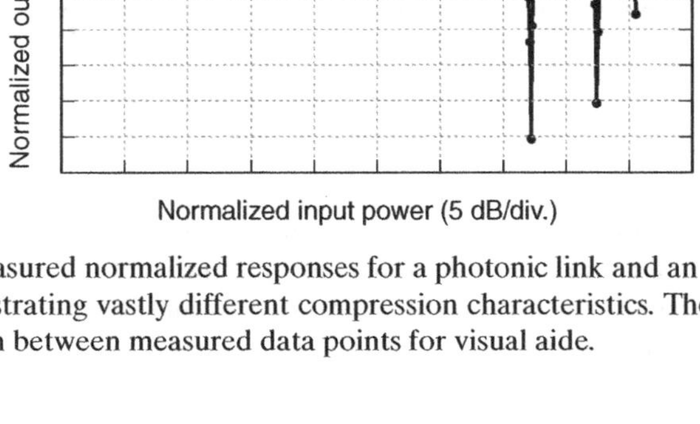

Figure 2.5. Measured normalized responses for a photonic link and an RF amplifier at 1 GHz, demonstrating vastly different compression characteristics. The solid lines are an interpolation between measured data points for visual aide.

RF components and systems are typically operated below the P_{1dB} point, in which case, the CDR_{1dB} adequately quantifies the dynamic range in terms of fundamental response. Harmonic distortions of a single-tone drive are problematic in that they represent frequency content not present in the input signal that is generated by the nonlinearity. The spurious signals arising from harmonic distortion can reduce the operational dynamic range below that specified by CDR_{1dB} if the distortion cannot be tolerated. Harmonic distortion is primarily a concern in multioctave systems (where the highest frequency is greater than twice the lowest frequency), but nonlinearities will also create IMD when more than one signal is present at the input. IMD, which impacts single-octave systems, is a major concern as described in the following section.

2.3.2 Spurious-Free Dynamic Range

The single-tone test discussed in the previous section quantifies a nonlinear system in the Taylor series description. The single-tone test is commonly used to determine the fundamental response and specify the CDR_{1dB}. Measuring the responses at the harmonics of the single-tone fundamental would allow for the coefficients in Equation (2.13) to be determined; with the coefficients known, the output could be calculated for an arbitrary input waveform, given the restrictions in applying a Taylor series expansion. In practice, however, a single-tone test is not commonly employed to determine the distortion response because of the multioctave nature of harmonic distortion. For example, determining the second- and third-order coefficients in Equation (2.13) for a 40-GHz device under test would require measuring the output power responses at 80 and 120 GHz, respectively. Rather, excitation with two closely spaced equal-amplitude sinusoids is a widely accepted method to determine the linearity of a system. Such a two-tone test will result in IMD as described in this following section.

Consider a signal of the form $V_{in}(t) = V_b + V\sin(\Omega_1 t) + V\sin(\Omega_2 t)$ input into a nonlinear circuit described by Equation (2.11). Carrying out the expansion yields

$$
\begin{aligned}
V_{out} = {} & (a_0 + a_2V^2) + \left(a_1V + \frac{9a_3V^3}{4}\right)\sin(\Omega_1 t) + \left(a_1V + \frac{9a_3V^3}{4}\right)\sin(\Omega_2 t) \\
& - \frac{a_2V^2}{2}\cos(2\Omega_1 t) - \frac{a_2V^2}{2}\cos(2\Omega_2 t) \\
& + a_2V^2\cos[(\Omega_1 - \Omega_2)t] - a_2V^2\cos[(\Omega_1 + \Omega_2)t]
\end{aligned}
$$

$$
\begin{aligned}
&-\frac{a_3V^3}{4}\sin(3\Omega_1 t)-\frac{a_3V^3}{4}\sin(3\Omega_2 t)\\
&+\frac{3a_3V^3}{4}\sin[(2\Omega_1-\Omega_2)t]+\frac{3a_3V^3}{4}\sin[(2\Omega_2-\Omega_1)t]\\
&-\frac{3a_3V^3}{4}\sin[(2\Omega_1+\Omega_2)t]-\frac{3a_3V^3}{4}\sin[(2\Omega_2+\Omega_1)t]+\cdots.
\end{aligned}
\tag{2.15}
$$

As compared to Equation (2.13), Equation (2.15) exhibits a different second-order component to the DC term, different third-order terms in the fundamentals, the same amplitudes for the second and third harmonics, and additional frequency content. One consequence of these differences is that a two-tone stimulus will result in a smaller compression power at the individual tones than a single-tone test. The additional frequency content in Equation (2.15) is the IMD at the frequencies f_1-f_2, f_2+f_1, $2f_1-f_2$, $2f_2-f_1$, $2f_1+f_2$, and $2f_2+f_1$. The IMD terms at $2f_1-f_2$ and $2f_2-f_1$ are problematic because they cannot be filtered when the two excitation signals are closely spaced. These two terms are typically used for quantifying the linearity of a single-octave system. The multioctave nonlinearity will usually be cast in terms of the IMD at f_1-f_2 and f_2+f_1. Comparing to Equation (2.13), these second-order IMDs are a factor of two larger (in voltage) than the second-order harmonic distortions. Likewise, the third-order IMDs in Equation (2.15) are three times larger than the third-order harmonic distortions.

The two-tone test therefore allows for determination of the third-order Taylor coefficients by making measurements at frequencies close to the fundamentals. However, a two-tone signal is not necessarily representative of a waveform that would be encountered in practice. Therefore, it is sometimes desirable to use signals with more spectral content. For example, Heiter (1973) presents results for a three-tone test and a noise waveform at the input (including phase nonlinearities). Multitones ($\geq$three tones) have been analyzed including early work by Sea (1968) and articles by Pedro and Carvalho (1999) and Gharaibeh et al. (2006) in the context of modern telecommunications signals. A further discussion of distortion measures is provided in Chapter 4. The remaining development in this case is in terms of intercept points, a concept that can be applied to many types of test signals. Specific examples throughout the book concentrate largely on two-tone excitations, as the results of a two-tone test remain a widely accepted measure of nonlinearity in microwave photonics.

The impact of new frequencies generated by nonlinearity is often quantified with the SFDR. The SFDR can be defined as the range of input powers (or output powers) over which the output signal is above the output noise floor and all spurious signals are less than or equal to the output noise floor. Alternatively, the SFDR can be viewed as being equal to the output SNR when the output SNR and the ratio of the fundamental power to the largest spurious signal are equivalent. When dealing with IMD, the ratio of the fundamental to the spurious signal is often termed the carrier-to-intermodulation ratio (CIR). Shown in Figure 2.6 is a graphical representation of the SFDR definition where the fundamental output power and the accompanying distortion are plotted as a function of RF input power. When using a dBm-scale for such a plot, the small-signal response of the fundamental will have a slope of one, whereas the nth-order distortion will exhibit a slope of n in the small-signal regime. The SFDR as shown in Figure 2.6 should be fairly clear, given the aforementioned definition. The fundamental and distortion responses are either measured to the noise floor or, as shown in Figure 2.6, extrapolated back to the noise floor. Extrapolating

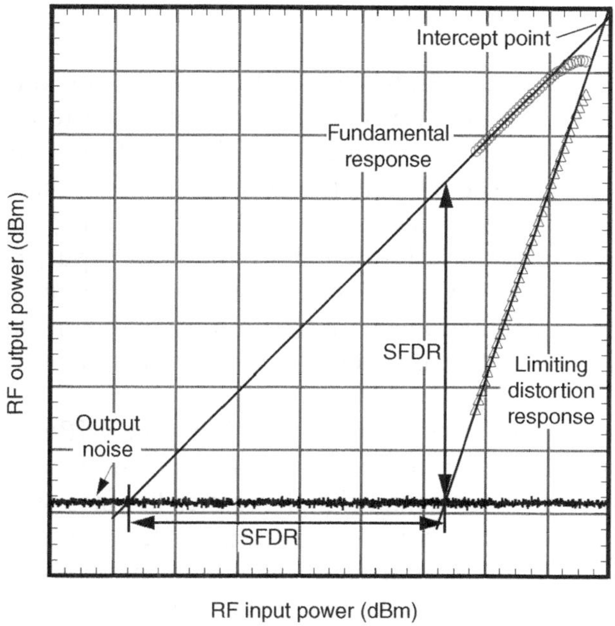

Figure 2.6. Definition of spurious-free dynamic range (SFDR) and intercept point. The solid lines are an extrapolation of the linear fundamental response with slope of one and that for the limiting distortion with a slope of n in the logarithmic space shown, where n is the order of the limiting distortion.

back to the noise floor from a measured data set that exhibits properly ordered small-signal responses assumes that the small-signal response is maintained over the range of input powers from the last measured point to the noise floor. This is usually a safe assumption, but some examples given in the following sections demonstrate that this is not always the case.

A convenient metric used to define the *n*th-order SFDR is the *n*th-order output intercept point (OIP*n*). As shown in Figure 2.6, the OIP*n* is the output power at the intersection of the extrapolated small-signal responses of the fundamental and the *n*th-order distortion. Note that the OIP*n* is a constructed output power that is not actually achievable in practice. The *n*th-order SFDR in terms of the OIP*n* is given as

$$\mathrm{SFDR}_n = \left(\frac{\mathrm{OIP}n}{N_{\mathrm{out}}B}\right)^{(n-1)/n}. \tag{2.16}$$

Equation (2.16) is often presented in dB-form as

$$\mathrm{SFDR}_n[\mathrm{dB}] = \frac{n-1}{n}\left\{\mathrm{OIP}n\,[\mathrm{dBm}] - N_{\mathrm{out}}\left[\frac{\mathrm{dBm}}{\mathrm{Hz}}\right] - 10\log(B[\mathrm{Hz}])\right\}. \tag{2.17}$$

The SFDR_n as given by Equations (2.16) and (2.17) is dimensionless. However, these two equations are often normalized to a 1-Hz bandwidth in units of $[\mathrm{Hz}^{(n-1)/n}]$ and $[\mathrm{dB}\cdot\mathrm{Hz}^{(n-1)/n}]$ for Equations (2.16) and (2.17), respectively. The presumption when doing so is that the unitless SFDR_n can then be obtained for an arbitrary *B*. This normalization invokes two assumptions: (i) that the output (or input) powers over which the analysis is being conducted pertain to the small-signal regime where the fundamental is linear and the *n*th-order distortion remains of order *n* and (ii) that N_{out} is independent of frequency over all *B* being analyzed.

A widely accepted practice for determining the OIP*n* is to measure the fundamental and *n*th-order distortion response over a wide range of input powers in the small-signal regime, fit lines with slopes of one and *n* to the data when expressed in dBm, and then extrapolate the lines to their intersection. (As mentioned previously, these lines could be extrapolated back to the noise floor in order to determine the SFDR_n directly. However, the OIP*n* metric is useful for cascading components as shown in Section 2.4.) The common point-slope form for the equation of a line suggests that a single measurement can predict the OIP*n*. This is true only if the device or system being characterized exhibits the expected fundamental (linear) and distortion (order n) response. Although it is not best practice to characterize a component or system at a single input power, a single-point OIPn expression is instructive to consider. If the output power at the fundamental frequency is denoted P_Ω and that for the nth-order distortion is P_n, both measured at the same input power, then the OIPn is

$$\text{OIP}n = \left(\frac{P_\Omega^n}{P_n}\right)^{1/(n-1)}. \tag{2.18}$$

In dB form,

$$\text{OIP}n[\text{dBm}] = \frac{1}{n-1}(n \cdot P_\Omega[\text{dBm}] - P_n[\text{dBm}]). \tag{2.19}$$

Again, Equations (2.18) and (2.19) are only valid in the small-signal regime.

A particularly useful observation can be made from Equations (2.18) and (2.19) concerning harmonic and IMD. As noted in the discussion following Equation (2.15), the small-signal second-order IMD is two times larger in voltage than the second-harmonic distortion, and the third-order IMD is three times larger in voltage than the third-harmonic. This is a general result for a system that can be described with a Taylor series [Equation (2.11) with the preceding restrictions]. Therefore, the second-order IMD is four times larger ($\approx$6.0 dB) in power than the second-harmonic distortion, and the third-order IMD is nine times larger ($\approx$9.5 dB) in power than the third-harmonic. Applying these relationships to Equations (2.18) and (2.19) determines that the harmonic OIP2 is four times larger ($\approx$6.0 dB) than the IMD OIP2 and that the harmonic OIP3 is three times larger ($\approx$4.8 dB) than the IMD OIP3.

The OIPn can be referenced back to the input by dividing by the gain factor. The result is termed the nth-order input intercept point:

$$\text{IIP}n = \frac{\text{OIP}n}{g}. \tag{2.20}$$

The SFDR_n can then be written in terms of the IIPn by inserting Equation (2.20) into Equation (2.16) and also using the definition of noise factor [Equation (2.6)]:

$$\text{SFDR}_n = \left(\frac{\text{IIP}n}{Fk_\text{B}T_\text{s}B}\right)^{(n-1)/n} \tag{2.21}$$

or in dB form as

$$\mathrm{SFDR}_n\,[\mathrm{dB}] = \frac{n-1}{n}\left\{174 + \mathrm{IIP}n\,[\mathrm{dBm}] - \mathrm{NF}\,[\mathrm{dB}] - 10\log\left(B\,[\mathrm{Hz}]\right)\right\}. \tag{2.22}$$

Finally, a third expression for SFDR_n can be obtained in terms of the carrier-to-distortion ratio at the output, the input power at the fundamental (P_Ω^{in}), and the MDS [Equation (2.9)]:

$$\mathrm{SFDR}_n = \left(\frac{P_\Omega}{P_n}\right)^{1/n}\left(\frac{P_\Omega^{\mathrm{in}}\cdot \mathrm{SNR}_{\mathrm{min}}}{\mathrm{MDS}}\right)^{(n-1)/n}. \tag{2.23}$$

Equations (2.16) and (2.21) are commonly used in microwave photonics, but Equation (2.23) is sometimes useful when dealing with receiver systems. The widely used SFDR and OIP metrics have associated subtleties that are best described by analyzing measured data.

Shown in Figure 2.7 is a measured data set for a fiber optic link of the type described in Chapter 6. When properly biased, this link is limited by third-order IMD when excited with a two-tone test. For test frequencies of 16.7 and 16.9 GHz producing IMD at 17.1 GHz and 16.5 GHz, the OIP3 = 7 dBm. With a measured noise floor of −164 dBm/Hz,

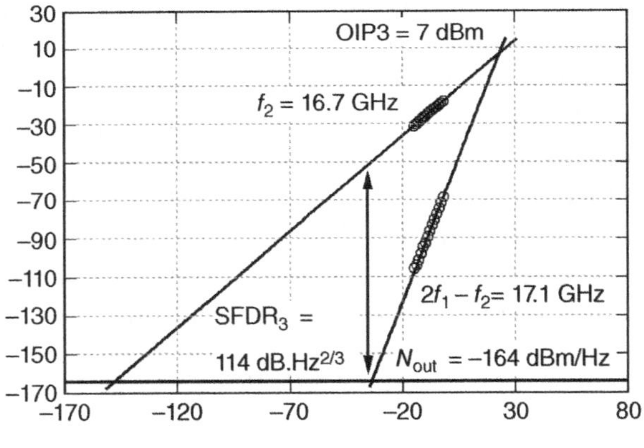

Figure 2.7. Measured third-order-limited spurious-free dynamic range (SFDR_3) of 114 dB · $\mathrm{Hz}^{2/3}$ for a photonic link as obtained via an equal-amplitude two-tone test employing frequencies $f_1 = 16.9$ and $f_2 = 16.7$ GHz. The limiting distortion is a third-order intermodulation product at $2f_1-f_2$ (and $2f_2-f_1$). The third-order output intercept point (OIP3) is 7 dBm, and the measured output noise power spectral density (N_{out}) is −164 dBm/Hz.

Equation (2.17) yields $SFDR_3 = 114$ dB for $B = 1$ Hz. This particular link is an example of a multioctave system that remains third order limited. This characteristic is usually not the case in microwave engineering and is an inherent advantage of this type of fiber optic link. Other RF components and systems will typically exhibit both even- and odd-order nonlinearity.

The measured second- and third-order IMDs for a wideband (2–18 GHz) RF amplifier are shown in Figure 2.8(a). Of first note is that both distortion terms deviate from the properly ordered curves (lines having slopes of two and three for the second- and third-order distortion, respectively) as the RF input passes −20 dBm; this must be taken into consideration when analyzing the data set in terms of intercept points. Assuming that the amplifier will be operated at input powers below −20 dBm, the measured OIP2 = 58 dBm and OIP3 = 29 dBm can be employed to specify the SFDR. With a measured $N_{out} = -148$ dBm/Hz and normalizing to $B = 1$ Hz, these intercept points yield $SFDR_2 = 103\,\text{dB} \cdot \text{Hz}^{1/2}$ and $SFDR_3 = 118\,\text{dB} \cdot \text{Hz}^{2/3}$. The $SFDR_3$ will dictate the single-octave performance of the amplifier, but the multioctave SFDR requires further analysis. Clearly, the second order IMD will dominate when $B = 1$ Hz, as it is the first distortion term to exceed the 1-Hz bandwidth noise level. However, the system bandwidth may be much larger in practice. Shown in Figure 2.8(b) is the SFDR(B) obtained using the data in Figure 2.8(a). The SFDR

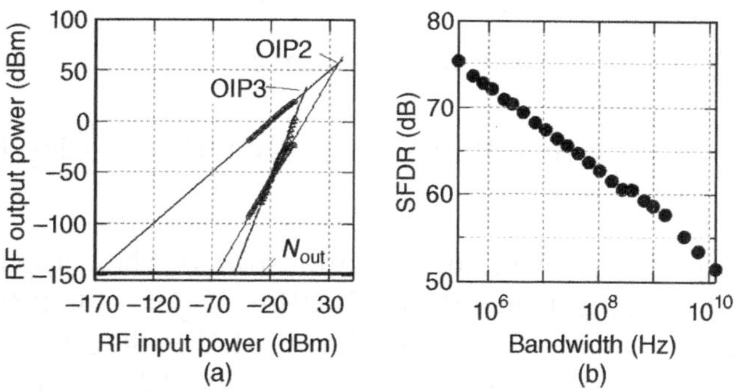

Figure 2.8. (a) Measured distortion response for a RF amplifier as obtained via an equal-amplitude two-tone test employing frequencies $f_1 = 16.9$ and $f_2 = 16.7$ GHz. Intermodulation products at $2f_2-f_1$, $2f_1-f_2$, and f_2-f_1 were measured. The third-order output intercept point (OIP3) is 29 dBm, and the second-order output intercept point (OIP2) is 58 dBm. Note that the spurious responses exhibit deviations from the fit near the top end of the data set. The measured output noise power spectral density (N_{out}) is −148 dBm/Hz. (b) The spurious-free dynamic range obtained from (a) as a function of noise bandwidth.

in Figure 2.8(a) is defined as the CIR when the highest-power IMD at a particular fundamental output power is equal to $N_{out} \times B$. This is a restatement of the previously used definition for SFDR as being equal to the output SNR when the output SNR and CIR are equivalent. At lower input powers in the data set, the second-order IMD limits the SFDR, and the curve SFDR(B) exhibits a slope of −1/2 in logarithmic space (the SFDR decreases 5 dB for every 10× increase in frequency). The third-order IMD becomes dominant as the input power is increased, where SFDR(B) has a slope of −2/3 in logarithmic space (−6.67 dB/decade). In the general small-signal limit, the plot of SFDR(B) will have a slope of $(1-n)/n$, where n is the order of the distortion. As shown in Figure 2.8(b), the transition from a slope of −1/2 to −2/3 (second- to third-order limited) occurs near 1 GHz bandwidth.

A useful expression for analyzing data sets as shown in Figure 2.8 can be obtained by setting $\mathrm{SFDR}_n = \mathrm{SFDR}_m$ and solving for OIPn(OIPm):

$$\mathrm{OIP}n = (\mathrm{OIP}m)^{[n(m-1)]/[m(n-1)]}(N_{out}B)^{1+[n(1-m)]/[m(n-1)]}. \tag{2.24}$$

Equation (2.24) is the nth-order output intercept point required to match the SFDR afforded by the mth-order output intercept point for a given N_{out} and B. Comparing OIP2 and OIP3 is a common problem; in the case that $n = 2$ and $m = 3$, Equation (2.24) becomes

$$\mathrm{OIP2} = (\mathrm{OIP3})^{4/3}(N_{out}B)^{-1/3}. \tag{2.25}$$

Equation (2.25) can be written in dB form as

$$\mathrm{OIP2[dBm]} = \frac{4}{3}\mathrm{OIP3[dBm]} - \frac{1}{3}N_{out}[\mathrm{dBm/Hz}] - \frac{10}{3}\log(B[\mathrm{Hz}]). \tag{2.26}$$

The utility of Equation (2.26) is demonstrated when applied to the data in Figure 2.8. Inserting OIP3 = 29 dBm, $N_{out} = -148$ dBm/Hz, and $B = 1$ Hz into Equation (2.26) results in OIP2 = 88 dBm. This is the OIP2 required to remain third-order limited at all $B \geq 1$ Hz for the RF amplifier under consideration. Obviously, the 58-dBm OIP2 falls short of this requirement, and therefore, the second-order distortion affects multioctave performance. A more useful quantity in this case is the B at which the third-order IMD will be larger than the second-order IMD. Solving Equation (2.26) for B and inserting OIP2 = 58 dBm, OIP3 = 29 dBm, and $N_{out} = -148$ dBm/Hz gives $B = 1.0$ GHz. Therefore, for $B \leq 1.0$ GHz, $\mathrm{SFDR} = 102\,\mathrm{dB}-(10/2)\log(B[\mathrm{Hz}])$, and for $B \geq 1.0$ GHz, SFDR = 118 dB−(20/3) log(B[Hz]). As described previously, this statement is consistent with Figure 2.8(b).

The OIPn metric is quite useful for the types of analysis exampled previously and for cascading a series of components as described in Section 2.4. However, an OIPn can sometimes be an inadequate or misleading metric for describing the actual system performance. Such an example is shown in Figure 2.9(a), where the measured results for a specialty fiber optic link are shown (McKinney et al. 2009). Over a range of input powers where the fundamental output is very linear (small-signal regime), the IMD exhibits a slope of three in a dB scale on the lower end of the range that transitions to a slope of five as the input power increases. Note that the distortion plotted in Figure 2.9(a) exhibiting this behavior remains at the same frequency. The two different regimes that correspond to third- and fifth-order nonlinearity can be described by OIP3 and OIP5, respectively. However, it would need to be explicitly specified over what range of input powers each OIPn was valid. Furthermore, neither an OIP3 nor an OIP5 will adequately describe the distortion over the range of powers where the transition from a slope of five to three occurs. Finally, these data exemplify situations where single-point measurements using Equation (2.18) can be disastrous. In these types of cases, it is better to plot SFDR as a function of noise power over the bandwidth using the definition of SFDR as being equal to the output SNR when the output SNR and CIR are equivalent. As shown in Figure 2.9(b), this type of analysis has been applied to the data in Figure 2.9(a). Plotted as a function of arbitrary noise power in dBm,

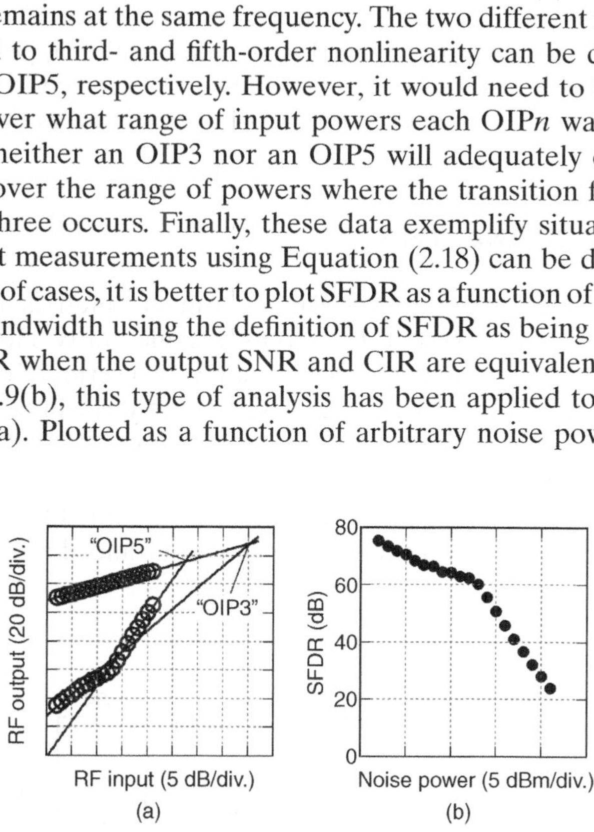

Figure 2.9. (a) Measured data for a linearized photonic link similar to that reported by McKinney et al. (2009). Shown are the fundamental response at $f_1 = 1.01$ GHz($f_2 = 1.02$ GHz) and the dominant intermodulation distortion response at $2f_1 - f_2 = 1.00$ GHz. Although a third- and fifth-order OIP are given, neither alone adequately describes the performance of the link. (b) The spurious-free dynamic range for the data in (a) plotted as a function of arbitrary noise power.

the SFDR in dB exhibits a slope of −2 when third-order limited and −4 when fifth-order limited. In general, for the small-signal regime, the slope of SFDR in dB as a function of noise power in dBm will be $(1 - n)$, where n is the order of the distortion.

The SFDR can be defined in a number of equivalent ways, and it should only follow that the measurement of SFDR is equally diverse. Most importantly, it must be ensured that the measurement apparatus is much more linear than the system being measured. One of the most commonly used procedures is to use filtered signal generators and a well-calibrated electrical spectrum analyzer to determine the OIPn. Equation (2.16) then gives the SFDR_n once N_{out} is obtained via manners described previously. Equation (2.23) prescribes an alternate method that perhaps is more general in that it does not rely on the OIPn. The best practice is to present as much data and description as possible when presenting SFDR results because, as shown in this section, the definitions can sometimes mislead those without experience in the matter.

2.4 CASCADE ANALYSIS

To this point, the presentation of RF performance metrics has been for a singular device or a system treated as a single entity. However, it is often required to determine the performance of a series of components given their individual metrics. Such cascade analysis will be the subject of this section. An architecture as shown in Figure 2.10 will be the concentration, a serial cascade of N stages where each stage has a known g, F, OIPn, and $P_{1\text{dB}}$. In all that follows, it is assumed that each stage is impedance matched on the input and output and that each stage is under small-signal excitation.

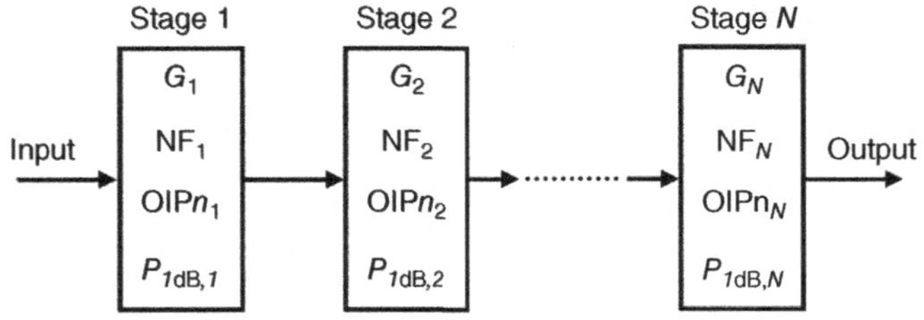

Figure 2.10. Block diagram for a cascade of N stages, each having RF gain (G), RF noise figure (NF), n^{th}-order output intercept point (OIPn), and output power at 1-dB compression (P1dB).

The gain factor is the simplest metric to cascade when the impedances are matched. The cascaded gain factor is the product of the individual gain factors,

$$g = \prod_{p=1}^{N} g_p, \tag{2.27}$$

where it is assumed that all stages are operated in the linear regime. Under the same assumptions, the dB form for the cascaded gain is the sum of the individual gains:

$$G[\text{dB}] = \sum_{i=1}^{N} G_i[\text{dB}]. \tag{2.28}$$

Equations (2.27) and (2.28) require that the $(i-1)$th stage does not compress the ith stage or

$$(P_{1\text{dB}}[\text{dBm}])_{i-1} \ll (P_{1\text{dB}}[\text{dBm}])_i - G_i[\text{dB}] + 1 \tag{2.29}$$

for all $i \geq 2$, where G_i is the small signal gain of the ith stage. The right-hand side of Equation (2.29) is the input power required to compress the ith stage by 1 dB. The ($\ll$) condition is a bit ambiguous and is intended to ensure that the small-signal gain is the actual gain for every stage in the cascade. Obviously, when some stages are near or in compression, Equations (2.27) and (2.28) must be adjusted accordingly.

The first presentation of the formula for cascaded noise factor is attributed to Friis (1944). The equation is most easily written in linear form as

$$F = F_1 + \sum_{\substack{i=2 \\ N\geq 2}}^{N} \left(\frac{F_i - 1}{\prod_{p=1}^{i-1} g_p} \right) \tag{2.30a}$$

or

$$F = F_1 + \frac{F_2 - 1}{g_1} + \frac{F_3 - 1}{g_1 g_2} + \cdots + \frac{F_N - 1}{g_1 \cdots g_{N-1}}. \tag{2.30b}$$

When combining a series of stages with high g_i and relatively low F_i (e.g., low noise amplifiers), the cascaded F is dominated by that of the first stage.

The final and most complicated cascaded performance metric treated in this chapter is the SFDR. The best method to cascade SFDRs is to cascade the OIPns and then use the cascaded F and g to determine the cascaded SFDR by way of Equations (2.16) or (2.21). The cascaded OIPn is given by (Wilson 1981)

$$\mathrm{OIP}n = \left\{ \sum_{\substack{i=2\\N\geq 2}}^{N-1} \left[\left(\mathrm{OIP}n_i \prod_{p=i+1}^{N} g_p \right)^{(1-n)/2} \right] + \mathrm{OIP}n_{\mathrm{N}}^{(1-n)/2} \right\}^{2/(1-n)} \tag{2.31a}$$

or in a more common form (Maas 2003)

$$(\mathrm{OIP}n)^{(1-n)/2} = (\mathrm{OIP}n_{\mathrm{N}})^{(1-n)/2} + (g_{\mathrm{N}} \times \mathrm{OIP}n_{\mathrm{N-1}})^{(1-n)/2}$$
$$+ (g_{\mathrm{N-1}} g_{\mathrm{N}} \times \mathrm{OIP}n_{\mathrm{N-2}})^{(1-n)/2} + \cdots + (g_2 \cdots g_{\mathrm{N}} \times \mathrm{OIP}n_1)^{(1-n)/2}. \tag{2.31b}$$

Equation (2.31) represents the minimum OIPn for a cascade (worst-case scenario) with the assumption that the distortions of each consecutive stage are combined in phase. These equations are therefore the most conservative approach for cascaded OIPn analysis. Kanaglekar et al. (1988) has presented a more-detailed analysis of OIP3 cascades including the minimum OIP3 (Equation (2.31) with $n = 3$), the maximum OIP3, and a statistical analysis assuming that the relative phases can be treated as random variables. Extending this analysis to an arbitrary order distortion is beyond the scope of this work. In fact, Maas (1995) has demonstrated that the worst-case OIP3 (Equation (2.31) with $n = 3$) is closer to real-world performance than was previously expected. Equation (2.31) should be used as a conservative guide in system design and analysis, noting that carefully measuring the cascaded OIPn is the best practice.

REFERENCES

Agilent, "Amplifier linear and gain compression measurements," Application Note 1408–7 (2006).

Agilent, "Fundamental of RF and microwave noise figure measurements," Application Note 57–1 (2010).

Bedrosian, E. and Rice, S. O., "The output properties of Volterra systems (non-linear systems with memory) driven by harmonic and Gaussian inputs," *Proceedings of the IEEE*, 59(12), 1688–1707 (1971).

Bussgang, J. J., Ehrman, L., and Graham, J. W., "Analysis of nonlinear systems with multiple inputs," *Proceedings of the IEEE*, 62(8), 1088–1119 (1974).

Feigenbaum, L., "Brook Taylor and the method of increments," *Archive for History of Exact Sciences*, 34(1), 1–140 (1985).

Friis, H. T., "Noise figures of radio receivers," *Proceedings of the IRE*, 32(7), 419–422 (1944).

Gharaibeh, K. M., Gard, K. G., and Steer, M. B., "In-band distortion of multisines," *IEEE Transactions on Microwave Theory and Techniques*, 54(8), 3227–3236 (2006).

Haus, H. A., "IRE standards on methods of measuring noise in linear twoports, 1959," *Proceedings of the IRE*, 48(1), 60–68 (1960).

Heiter, G. L., "Characterization of nonlinearities in microwave devices and systems," *IEEE Transactions on Microwave Theory and Techniques*, 21(12), 797–805 (1973).

Johnson, J. B., "Thermal agitation of electricity in conductors," *Physical Reviews*, 32, 97–109 (1928).

Kanaglekar, N. G., McIntosh, R. E., and Bryant, W. E., "Analysis of two-tone, third-order distortion in cascaded two-ports," *IEEE Transactions on Microwave Theory and Techniques*, 36(4), 701–705 (1988).

Maas, S. A., "Third-order intermodulation distortion in cascaded stages," *IEEE Microwave and Guided Wave Letters*, 5(6), 189–191 (1995).

Maas, S. A., *Nonlinear Microwave and RF Circuits*, Artech House, Boston (2003).

Martin, W. H., "Decibel—The name for the transmission unit," *Bell System Technical Journal*, 8(1), 1–2 (1929).

McKinney, J. D., Colladay, K., and Williams, K. J., "Linearization of phase-modulated analog optical links employing interferometric detection," *Journal of Lightwave Technology*, 27(9), 1212–1220 (2009).

Nyquist, H., "Thermal agitation of electric charge in conductors," *Physical Review*, 32, 110–113 (1928).

Pedro, J. C. and de Carvalho, N. B., "On the use of multitone techniques for assessing RF components' intermodulation distortion," *IEEE Transactions on Microwave Theory and Techniques*, 47(12), 2393–2402 (1999).

Pozar, D. M., *Microwave Engineering*, Wiley, USA (2005).

Ramo, S., Whinnery, J. R., and Duzer, T. V., *Fields and Waves in Communications Electronics*, Wiley, New York (1994).

Sea, R. G., "An algebraic formula for amplitudes of intermodulation products involving an arbitrary number of frequencies," *Proceedings of the IEEE*, 56(8), 1388–1389 (1968).

Urick, V. J., McKinney, J. D., Diehl, J. F., and Williams, K. J., "Fiber-optic links with all-photonic RF gain and low RF noise figure," *IEEE International Microwave Symposium Digest* (2011).

Webber, J. C. and Pospieszalski, M. W., "Microwave instrumentation for radio astronomy," *IEEE Transactions on Microwave Theory and Techniques*, 50(3), 986–995 (2002).

Wiener, N., "Response of a non-linear device to noise," MIT Radiation Lab Report V-16S, R-129 (1942).

Wilson S. E., "Evaluate the distortion of modular cascades," *Microwaves*, 20, 67–70 (1981).

CHAPTER 3

SOURCES OF NOISE IN FIBER OPTIC LINKS

The concept of electrical noise power spectral density (PSD) was introduced in the previous chapter to define radio-frequency (RF) performance metrics. The present chapter is devoted to the description of the noise processes relevant to microwave photonics, therefore spanning electrical and optical domains. Optical noise over a wide bandwidth must be considered with carriers commonly used from 1310 nm (228.9 THz) to 1550 nm (193.4 THz). Modulating signals span less bandwidth, ranging from audio signals (<20 kHz) to the millimeter wave bands (>110 GHz). In addition, advancing technology may lessen the gap between carrier and modulation signals in microwave photonics. The convention in this case for designating frequencies is as follows: Ω, microwave angular frequency; $f = \Omega/2\pi$, microwave Hertzian frequency; ω, optical angular frequency; $\nu = \omega/2\pi$, optical Hertzian frequency. When the distinction is not clear and/or in generalized cases, the optical frequencies will be employed.

An entire book could be dedicated to a complete analysis of the noise in each domain. A particularly useful volume that succeeds in such an endeavor employing basic physics was provided by Haus (2000a). Other useful applied books on electronic noise include those by Robins (1982) and Schiek et al. (2006). This chapter provides a

Fundamentals of Microwave Photonics, First Edition.
Vincent J. Urick Jr, Jason D. McKinney, and Keith J. Williams.

focused treatment of the topic as it pertains to microwave photonics. Not every type of noise in electromagnetism is covered in this work. Rather, the attention is on those sources of noise that are most important in applications of microwave photonics. The treatment in this chapter begins with an introduction to basic noise concepts and how they are tied to the RF performance. Individual noise components are then treated separately starting with fundamental sources: thermal and shot noise. Additional noise arising from basic components in a fiber optic link is then analyzed including excess noise from lasers, optical amplifiers, and photodiodes. Optical propagation effects will be excluded from the analysis in this chapter, as they are covered in Chapter 5.

3.1 BASIC CONCEPTS

The RF performance metrics F, CDR, and SFDR are all defined in terms of the total output noise PSD, N_{out} (Chapter 2). The interest in this case is therefore to obtain expressions for N_{out} of a general analog fiber optic link such as shown in Figure 3.1. An electrical signal is modulated onto an optical source; such a modulation can be achieved with an external component or by modulating the laser directly. Optical amplification may be employed in the link. Recovery of the electrical signal employs a photodetector, which requires modulation of the optical intensity. The optical intensity modulation is achieved directly or, if another modulation was employed (phase, frequency, or polarization), indirectly via a conversion to optical intensity modulation before illuminating the photodetector. The photodiode circuit shown in Figure 3.1 will be used, which is equivalent to that described in Figure 1.17 for $R_{sh} \rightarrow \infty$ and $R_{series} \rightarrow 0$.

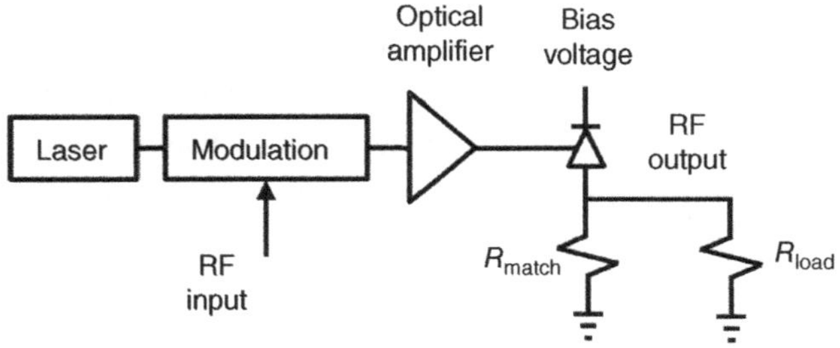

Figure 3.1. Photonic link block diagram.

Photodetectors convert photons into electrons and will therefore transfer intensity variations in the optical power to amplitude variations in the output photocurrent. Likewise, photodetectors do not respond to optical phase changes unless there is a conversion from phase variations to intensity variations in the optical domain, such as with an interferometer. Following the development given by Coldren (2012), it is convention to express optical intensity variations in terms of relative intensity noise:

$$\mathrm{RIN} \equiv \frac{\left\langle P_{\mathrm{opt}}^2(t) \right\rangle}{\left\langle P_{\mathrm{opt}} \right\rangle^2}, \tag{3.1}$$

where the numerator is the mean-square of the optical intensity noise distribution, and the denominator is the squared average optical power. Assuming that the noise is completely random, a Fourier transform (see Appendix V) allows for Equation (3.1) to be expressed in the frequency domain:

$$\left\langle P_{\mathrm{opt}}^2(t) \right\rangle = \frac{1}{2\pi} \int_{-\infty}^{\infty} S_{\mathrm{opt}}(\omega) |F(\omega)|^2 d\omega, \tag{3.2}$$

where $S_{\mathrm{opt}}(\omega)$ is the two-sided spectral density of the mean-square of the optical intensity noise distribution with units of W^2/Hz, and $F(\omega)$ is the filter transfer function used to measure the noise. If the measurement bandwidth B about frequency $f = \omega/2\pi$ is much narrower than changes in the spectral density, then $F(f) = 1$ and $\langle P_{\mathrm{opt}}^2(t)\rangle = 2S_{\mathrm{opt}}(f)B$. A two-sided spectral density consisting of positive and negative frequencies is typically used, because the Fourier transform of such a spectral density is the autocorrelation function according to the Wiener–Khintchine Theorem (McQuarrie 2000). However, the one-sided spectral density having only positive frequency content is more significant physically. The one-sided spectral density is twice the two-sided spectral density because $S_{\mathrm{opt}}(f) = S_{\mathrm{opt}}(-f)$ (excluding $f = 0$). Now, a photodetector is required to measure $\langle P_{\mathrm{opt}}^2(t)\rangle$ and $\langle P_{\mathrm{opt}}\rangle$; the metric used to quantify the conversion of photons to electrons is the photodetector responsivity $\mathfrak{R} = \eta q/(h\nu)$, where η is the photodetector quantum efficiency, q is the magnitude of the electron charge, h is Planck's constant, and ν is the optical frequency. The DC photocurrent is then $I_{\mathrm{dc}} = \mathfrak{R}\langle P_{\mathrm{opt}}\rangle$. The mean-squared photocurrent due to the optical intensity noise is $\langle i_{\mathrm{opt}}^2(t)\rangle = 2\mathfrak{R}^2|H_{\mathrm{pd}}(f)|^2 S_{\mathrm{opt}}(f) = \mathfrak{R}^2|H_{\mathrm{pd}}(f)|^2\langle P_{\mathrm{opt}}^2(t)\rangle/B$, where the one-sided spectral density of the mean-square of the optical intensity noise distribution was used, and H_{pd} is the photodetector filter transfer function normalized to the responsivity. These relations allow for Equation (3.1) to be rewritten in terms of the photocurrent as

$$\mathrm{RIN} = \frac{\left\langle i_{\mathrm{opt}}^2(t) \right\rangle B}{I_{\mathrm{dc}}^2 \left| H_{\mathrm{pd}}(f) \right|^2}. \tag{3.3}$$

The noise PSD delivered to a load with resistance R_{load} at the output of the link due to optical intensity fluctuations is $N_{\mathrm{opt}} = \langle i_{\mathrm{opt}}^2(t) \rangle R_{\mathrm{load}}$. With all of the photocurrent delivered to the load, R_{match} is an open circuit, and a maximum value of $H_{\mathrm{pd}} = 1$ can be obtained. As shown in Figure 3.1 when $R_{\mathrm{match}} = R_{\mathrm{load}}$, $H_{\mathrm{pd}} = 1/2$, accounting for the noise current being split evenly between R_{load} and R_{match}. Inserting the aforementioned expression for N_{opt} into Equation (3.3) and dropping the subscript on R yields

$$\mathrm{RIN} = \frac{N_{\mathrm{opt}} B}{I_{\mathrm{dc}}^2 R \left| H_{\mathrm{pd}}(f) \right|^2}. \tag{3.4}$$

Equation (3.4) is a commonly used metric to describe the effect of optical intensity fluctuations in terms of measureable quantities in the electrical domain: the average current sourced by the photodetector (I_{dc}) and the output noise PSD (N_{opt}) measured in a bandwidth B delivered to the load with resistance R. The RIN given by Equation (3.4) is typically normalized to $B = 1$ Hz and expressed in a decibel scale; for $R = 50\,\Omega$,

$$\mathrm{RIN[dBc/Hz]} = 13 + N_{\mathrm{opt}}[\mathrm{dBm/Hz}] - 20\log(I_{\mathrm{dc}}[\mathrm{mA}]) - 2H_{\mathrm{pd}}[\mathrm{dB}], \tag{3.5}$$

where the units dBc/Hz represent the noise relative to the carrier. The RIN will usually be a function of frequency in the RF domain. Typical values for RIN are very small (large negative numbers in decibels), and those encountered in microwave photonics usually range from −100 to −175 dBc/Hz.

Intensity noise in the optical domain will contribute N_{opt} to the total noise PSD N_{out}, which plays a critical role in the RF performance (see Chapter 2). As described previously, N_{out} in this context is to be taken as the single-sided noise PSD, as it represents what would be physically measured. The definition of noise factor [Equation (2.6)] implies that N_{out} is to be measured in the absence of a signal with the input to the system terminated with a matched impedance. In many applications, however, noise on a signal is of particular interest and may not necessarily be completely quantified by the noise factor. Such applications include time and frequency metrology (Diddams et al. 2004) and radar systems (Skolnik 2008). Noise present on a signal that is not captured by the noise factor can include noise associated with the signal source itself or up-converted baseband noise. The latter can occur in systems employing mixers (see Figure 1.18) or in fiber optic systems where baseband RIN can degrade the RF signal purity (see Figure 3.15). In these instances, the signal-to-noise ratio (SNR) at the output is often used to describe the performance:

$$\mathrm{SNR}(f') = \frac{P_{\mathrm{rf}}}{N_{\mathrm{SSB}}(f')}, \tag{3.6}$$

where N_{SSB} is the "single-sideband" (SSB) noise power, that is, the total noise power in a 1-Hz bandwidth at some offset frequency from the carrier with average power P_{rf}. This N_{SSB} is not equivalent to the single-sided noise PSD mentioned previously. However, in the absence of oscillator and up-converted noise, the situation $N_{\mathrm{SSB}}(f') = N_{\mathrm{out}}(f') \times (1\,\mathrm{Hz})$ is possible and can be used to specify the SNR as a function of offset frequency from the carrier, f'. The usage of SNR in Equation (3.6) should not be confused with that in Equation (2.8); the latter was used to describe the noise factor, and implicitly requires that the signal does not carry additional noise. In addition, the definitions of SNR in Chapter 2 integrate the noise PSD over the operational bandwidth, whereas Equation (3.6) gives the SNR as a function of offset frequency by normalizing the noise to a 1-Hz bandwidth.

Phase noise is of particular interest for many applications. There are a variety of measures employed to quantify the stability in time and frequency of a signal, but one particular metric that is widely used is "script L" (Allan et al. 1988). "Script L" is defined as

$$\mathcal{L}(f') \equiv \frac{\text{SSB PSD due to phase-noise modulation [W/Hz]}}{\text{total signal power [W]}}, \tag{3.7}$$

where f' is the offset frequency from the carrier. In the case where pure noise N_{out} is superposed with a noise-free signal (Robins 1982),

$$\mathcal{L}(f') = \frac{N_{\mathrm{out}}(f')}{2P_{\mathrm{rf}}} \quad \text{(pure noise superposed).} \tag{3.8}$$

Amplified thermal noise and shot noise present situations in microwave photonics where Equation (3.8) applies, and this will be analyzed in the following two sections. Up-converted laser noise can also degrade the SNR [Equation (3.6)] in a microwave photonic link as exemplified in Section 3.4. Aside from these specific cases and some other scattered examples, the main emphasis in the remainder of this work will be to provide expressions for the RF noise factor in terms of measureable quantities. Namely, the total output noise PSD will be analyzed in terms of its constituent components.

3.2 THERMAL NOISE

The concept of thermal noise was introduced in Section 2.2 in order to define the RF noise factor. The term thermal noise is meant to describe the random voltage fluctuations across the terminals of a conductor in thermal equilibrium. These fluctuations are caused by the random motion of electrons in the conductor due to thermal agitation. The effect was quantified experimentally by Johnson (1928) and described theoretically by Nyquist (1928) in two sequential articles. Therefore, thermal noise is sometimes referred to as Johnson or Nyquist noise.

The Equipartition Theorem (Reif 1965)—a result of classical statistical mechanics—states that the average energy per degree of freedom is $k_B T/2$, where k_B is Boltzmann's constant and T is the absolute temperature. A one-dimensional oscillator has two degrees of freedom resulting in an average energy of $k_B T$ for an electromagnetic mode, with equal contributions from the electric and magnetic field. A conductor in thermal equilibrium can be modeled as a one-dimensional oscillator. As was first described by Nyquist (1928) and later by others such as Gordon (1964) and Haus (2000a), a circuit consisting of matched parallel resistors in thermal equilibrium will propagate power in both directions in the amount of $k_B TB$, where B is the bandwidth of the circuit. It is common to quantify thermal noise in terms of the mean-squared voltage across a single resistor with resistance R,

$$\langle V^2 \rangle = 4k_B TRB. \tag{3.9}$$

As described in Equation (2.4) and shown by Gordon (1964) using formal network theory, this voltage results in a maximum available power of $k_B TB$.

As noted by Nyquist, if the energy of the mode is taken as $hf/(e^{hf/k_BT} - 1)$, then the mean-squared voltage due to thermal noise in a resistor is

$$\langle V^2 \rangle = \frac{4RBhf}{e^{hf/k_BT} - 1}, \tag{3.10}$$

where f is the frequency and h is Planck's constant. The available PSD from Equation (3.10), $\langle V^2 \rangle/(4RB) = hf/(e^{hf/k_BT} - 1)$, is the original form given by Planck (1901) for the energy of a one-dimensional blackbody. This is a quantum-mechanical result but, as discussed in the following section, does not necessarily provide a complete quantum description. When $k_BT \gg hf$, a situation most likely encountered in electronics, Equations (3.9) and (3.10) are approximately equivalent. Consider the Taylor Series expansion for the exponential in the denominator of Equation (3.10),

$$e^{hf/k_BT} = \sum_{n=0}^{\infty} \frac{1}{n!} \left(\frac{hf}{k_BT} \right)^n \approx 1 + \frac{hf}{k_BT} \quad \text{for} \quad k_BT \gg hf \tag{3.11}$$

which when inserted into Equation (3.10) returns Equation (3.9). In the same manner, Planck's law for the energy in a one-dimensional blackbody, $U = hf/(e^{hf/k_BT} - 1)$, is approximately equal to the Rayleigh-Jeans law, $U = k_BT$, when $k_BT \gg hf$. The problem with the latter frequency-independent expression is that the thermal noise power is infinite at an infinite bandwidth. This so-called ultraviolet catastrophe is corrected by Planck's law. Note that both Planck's law as given previously and the Rayleigh–Jeans approximation yield zero energy at absolute zero temperature. However, quantum mechanics predicts a zero-point energy as described in the following section.

The equations previously stated describe the topic of this section, thermal noise. At the standard noise temperature (T_s = 290 K) and for microwave frequencies, the Nyquist equation [Equation (3.9)] adequately describes the output noise "engendered at the input frequency by the input termination" (Haus 1960) for any component or system. Therefore, the RF noise factor requires no further discussion. However, situations do arise in microwave photonics where other considerations may become important (when $k_BT \gg hf$ no longer holds). A few examples include cooled RF receivers where $T \ll T_s$, technology advances far into the millimeter wave range (>100 GHz), and the fact that the optical carrier frequency is hundreds of terahertz. For these reasons and for completeness, zero-point energy is briefly described in this chapter. The complete form for the energy of a blackbody is

$$U = \frac{\hbar\omega}{e^{\hbar\omega/k_B T} - 1} + \frac{\hbar\omega}{2}, \tag{3.12}$$

where $\hbar = h/(2\pi)$ and $\omega = 2\pi\nu$ is the angular frequency. Equation (3.12) is the conclusion of Planck's second formulation of blackbody radiation in 1912, although Planck did not employ modern quantum mechanics to arrive at the result (Milonni and Shih 1991). Today, the term $\hbar\omega/2$ is referred to as the zero-point energy. Zero-point energy was a consequence of the quantum theory firmly in place by the 1930s, where the mean square field fluctuations are non-zero in the vacuum state. Decades later, Callen and Welton (1951) arrived at the conclusion that a generalized resistance in thermal equilibrium exhibits an electromotive force including a term arising from the zero-point energy. Weber (1953) stated that there is no zero-point contribution to the maximum available power from a resistor; rather, the zero-point energy is observable only as the mean-squared value of the voltage across the resistor. Harris (1971) attributes the zero-point energy to the detection process and therefore discounts its use in thermal noise standards such as noise factor. However, Kerr (1999) suggests that the zero-point energy should be included in the thermal noise of a resistor. Before discussing this controversial topic further, it is instructive to consider numerical values of the noise PSDs associated with each description.

Let the available noise PSDs associated with Equations (3.9), (3.10), and (3.12) be

$$N_1 = k_B T, \tag{3.13}$$

$$N_2 = \frac{hf}{e^{hf/k_B T} - 1}, \tag{3.14}$$

$$N_3 = \frac{hf}{e^{hf/k_B T} - 1} + \frac{hf}{2}, \tag{3.15}$$

respectively. At the standard noise temperature of T_s = 290 K, N_1 = −173.9753 dBm/Hz, N_2 = −174.0112 dBm/Hz, and N_3 = −173.9751 dBm/Hz. Shown in Figure 3.2 are plots of these three equations for $f = 100$ GHz as a function of temperature from $T = 0$ to 50 K. At 50 K, the three curves are within 0.22 dB of each other. The curves diverge further as the temperature decreases. As $T \to 0$, N_1 and N_2 approach zero, with N_2 doing so faster than N_1 for $f = 100$ GHz. The

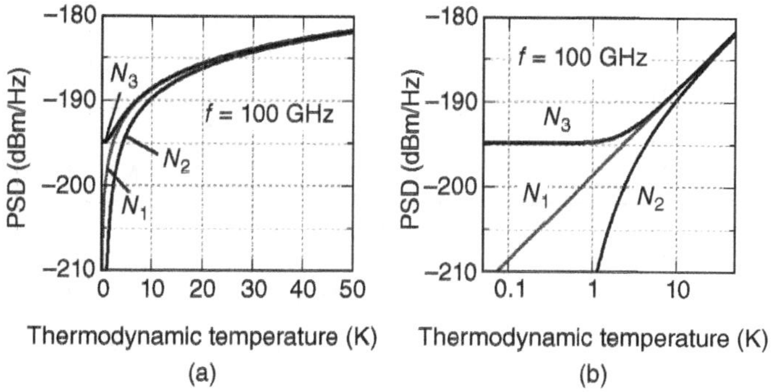

Figure 3.2. Calculations of the power spectral densities (PSDs) from Equations (3.13), (3.14), and (3.15) for $f = 100$ GHz.

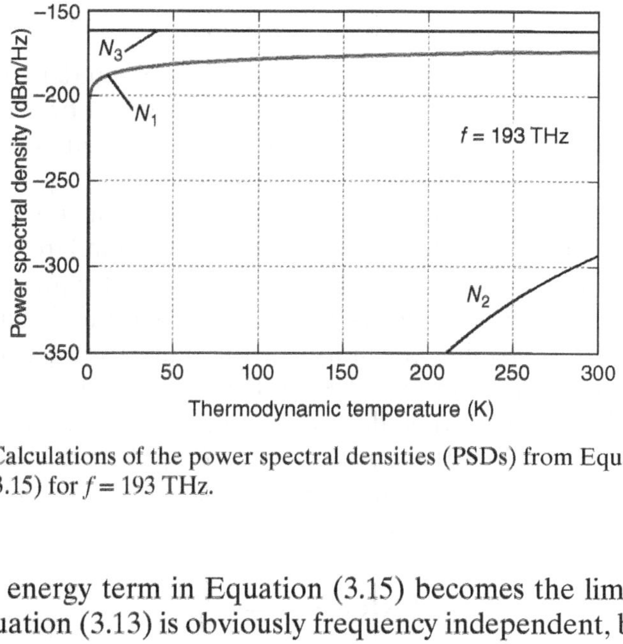

Figure 3.3. Calculations of the power spectral densities (PSDs) from Equations (3.13), (3.14), and (3.15) for $f = 193$ THz.

zero-point energy term in Equation (3.15) becomes the limit of N_3 as $T \rightarrow 0$. Equation (3.13) is obviously frequency independent, but the differences between N_1, N_2, and N_3 become more drastic as the frequency is increased. Consider the case where $f = 193$ THz, the optical carrier frequency for a 1550-nm fiber optic link, as shown in Figure 3.3. The value given by Planck's law, N_2, is below −290 dBm/Hz (1×10^{-59} W/Hz) for $T < 300$ K. The zero-point term represents the main contribution to N_3 and is larger than Nyquist's equation, N_1. The zero-point energy is therefore more important than thermal contributions at optical frequencies and can be considered the cause of shot noise as discussed in the following section. Finally, Equations (3.13)–(3.15) are plotted

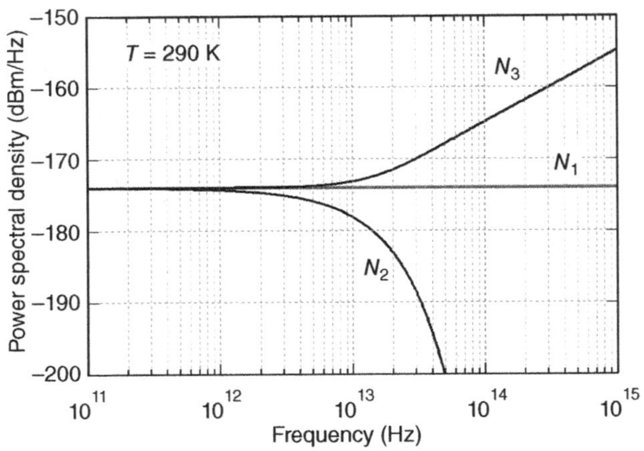

Figure 3.4. Calculations of Equations (3.13), (3.14), and (3.15) for $T = 290$ K.

as a function of frequency in Figure 3.4. At the standard noise temperature $T_s = 290$ K, the three curves are nearly identical until about 1 THz, after which N_1 remains constant, while N_2 eventually decreases exponentially and N_3 eventually increases linearly.

The primary metric for noise performance is the noise factor as defined in Chapter 2, which is defined at the standard noise temperature $T_s = 290$ K. Therefore, for electronic frequencies below 100 GHz, the PSD associated with Nyquist's equation, Equation (3.9), is adequate in describing the noise factor. Furthermore, at thermodynamic temperatures above 50 K and frequencies below 100 GHz, Equations (3.13)–(3.15) yield nearly the same value. For most instances in microwave photonics these conditions will apply, and therefore, the value $k_B T$ will be used for the thermal noise PSD from this point on. The associated SNR due to thermal noise is then

$$\mathrm{SNR}_{\mathrm{th}} = \frac{P_{\mathrm{rf}}}{k_B T B}, \tag{3.16}$$

which is equivalent to the input SNR defined in Equation (2.8) when $T = T_s$. Equation (3.16) can be normalized to $B = 1$ Hz, in which case, an expression in the form of Equation (3.6) is obtained. The thermal noise will contribute equally to amplitude and phase noise when superposed with an RF signal; the SSB phase noise PSD due to superposed thermal noise is

$$\mathcal{L}_{\mathrm{th}}(f') = \frac{k_B T}{2P_{rf}}. \tag{3.17}$$

Thermal-noise-limited signals as described by Equations (3.16) and (3.17) seldom exist in practice. Rather, added noise defined by the system noise factor will decrease the SNR and increase $\mathcal{L}_{\text{th}}(f')$. If the system noise factor and gain are frequency independent over the measurement bandwidth, Equations (3.16) and (3.17) can be adjusted to yield

$$\text{SNR} = \frac{P_{\text{rf}}}{Fgk_{\text{B}}T_sB} \tag{3.18}$$

$$\mathcal{L}(f') = \frac{Fgk_{\text{B}}T_{\text{s}}}{2P_{\text{rf}}}. \tag{3.19}$$

It is assumed that F relates to white noise in Equations (3.18) and (3.19). Equation (3.18) is the SNR at the output of a system at the standard noise temperature with noise factor F and gain factor g; Equation (3.19) is $\mathcal{L}_{\text{th}}(f')$ at the output of the same system.

Shown in Figure 3.5 are the measured $\mathcal{L}_{\text{th}}(f')$ for a 10.24-GHz oscillator (Poseidon Scientific Instruments SLCO-BCS) with and without an RF amplifier (Urick et al. 2012), the former demonstrating a significant increase in noise that is described by Equation (3.19). An Agilent E5500A/B phase noise test system was employed for the collection of these data. For both data sets in Figure 3.5, the RF output power was $P_{\text{rf}} = 5.3$ dBm. The thermal noise limit in this case is given by Equation (3.17) as −182 dBc/Hz. The oscillator's phase noise is

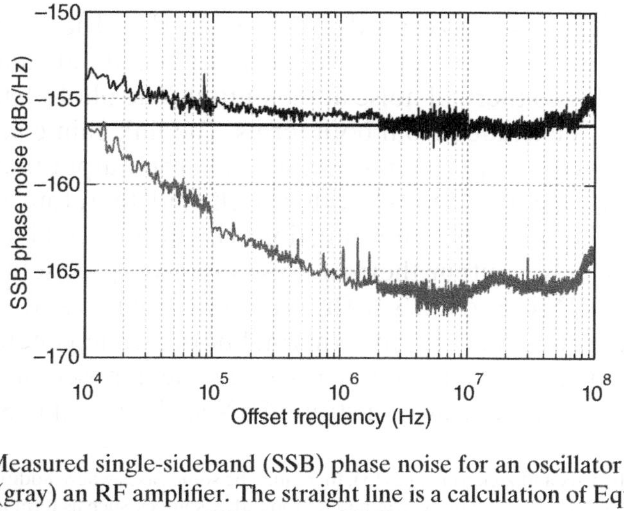

Figure 3.5. Measured single-sideband (SSB) phase noise for an oscillator with (black) and without (gray) an RF amplifier. The straight line is a calculation of Equation (3.19) for the experimental parameters (Urick et al. 2012).

shown by the grey curve and is well above the thermal noise limit. The black curve is the measured phase noise after passing through an RF amplifier with G = 19.5 dB and NF = 6.3 dB at and about 10.24 GHz. The solid black line is the result of Equation (3.19) with these parameters inserted, $\mathcal{L}_{\mathrm{th}}(f') = -156.5\,\mathrm{dBc/Hz}$. The amplified oscillator phase noise does contribute to the black curve, but the noise floor at −156.5 dBc/Hz is dominant from about 1 to 70 MHz.

Thermal noise exists at the input and output of an analog fiber optic link such as shown in Figure 3.1. The total noise PSD due to thermal noise at the output of such a link can be written as

$$N_{\mathrm{th}} = k_{\mathrm{B}}T(g+1), \tag{3.20}$$

where g is the gain factor for the link. The first term, $gk_{\mathrm{B}}T$, accounts for thermal noise generated at the input of the link that experiences the link transfer function as described by g.[1] Note that g contains the modulation response and the photodetector filter function. The second term in Equation (3.20), $k_{\mathrm{B}}T$, is due to thermal noise at the link output, assuming that the impedance of the photodetector matches that of the load and therefore transfers a noise PSD of $k_{\mathrm{B}}T$.

Equation (3.20) can be inserted into Equation (3.4) to obtain a RIN associated with thermal noise:

$$\mathrm{RIN}_{\mathrm{th}} = \frac{gk_{\mathrm{B}}T}{I_{\mathrm{dc}}^2 R} + \frac{k_{\mathrm{B}}T}{I_{\mathrm{dc}}^2 R}. \tag{3.21}$$

Equation (3.21) is normalized to B = 1 Hz. The first term in Equation (3.21) is due to thermal noise at the input, and the second term is due to thermal noise at the output. The photodetector filter function H_{pd} is excluded from both terms. The link gain g accounts for the photodetector response in the first term. The output thermal noise is generated after the photodetector itself and therefore should not contain the photodetector filter function. This latter point indicates that defining a RIN for output thermal noise does not precisely follow the definition of RIN in Section 3.1. The output thermal noise does not exist in the optical domain. The input thermal noise, while generated in the electrical domain at the link input, does become imposed on the optical carrier and would therefore constitute RIN as described in Section 3.1.

[1] As described by Cox and Ackerman (2013), there may be some cases where additional terms are required to account for thermal noise originating from other sources, such as termination resistors and lossy electrodes.

Equation (3.21) is included in this chapter for completeness and will not be used for the remainder of this text owing to the unconventional usage of the RIN definition for output thermal noise. The term RIN will be reserved to account for excess noise originating in the optical domain. However, as noted previously (Urick et al. 2011), generalizing the definition of RIN such as in Equation (3.21) allows for a compact formulation of the performance for some microwave photonic link architectures.

3.3 SHOT NOISE

Shot noise was discovered by Schottky (1918) while studying electron emission in vacuum tubes. Haus (2000a) gives a thorough mathematical description of shot noise from a few perspectives pertinent to microwave photonics. Following his work, this section gives brief, largely qualitative explanations of shot noise using three methods: a statistical analysis using the original assumptions made by Schottky, a semiclassical description of detection with a photodiode, and a purely quantum mechanical treatment of photodetection. For the purposes of this work, all three descriptions correspond.

In his original work, Schottky assumed that the emission of electrons from the cathode in a vacuum tube was completely random, resulting in current fluctuations with a mean squared value of

$$\left\langle i_{\text{sh}}^2 \right\rangle = 2qI_{\text{dc}}B, \tag{3.22}$$

where q is the elementary charge constant, I_{dc} is the DC current, and B is the bandwidth. Equation (3.22) is the commonly cited expression associated with shot noise. Furthermore, it can be shown that the emission of electrons from the cathode follows a Poisson probability distribution, which is an interesting result as it pertains to photodiodes.

As described in Chapter 1, a p–i–n junction can be used as a photodiode. Such a photodiode acts as a classical square-law device and will, with some efficiency, convert optical power into electrical power. For an incident optical field $E(t)$, the resulting photocurrent is $I(t) = \Re A(\varepsilon/\mu)^{1/2}|E(t)|^2/2$, where $\Re$ is the photodiode responsivity with units of A/W, A is the cross-sectional area of the fiber mode, ε is the permittivity of the fiber core, and μ is the permeability of the fiber core.[2] The

[2]This expression assumes that all of the power in the fiber core is transferred to the photodiode with a particular quantum efficiency. If power is lost between the fiber and the photodiode absorption region, the responsivity becomes an "effective responsivity," $\Re_{\text{eff}}$, where $\Re_{\text{eff}} \neq \eta q/(\hbar\omega)$.

resulting DC photocurrent sourced from a p–i–n junction illuminated by an average optical power $\langle P_{opt} \rangle$ is

$$I_{dc} = \Re \left\langle P_{opt} \right\rangle = \frac{\eta q}{\hbar\omega} \left\langle P_{opt} \right\rangle . \tag{3.23}$$

Equation (3.23) can be inserted into Equation (3.22) to determine the mean-squared photocurrent fluctuations due to shot noise. Equation (3.23) is semiclassical in the sense that it employs quantization of the electron charge (q) and the photon energy ($\hbar\omega$).

From a purely quantum-mechanical view, the photodiode absorbs photons and generates electron–hole pairs. The incident optical field can be represented by coherent quantum states, which exhibit fluctuations described by a Poisson process and demonstrate shot noise. The resulting photocurrent will therefore also have a shot noise component (Haus 2000a). Alternatively, the origin of shot noise can be related to the zero-point fluctuations discussed in the previous section. Shot noise in the optical domain may be viewed as fluctuations of carrier emission induced by zero-point fluctuations of the field. However, these zero-point fluctuations of the field "only contribute to the fluctuations" in the photocurrent; "zero-point fluctuations by themselves produce no photocurrent" (Haus 2000a).

The PSD at the output of a link such as shown in Figure 3.1 due to shot noise is

$$N_{sh} = 2qI_{dc}R\left|H_{pd}(f)\right|^2 . \tag{3.24}$$

Equation (3.24) is the PSD associated with Equation (3.22) including the response of the photodetector. Equation (3.24) can be written in a useful dB form as

$$N_{sh}[\text{dBm/Hz}] = -168 + 10\log(I_{dc}[\text{mA}]) + 2H_{pd}[\text{dB}]. \tag{3.25}$$

The shot noise is sometimes expressed as RIN by inserting Equation (3.24) into Equation (3.4):

$$\text{RIN}_{sh} = \frac{2q}{I_{dc}}. \tag{3.26}$$

Logarithmically, Equation (3.26) becomes

$$\text{RIN}_{sh}[\text{dBc/Hz}] = -155 - 10\log(I_{dc}[\text{mA}]). \tag{3.27}$$

The usage of RIN to quantify shot noise is more commonly used than for thermal noise [Equation (3.21)]. As described previously, shot noise does exist as fluctuations in the optical power, and the strict RIN definition therefore applies. However, as described in the following section, RIN will be used primarily to quantify noise in excess of thermal and shot noise that originates in the optical domain. Two examples of such excess noise are laser noise (Section 3.4) and noise introduced by optical amplification (Section 3.5).

Equations (3.25) and (3.27) are plotted as functions of I_{dc} in Figure 3.6 for $H_{pd} = 1$ (0 dB). As will be discussed in later chapters, microwave photonic links operating in the shot noise limit are desirable because the N_{sh} increases linearly with I_{dc}, whereas the signal level for many links increase quadratically with I_{dc}. In such cases, the noise factor will decrease as the received optical power increases. By the same token, RIN_{sh} decreases linearly with increasing I_{dc}. The is an important property of shot noise as will be demonstrated in Section 3.4, where the sensitivity of a laser RIN measurement is shown to increase with I_{dc}.

Shown in Figure 3.7 are the measured N_{sh} for $I_{dc} = 5$ and 10 mA (Urick et al. 2012). The data were obtained with a low-noise solid-state laser at 1.319 μm implemented in a crystal of $Y_3Al_5O_{12}$ doped with Nd^{3+} ions, commonly referred to as a Nd:YAG laser, and a p–i–n photodiode with $H_{pd} = 1$ over the frequency range shown. The particular Nd:YAG laser employed is shot-noise-limited over the frequency range of 100 MHz–1 GHz for the photocurrents employed. As depicted in

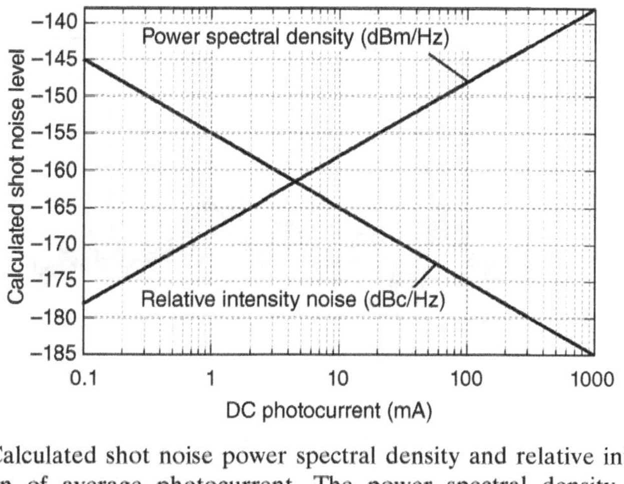

Figure 3.6. Calculated shot noise power spectral density and relative intensity noise as a function of average photocurrent. The power spectral density is given by Equation (3.25) with $H_{pd} = 1$; the relative intensity noise is given by Equation (3.27).

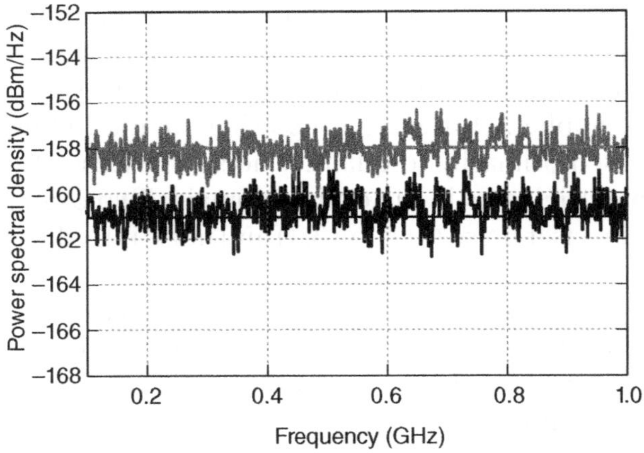

Figure 3.7. Measured shot noise power spectral density for average photocurrents of 5 mA (black) and 10 mA (gray). The straight lines are calculations of Equation (3.25) (Urick et al. 2012).

Figure 3.7, the calculated N_{sh} = −161 and −158 dBm/Hz for I_{dc} = 5 and 10 mA, respectively. The measured values vary about these calculated levels. A similar Nd:YAG laser was employed for the data set shown in Figure 3.8. Depicted there is RIN_{sh} measured with a high power partially depleted absorber photodiode (Tulchinsky et al. 2004) operated at I_{dc} = 80 mA (Urick et al. 2007). The sharp peak between 8 and 9 GHz is due to a sidemode in the Nd:YAG laser, but the laser is otherwise shot noise limited over the 1–10 GHz range for I_{dc} = 80 mA.

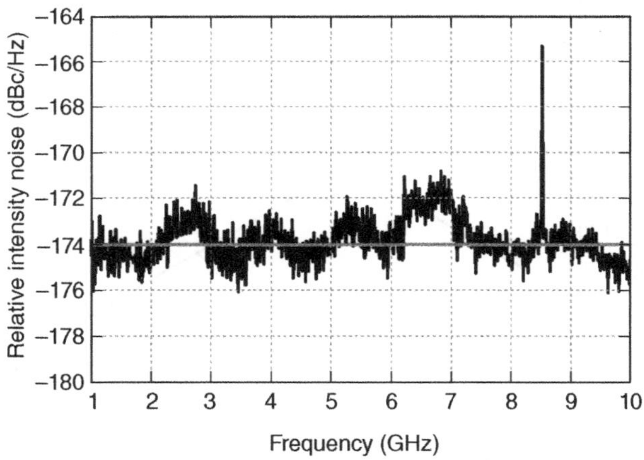

Figure 3.8. Measured relative intensity noise for a Nd:YAG laser at 80 mA average photocurrent. The bold line is a calculation of Equation (3.27) (Urick et al. 2007).

Equation (3.27) predicts $\text{RIN}_{\text{sh}} = -174$ dBc/Hz for $I_{\text{dc}} = 80$ mA. As shown in Figure 3.8, the measured RIN varied significantly about that value with some periodic structure. The cause of this structure was impedance mismatch between the photodiode and the measurement apparatus.

The SNR for a shot-noise-limited signal is

$$\text{SNR}_{\text{sh}} = \frac{P_{\text{rf}}}{2qI_{\text{dc}}RB\left|H_{\text{pd}}(f)\right|^2}, \tag{3.28}$$

where it is assumed that the photodetector filter function is already accounted for in the average RF power P_{rf}; the photodetector filter function is therefore included as part of the shot noise power in the denominator. Similarly to thermal noise, one-half of the shot noise PSD will contribute to the SSB phase noise when superposed with an RF signal. The expression for a shot-noise-limited $\mathcal{L}_{\text{th}}(f')$ is

$$\mathcal{L}_{\text{sh}}(f') = \frac{qI_{\text{dc}}R\left|H_{\text{pd}}(f)\right|^2}{P_{\text{rf}}}. \tag{3.29}$$

Shown in Figure 3.9 is the measured $\mathcal{L}_{\text{th}}(f')$ for a shot-noise-limited fiber optic at $I_{\text{dc}} = 10$ mA (Urick et al. 2012). For these experiments, a photodiode with an impedance-matching circuit was

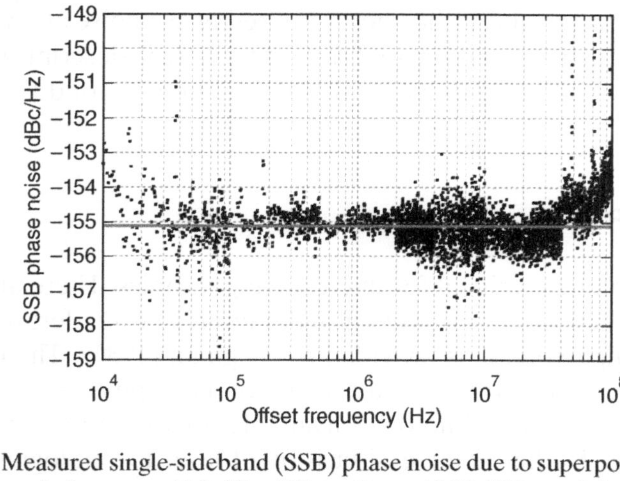

Figure 3.9. Measured single-sideband (SSB) phase noise due to superposed shot noise. The data are relative to a −11.9 dBm RF carrier at 10.24 GHz, and the average photocurrent is 10 mA. The bold line is a calculation of Equation (3.29) (Urick et al. 2012).

employed such that $|H_{pd}(f)|^2 = 1/4$ (−6 dB) and $R = 50\ \Omega$. With this photodiode, Equation (3.29) yields $\mathcal{L}_{sh}[\text{dBc/Hz}] = -177 + 10\log(I_{dc}[\text{mA}]) - P_{rf}[\text{dBm}]$. The experiment was arranged such that $P_{rf} = -11.9$ dBm, which results in $\mathcal{L}_{th}(f') = -155.1$ dBc/Hz. As depicted in Figure 3.9, the measured values vary about this level with exception to the high and low end of the frequency range shown; the oscillator phase noise contributes at those frequency offsets.

In all of the previous discussions concerning shot noise, it was assumed that the field could be represented by coherent states of light where zero-point fluctuations give rise to shot noise. Other "squeezed states" of light exist such as quadrature-squeezed states or states exhibiting sub-Poissonian statistics (Teich and Saleh 1989). It is beyond the scope of this book to treat squeezed states in any detail, but a brief mention in this chapter is included for completeness. As described by Walls (1983), a single-frequency field can be decomposed into two quadrature components. In coherent states, each quadrature exhibits the same noise due to the zero-point fluctuations. However, in a quadrature-squeezed state, the noise in one quadrature can be reduced at the cost of increased noise in the other quadrature. Such quadrature-squeezed states have been demonstrated in fiber systems to reduce noise below the shot noise limit (Bergman et al. 1993). These and other exotic states of light are rarely encountered in microwave photonics where the shot noise limits described previously are valid. However, a recent experimental demonstration by Quinlan et al. (2013a) has shown that classical states of light consisting of short optical pulses can exhibit shot-noise-limited phase noise at levels below that described by Equation (3.29). A theoretical description of that effect is provided in (Quinlan et al. 2013b) but is beyond the scope of this work.

3.4 LASERS

Laser technology is critical for microwave photonics. There are numerous pedagogical works on lasers for the interested reader; one particularly thorough volume is provided by Siegman (1986). The main purpose in this case is not to give a tutorial on laser physics nor detail all of the possible laser technologies available for use in microwave photonic systems. Rather, a terse discussion of laser technologies relevant to microwave photonics is provided followed by a description of laser noise, primarily in terms of RIN.

There are a variety of lasers that can be used for microwave photonics. Semiconductor lasers (Coldren et al. 2012) were widely used in early direct-modulation analog links and later as continuous-wave (CW) sources in externally modulated links and are staples in optical fiber communications (Suematsu 2014). Using a wide variety of materials, such as GaN, GaAs, InAlAs, and InGaAsP, semiconductor lasers are available over a very wide range of wavelengths from approximately 400 to 1600 nm and are pumped directly with a constant current. A variety of material structures can be employed to form a semiconductor laser. The most basic structure is a simple forward-biased diode using a p- and n-type semiconductor of the same material to form a homojunction. Homojunction lasers lack both carrier and optical mode confinement, requiring large threshold currents for inversion. These devices are much less efficient than heterojunction structures that use two or more different semiconductor materials to form the laser cavity (Delfyett and Lee 1991). Various styles of heterostructure lasers are commonly used today. A double-heterostructure structure design employs an active region surrounded with semiconductor material having a larger bandgap, thus minimizing absorption losses in unpumped material. When the active region layer thickness is comparable to the de Broglie wavelength ($\lambda = h/p$, where h is Planck's constant and p is the electron momentum), a quantum-well laser can be formed. The use of quantum wells discretizes the density of states, leading to lower threshold currents and higher efficiencies. Quantum-well laser designs make up the majority of semiconductor lasers today.

Many of the first semiconductor lasers exhibited multiple longitudinal modes. Multiwavelength emission can lead to various problems in both analog and digital links, including effects associated with fiber chromatic dispersion (Esman et al. 1988). Therefore, single-wavelength designs were quickly pursued. There are a few prevalent approaches to obtaining single-wavelength oscillations in semiconductor lasers, where feedback can be used to restrict the oscillating wavelength. Edge-emitting cavities can be coupled to a distributed Bragg reflector (DBR) or, alternatively, the reflector can be designed directly into the laser itself as in a distributed-feedback (DFB) structure. The DFB laser is commonly used because it is more robust to temperature fluctuations than the DBR design and it is widely available at the important optical fiber wavelengths of 1.32- and 1.55-μm. The vertical cavity surface emitting laser (VCSEL) also offers a small-scale efficient laser, but its use has been limited primarily to integrated chip-to-chip communications applications (Towe et al. 2000).

An alternative to the semiconductor laser for analog links is the solid state laser—that is, a laser based on active ions in a glass or crystal host (Siegman 1986). Solid state lasers have significantly slower gain dynamics than semiconductor materials leading to better high frequency RIN characteristics as well as narrower linewidths. Narrow laser linewidths can yield lower optical phase noise levels, which can be important in coherent- or angle-modulation formats. Solid state lasers are generally less efficient than their semiconductor counterparts, requiring an electrical current to pump a semiconductor laser that is then used as a pump laser to excite the ions.

A common solid-state laser is the $Y_3Al_5O_{12}$:Nd^{3+} laser, which has an active region comprising the yttrium aluminum garnet crystal ($Y_3Al_5O_{12}$ or YAG) doped with triply-ionized neodymium, Nd^{3+} (Powell 1998). Commonly referred to as Nd:YAG, this laser is pumped primarily at 730 or 808 nm. The main Nd:YAG emission line is at 1.064 μm, but it will also lase at other wavelengths, including the 1.32-μm line, which is near the zero-dispersion wavelength window of silica optic fiber. As described in Chapter 1, the advent of the erbium-doped fiber amplifier (EDFA) significantly advanced the capabilities of fiber optic systems operating around the low-loss wavelength region in fiber near 1.55 μm. The same dopant in an EDFA, Er^{3+}, can be doped into glasses or other materials to construct lasers matching the EDFA emission wavelengths (Digonnet 2001). Oftentimes, Yb^{3+} will be co-doped with Er^{3+} to increase the pump absorption, especially at 980 nm, offering the potential for short-cavity lasers. The list of lasers in this case is certainly not inclusive but does represent the main technologies used for microwave photonics. Other, less-prevalent laser technologies for microwave photonics worth mentioning include Brillouin (Geng et al. 2006) and Raman (Dianov and Prokhorov 2000) fiber lasers, where the latter is primary used as a pump laser.

Whatever laser technology is employed, it is desirable to have high optical power with low noise. Because photodetectors will only respond to changes in optical intensity, laser intensity fluctuations are of primary importance. However, as will be described later, laser phase fluctuations can be problematic in systems when certain mechanisms convert phase noise into intensity noise. Laser noise is typically cast in terms RIN as given by Equation (3.4):

$$\mathrm{RIN}_{\mathrm{laser}} = \frac{N_{\mathrm{laser}}B}{I_{\mathrm{dc}}^2 R|H_{\mathrm{pd}}(f)|^2}, \tag{3.30}$$

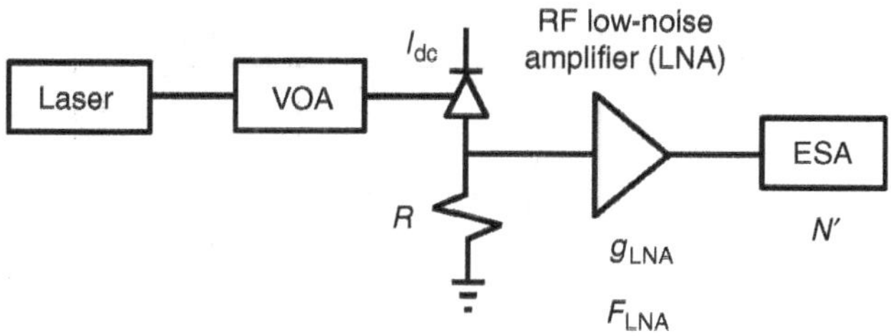

Figure 3.10. Experimental setup for laser relative intensity noise (RIN) measurements. VOA: variable optical attenuator, ESA: electrical spectrum analyzer.

where N_{laser} is the noise PSD at the output of the photodiode due to laser noise. As described in the text following Equation (3.4), Equation (3.30) is typically normalized to $B = 1$ Hz and expressed on a decibel scale with units dBc/Hz. An apparatus to measure intrinsic laser intensity noise is shown in Figure 3.10. Figure 3.10 is very similar to the link shown in Figure 3.1, with the exception that the measurement apparatus does not include an external modulation mechanism. The output of a laser is directed to a photodiode, typically via fiber for microwave-photonics applications. The laser light may need to be attenuated depending on the relationship between the laser output power and the photodiode power handling in order to keep the photodiode in its linear operating regime. Therefore, a variable optical attenuator may be placed between the laser and photodetector. Because the RIN is a relative measure, attenuation between the laser and the photodetector does not affect the laser RIN. However, as described by Equation (3.26), the shot-noise-limited RIN decreases with increasing photocurrent. Therefore, the sensitivity of a RIN measurement is determined by the amount of average laser power that is detected as shown in Figure 3.6. Although it is possible to measure RIN a few decibels below that determined by Equation (3.26), the accuracy of the measurement (Obarski and Splett 2000) is determined by many factors including the underlying statistics of the RIN and the methodology used in the measurement process. A common method, as depicted in Figure 3.10, allows the output noise PSD to be analyzed on an electrical spectrum analyzer. The laser noise will usually be at a level below the electrical spectrum analyzer noise floor. Therefore, a low noise amplifier (LNA) can be inserted between the photodetector and spectrum analyzer to allow the optically generated noise to be measured. Assuming that the noise floor of the spectrum analyzer is negligible with respect to the noise at the output of the LNA (N'), the RIN is given by

$$\text{RIN} = \frac{4\left(N'/g_{\text{LNA}} - F_{\text{LNA}}k_{\text{B}}T\right)}{I_{\text{dc}}^2 R}, \tag{3.31}$$

where g_{LNA} is the gain factor for the LNA, and F_{LNA} is the noise factor for the LNA. Additional assumptions pertaining to Equation (3.31) are that output thermal noise does not contribute to the RIN [see the discussion below Equation (3.21)], $H_{\text{pd}} = 1/2$ over the RF bandwidth, $B = 1$ Hz, g_{LNA} and F_{LNA} are constant over the measurement bandwidth, and the apparatus in Figure 3.10 is employed. Neglecting any mechanisms to modulate input thermal noise onto the laser, the RIN as given by Equation (3.31) should include contributions from laser noise and shot noise only.

In applied RF photonics applications, the best practice is to measure the RIN spectrum of any laser under consideration carefully over the frequency range of interest using an architecture such as that shown in Figure 3.10. Shown in Figure 3.11(a) are the measured noise PSDs for a Nd:YAG laser with $I_{\text{dc}} = 5$, 10, and 20 mA (Urick et al. 2012). For these data, laser noise dominates below 3 MHz and shot noise is the principle noise source above 300 MHz. As shown in Figure 3.11(a), the PSD scales as I_{dc} (3 dB change for a doubling of the average current) in the shot-noise-limited regime and as I_{dc}^2 (6 dB for a 2× increase in average photocurrent) when laser noise dominates. The RIN spectra corresponding to the measured noise PSDs in Fig. 3.11(a) are shown

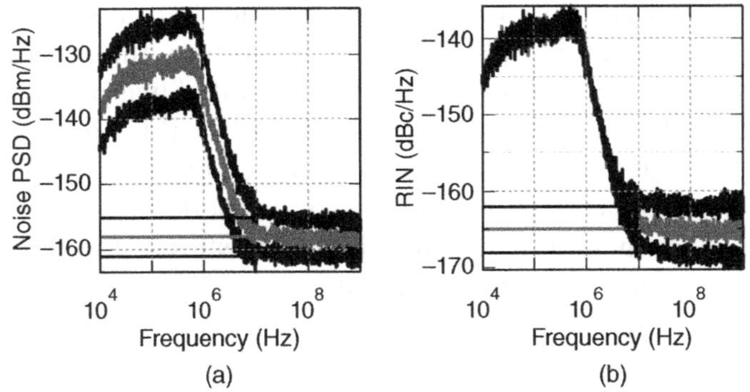

Figure 3.11. (a) Measured noise power spectral density (PSD) using a Nd:YAG laser with average photocurrents of 5 mA (bottom black), 10 mA (gray), and 20 mA (top black). The bold lines are calculations of the shot noise limits from Equation (3.25). (b) Measured relative intensity noise (RIN) using a Nd:YAG laser with average photocurrents of 5 mA (top black), 10 mA (gray), and 20 mA (bottom black). The bold lines are calculations of the shot noise limits from Equation (3.27) (Urick et al. 2012).

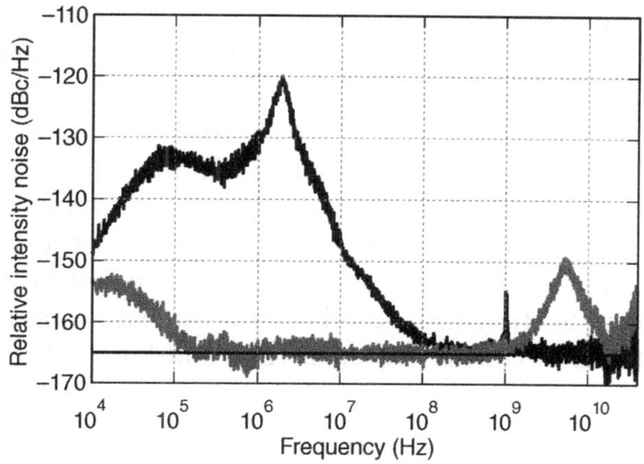

Figure 3.12. Measured relative intensity noise for a semiconductor distributed feedback laser (gray) and an erbium-doped fiber laser (black). Both data sets are at an average photocurrent of 10 mA, which results in a shot-noise limit of −165 dBc/Hz designated with the bold line (Singley et al. 2011a).

in Figure 3.11(b). Note that where laser noise dominates ($f < 3$ MHz), the RIN spectra are independent of the average photocurrent. In contrast, in the shot-noise-limited regime ($f > 300$ MHz), the measured RIN decreases with increasing photocurrent as $\mathrm{RIN}_{\mathrm{shot}} \propto 1/I_{\mathrm{dc}}$. In practice, converting between RIN and PSD is often required when translating laser noise performance, which is usually cast as RIN, into RF system metrics.

Shown in Figure 3.12 are two RIN spectra measured with an apparatus such as that in Figure 3.10. Shown are spectra for a solid-state erbium-doped fiber laser and a semiconductor DFB laser, both for $I_{\mathrm{dc}} = 10$ mA, which sets the shot-noise-limited RIN level at −165 dBc/Hz (Singley et al. 2011a). The solid-state laser is shot noise limited at this level from about 1 to 20 GHz; the semiconductor laser is shot noise limited from just above 100 kHz to about 1 GHz. Peaks in the RIN spectrum of a laser can be caused by numerous sources, including competing modes or noise external to the cavity. The sharp peak near 1 GHz in the fiber laser spectrum shown in Figure 3.12 is due to a second longitudinal mode. However, the broader peaks at a few megahertz for the fiber laser and near 6 GHz for the semiconductor laser are due to relaxation oscillations. Most lasers encountered in microwave photonics will exhibit relaxation oscillations; the relaxation oscillation frequency will sometimes be referred to as "the RIN peak" for a particular laser. Relaxation oscillations generally occur in lasers when the excited-state lifetime is longer than the cavity damping time, which is a condition met by most solid-state and semiconductor lasers. Relaxation oscillations occur about the steady-state amplitude and are triggered by even the smallest perturbations in the laser cavity or pump intensity. As shown in Figure 3.12, the relaxation oscillation frequencies can vary significantly among different lasers.

One can find numerous expressions for relaxation oscillation frequencies in the literature, and two will be mentioned in this chapter. As described in Section 1.3, the relaxation oscillation frequency for a semiconductor laser can be written as (Lau and Yariv 1985)

$$f_r = \frac{1}{2\pi}\sqrt{\frac{Ap_0}{\tau_p}}, \tag{3.32}$$

where A is the differential optical gain coefficient, p_0 is the steady-state photon density, and τ_p is the photon lifetime. Equation (3.32) is important in semiconductor laser design not only because it describes a peak in the noise spectrum, but also because it gives the direct modulation bandwidth of the laser (also see Section 8.1). Siegman (1986) employed the rate equations for the four-level Nd:YAG system to derive the relaxation frequency as

$$f_r = \frac{1}{2\pi}\sqrt{\frac{(W/W_{th} - 1)}{\tau_a \tau_c}}, \tag{3.33}$$

where W is the pump intensity, W_{th} is the threshold pump intensity, τ_a is the atomic lifetime, and τ_c is the cavity lifetime. In a Nd:YAG $\tau_a \gg \tau_c$, dictating that f_r is determined by the pump intensity and atomic lifetime for a fixed threshold pump intensity.

Relaxation oscillations can sometimes be suppressed in lasers. Shown in Figure 3.13 (Urick et al. 2007) is a particular example of the effectiveness of a feedback circuit used to suppress the relaxation oscillation in a Nd:YAG laser (Lightwave Electronics Model No. M125N-1319-200). In this case, the RIN peak near 300 kHz was reduced by more than 40 dB at the cost of slightly increased noise in the region between 800 kHz and 3 MHz. The spectra in Figure 3.13 were obtained with a high power photodiode (Tulchinsky et al. 2004) operated at $I_{dc} = 80\,\text{mA}$, which sets the shot noise limit at −174 dBc/Hz. The photodiode did not have an impedance-matching circuit leading to a mismatch causing the ripple seen above 30 MHz. The spike near 8.5 GHz corresponds to the laser mode spacing. These two features are more easily seen in Figure 3.8,

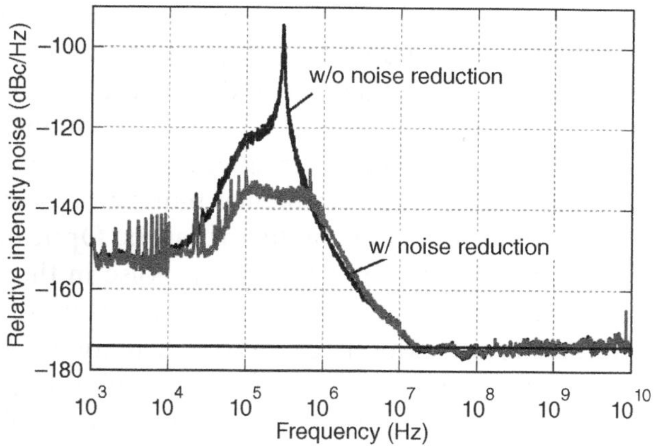

Figure 3.13. Measured relative intensity noise spectra for a Nd:YAG laser with and without noise reduction circuitry employed. The bold line is the shot noise limit for 80 mA average photocurrent as given by Equation (3.27) (Urick et al. 2007).

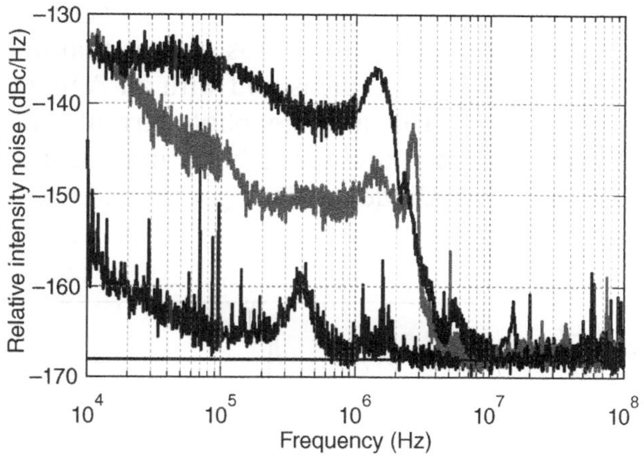

Figure 3.14. Measured relative intensity noise spectra for a semiconductor distributed feedback laser employing three different current sources. The bold line is the shot-noise limit for 20 mA average photocurrent (Diehl et al. 2007).

which is a plot of a portion the data in Figure 3.13 with a linear frequency axis.

The drive electronics can also significantly affect the noise performance of a laser. Shown in Figure 3.14 are three RIN spectra for the same semiconductor DFB laser operated with three different electronic drivers. The details for that data set are provided by Diehl et al. (2007) but suffice it to say that the choice of driver can significantly change the noise spectrum for a laser. In the case of a semiconductor DFB, noise in the pump current can directly modulate the laser intensity and manifest as excess RIN. The efficiency and bandwidth of this process depend on the particular laser. Generally, lasers will respond to changes in the pump level most efficiently at frequencies below the relaxation oscillation. Therefore, even high frequency pump noise on the order of gigahertz can couple into semiconductor lasers. Optically pumped solid-state lasers can be more robust to pump noise in the microwave regime. For example, solid-state lasers including erbium-doped fiber lasers and Nd:YAG lasers exhibit relaxation oscillations below a few megahertz (see Figures 3.12 and 3.13). For links operating above the relaxation oscillation, such low frequency noise is only problematic in that it can up-convert to the RF carrier.

Shown in Figures 3.15 and 3.16 are examples of laser noise that has been up-converted onto an RF carrier. Shown in Figure 3.15 are three measured spectra for an erbium-doped fiber laser: the baseband RIN spectrum, the SSB phase, and SSB amplitude noise for an RF carrier in a photonic link employing the same laser (Urick et al. 2012). The RIN was measured with an apparatus such as that in Figure 3.10, and the up-converted noise was measured with an Agilent E5500A/B phase noise test system. The average photocurrent was 5 mA for both the RIN and link measurements; both architectures employed impedance-matched photodiodes with $H_{\text{pd}} = 1/2$ and $R = 50\ \Omega$. The

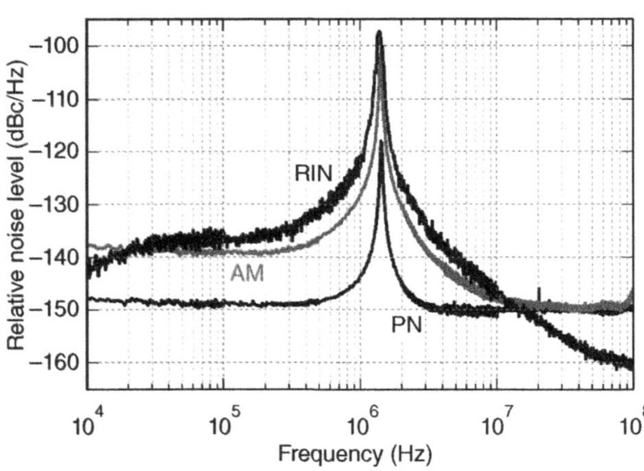

Figure 3.15. Measured baseband relative intensity noise (RIN), up-converted single-sideband (SSB) amplitude noise (AM), and up-converted SSB phase noise (PN) for a link employing an erbium-doped fiber laser. The up-converted spectra are relative to a RF carrier at 10.24 GHz (Urick et al. 2012).

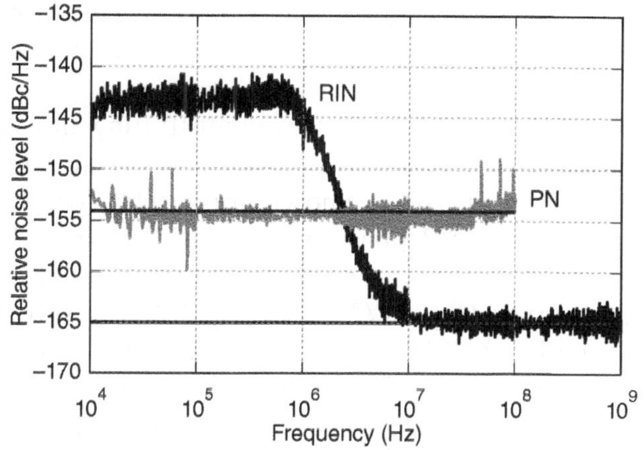

Figure 3.16. Measured baseband relative intensity noise (RIN) and up-converted single-sideband phase noise (PN) for a link employing a Nd:YAG laser. The shot-noise limits are shown with bold lines for each data set. The up-converted spectra are relative to a RF carrier at 10.24 GHz (Urick et al. 2012).

frequency axis for the up-converted noise is relative to the 10.24-GHz carrier frequency at an absolute amplitude of P_{rf} = 4 dBm. An LNA with G = 20 dB and NF = 6 dB was employed for the up-converted noise measurements. The noise floor for the up-converted measurements is determined by superposed shot noise [Equation (3.29)] and superposed noise from the LNA [Equation (3.19)]. Given the aforementioned parameters, Equation (3.19) yields $\mathcal{L}_{th}(f') = -155\,\text{dBc/Hz}$; the shot-noise limit is $\mathcal{L}_{sh}(f') = -154\,\text{dBc/Hz}$ noting that $P_{rf} = -16$ dBm was inserted in Equation (3.29), which is the RF power level before the LNA. The combination of these two sources sets the floor at −151 dBc/Hz, which agrees with the measured values above 10 MHz in Figure 3.15. The shot noise floor for the RIN spectrum is given by Equation (3.27) as −162 dBc/Hz. The relaxation-oscillation peak near 1.4 MHz is the most notable feature in Figure 3.15. The peak is at a level of −97 dBc/Hz in terms of RIN and near −101 dBc/Hz in terms of SSB amplitude noise. The relaxation oscillation can be considered a modulating signal for the RF carrier; the expected level per sideband in this case would be 3 dB below the modulating signal power (Robins 1982). The observed 4-dB difference in the RIN and SSB amplitude noise spectra is in agreement, with the 1-dB discrepancy being within the experimental error. The peak in the SSB phase noise spectrum is at a level of −118 dBc/Hz, 17 dB below the amplitude noise, and is attributed to the measurement system's limited ability to reject amplitude noise. Therefore, the laser noise in this case contributes to the up-converted amplitude noise but not to the up-converted phase noise.

Shown in Figure 3.16 are two spectra demonstrating another instance where laser intensity noise does not contribute to up-converted RF phase noise (Urick et al. 2012). Shown are spectra for the RIN of a Nd:YAG laser and the SSB phase noise relative to a 10.24-GHz RF carrier at the output of a link using the same Nd:YAG laser. For both data sets, $I_{dc} = 10\,\text{mA}$, $H_{pd} = 1/2$, and $R = 50\,\Omega$. The shot noise RIN is given by Equation (3.27) as −165 dBc/Hz. The RIN spectrum is shot noise limited above 10 MHz, but laser RIN well above shot noise is observed at lower frequencies. The phase noise data are relative to a −13-dBm carrier, setting the shot noise limit at −154 dBc/Hz [Equation (3.29)]. The measured phase noise is shot noise limited over the entire frequency range shown. Clearly, the laser RIN, as high as −144 dBc/Hz, does not contribute to up-converted phase noise in this case.

As mentioned previously, photodiodes convert fluctuations in optical intensity to fluctuations in current. Therefore, laser intensity noise is the main concern for RF photonic links. Optical phase fluctuations do not affect the RF performance of photonic links unless there is some mechanism to convert phase fluctuations to intensity fluctuations. Unfortunately, there are numerous such mechanisms. Unintended reflections in a link can create a multipath environment in which phase noise converts to intensity noise before photodetection. Fiber effects such as double Rayleigh scattering and chromatic dispersion (see Chapter 5) can also convert optical phase fluctuations to intensity fluctuations in direct-detection systems. Phase-to-intensity conversion is intentional in systems where information is encoded onto the optical phase, and such systems will certainly be affected by optical phase noise as well as intensity noise. The remainder of this section will therefore give an introduction into the concepts related to laser phase noise and its measurement.

There are numerous sources of phase noise in a laser system. The concentration in this section is on fluctuations about the laser center line described by a linewidth. Many mechanisms can contribute to the broadening of a laser line. Such mechanisms are classified into two categories: homogenous broadening and inhomogeneous broadening. Homogenous broadening occurs when the broadening mechanism affects all atoms with the same probability and is described by a

Lorentzian intensity profile (Powell 1998):

$$I_o(\omega) \approx \frac{\Delta\omega}{(\omega-\omega_0)^2+(\Delta\omega/2)^2}, \tag{3.34}$$

where ω_0 is the angular frequency of the peak and $\Delta\omega$ is the full width at half-maximum (FWHM) linewidth. Mechanisms that cause homogenous broadening include the natural broadening of laser lines governed by quantum mechanics, Raman scattering and lifetime broadening in solid-state laser systems, and collision broadening in gas lasers. Homogeneously broadened lasers can exhibit spatial hole burning. Spatial hole burning occurs when a laser gain medium is spatially saturated by one cavity mode. Other longitudinal modes can still experience gain in spatial regions where the gain remains unsaturated (Tang et al. 1963).

Inhomogeneous broadening is caused by processes affecting each atom with random probability. A Gaussian function can be used to describe inhomogeneous broadening (Powell 1998):

$$I_o(\omega) = \frac{4\sqrt{\pi \ln 2}}{\Delta\omega} \exp\left[\frac{-4\left(\omega-\omega_0\right)^2 \ln 2}{(\Delta\omega)^2}\right]. \tag{3.35}$$

Doppler broadening in gas systems is a classic example of inhomogeneous broadening. Gaussian functions have also been used to quantify $1/f$ noise in lasers (Mercer 1991). Spectral hole burning is a phenomenon associated with inhomogeneous broadening. Spectral hole burning occurs when different excited modes independently saturate, thus producing "holes" in the gain with respect to wavelength.

The natural linewidth of a laser, determined by the Heisenberg uncertainty principle, was first described by Schawlow and Townes (1958). Early measurements of the quantum-limited Schawlow–Townes linewidth were conducted by Hinkley and Freed (1969) and follow the expected Lorentzian profile. Henry (1982) later introduced a modified Schawlow–Townes linewidth, including the "alpha parameter," to describe broadening observed in semiconductor lasers. The alpha parameter is the ratio of deviations from the steady-state value of the real part of the index of refraction to that of the imaginary part. The alpha parameter is described thoroughly in two later review articles by Henry (1986) and Osinski and Buus (1987) and is still employed for semiconductor laser analysis (Chuang et al. 2014). In some cases, both homogenous and inhomogeneous processes can determine the line

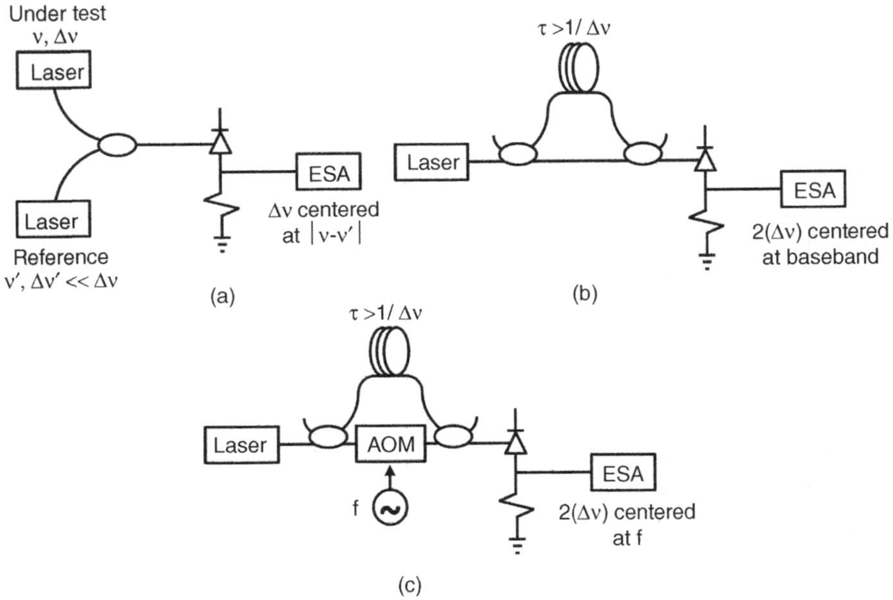

Figure 3.17. Experimental setups for laser linewidth characterization. Shown are (a) heterodyne setup, (b) homodyne setup, and (c) self-heterodyne setup. AOM: acousto-optic modulator.

shape of a laser. In such cases, neither a Lorentzian nor Gaussian by itself will reproduce the line shape. Rather, a Voigt function containing contributions from both Lorentzian and Gaussian distributions can be employed to describe the measured spectrum (Mercer 1991).

Measurement of laser linewidth can be conducted in a variety of manners. The three most prevalent methods are shown in Figure 3.17. Depicted are the conceptual diagrams for the (a) heterodyne, (b) homodyne, and (c) self-heterodyne techniques. It is beyond the scope of this work to detail each extensively, and only the basics will be described in this chapter. Baney and Sorin (1998) give a complete explanation of these and other techniques. The heterodyne technique uses a reference laser to measure the laser of interest. The reference laser and laser under test are combined in the optical domain with, for example, a fiber optic coupler. The combined field is detected with a photodetector and observed in the electrical domain typically with an electrical spectrum analyzer. If the linewidth of the reference laser is much smaller than the laser being characterized, then the line shape in the electrical domain is a copy of the laser under test centered at the frequency difference between the two lasers. The heterodyne technique is useful because it will display asymmetries in the laser under test. However, the method is difficult to implement for state-of-the-art lasers, because a reference laser may not be attainable with sufficiently narrow linewidth.

Neither the self-homodyne nor the self-heterodyne technique requires a reference laser. Rather, the laser under test is mixed with itself. This is achieved by splitting the laser, delaying one signal with respect to the other, recombining the signals, and then passing them to a photodetector. In the self-homodyne technique, the measurement is performed at baseband, whereas one signal is frequency shifted in the self-heterodyne method, thus producing a beat note at the shift frequency in the electrical domain. This is advantageous because it moves the measurement away from baseband in the electrical domain where significant system noise can be present. A stable measurement is difficult to obtain using either technique when the delayed copy of the laser remains coherent with the undelayed copy. Therefore, it is desirable to use a delay much longer than the coherence length of the laser. For a linewidth of $\Delta\nu = \Delta\omega/(2\pi)$, the time delay must be

$$\Delta\tau > \frac{1}{\Delta\nu} \tag{3.36}$$

to be outside the coherence time of the source. From Equation (3.36), the required delay length in a medium with index of refraction n is

$$\Delta L > \frac{c}{n\Delta\nu}, \tag{3.37}$$

where c is the speed of light in vacuum. Equation (3.37) is plotted in Figure 3.18 for $n = 1.5$, a value typical for optical fiber. For both the self-homodyne and the self-heterodyne techniques, the spectrum in the electrical domain represents a convolution of the laser spectrum with itself. Two consequences of this are that any asymmetries in the line shape are lost and the linewidth in the electrical domain is larger than the actual linewidth of the laser. For a pure Lorentzian line shape, the measured FWHM linewidth is twice the actual linewidth; a pure Gaussian line shape will be broadened by a factor of $\sqrt{2}$ (Baney and Sorin 1998). If the Lorentzian line shape given by Equation (3.34) is shifted to baseband ($\omega_0 = 0$), changed to a function of Hertzian frequency ($\nu = \omega/2\pi$), and converted to a Lorentzian with twice the width ($\Delta\omega/2\pi = 2\Delta\nu$),

$$I_0(\nu) \approx \frac{\Delta\nu}{\pi[\nu^2 + (\Delta\nu)^2]}. \tag{3.38}$$

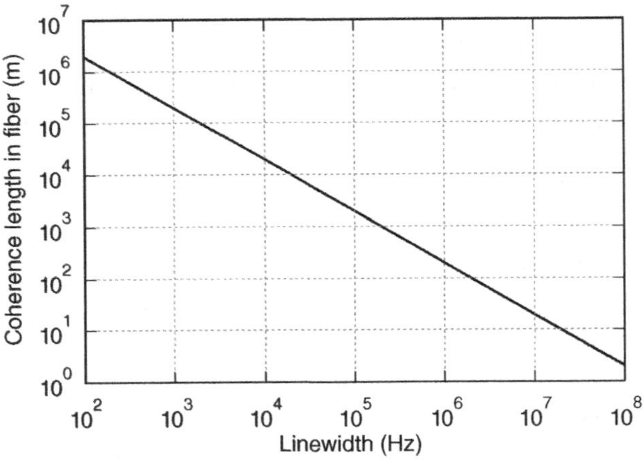

Figure 3.18. Plot of the minimum coherence length as given by Equation (3.37) with $n = 1.5$ (optical fiber) as a function of laser linewidth.

Equation (3.38) can be used to fit the intensity profile in the electrical domain to determine the FWHM Lorentzian linewidth in the optical domain, $\Delta\nu$.

The self-heterodyne technique is often employed for characterization of lasers used in microwave photonics. The first demonstration of the self-heterodyne technique was conducted by Okoshi et al. (1980) using a fiber Mach–Zehnder interferometer. One arm of the interferometer comprised a fiber delay, and the other provided a frequency shift via an acousto-optic modulator (AOM). Other types of asymmetric interferometers could be employed to the same end, such as a Michelson interferometer. The resolution for such an architecture has been given as (Tsuchida 1990)

$$\Delta\nu_{\text{min}} = \frac{c\left[-0.4\log\left(\text{RBW}\cdot n\Delta L/c\right)+0.6\right]}{n\Delta L}, \tag{3.39}$$

where RBW is the resolution bandwidth of the electrical spectrum analyzer, and the condition $(\text{RBW}\cdot n\Delta L/c) < 0.3$ must be upheld. Equation (3.39) sets the fiber length for a self-heterodyne architecture and agrees with Equation (3.37) when $(\text{RBW}\cdot n\Delta L/c) = 0.1$. To reduce the need for long lengths of fiber, recirculating optical delay lines have also been employed for laser characterization. Tsuchida (1990) was the first to propose a recirculating fiber optic delay line for the self-heterodyne technique, with other variants following such as those with loss compensation (Dawson et al. 1992).

Shown in Figure 3.19 is a recirculating fiber optic delay line with some unique features. The laser under test is passed through a fiber optic isolator and split equally into two paths. The first path goes directly to a photodiode, while the other is launched into a custom fiber span. The span consists of twelve 3.125-km lengths, six each of two types of fiber spliced in an alternating manner, for a total length of 37.5 km (187.5 μs delay). Such an arrangement mitigates the deleterious effects of stimulated Brillouin scattering (SBS) and chromatic dispersion, fiber effects that are detailed in Chapter 5. After propagating the fiber span, the signal frequency is shifted by 55 MHz via an AOM and then passed to a polarization controller. The polarization controller is used to align the delayed signal with the undelayed signal, thus maximizing the beat note. Half of the delayed signal is coupled to the photodiode, with the other half being recirculated. If the input electric field is assumed to be of the form, $E_{in}(t) = E_0 \exp[i\omega_0 t + i\phi(t)]$, the ideal output electric field is given by

$$E_{out}(t) = \frac{E_0}{\sqrt{2}} \sum_{m=0}^{\infty} \left(\sqrt{\alpha/2}\right)^m \exp[i(\omega_0 + m\Omega)(t - m\tau) + i\phi(t - m\tau)], \tag{3.40}$$

where E_0 is the electric field amplitude, ϕ is the time-dependent phase, m corresponds to the number of circulations through the fiber loop, α is the optical power loss in the fiber loop, Ω is angular drive frequency for the AOM, and τ is the time delay through the loop. Square-law detection of the field given by Equation (3.40) yields a delayed self-heterodyne

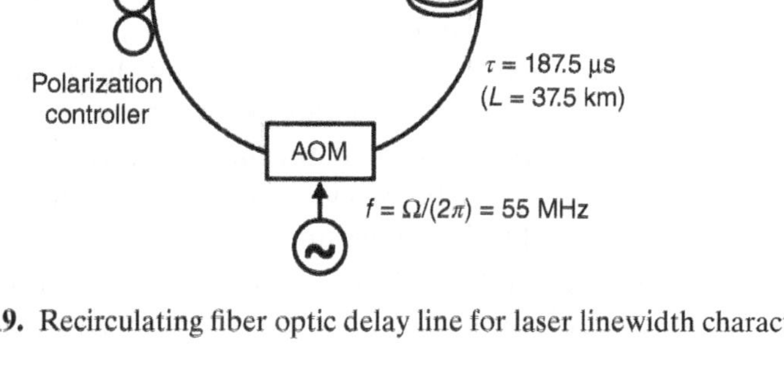

Figure 3.19. Recirculating fiber optic delay line for laser linewidth characterization.

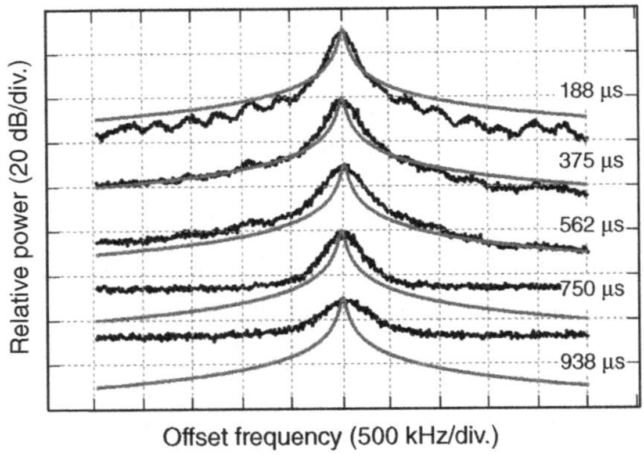

Figure 3.20. Measured lineshape as a function of recirculation time of an erbium-doped fiber laser for one to five passes through the loop shown in Figure 3.19. The amplitudes are offset 30 dB and normalized in frequency for comparison purposes. The lines are a calculated Lorentzian [Equation (3.38)] with a 2.5-kHz full-width half-maximum linewidth (Urick 2004).

measurement of the laser linewidth at time delays $m\tau$ centered about frequencies $m\Omega/(2\pi)$.

The architecture shown in Figure 3.19 was employed to characterize an erbium fiber laser. The results are shown in Figure 3.20 for up to five passes through the fiber delay line (Urick 2004). With RBW = 1 kHz, the resolution given by Equation (3.39) for these data are 4.747, 2.052, 1.243, 0.866, and 0.651 kHz for $m = 1$–5, respectively. Each data set in Figure 3.20 is normalized to the center frequency of the beat note ($m \cdot 55$ MHz) and the maximum power. Each spectrum is compared to a Lorentzian line shape [Equation (3.38)] with $\Delta\nu = 2.5$ kHz. For the first pass with a 188-μs delay, the laser is still coherent and interference fringes can be seen in the skirts of the spectrum. By Equation (3.36), this indicates that the linewidth is smaller than 5.4 kHz. At a 375-μs delay ($m = 2$), the 2.5-kHz Lorentzian provides a good fit to the skirts, but there appears to be some Gaussian broadening near the peak. In addition, there is very weak interference observed pointing to a linewidth on the order of 2.6 kHz according to Equation (3.36). On subsequent passes, the skirts of the spectra become masked by the noise floor of the measurement system and the broadening worsens. A conclusion from the data set in Figure 3.20 is that the laser exhibits a linewidth of about 2.5 kHz, but there may be broadening mechanisms other than homogeneous ones. In addition, environmental variations in the fiber (thermal or acoustic) may contribute to phase variations. In the aforementioned analysis, fiber lengths are assumed to be constant, which may

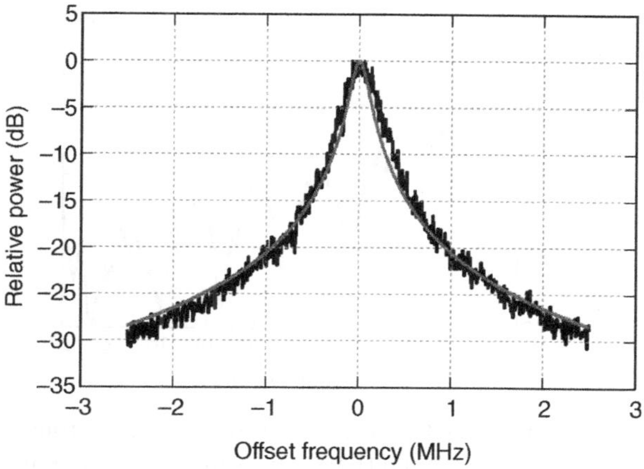

Figure 3.21. Delayed self-heterodyne measurement of a semiconductor distributed feedback laser. The solid line designates a calculated Lorentzian [Equation (3.38)] with a 95 kHz full-width half-maximum linewidth (Singley 2011b).

not always be the case. Further investigation would be required, demonstrating how challenging laser linewidth characterization can become for narrow-linewidth lasers. A semiconductor DFB laser was also measured with the architecture shown in Figure 3.19 to produce the data shown in Figure 3.21 (Singley et al. 2011b). Depicted are measured data for a single pass through the loop, normalized to the peak measured power and 55 MHz center frequency. Equation (3.38) was employed to fit the data with $\Delta\nu = 95\,\text{kHz}$. Good agreement is observed between the fit and the measurement, indicating that homogenous broadening dominates the lineshape of this particular laser.

Laser phase noise can manifest in a variety of ways in microwave photonics applications. Any process that introduces a differential time delay between the signal and a copy of itself has the potential to convert laser phase noise to intensity noise. Interferometers are often used for sensors or to demodulate phase-encoded signals (see Chapter 7). The theory for laser phase-to-intensity noise conversion in two-beam interferometers was established in two sequential articles by Moslehi (1986) and Tkach and Chraplyvy (1986). The equations from those articles can be used to write the laser phase noise that has been converted to RIN as

$$\text{RIN} = \frac{\Delta\nu \cdot e^{-2\pi\tau\Delta\nu}}{\pi(\Delta\nu^2 + f^2)} \left[\cosh(2\pi\tau \cdot \Delta\nu) - \cos(2\pi f\tau)\right], \qquad (3.41)$$

where $\Delta\nu$ is the linewidth assumed to be Lorentzian, f is the frequency in the electrical domain, and τ is the differential delay between the two

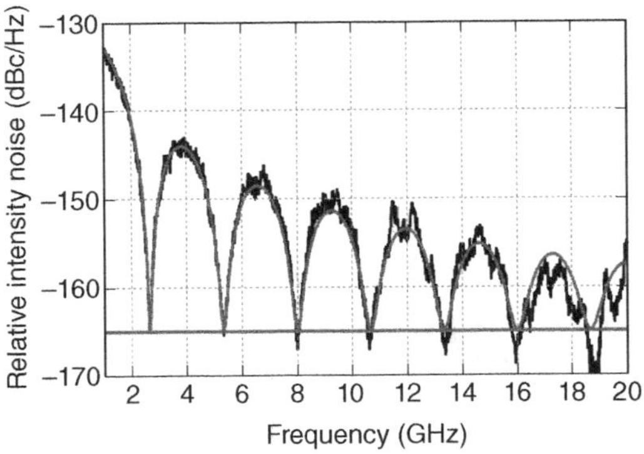

Figure 3.22. Measurement of relative intensity noise of the laser from Figure 3.21 after passing through a Mach–Zehnder interferometer with a 375-ps differential delay. The solid lines are for Equation (3.41) and for the shot noise limit of −165 dBc/Hz (Singley et al., 2011b).

arms in the interferometer. Equation (3.41) further assumes that the two-beam interferometer is biased at quadrature (equal optical power on both output ports). Shown in Figure 3.22 (Singley et al. 2011b) are a measured noise spectrum and a calculation of Equation (3.41). For the experiment, the same 95-kHz linewidth laser used to produce the data in Figure 3.21 was passed through a quadrature-biased fiber Mach–Zehnder interferometer with $\tau = 375\,\text{ps}$. The detected photocurrent was $I_{\text{dc}} = 10\,\text{mA}$. The calculation of Equation (3.41) employs these parameters along with $\Delta\nu = 95\,\text{kHz}$, the result from analysis of Figure 3.21. The shot-noise-limited RIN [Equation (3.26)] was added to Equation (3.41) to produce the curve in Figure 3.22. Excellent agreement between measurement and theory is shown in Figure 3.22. The Lorentzian linewidth $\Delta\nu = 95\,\text{kHz}$ is a good measure for the DFB laser in this case, where an incoherent ($\tau \gg 1/\Delta\nu$) self-heterodyne measurement agrees nicely with this coherent ($\tau \ll 1/\Delta\nu$) self-homodyne experiment.

Equation (3.41) can therefore be employed to quantify the impact of laser phase noise, when described by a Lorentzian line shape, on a system employing a two-beam interferometer. In practice, there are other mechanisms that can convert laser phase noise into RIN. Coherent detection techniques (Chapters 7 and 8), chromatic dispersion (Chapter 5), and Rayleigh scattering (Chapter 5) will be treated elsewhere in this text. There are also numerous mechanisms that can cause optical phase noise not easily described by a linewidth parameter, including undesired frequency or phase modulation caused by external signals and/or noise.

The most important aspects of laser noise for microwave photonics have been treated in this section. Intensity fluctuations on the optical field detected by a photodiode will result in noise in the RF domain. Laser intensity fluctuations are of the utmost importance in RF photonics, whether they are intrinsic laser intensity noise or phase noise that has been converted to intensity noise. As the power handling of RF photodiodes continually increases (Chapter 9), demand for lower-noise lasers will also rise. When concerning laser noise for RF photonics applications, the best practice is to characterize the laser noise carefully as it manifests in the RF domain for a particular architecture.

3.5 OPTICAL AMPLIFIERS

Optical amplifiers are often used in microwave photonic links. They can be employed in short and long links alike to increase or maintain signal power. In long-haul applications, discrete or distributed optical amplification throughout a link can offset propagation losses in fiber. Discrete amplifiers can be used in short fiber optic links to compensate for component loss or used in conjunction with a laser to boost power in a master-oscillator power amplifier (MOPA) configuration. In addition, optical amplifiers can be placed just before photodetection in a link to boost the output signal level to help optimize dynamic range. Regardless of the specific application, the process of amplification will add noise to the signal. There are circumstances and some amplifier designs where this is not necessarily the case (see Haus 2000a), but those devices and systems are beyond the scope of this work. The noise added by the amplification process can be treated in a variety of manners (Yamamoto and Inoue 2003); the treatment in this case concentrates on analog link performance and design.

It is beyond the scope of this work to cover all possible optical amplifiers for microwave photonics. The most prevalent are rare-earth-doped fiber amplifiers, most notably the EDFA. As silica fibers doped with Er^{3+} became readily available (Mears et al. 1987), the EDFA quickly formed the basis of modern telecommunications systems due to the broad emission spectrum in the low loss 1.55-μm window of silica fibers. The Yb^{3+} ion can also be co-doped with Er^{3+} into silica fiber to increase the absorption cross-section compared to an amplifier employing Er^{3+} alone. Rare earth amplifiers other than those employing Er^{3+} are possible as well (Kenyon 2002). For example, optical fibers have been doped with terbium (Tb^{3+}) and neodymium (Nd^{3+}) ions, the former having a fluorescence peak near 600 nm and that for the latter being near 1.06 μm

(Poole et al. 1986). The ytterbium-doped silica fiber amplifier (YDFA) has also been shown to be very efficient and achieves gain in the 1-μm region (Paschotta et al. 1997). The praseodymium-doped (Pr^{3+}) fiber amplifier (PDFA) offers options for gain in the 1.3-μm band (Whitley 1995) but requires fluoride-based optical fiber hosts that are difficult to couple with silica fibers. Given all these options, however, the EDFA remains the most popular choice among digital- and analog-photonic system designers. The impact and noise performance of using an EDFA within an analog link will be treated extensively in Section 3.5.1.

There are other optical amplifier technologies available besides the rare-earth-doped fiber amplifiers listed previously. Fiber amplifiers based on the Raman and Brillouin effects have been used as described in Section 3.5.2. Other possible mechanisms for optical gain in fibers include parametric processes. Four-photon mixing due to the third-order susceptibility ($\chi^{(3)}$) in a nonlinear fiber can be exploited to implement an optical parametric amplifier (OPA). Although not covered in this book, an OPA (Stolen and Bjorkholm 1982) has been demonstrated for numerous applications (Hansryd et al. 2002) and is promising as a very low-noise approach but requires high pump power. There are also optical amplifiers that are not based in fiber, such as semiconductor optical amplifiers (SOAs). The SOA is very attractive for its integration potential and is covered in Section 3.5.3.

3.5.1 Erbium-Doped Fiber Amplifiers

A schematic for an EDFA is shown in Figure 3.23. The gain medium consists of a length of fiber doped with Er^{3+}. A partial energy diagram for this ion is shown in Figure 3.24. The transition from the first excited state to the ground state, ${}^4I_{13/2} \rightarrow {}^4I_{15/2}$, corresponds to the 1.53-μm emission band that overlaps with the low propagation loss window of silica fiber optic systems. A signal near 1.55 μm can

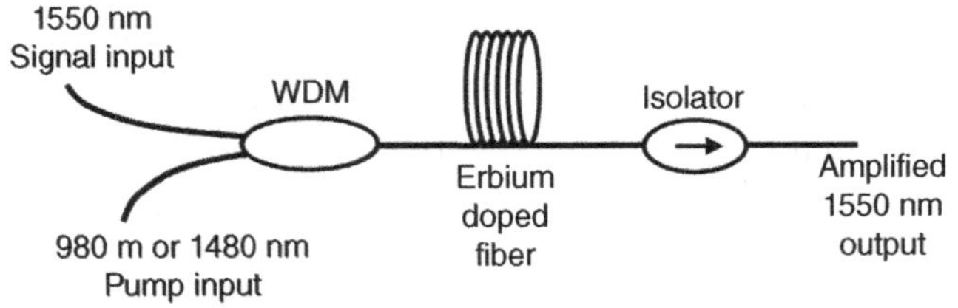

Figure 3.23. Basic components of an Erbium-doped fiber amplifier (EDFA). WDM: wavelength division multiplexer.

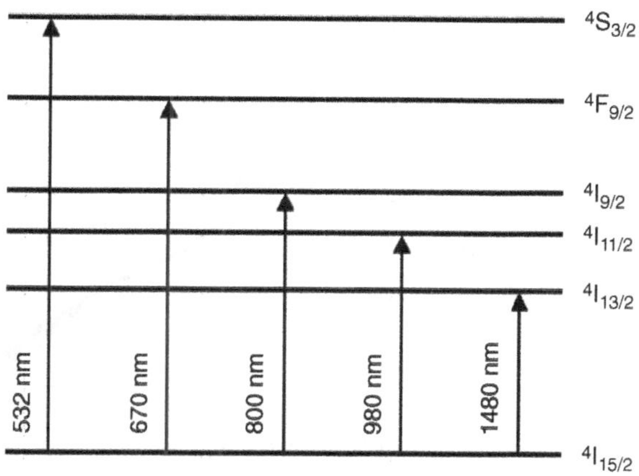

Figure 3.24. Partial energy level diagram for Er^{3+}. The wavelengths shown are for absorption. The notation on the right is in the form ${}^{2S+1}L_j$, where S is the spin quantum number, L is the orbital angular momentum quantum number, and j is the total angular momentum quantum number.

be coupled with a pump beam into an erbium-doped fiber (EDF) using a WDM. The input optical signal stimulates emission from the ${}^4I_{13/2} \rightarrow {}^4I_{15/2}$ transition resulting in coherent signal gain. Spontaneous emission events also occur from the excited state to the ground state and result in noise. As will be described in the following section, the optical spontaneous emission translates to different types of noise in the electrical domain. As shown in Figure 3.24, there are numerous possibilities for pump wavelengths, and co-doping with Yb^{3+} offers additional options. A pump near 1480 nm directly excites the ${}^4I_{13/2}$ state and creates a two-level system. A three-level system results when pumped with light at 980 nm, which can excite the medium to the ${}^4I_{11/2}$ state that will in turn decay nonradiatively to the metastable first excited state. Both of these wavelengths are often employed, and single-spatial-mode semiconductor lasers are readily available with Watt-level or higher powers suitable for core pumping. The 1480-nm band offers a wider absorption spectrum, but 980-nm pumping is more often employed because it offers better noise performance (Yamada et al. 1990). Another advantage of pumping at 980 nm is that reliable fused-coupler fiber optic WDMs are readily available due to the larger wavelength difference between pump and signal wavelengths.

Shown in Figure 3.25 is the measured spontaneous emission spectrum (no input signal) at the output of an EDFA pumped at 980 nm

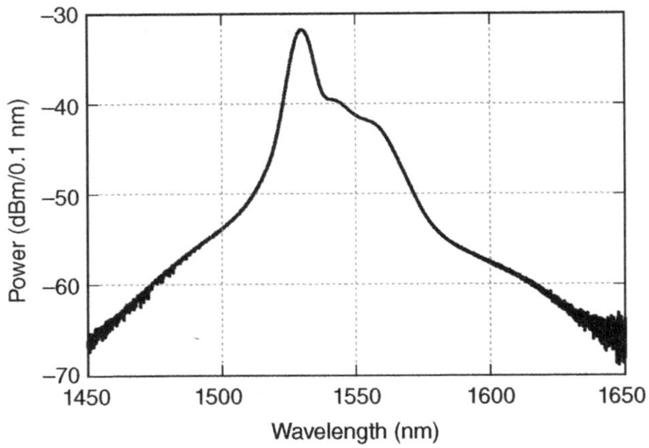

Figure 3.25. Measured amplified spontaneous emission spectrum for an erbium-doped fiber amplifier (Urick 2007).

(Urick 2007). The gain peak near 1535 nm is evident there, noting that significant emission spans the 1500–1575-nm range. The lifetime of the $^4I_{13/2}$ state is typically on the order of 10 ms but does depend on the host material. Aluminum co-doping in germanosilicate fiber can be employed to flatten the emission spectrum in the 1550-nm range (Kikushima and Yoshinaga 1991) and results in a 9.5–10.0 ms lifetime for the $^4I_{13/2}$ state (Becker et al. 1999). However, a variety of lifetimes are possible, ranging from 4 ms for a tellurite host to 14.7 ms for silicate glass (Becker et al. 1999). As discussed in Chapter 4, the relatively long lifetime of the EDFA gain medium is advantageous, because microwave signals will not experience distortion due to the slow EDFA gain dynamics.

The spontaneous emission from an EDFA manifests in three prevalent types of noise in the electrical domain. The spontaneous emission can beat with itself or the signal during photodetection, resulting in what are termed spontaneous-spontaneous noise and signal-spontaneous noise, respectively. There is also shot noise associated with the average spontaneous emission noise power. Desurvire (2002) provides a thorough treatment of noise in EDFAs, and the equations there will provide the starting points for the analysis in this chapter. The treatment of noise from an EDFA begins with the introduction of the spontaneous emission factor (Desurvire 2002):

$$n_{sp} = \frac{N_2}{N_2 - (\sigma_a/\sigma_e)N_1}, \tag{3.42}$$

where N_1 and N_2 are the densities of atoms in the ground and excited states, respectively, and σ_a and σ_e are the absorption and emission cross-sections, respectively. Equation (3.42) is for a two-level system, but a similar expression can also be derived for a three-level system. The spontaneous emission factor is employed to write the average optical power at an EDFA output due to spontaneous emission as (Desurvire 2002)

$$P_{sp} = n_{sp}h\nu(g_o - 1)B_o, \tag{3.43}$$

where h is Planck's constant, ν is the optical frequency, g_o is the optical power gain factor for the EDFA, and B_o is the optical bandwidth. The shot noise due to spontaneous emission is given by Equation (3.24) with $I_{dc} = M_{sp}l_o\Re P_{sp}$,

$$N_{sh,sp} = 2M_{sp}\Re l_o q h\nu n_{sp}(g_o - 1)B_o R|H_{pd}(f)|^2, \tag{3.44}$$

where $\Re$ is the photodiode responsivity, l_o is the optical loss factor between the EDFA and the photodiode, and M_{sp} is the number of amplified spontaneous emission (ASE) modes ($M_{sp} = 2$ for a typical single-mode-fiber EDFA configuration).

The derivation of expressions pertaining to spontaneous-spontaneous noise starts with Desurvire's 2002 expression for the spontaneous-spontaneous spectral density at the EDFA output:

$$\hat{\sigma}^2_{sp\text{-}sp}(f) = 2q^2N^2B_o\left(1 - \frac{f}{B_o}\right), \tag{3.45}$$

where $N = n_{sp}(g_o-1)$ is the amplified spontaneous emission (ASE) photon number at the EDFA output, and f is the electronic frequency. The mean-squared current spectral density after detection with a photodiode is

$$\langle i^2_{sp\text{-}sp}(f)\rangle = \eta^2 l_o^2 M_{sp}\hat{\sigma}^2_{sp\text{-}sp}. \tag{3.46}$$

Inserting Equation (3.45) and $\eta = \Re h\nu/q$ for the quantum efficiency into Equation (3.46) yields

$$\langle i^2_{sp\text{-}sp}(f)\rangle = 2M_{sp}(\Re h\nu l_o n_{sp})^2(g_o - 1)^2B_o\left(1 - \frac{f}{B_o}\right). \tag{3.47}$$

The mean-squared current fluctuations over a specified RF bandwidth B are calculated from Equation (3.47),

$$\langle i^2_{\text{sp-sp}} \rangle = \int_0^B \langle i^2_{\text{sp-sp}}(f) \rangle \, df = 2M_{\text{sp}}(\Re h\nu l_o n_{\text{sp}})^2 (g_o - 1)^2 B_o B \left(1 - \frac{B}{2B_o}\right). \tag{3.48}$$

Equation (3.48) can then be employed to calculate the total noise power due to spontaneous-spontaneous beat noise in a given electrical bandwidth. The RF noise PSD is obtained from Equation (3.47):

$$N_{\text{sp-sp}}(f) = 2|H_{\text{pd}}(f)|^2 M_{\text{sp}}(\Re h\nu l_o n_{\text{sp}})^2 (g_o - 1)^2 B_o \left(1 - \frac{f}{B_o}\right) R. \tag{3.49}$$

The procedure for arriving at the RF noise PSD resulting from signal-spontaneous beat noise is similar to that in arriving at Equation (3.49). The signal-spontaneous spectral density at the EDFA output as given by Desurvire (2002) is

$$\hat{\sigma}^2_{\text{sig-sp}}(f) = \frac{4q^2 g_o P_{\text{sig}} N}{h\nu}, \tag{3.50}$$

where P_{sig} is the optical power of the signal at the input to the EDFA. The signal-spontaneous mean squared current spectral density is then

$$\left\langle i^2_{\text{sig-sp}}(f) \right\rangle = 4h\nu P_{\text{sig}} n_{\text{sp}} (\Re l_o)^2 g_o (g_o - 1). \tag{3.51}$$

The signal-spontaneous mean squared current fluctuations are obtained by multiplying Equation (3.51) with the bandwidth:

$$\left\langle i^2_{\text{sig-sp}} \right\rangle = 4h\nu P_{\text{sig}} n_{\text{sp}} (\Re l_o)^2 g_o (g_o - 1) B. \tag{3.52}$$

Finally, the signal-spontaneous RF noise PSD is

$$N_{\text{sig-sp}}(f) = 4|H_{\text{pd}}(f)|^2 h\nu P_{\text{sig}} n_{\text{sp}} (\Re l_o)^2 g_o (g_o - 1) R. \tag{3.53}$$

Data stressing the frequency dependence of spontaneous-spontaneous beat noise and signal-spontaneous beat noise are shown in Figure 3.26 (Urick 2007). Depicted are measured $N_{\text{sp-sp}}$ and $N_{\text{sig-sp}}$ ranging from 2 to 16 GHz for an analog link employing an EDFA. The $H_{\text{pd}} \approx 1/2$

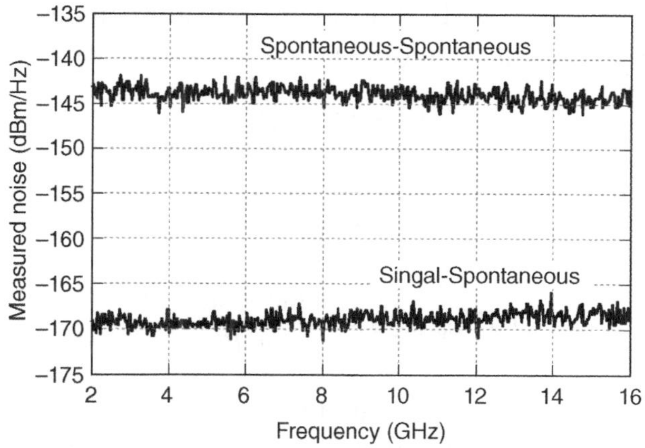

Figure 3.26. Measured noise at the output of two EDFA configurations showing spontaneous-spontaneous beat noise at 1 mA average photocurrent (top) and signal-spontaneous beat noise at 2 mA average photocurrent (bottom) (Urick 2007).

for each data set. For the $N_{\text{sp-sp}}$ curve, a DC photocurrent of 1 mA was detected and there was no ASE filter used. Therefore, $B_o \gg f$ and the RF dependence of $N_{\text{sp-sp}}$ is negligible as given by Equation (3.49), which agrees with the data in Figure 3.26. The DC photocurrent was 2 mA for the measured $N_{\text{sig-sp}}$ in Figure 3.26. Note that the $N_{\text{sig-sp}}$ is also flat across frequency, in agreement with the RF independence of Equation (3.53).

Equations (3.44), (3.49), and (3.53) with Equation (3.4) can be used to give expressions for the RIN associated with ASE shot noise, spontaneous-spontaneous beat noise, and signal-spontaneous beat noise, respectively. To that end, a full expression for the DC photocurrent is

$$I_{dc} = \Re l_o g_o P_{sig} + M_{sp} \Re h\nu l_o n_{sp} (g_o - 1) B_o, \qquad (3.54)$$

where the first term accounts for the average photocurrent created by the optical signal, and the second term is due to the ASE. In many practical occurrences, the DC current due to the optical signal dominates and, if $g_o \gg 1$, then the inequality $P_{sig} \gg M_{sp} h\nu n_{sp} B_o$ defines this special case. For example, if $M_{sp} = 2$, $\nu = 193$ THz, $n_{sp} = 1$, and $B_o = 200$ GHz, then $P_{sig} \gg 50$ nW. For microwave photonics applications, $B_o = 200$ GHz will allow for 100 GHz of dual-sideband RF modulation, and the resulting restriction on $P_{sig} \gg 50$ nW is representative of the power levels typically employed. In this case, the signal-spontaneous noise will be much larger than the ASE shot noise and spontaneous-spontaneous noise. In this limit, Equation (3.54) can be rewritten as

$$I_{dc} \approx \Re I_o g_o P_{sig}. \tag{3.55}$$

The RIN for signal-spontaneous beating is obtained by inserting Equations (3.53) and (3.55) into Equation (3.4):

$$\mathrm{RIN}_{sig\text{-}sp}(f) \approx \frac{4 n_{sp} h\nu (g_o - 1)}{g_o P_{sig}}, \tag{3.56}$$

which has been normalized to a unit bandwidth.

A cursory inspection of Equation (3.56) might suggest that there is no noise present when $g_o = 1$. However, the value for the spontaneous emission factor prevents this from happening as described in the following. The gain for an EDFA can be written as a function of the EDFA length z as (Desurvire 2002)

$$g_o(z) = \exp[\sigma_e N_2(z) - \sigma_a N_1(z)]. \tag{3.57}$$

Inserting Equation (3.42) into Equation (3.57) and solving for the spontaneous emission factor yield

$$n_{sp} = \frac{\sigma_e N_2 z}{\ln(g_o)}, \tag{3.58}$$

where N_2 and σ_e are assumed to be constant over the amplifier length z. Equation (3.58) can then be inserted into Equation (3.56) to obtain

$$\mathop{\mathrm{RIN}_{sig\text{-}sp}}_{\lim g_o \to 1} \approx \frac{4\sigma_e N_2 z h\nu}{P_{sig}}. \tag{3.59}$$

Therefore, as given by Equation (3.59), the signal-spontaneous beat noise does not vanish in the limit $g_o \to 1$. A similar analysis can be applied to Equation (3.49).

The optical noise factor (F_o) and optical noise figure, $\mathrm{NF}_o \equiv 10\log(F_o)$, are commonly employed to quantify the noise performance of an EDFA and are defined relative to a shot-noise-limited input. The optical noise factor for an EDFA can be defined as

$$F_o \equiv \frac{N_{sh,sig} + N_{sh,sp} + N_{sp\text{-}sp} + N_{sig\text{-}sp}}{N_{sh,in}}, \tag{3.60}$$

where $N_{\text{sh,sig}}$ is shot noise due to the signal at the EDFA output and $N_{\text{sh,in}}$ is shot noise due to the signal at the EDFA input. The numerator in Equation (3.60) captures all of the optical noise at the EDFA output assuming a shot-noise-limited input. The four terms there are adequate for this work, noting that other sources of noise may arise such as multipath interference, amplified double Rayleigh backscatter, and shot noise due to unabsorbed pump power (Baney et al. 2000). By definition, all terms in Equation (3.60) are assumed to be measured with an ideal detector having $\eta = 1$. Therefore, Equations (3.44), (3.49), and (3.53) with $H_{\text{pd}} = 1$, $\Re = q/(h\nu)$ and $l_{\text{o}} = 1$ yield expressions for $N_{\text{sh,sp}}$, $N_{\text{sp-sp}}$, and $N_{\text{sig-sp}}$, respectively. Under the same conditions, $N_{\text{sh,sig}} = 2q^2 g_{\text{o}} P_{\text{sig}} R/(h\nu)$ and $N_{\text{sh,in}} = 2q^2 P_{\text{sig}} R/(h\nu)$.

In the high gain limit ($g_{\text{o}} \gg 1$), another expression for n_{sp} can be employed. When $g_{\text{o}} \gg 1$, n_{sp} can be written in terms of the optical noise factor for the EDFA (Desurvire 2002):

$$n_{\text{sp}} = \frac{F_{\text{o}}}{2} \quad (\text{for}\, g_{\text{o}} \gg 1). \tag{3.61}$$

Equation (3.61) restricts $F_{\text{o}} \geq 2$ ($\text{NF}_{\text{o}} \geq 3\,\text{dB}$) for a high gain EDFA because the minimum absolute value of $n_{\text{sp}} = 1$. A useful expression for the signal-spontaneous RIN can be written in terms of the optical noise factor given some assumptions: the aforementioned $g_{\text{o}} \gg 1$, which leads to Equation (3.61) and $P_{\text{sig}} \gg M_{\text{sp}} h\nu n_{\text{sp}} B_{\text{o}}$, that is, the DC photocurrent due to ASE is negligible compared to the signal photocurrent. Under these conditions, Equation (3.61) can be inserted into Equation (3.56) to yield

$$\text{RIN}_{\text{sig-sp}} \approx \frac{2h\nu F_{\text{o}}}{P_{\text{sig}}}. \tag{3.62}$$

Equation (3.62) is a useful system design equation for the RIN at the output of an EDFA in the signal-spontaneous limit, a regime often encountered in microwave photonics. Note that if $N_{\text{sig-sp}}$ is the dominate term in the numerator of Equation (3.60), Equation (3.62) can be obtained by writing F_{o} in terms of RIN and using $\text{RIN}_{\text{sh,in}} = 2h\nu/P_{\text{sig}}$ in the denominator. Many EDFA manufacturers will specify NF_{o} in decibels as a function of input power to the EDFA. Therefore, a useful decibel form of Equation (3.62) can be written as

$$\text{RIN}_{\text{sig-sp}}[\text{dBc/Hz}] \approx -155.9 + \text{NF}_{\text{o}}[\text{dB}] - P_{\text{sig}}[\text{dBm}], \tag{3.63}$$

where $\nu = 193$ THz (1550 nm wavelength) was used.

The optical noise figure is the conventional metric to describe the noise performance of an EDFA, although debates on its usage have arisen (Haus 2000b). However, a noise penalty metric (Urick et al. 2006) can be more useful for system design as opposed to the component-level NF_o. A noise penalty can be defined as

$$\mathrm{NP} \equiv \frac{N_{\mathrm{sh,sig}} + N_{\mathrm{sh,sp}} + N_{\mathrm{sp\text{-}sp}} + N_{\mathrm{sig\text{-}sp}}}{N_{\mathrm{sh,sig}}}, \tag{3.64}$$

where the quantities are assumed to be measured in the system under consideration, including post-EDFA losses and the nonideality of the photodiode. Less this important caveat, the numerator in Equation (3.64) is the same as that in Equation (3.60). However, the reference (denominator) for the NP is signal shot noise at the output, as opposed to shot noise at the input for NF_o. This important difference allows the link equations to be modified directly to include a multiplicative NP term (Urick et al. 2006). Thus, the NP allows the link designer to assess quickly what the impact an optical amplifier will have on link performance. Equations (3.44), (3.49), and (3.53) can be inserted into Equation (3.64) for $N_{\mathrm{sh\text{-}sp}}$, $N_{\mathrm{sp\text{-}sp}}$, and $N_{\mathrm{sig\text{-}sp}}$, respectively. In the case of the NP definition, $N_{\mathrm{sh,sig}} = 2q^2|H_{\mathrm{pd}}(f)|^2\eta l_o g_o P_{\mathrm{sig}} R/(h\nu)$. An interesting expression for the NP is obtained in the signal-spontaneous noise limit, where $N_{\mathrm{sig\text{-}sp}}$ is much greater than both $N_{\mathrm{sp\text{-}sp}}$ and $N_{\mathrm{sp\text{-}sp}}$. Applying this condition to Equation (3.64) and also writing the noise in terms of RIN yield

$$\mathrm{NP} \approx 1 + \frac{\mathrm{RIN}_{\mathrm{sig\text{-}sp}}}{\mathrm{RIN}_{\mathrm{sh,sig}}}, \tag{3.65}$$

where $\mathrm{RIN}_{\mathrm{sh,sig}} = 2h\nu/(\eta l_o g_o P_{\mathrm{sig}})$ is the RIN associated with the signal shot noise at the output of the link (as opposed to the output of the EDFA for the NF_o definition). According to Equation (3.62), the $\mathrm{RIN}_{\mathrm{sig\text{-}sp}}$ is independent of post-EDFA losses, and therefore, $\mathrm{RIN}_{\mathrm{sig\text{-}sp}}$ is fixed at the EDFA output. However, $\mathrm{RIN}_{\mathrm{sh,sig}}$ is a function of post-EDFA losses. Therefore, if ηl_o is very small (very large losses), then $\mathrm{RIN}_{\mathrm{sh,sig}}$ can be much larger than $\mathrm{RIN}_{\mathrm{sig\text{-}sp}}$ and NP $\to$ 1 (0 dB). This can be seen more explicitly by inserting the aforementioned expression for $\mathrm{RIN}_{\mathrm{sh,sig}}$ and Equation (3.62) into Equation (3.65),

$$\mathrm{NP} \approx 1 + \eta l_o g_o F_o. \tag{3.66}$$

Therefore, in the signal-spontaneous limit when $\eta l_o \ll g_o F_o$, NP $\to$ 1 (0 dB) and the link is essentially shot noise limited. The condition

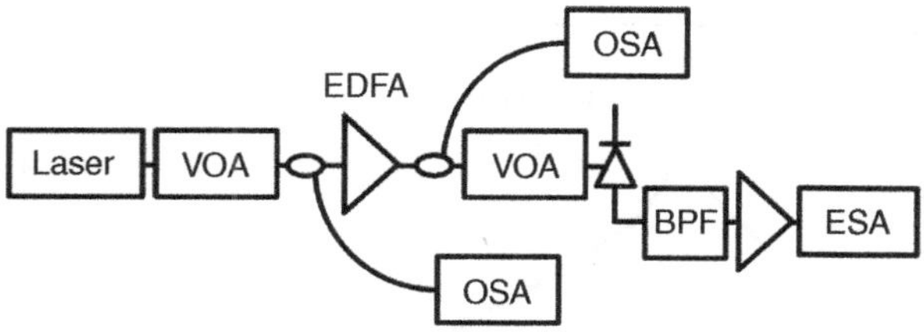

Figure 3.27. Measurement apparatus to characterize noise from an erbium-doped fiber amplifier.

$\eta l_o \ll g_o F_o$ encompasses many useful RF link situations, where EDFAs can be employed to offset distribution splitting, propagation losses, or poor photodiode quantum efficiency.

An architecture to demonstrate the difference between NP and NF_o is shown in Figure 3.27. A laser is passed through a variable optical attenuator (VOA) and then to the input of an EDFA under test. A calibrated optical spectrum analyzer (OSA) and fiber optic coupler are employed to monitor the input power to the EDFA. A second OSA and coupler pair is used on the opposite side of the EDFA to record the output power. After photodetection, an RF bandpass filter (BPF), amplifier, and electrical spectrum analyzer (ESA) are used to measure the output noise PSD. An experiment was conducted with this apparatus including an Er–Yb glass laser (CLR Photonics), a commercial EDFA (Avanex), and an InGaAs photodiode (Discovery Semiconductor) (Urick et al. 2006). The RIN of the laser alone was measured without the RF filter and is shown in Figure 3.28 (Urick et al. 2006). For this measurement, $I_{dc} = 20\,\text{mA}$, which by Equation (3.27), sets $\text{RIN}_{sh} = -168\,\text{dBc/Hz}$. For this photocurrent, the laser is shot noise limited at 1 GHz; assuming that the shape of the laser RIN spectrum follows a Lorentzian curve, the expected RIN approaches −190 dBc/Hz at 1 GHz. The maximum output power of the laser is 70 mW at 1550 nm; $\text{RIN}_{sh} = -174.4\,\text{dBc/Hz}$ for detection of that power with an ideal photodiode ($I_{dc} = 87.5\,\text{mA}$). Therefore, characterization of the EDFA at 1 GHz with this laser upholds the assumption for the definition of NF_o that the input is shot noise limited.

Shown in Figure 3.29 are three data sets obtained from the apparatus in Figure 3.27, all as a function of optical input power into the EDFA (P_{sig}) ranging from just below −20 dBm to just below +20 dBm (Urick et al. 2006). As shown by the optical output power curve, the EDFA is in compression for much of this regime with a small range of linear operation where $G_o \approx 35$ dB. The solid line designates the calculated RIN due to the signal shot noise at the input of the EDFA,

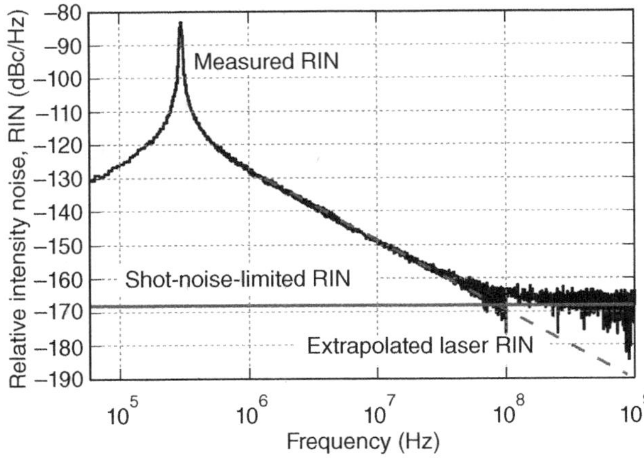

Figure 3.28. Measured laser noise for an erbium-doped fiber laser employed in amplifier noise characterization. Shown are the shot noise limit for 20-mA average current and extrapolated noise assuming a Lorentzian lineshape (Urick et al. 2006).

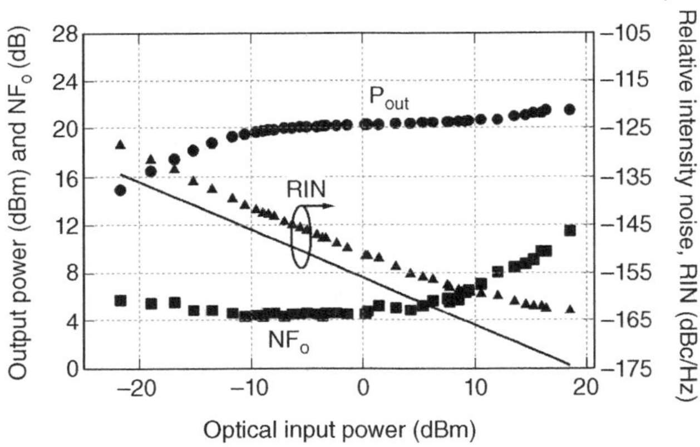

Figure 3.29. Measured optical output power (P_{out}), optical noise figure (NF_o) and relative intensity noise (RIN) at the output of an EDFA as a function of optical input power level, seeded with the laser used in Figure 3.28. The solid line is the calculated shot noise limit of the input signal (Urick et al. 2006).

$RIN_{sh,in} = 2hv/P_{sig}$, which assumes an ideal detector with $\eta = 1$. The measured RIN at the EDFA output is also shown; this level divided by $RIN_{sh,in}$ yields the plotted $F_o(P_{sig})$ as defined by Equation (3.60). Note that NF_o is about 5 dB until about $P_{sig} = 5$ dBm, after which NF_o rises significantly as the EDFA enters into deep compression. For comparison, consider the data in Figure 3.30, the measured NP for $I_{dc} = 5$, 10, and 20 mA (Urick et al. 2006). For the data in this case, the second VOA was adjusted to set the DC photocurrent, and the RIN was

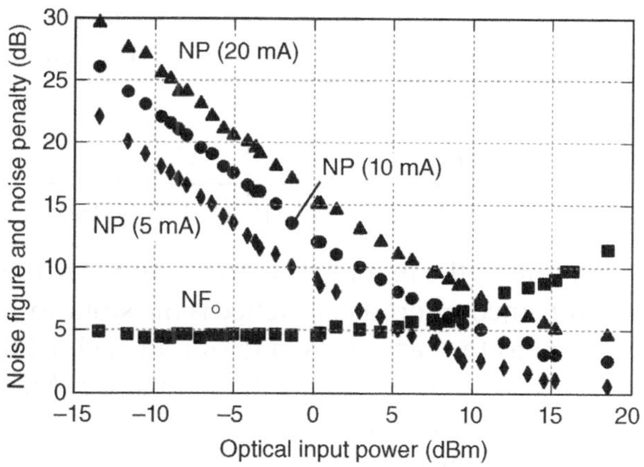

Figure 3.30. Measured optical noise figure (NF_o) and noise penalty (NP) from the output of an EDFA as a function of input power level, seeded with the laser used in Figure 3.28. The noise penalty was measured at average photocurrents of 5, 10, and 20 mA (Urick et al. 2006).

measured. As prescribed by Equation (3.64), the total RIN was divided by $RIN_{sh,sig} = 2h\nu/(\eta l_o g_o P_{sig}) = 2q/I_{dc}$ to obtain NP. As can be seen in Figure 3.30, the NP reduces as the EDFA goes deeper into compression, although the NF_o rises. The utility of the NP metric is that it describes the penalty incurred by employing an EDFA from the system level as compared to having a system with shot-noise-limited light at the same output power level.

Both the F_o and NP metrics can be cascaded to describe the performance of a chain of EDFAs. Consider the architecture shown in Figure 3.31, a chain of N EDFAs each having an optical gain factor g_i,

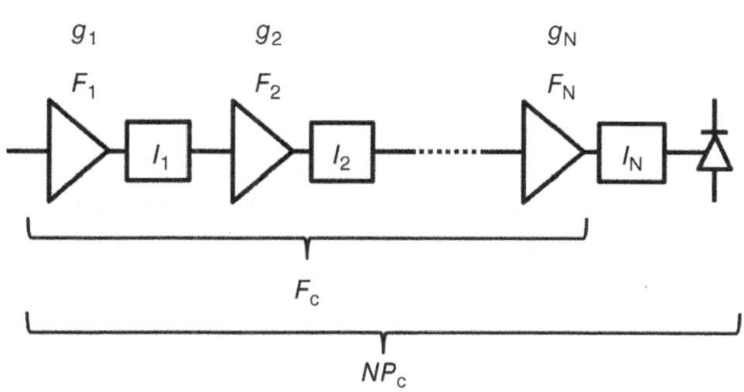

Figure 3.31. Architecture for an optical link with a cascade of N erbium-doped fiber amplifiers (EDFAs). The cascaded optical noise factor (F_c) is defined from the input of the first EDFA to the output of the last EDFA; the cascaded noise penalty (NP_c) is defined from the input of the first EDFA to the output of the photodiode.

an optical noise factor F_i, and a post-EDFA loss factor l_i $(1 \leq i \leq N)$. The cascaded noise factor for such a chain of EDFAs can be written as (Becker et al. 1999)

$$F_c = F_1 + \frac{F_2}{g_1 l_1} + \cdots + \frac{F_N}{g_1 l_1 g_2 l_2 \cdots g_{N-1} l_{N-1}}, \tag{3.67}$$

defined from the input of the first EDFA to the output of the last EDFA. A special case often encountered in practice is the situation where every EDFA in the chain has the same noise factor and each EDFA compensates precisely for loss between it and the next stage. Mathematically, this can be written as $F_n = F_o$ and $g_n l_n = 1$ for all n. Under these conditions, Equation (3.67) reduces to

$$F_c = N \cdot F_o \qquad \text{(loss compensated)}. \tag{3.68}$$

The cascaded noise penalty is defined from the input of the first EDFA to the output of the link, as shown in Figure 3.31. In the signal-spontaneous limit [Equation (3.66)], the cascaded noise penalty can be written as

$$\mathrm{NP_c} \approx 1 + \eta l_N g_c F_c, \tag{3.69}$$

where $g_c = g_1 l_1 g_2 l_2 \cdots g_{N-1} l_{N-1} g_N$ is the cascaded optical gain factor. For a loss-compensated system as defined previously, Equation (3.69) becomes

$$\mathrm{NP_c} \approx 1 + \eta l_N g_N N \cdot F_N \quad \text{(loss compensated)}. \tag{3.70}$$

Equation (3.70) can be rewritten as

$$\mathrm{NP_c} \approx 1 + N(\mathrm{NP_N} - 1) \approx N \cdot \mathrm{NP_N} \ \text{for} \ \mathrm{NP_N} \gg 1, \tag{3.71}$$

which is a useful "rule of thumb" for system design.

Shown in Figures 3.32 and 3.33 are measured SSB phase and amplitude noise, respectively, for a photonic link employing two EDFAs in a loss-compensated configuration (Urick et al. 2012). For these data, the RF frequency was 10.24 GHz with a power level of −15 dBm, $I_{dc} = 10\,\mathrm{mA}$, $R = 50\,\Omega$, and $H_{pd} = 1/2$. Inserting these parameters into Equation (3.29) yields a shot noise limit at −152 dBc/Hz. The main contribution to the phase and amplitude noise above 1 MHz is superposed noise arising from the EDFAs. The data in Figures 3.32 and 3.33 demonstrate two properties of the EDFA noise. Firstly,

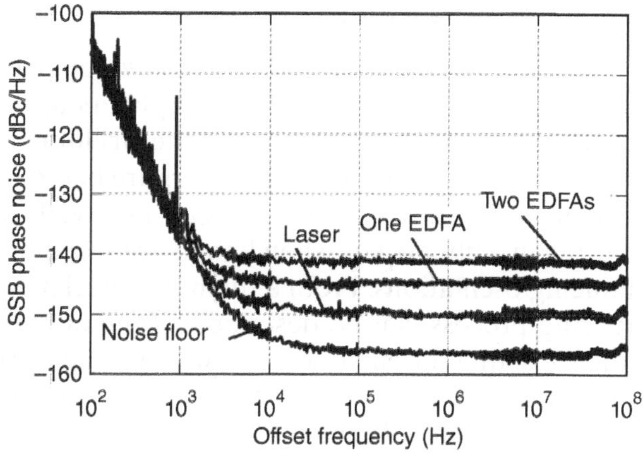

Figure 3.32. Measured single-sideband (SSB) phase noise for a link employing a laser alone and the same laser with one and two erbium-doped fiber amplifiers (EDFAs). Also shown is the measurement noise floor limited by the oscillator at 10.24 GHz (Urick et al. 2012).

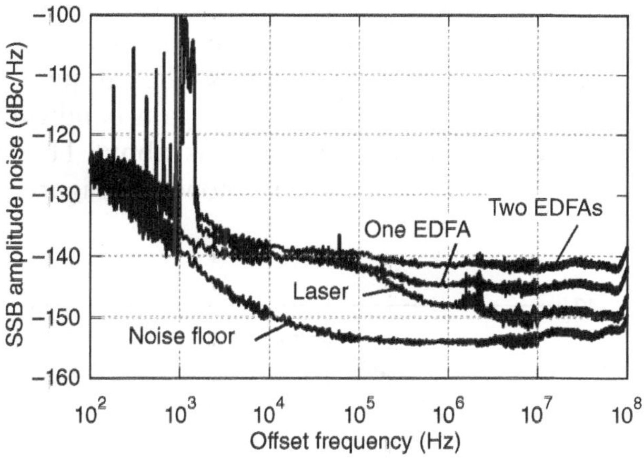

Figure 3.33. Measured single-sideband (SSB) amplitude noise for a link employing a laser alone and the same laser with one and two erbium-doped fiber amplifiers (EDFAs). Also shown is the measurement noise floor limited by the oscillator at 10.24 GHz (Urick et al. 2012).

it is equally distributed among the RF phase and amplitude noise. Secondly, the change in noise level from one EDFA to two EDFAs in a loss-compensated configuration is about 3 dB, which agrees with the aforementioned analysis.

3.5.2 Raman and Brillouin Fiber Amplifiers

Scattering processes in fiber can be used to achieve optical amplification. Two common types of scattering fiber amplifiers are the Raman fiber amplifier (Rottwitt and Stentz 2002) and the Brillouin fiber amplifier (Tkach and Chraplyvy 1989). Both amplifiers are based on the same scattering mechanism but differ in the energies involved. Raman and Brillouin scattering in a fiber occur when light is scattered by molecules in the fiber inducing oscillations. From a quantum-mechanical point of view, the scattering process can be described as the annihilation of a photon with the subsequent emission of a lower frequency photon and a phonon (Stokes scattering), or annihilation of a photon and a phonon with the subsequent emission of a higher frequency photon (anti-Stokes scattering). For the applications in this case, the Stokes process is much more probable than the anti-Stokes scattering. An optical phonon is created in Raman scattering, whereas an acoustic phonon results from Brillouin scattering.

The Raman and Brillouin scattering processes can be harnessed to achieve optical amplification by pumping fiber with the wavelength that corresponds to the Stokes shift relative to the signal wavelength. The signal wavelength can then stimulate the scattering, resulting in coherent optical gain. In typical fibers, the frequency shift for Raman scattering is about 13 THz, and the gain bandwidth is about 7 THz. For Brillouin scattering, the shift is about 11 GHz with a gain bandwidth in the 20 MHz range. Shown in Figure 3.34 is the

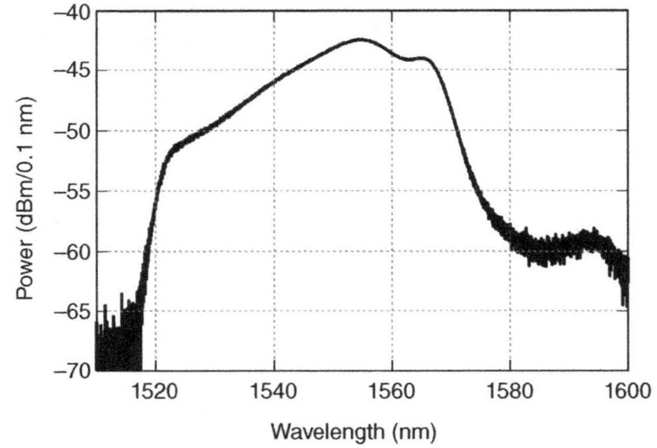

Figure 3.34. Amplified spontaneous emission spectrum for a Raman amplifier pumped at 1455 nm. (Urick 2007).

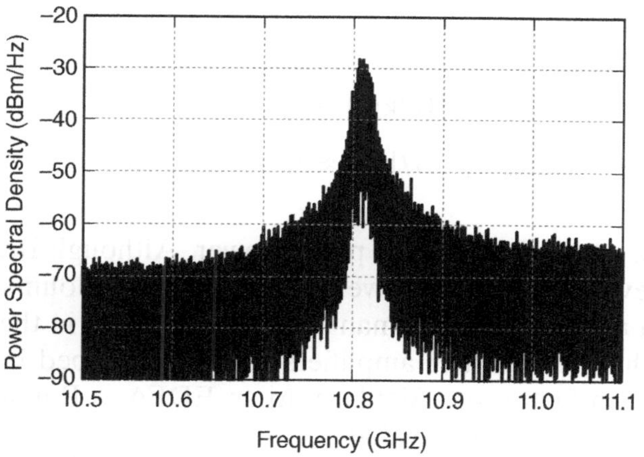

Figure 3.35. Measured stimulated Brillouin scattering noise at the output of 20 km of standard single mode fiber pumped at 1551 nm (Urick 2007).

measured spontaneous emission spectrum due to Raman scattering for a 105-km length of fiber (Urick 2007), comprising a combination of positive- and negative-dispersion fiber (Urick et al. 2004). For these data, the pump power was about 500 mW at 1455 nm. The peak in the spontaneous emission spectrum is at about 1555 nm, corresponding to an increase in wavelength of $\Delta\lambda = 100\,\text{nm}$. The associated frequency shift is $\Delta f = 13.2\,\text{THz}$. As noted previously, the expected gain bandwidth of the Raman process is about $\Delta f = 7\,\text{THz}$. The corresponding wavelength range can be calculated as $|\Delta\lambda| = |\Delta f|\lambda^2/c$. Inserting $\lambda = 1555\,\text{nm}$ into this expression yields $|\Delta\lambda| = 56\,\text{nm}$, which is in line with the emission shown in Figure 3.34. Shown in Figure 3.35 is a measured spectrum demonstrating some properties of the Brillouin scattering process (Urick 2007). The data shown in this case were obtained with an electrical spectrum analyzer used to measure the beat note between the Brillouin scattered signal and the pump signal. For these data, 20 km of standard single mode fiber was pumped at about 10 mW with a laser at 1551 nm having a linewidth less than 1 MHz. As shown in this case, the Brillouin frequency is just above 10.8 GHz, and the spectrum is about 25 MHz wide 10 dB down from the peak.

The pump power required to achieve Raman gain in typical fiber is about three orders of magnitude larger than that for an equivalent amount of Brillouin gain. For non-polarization-maintaining silica fiber at 1.5 μm, the Raman and Brillouin gain can be approximated by

(Desurvire 2002)

$$G_{\text{Raman}}[\text{dB/km}] \approx 6.6 \times P_{\text{pump}}[\text{W}], \tag{3.72}$$

$$G_{\text{Brillouin}}[\text{dB/km}] \approx 4.8 \times P_{\text{pump}}[\text{mW}], \tag{3.73}$$

respectively, where P_{pump} is the pump power. Although Brillouin gain can be achieved with a much lower power level, a Brillouin fiber amplifier is much noisier than a Raman fiber amplifier. As in the case of an EFDA, both of these fiber amplifiers can be described by a spontaneous emission factor, n_{sp}. Similarly to an EDFA, a Raman amplifier can approach the minimum value $n_{\text{sp}} \to 1$; however, a typical Brillouin amplifier will exhibit $n_{\text{sp}} \approx 50\text{–}500$ (Desurvire 2002). Given the amount of added noise and limited gain bandwidth, the Brillouin fiber amplifier alone has few applications. Therefore, the remainder of this section concentrates on Raman fiber amplifiers.

A Raman fiber amplifier offers numerous benefits for photonic systems. Firstly, optical gain can be achieved at any wavelength where a pump laser is available at the associated Stokes shift wavelength. Secondly, a Raman-amplified link can be pumped from a remote location. This is advantageous in links where placement of discrete in-line amplifiers can be difficult. Furthermore, the link can be pumped from either side, that is, the pump can be co- or counter-propagated with the signal. Finally, there are some applications where distributed Raman amplification can provide SNR benefits as opposed to discrete amplifiers.

The benefits of Raman amplifiers do come at a cost. In addition to the high pump power, there are other concerns for microwave photonic systems employing Raman fiber amplifiers. The speed of the Raman process, on the order of 100 fs, can be problematic in two regards. Firstly, the gain medium will respond to microwave and millimeter-wave modulation, which can lead to distortion for analog applications if the amplifier is operated near compression (see Section 4.3). Secondly, pump noise can be transferred to the signal at the frequencies of interest. Neither of these concerns is as important in an EDFA because the gain dynamics are much slower (~10 ms), which corresponds to an RF frequency typically below those of interest. Unlike an EDFA, the pump and the signal in Raman (and Brillouin) fiber amplifiers must be aligned in polarization. This can present a problem especially in links where polarization dependent chromatic dispersion will cause the pump and the signal polarizations to evolve differently.

While both an EDFA and a Raman fiber amplifier can approach $n_{sp} \to 1$, there are other possible sources of noise in a Raman amplifier that are not as important in an EDFA because of the lower power levels involved in the latter (Bromage 2004). For example, four-wave mixing can occur in a Raman fiber amplifier due to pump–signal interactions or pump–pump interactions when multiple pump wavelengths are employed. In addition, multipath interference due to double Rayleigh scattering is prevalent in Raman amplifiers. It is beyond the scope of this work to treat all of these effects analytically, but these mentioned propagation effects are covered generally in Chapter 5.

Two important measures employed for distributed amplifiers are effective gain and effective noise factor. Let the performance of a Raman amplifier be quantified by the gain factor and noise factor, g_o and F_o, respectively. Both of these metrics are defined as they were for an EDFA in Section 3.5.1. The ratio of the output power to the input power is g_o; F_o is defined relative to a shot-noise-limited input as given by Equation (3.60). The effective gain of a Raman fiber amplifier is (Rottwitt and Stentz 2002)

$$F_{o,\text{eff}} = F_o e^{-\alpha_{\text{sig}} L}, \tag{3.74}$$

where α_{sig} is the fiber attenuation constant for the signal, and the length of the amplifier is L. The attenuation constant is typically expressed in units of dB/km, and a formula for conversion from linear to dB form is

$$\alpha_{\text{sig}}[\text{dB}] = \frac{10 \cdot \alpha_{\text{sig}}[\text{linear}]}{\ln(10)} \approx 4.343 \cdot \alpha_{\text{sig}}[\text{linear}]. \tag{3.75}$$

Normalizing α_{sig} to the loss per km allows Equation (3.74) to be written as

$$\text{NF}_{o,\text{eff}}[\text{dB}] = \text{NF}_o[\text{dB}] - \alpha_{\text{sig}}[\text{dB/km}] \cdot L[\text{km}], \tag{3.76}$$

where $\text{NF}_{o,\text{eff}}$ is the effective optical noise figure and NF_o is the optical noise figure for the Raman amplifier. As described by Equation (3.76), the effective noise figure is a reduction of the noise figure by the total fiber loss in the Raman amplifier. The effective gain factor is cast in a similar manner:

$$g_{o,\text{eff}} = g_o e^{\alpha_{\text{sig}} L}, \tag{3.77}$$

which can be written in dB form as

$$G_{o,\text{eff}}[\text{dB}] = G_o[\text{dB}] + \alpha_{\text{sig}}[\text{dB/km}] \cdot L[\text{km}]. \tag{3.78}$$

The effective gain is therefore an increase of the Raman amplifier gain by the total fiber loss. The effective gain is sometimes termed the "on-off gain," as it quantifies the change in output power when the amplifier is on compared to the propagation loss in the fiber comprising the amplifier.

An understanding of all of the metrics for fiber amplifiers allows for the informed design of photonic systems. Whether effective noise figure, noise figure, noise penalty, or RIN is used to describe the noise performance, it is most important to choose an amplification scheme carefully. Hybrid fiber amplifiers can take advantage of the useful properties from different techniques. For example, a hybrid Brillouin/EDFA was demonstrated to provide better noise performance than an EDFA alone (Strutz and Williams 2000). Hybrid Raman/EDFAs were employed to achieve 3.28 Tb/s signal transmission (Nielsen et al. 2000)—a record at the time of publication. The choice of what type of fiber amplifier to use, if any, will be driven by system requirements. However, as mentioned previously, fiber amplifiers do not present the only option for optical amplification.

3.5.3 Semiconductor Optical Amplifiers

A SOA can be coupled to fiber but, as opposed to the other amplifiers treated previously, does not employ the fiber itself as the gain medium. Rather, the semiconductor material serves as the active region for a SOA. Similarly to semiconductor lasers described in Section 3.4, a variety of materials can be used to construct SOAs to achieve optical gain over a wide bandwidth. For example, GaInAsP is commonly used for amplification in the 1.5-μm region (Simon, 1987). There are two main types of SOAs: a Fabry–Perot amplifier and a traveling-wave amplifier (Olsson 1992). In a Fabry–Perot design, the facets of the semiconductor chip are partially reflective, setting up a resonant amplifier scheme. These resonant designs tend to be sensitive to temperature and other environmental conditions. A traveling-wave SOA is a single-pass amplifier with facets having very low reflectivity. Regardless of the design, there are a number of issues to consider when evaluating a SOA for microwave photonics applications.

Similarly to the EDFA, a SOA can be described by an optical gain factor and a spontaneous emission factor. However, there are several differences between an EDFA and SOA to consider. As mentioned previously, SOAs are available over a wide wavelength range, spanning approximately 400–1600 nm. SOAs also offer low power consumption, small size, and excellent integration potential. However, they typically exhibit lower gain and higher noise figure compared to an EDFA. Similarly to Brillouin and Raman fiber amplifiers though, SOAs exhibit polarization-dependent gain (O'Mahony 1988). While fiber amplifiers do not require any special interconnectivity, a SOA chip must be coupled to the fiber. This can sometimes be problematic, particularly in achieving proper antireflection coatings. Perhaps, the biggest issue with SOAs for microwave photonics applications is distortion of the analog signals. The speed of the gain recovery in a SOA will typically be on the order of a few hundred picoseconds, which is fast enough to distort signals in the gigahertz range when the amplifier is operated in or near compression (see Section 4.3). However, because of the integration potential, the SOA is receiving attention for integrated microwave photonic circuits, an area that will continue to evolve.

3.6 PHOTODETECTION

Shot noise is considered the primary noise contribution associated with the photodetection process for many high performance links. However, other types of noise can be generated within a photodetector in addition to shot noise. In this section, a treatment of excess amplitude noise due to the ionization (avalanche) process will be discussed in detail. Another important noise source is excess phase noise, but the discussion of photodetector phase noise is beyond the scope of this text. The reader is referred to references by Ivanov et al. (2005), Tulchinsky and Williams (2005), Taylor et al. (2010), and Fortier et al. (2013) for discussions of photodetector phase noise. Another noise contribution is associated with impedance matching photodiodes to output impedances higher than 50 Ω, which can be important for low frequency or narrowband systems (Helkey et al. 1997). Such circuits have been employed for lightwave analog cable television (CATV) systems but will not be addressed in this chapter.

High performance analog links use high photocurrents to improve link noise figure and dynamic range. If shot-noise-limited link performance is achieved, the SNR improves as the photocurrent increases due to the slower rise in shot noise power compared to the signal power. With the widespread availability of 50 mW or higher laser power and low optical loss, achieving link photocurrents well above 2 mA is common. Therefore, the output shot noise power will surpass output thermal noise in most high performance links [see Equations (3.20) and (3.24)].

However, some low power link designs could take advantage of the current gain provided by avalanche photodiodes (APDs) for chip-to-chip applications or other low power analog links.

Digital communication links have benefited from the use of APDs or optical amplifiers to improve output SNR. When output photocurrents are well below 500 μA and the laser RIN is low, output thermal noise dominates receiver performance. Photocurrent multiplication (avalanche gain) or optical preamplification in a receiver can be used to increase signal current (power) out of a photodiode so long as the excess noise from the particular gain process is manageable. In these low power applications where receivers operate with low levels of avalanche current gain, the SNR increases as the square of effective current gain so long as the noise associated with the avalanche process remains below the output thermal noise. The point of maximum SNR occurs when the noise associated with the gain process equals or slightly exceeds the output thermal noise. In APDs, this noise crossover point is determined by the input signal power, the multiplication factor, and the excess noise factor of the APD.

McIntyre (1966) developed a local-field avalanche theory to evaluate the contributions to the output noise from the avalanche process. In this work, both the noise and the gain-bandwidth product of the APD can be described by the electron (α) and the hole (β) ionization coefficients and, more specifically, by their ratio or k-factor, $k = \beta/\alpha$. McIntyre (1966) derived an expression for the noise spectral density of the output current with electron injection assumed:

$$2qI_{\text{in}}M^3 \left[1 - (1-k)\left(\frac{M-1}{M}\right)^2\right], \tag{3.79}$$

where M is the mean current gain and I_{in} is the input injected current for a device with $M = 1$. An excess noise factor for APDs has been defined by dividing the output noise spectral density by the output noise current that would be present if there were no added noise, specifically the multiplied input current shot noise, $2qI_{\text{in}}M^2$. Dividing Equation (3.79) by this output noise gives the common expression for the APD excess noise factor,

$$F(M) = kM + (1-k)\left(2 - \frac{1}{M}\right). \tag{3.80}$$

Equation (3.80) is plotted in Figure 3.36 for various k-factors as a function of M. For $k = 1$ (equal hole and electron ionization coefficients),

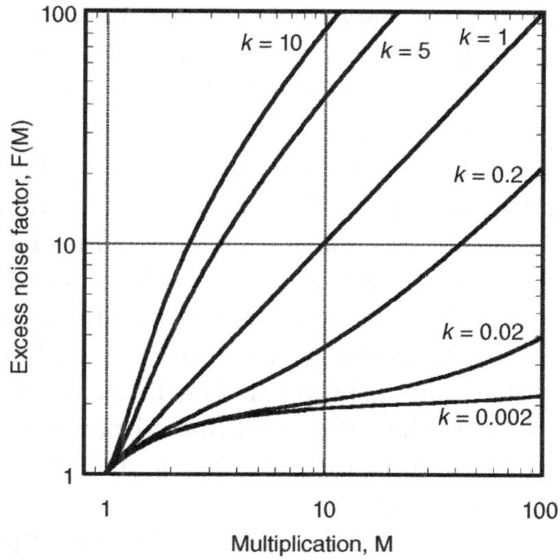

Figure 3.36. Plot of APD excess noise factor, Equation (3.80), as a function of avalanche gain, M, for various k-factors.

$F = M$, which represents a "good" APD design. Another trend that can be seen is that $F \to 2$ (3 dB) as $k \to 0$ for large M (high gain).

Equation (3.79) was derived under the assumption that the carriers (and thus their ionization coefficients) are in local equilibrium with the electric field and thus the ionization probabilities are determined solely by the electric field at that position. It is well known that the impact ionization is nonlocal, in the sense that the carriers that are injected "cold" require a certain distance to attain sufficient energy to ionize (Okuto and Crowell 1974). This distance is usually referred to as the "dead space". Likewise, there is also dead space between ionization events because ionized carriers require a certain distance to attain sufficient energy before they can ionize another carrier. The local-equilibrium condition is sufficient to explain APD behavior and noise tendencies in APDs with avalanche regions much greater than the dead space. However, the local-equilibrium condition can lead to over-estimations of $F(M)$ for thin APDs. In such cases, the nonlocal effects of the dead space cannot be excluded, and a further refinement of the aforementioned APD theory is required. Several treatments have been developed including using history-dependent ionization coefficients (McIntyre 1999) and calculating the spatial distribution of ionization events (Ramirez et al. 2014). The theory by McIntyre (1999) has been used to analyze ionization characteristics of thin avalanche region APDs (Yuan et al. 1999). It is beyond the scope of this text to explain the details of designing APDs with low k-factors. The interested reader is pointed toward Campbell (2007) for a good review on APD design. For the purposes in this case, it is sufficient to study the impact that a given k-factor has on the performance of a photodiode within an analog link. For this purpose, the excess noise factor will be used to touch upon the following two relevant scenarios: links that use APDs to achieve low noise current gain and links using high current photodiodes that exhibit some small levels of avalanche multiplication due to the high operating electric field conditions within the depletion region.

In the first scenario, a link using an APD will be considered. Normally, APDs are utilized only when the input optical power to a photodiode is small such that the shot noise is negligible and the output thermal noise dominates. In this case, the current gain directly causes the SNR to increase, in dB, by 20log(M) until the APD noise output exceeds the output thermal noise (Melchior and Lynch 1966). Thus, making APDs with low excess noise factors will allow the APD to be operated at a higher gain before the APD noise exceeds the output thermal noise.

In the second scenario, a link using a high current photodiode is considered. High current photodiodes are often employed with low

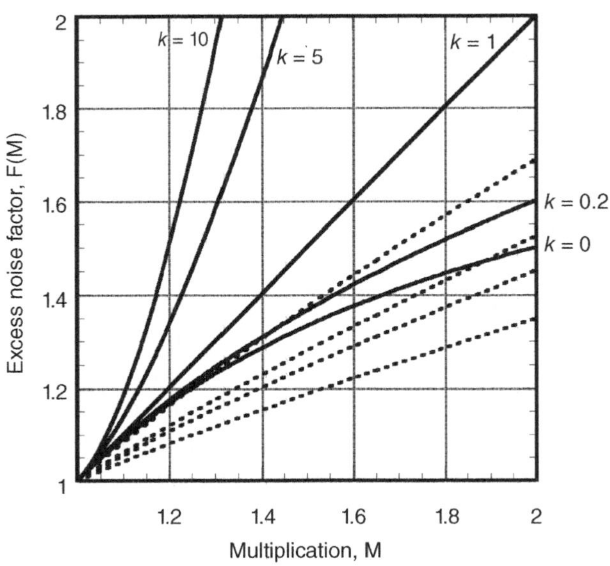

Figure 3.37. Plot of APD excess noise factor at low gain, Equation (3.80)-(solid), and an empirical function from Hu et al. (1996)-(dashed) that more closely predicts measurement results from thin APDs.

noise lasers to improve the noise figure in high performance links if shot-noise-limited performance can be obtained. Excess noise in high current photodiodes may be important, as these devices tend to operate at high-intrinsic-region electric fields, which helps to overcome space-charge effects and allow for linear operation (Williams et al. 1996). Photodiodes operating under these conditions have shown a small quantum efficiency increase with the higher applied voltages, which was attributed to the avalanche process (Hastings et al. 2009). In addition, many of these high current photodiode designs use short intrinsic or drift regions. Thus, they may show similar k-factor characteristics as compared to short drift region APDs at high bias voltages. Figure 3.37 shows a plot of Equation (3.80) at low M combined with a plot of an empirical function (Hu et al. 1996) used to calculate $F(M)$ with reasonable accuracy for thin APDs having low measured k-factors. It can be seen that thin APDs can yield excess noise lower than Equation (3.80) predicts for low M. The exact impact that excess noise has on high current photodiodes has received little attention, and more research is needed to minimize the impact of APD k-factors on high performance analog links operating at high photocurrents.

REFERENCES

Allan, D., Hellwig, H., Kartaschoff, P., Vanier, J., Vig, J., Winkler, G. M. R., and Yannoni, N. F., "Standard terminology for fundamental frequency and time metrology," in *42nd Annual Frequency Control Symposium*, 419–425 (1988).

Baney, D. M. and Sorin, W. V., "High resolution optical frequency analysis," in *Fiber Optic Test and Measurement*, D. Derickson (editor), Prentice Hall (1998).

Baney, D. M., Gallion, P., and Tucker, R. S., "Theory and measurement for the noise figure of optical amplifiers," *Optical Fiber Technology*, 6, 122–154 (2000).

Becker, P. C., Olsson, N. A., and Simpson, J. R., *Erbium-Doped Fiber Amplifiers Fundamentals and Technology*, Academic (1999).

Bergman, K., Doerr, C. R., Haus, H. A., and Shirasaki, M., "Sub-shot-noise measurement with fiber-squeezed optical pulses," *Optics Letters*, 18(8), 643–645 (1993).

Bromage, J., "Raman amplification for fiber communications systems," *Journal of Lightwave Technology*, 22(1), 79–93 (2004).

Callen, H. B. and Welton, T. A., "Irreversibility and generalized noise," *Physical Review*, 83(1), 34–40 (1951).

Campbell, J. C., "Recent advances in telecommunications avalanche photodiodes." *Journal of Lightwave Technology*, 25(1), 109–121 (2007).

Chuang, C. F., Liao, Y. H., Lin, C. H., Chen, S. Y., Grillot, F., and Lin, F. Y., "Linewidth enhancement factor in semiconductor lasers subject to various external optical feedback conditions," *Optics Express*, 22(5), 5651–5658 (2014).

Coldren, L. A., Corzine, S. W., and Mashanovitch, M. L., *Diode Lasers and Photonic Integrated Circuits*, Wiley (2012).

Cox, C. H. III and Ackerman, E. I., "Recent advances in high-frequency (>10 GHz) microwave photonic links," in *Optical Fiber Telecommunications VIB*, I. P. Kaminow, T. Li, and A. E. Willner (editors), Academic (2013).

Dawson, J. W., Park, N., and Vahala, K. J., "An improved delayed self-heterodyne interferometer for linewidth measurements," *IEEE Photonics Technology Letters*, 4(9), 1063–1065 (1992).

Delfyett, P. J., and Lee, C-H., "Semiconductor injection lasers," in *Encyclopedia of Lasers and Optical Technology*, Academic Press, San Diego (1991).

Desurvire, E., *Erbium-Doped Fiber Amplifiers Principles and Applications*, Wiley (2002).

Dianov, E. M. and Prokhorov, A. M., "Medium-power CW Raman fiber lasers," *IEEE Journal on Selected Topics in Quantum Electronics*, 6(6), 1022–1028 (2000).

Diddams, S. A., Berquist, J. C., Jefferts, S. R., and Oates, C. W., "Standards of time and frequency at the outset of the 21st century," *Science*, 306, 1318–1324 (2004).

Diehl, J. F., Swingen, L. A., Devgan, P. S., and Urick, V. J., "An evaluation of compact laser drivers for field-deployed, analog fiber-optic systems," Naval Research Laboratory Memorandum Report, NRL/MR/5652--07-9088 (2007).

Digonnet, M. J. F. (editor), *Rare-Earth-Doped Fiber Lasers and Amplifiers*, Marcel Dekker, New York (2001).

Esman, R. D., Goldberg, L., and Weller, J. F., "Feedback induced microwave signal dropout in 0.83 μm fire-optic links," *Electronics Letters*, 24(6), 347–349 (1988).

Fortier, T. M., Quinlan, F., Hati, A., Nelson, C., Taylor, J. A., Fu, Y., Campbell, J., and Diddams, S. A., "Photonic microwave generation with high-power photodiodes," *Optics Letters*, 38(10), 1712–1714 (2013).

Geng, J., Staines, S., Wang, Z., Zong, J., Blake, M., and Jiang, S., "Highly stable low-noise Brillouin fiber laser with ultranarrow spectral linewidth," *IEEE Photonics Technology Letters*, 18(17), 1813–1815 (2006).

Gordon, E. I., "Optical maser oscillators and noise," *Bell System Technical Journal*, 43, 507–539 (1964).

Hansryd, J., Andrekson, P. A., Westlund, M., Li, J., and Hedekvist, P., "Fiber-based optical parametric amplifiers and their applications," *IEEE Journal of Selected Topics in Quantum Electronics*, 8(3), 506–520 (2002).

Harris, I. A., "Zero-point fluctuations and thermal-noise standards," *Electronics Letters*, 7(7), 148–149 (1971).

Hastings, A. S., Tulchinsky, D. A., and Williams, K. J., "Photodetector nonlinearities due to voltage-dependent responsivity," *IEEE Photonics Technology Letters*, 21(21), 1642–1644 (2009).

Haus, H. A., "IRE standards on methods of measuring noise in linear twoports, 1959," *Proceedings of the IRE*, 48(1), 60–68 (1960).

Haus, H. A., *Electromagnetic Noise and Quantum Optical Measurements*, Springer, Germany (2000a).

Haus, H. A., "Noise figure definition valid from RF to optical frequencies," *IEEE Journal of Selected Topics in Quantum Electronics*, 6(2), 240–247 (2000b).

Helkey, R., Twichell, J. C., and Cox, C. I., "A down-conversion optical link with RF gain," *Journal of Lightwave Technology*, 15(6), 956–961 (1997).

Henry, C. H., "Theory of the linewidth of semiconductor lasers," *IEEE Journal of Quantum Electronics*, 18(2), 259–264 (1982).

Henry, C. H., "Phase noise in semiconductor lasers," *Journal of Lightwave Technology*, 4(3), 298–311 (1986).

Hinkley, E. D. and Freed, C., "Direct observation of the Lorentzian line shape as limited by quantum phase noise in a laser above threshold," *Physical Review Letters*, 23(6), 277–280 (1969)

Hu, C., Anselm, K. A., Streetman, B. G., and Campbell, J. C., "Noise characteristics of thin multiplication region GaAs avalanche photodiodes," *Applied Physics Letters*, 69(24), 3734–3736 (1996).

Ivanov, E. N., Diddams, S. A., and Hollberg, L., "Study of the excess noise associated with demodulation of ultra-short infrared pulses," *IEEE Trans. on Ultrasonics, Ferroelectrics and Frequency Control*, 52(7), 1068–1074 (2005).

Johnson, J. B., "Thermal agitation of electricity in conductors," *Physical Review*, 32, 97–109 (1928).

Kenyon, A. J., "Recent developments in rare-earth doped materials for optoelectronics," *Progress in Quantum Electronics*, 26, 225–284 (2002).

Kerr, A. R., "Suggestions for revised definitions of noise quantities, including quantum effects," *IEEE Transactions on Microwave Theory and Techniques*, 47(3), 325–329 (1999).

Kikushima, K. and Yoshinaga, H., "Distortion due to gain tilt of erbium-doped fiber amplifiers," *IEEE Photonics Technology Letters*, 3(10), 945–947 (1991).

Lau, K. Y. and Yariv, A., "Ultra-high speed semiconductor lasers," *IEEE Journal of Quantum Electronics*, 21(2), 121–138 (1985).

McIntyre, R. J., "Multiplication noise in uniform avalanche diodes," *IEEE Transactions on Electron Devices*, 13(1), 164–168 (1966).

McIntyre, R. J., "A new look at impact ionization-Part I: A theory of gain, noise, breakdown probability, and frequency response," *IEEE Transactions on Electron Devices*, 46(8), 1623–1631 (1999).

McQuarrie, D. A., *Statistical Mechanics*, University Science Books, pp. 553–561 (2000).

Mears, R. J., Reekie, L., Jauncey, I. M., and Payne, D. N., "Low-noise erbium-doped fibre amplifier operating at 1.54 μm," *Electronics Letters*, 23(19), 1026 (1987).

Melchior, H. and Lynch, W. T., "Signal and noise response of high speed Germanium avalanche photodiodes," *IEEE Transactions on Electron Devices*, 13(12), 829–838 (1966).

Mercer, L. B., "1/f Frequency noise effects on self-heterodyne linewidth measurements," *Journal of Lightwave Technology*, 9(4), 485–493 (1991).

Milonni, P. W. and Shih, M. L., "Zero-point energy in early quantum theory," *American Journal of Physics*, 59(8), 684–698 (1991).

Moslehi, B., "Noise power spectra of optical two-beam interferometers induced by the laser phase noise," *Journal of Lightwave Technology*, 4(11), 1704–1710 (1986).

Nielsen, T. N., Stentz, A. J., Rottwitt, K., Vengsarkar, D. S., Chen, Z. J., Hansen, P. B., Park, J. H., Feder, K. S., Strasser, T. A., Cabot, S., Stulz, S., Peckham, D. W., Hsu, L., Kan, C. K., Jurdy, A. F., Sulhoff, J., Park, S. Y., Nelson, L. E., and Gruner-Nielsen, L., "3.28-Tb/s (82x40 Gb/s) transmission over 3x100 km nonzero-dispersion fiber using dual C- and L-band hybrid Raman/erbium-doped inline amplifiers," in *Optical Fiber Communications Conference Post-Deadline Digest*, Paper PD23-1, 236 (2000).

Nyquist, H., "Thermal agitation of electric charge in conductors," *Physical Review*, 32, 110–113 (1928).

O'Mahony, M. J., "Semiconductor laser optical amplifiers for use in future fiber systems," *Journal of Lightwave Technology*, 6(4), 531–544 (1988).

Obarski, G. E. and Splett, J. D., "Measurement assurance program for the spectral density of relative intensity noise of optical fiber sources near 1550 nm," NIST Special Publication 250–57, US Government Printing Office (2000).

Okoshi, T., Kikuchi, K., and Nakayama, A., "Novel method for high resolution measurement of laser output spectrum," *Electronics Letters*, 16(16), 630–631 (1980).

Okuto, Y. and Crowell, C., "Ionization coefficients in semiconductors: A nonlocalized property," *Physical Review B*, 10(10), 4284–4296 (1974).

Olsson, N. A., "Semiconductor optical amplifiers," *Proceedings of the IEEE*, 80(3), 375–382 (1992).

Osinski, M. and Buus, J., "Linewidth broadening factor in semiconductor lasers-an overview," *IEEE Journal of Quantum Electronics*, 23(1), 9–29 (1987).

Paschotta, R., Nilsson, J., Tropper, A. C., and Hanna, D. C., "Ytterbium-doped fiber amplifiers," *IEEE Journal of Quantum Electronics*, 33(7), 1049–1056 (1997).

Planck M., "On the law of distribution of energy in the normal spectrum," *Annalen der Physik*, 309(4), 553–563 (1901).

Poole, S. B., Payne, D. N., Mears, R. J., Fermann, M. E., and Laming, R. I., "Fabrication and characterization of low-loss optical fibers containing rare-earth ions," *Journal of Lightwave Technology*, 4(7), 870–876 (1986).

Powell, R. C., *Physics of Solid-State Laser Materials*, Springer, New York (1998).

Quinlan, F., Fortier, T. M., Jiang, H., Hati, A., Nelson, C., Fu. Y., Campbell, J. C., and Diddams, S. A., "Exploiting shot noise correlations in the photodetection of ultrashort optical pulse trains," *Nature Photonics*, 7(4), 290–293 (2013a).

Quinlan, F., Fortier, T. M., Jiang, H., and Diddams, S. A., "Analysis of shot noise in the detection of ultrashort optical pulse trains," *Journal of the Optical Society of America B*, 30(6), 1775–1785 (2013b).

Ramirez, D. A., Hayat, M. M., Huntington, A. S., and Williams, G. M., "Non-local model for the spatial distribution of impact ionization effects in avalanche photodiodes," *IEEE Photonics Technology Letters*, 26(1), 25–28 (2014).

Reif, F., *Fundamentals of Statistical and Thermal Physics*, McGraw Hill, Boston (1965).

Robins, W. P., *Phase Noise in Signal Sources*, The Institute of Engineering and Technology, United Kingdom (1982).

Rottwitt, K. and Stentz, A. J., "Raman amplification in lightwave communication systems," in *Optical Fiber Telecommunications IVA*, I. Kaminow and T. Li (editors), Academic (2002).

Schawlow, A. L. and Townes, C. H., "Infrared and optical masers," *Physical Review*, 112(6), 1940–1949 (1958).

Schiek, B., Rolfes, I., and Siweris, H.-J., *Noise in High-Frequency Circuits and Oscillators*, Wiley, Hoboken (2006).

Schottky, W., "On spontaneous current fluctuations in various electricity conductors," *Annalen der Physik*, 57, 541–567 (1918).

Siegman, A. E., *Lasers*, University Science Books, Mill Valley (1986).

Simon, J. C., "GaInAsP semiconductor laser amplifiers for single-mode fiber communications," *Journal of Lightwave Technology*, 5(9), 1286–1295 (1987).

Singley, J. M., Diehl, J. F., and Urick, V. J., "Characterization of lasers for use in analog photonic links," Naval Research Laboratory Memorandum Report, NRL/MR/5650--11-9370 (2011a).

Singley, J. M., Diehl, J. F., and Urick, V. J., "Laser noise considerations for phase modulated links," *IEEE AVFOP Conf. Dig.*, paper ThD5 (2011b).

Skolnik, M., *Radar Handbook*, McGraw-Hill (2008).

Stolen, R. H. and Bjorkholm, J. E., "Parametric amplification and frequency conversion in optical fibers," *IEEE Journal of Quantum Electronics*, 18(7), 1062–1072 (1982).

Strutz, S. J. and Williams, K. J., "Low-noise hybrid erbium/Brillouin amplifier," *Electronics Letters*, 36(16), 1359–1360 (2000).

Suematsu, Y., "Dynamic single-mode lasers," *Journal of Lightwave Technology*, 32(6), 1144–1158 (2014).

Tang, C. L., Statz, H., and deMars, G., "Spectral output and spiking behavior of solid-state lasers," *Journal of Applied Physics*, 34(8), 2289–2295 (1963).

Taylor, J. A., Quinlan, F., Hati, A., Nelson, C., Diddams, S. A., Datta, S., and Joshi, A., "Phase noise in the photodetection of ultrashort optical pulses," *2010 IEEE International Frequency Control Symposium (FCS)*, pp. 684–688 (2010).

Teich, M. C. and Saleh, B. E. A., "Squeezed states of light," *Quantum Optics: Journal of the European Optical Society Part B*, 1(2), 153–191 (1989).

Tkach, R. W. and Chraplyvy, A. R., "Phase noise and linewidth in an InGaAsP DFB laser," *Journal of Lightwave Technology*, 4(11), 1711–1716 (1986).

Tkach, R. W. and Chraplyvy, A. R., "Fibre Brillouin amplifiers," *Optical and Quantum Electronics*, 21, S105–S112 (1989).

Towe, E., Leheny, R. F., and Yang, A., "A historical perspective of the development of the vertical-cavity surface-emitting laser," *IEEE Journal of Selected Topics in Quantum Electronics*, 6(6), 1458–1464 (2000).

Tsuchida, H., "Simple technique for improving the resolution of the delayed self-heterodyne method," *Optics Letters*, 15(11), 640–642 (1990).

Tulchinsky, D. A. and Williams, K. J., "Excess amplitude and excess phase noise of RF photodiodes operated in compression," *IEEE Photonics Technology Letters*, 17(3), 654–656 (2005).

Tulchinsky, D. A., Li, X., Li, N., Demigual, S., Campbell, J. C., and Williams, K. J., "High-saturation current wide-bandwidth photodetectors," *IEEE Journal of Selected Topics in Quantum Electronics*, 10(4), 702–708 (2004).

Urick, V. J., "Characterization of erbium-doped fiber lasers," Master's Thesis, George Mason University Library, Thesis Collection, Phys. 2004 U.75 (2004).

Urick, V. J., "Long-haul analog photonics principles with applications," Doctoral Dissertation, UMI No. 3255815 (2007).

Urick, V. J., Qiu, J. X., Bucholtz, F., McDermitt, C., and Williams, K. J., "4 Gbit/s transmission over single-channel Raman-amplified 105 km link," *Electronics Letters*, 40(8), 1–2 (2004).

Urick, V. J., Bucholtz, F., and Williams, K. J., "Noise penalty of highly saturated erbium-doped fiber amplifiers in analog links," *IEEE Photonics Technology Letters*, 18(6), 749–751 (2006).

Urick, V. J., Devgan, P. S., McKinney, J. D., and Dexter, J. L., "Laser noise and its impact on the performance of intensity-modulation with direct-detection analog photonic links," Naval Research Laboratory Memorandum Report, NRL/MR/5652--07-9065 (2007).

Urick, V. J., Bucholtz, F., McKinney, J. D., Devgan, P. S., Campillo, A. L., Dexter, J. L., and Williams, K. J., "Long-haul analog photonics," *Journal of Lightwave Technology*, 29(8), 1182–1205 (2011).

Urick, V. J., Singley, J. M., Sunderman, C. E., Diehl, J. F., and Williams, K. J., "Design and performance of Ka-band fiber-optic delay lines," Naval Research Laboratory Memorandum Report, NRL/MR/5650—12-9456, (2012).

Walls, D. F., "Squeezed states of light," *Nature*, 306(10), 141–146 (1983).

Weber, J., "Quantum theory of a damped electrical oscillator and noise," *Physical Review*, 90(5), 977–982 (1953).

Whitley, T. J., "A review of recent system demonstrations incorporating 1.3-μm praseodymium-doped fluoride fiber amplifiers," *Journal of Lightwave Technology*, 13(5), 744–760 (1995).

Williams, K. J., Esman, R. D., and Dagenais, M., "Nonlinearities in p-i-n microwave photodetectors," *Journal of Lightwave Technology*, 14(1), 84–96 (1996).

Yamada, M., Shimizu, M., Okayasu, M., Takeshita, T., Horiguchi, M., Tachikawa, Y., and Sugita, E., "Noise characteristics of Er^{3+}-doped fiber amplifiers pumped by 0.98 and 1.48 μm laser diodes," *IEEE Photonics Technology Letters*, 2(3), 205–207 (1990).

Yamamoto, Y. and Inoue, K., "Noise in amplifiers," *Journal of Lightwave Technology*, 21(11), 2895–2915 (2003).

Yuan, P., Anselm, K. A., Hu, C., Nie, H., Lenox, C., Holmes, A. L., Streetman, B. G., Campbell, J. C., and McIntyre, R. J., "A new look at impact ionization-Part II: Gain and noise in short avalanche photodiodes." *IEEE Transactions on Electron Devices*, 46(8), 1632–1639 (1999).

CHAPTER 4

DISTORTION IN FIBER OPTIC LINKS

As discussed in Chapter 2, the signal fidelity of analog photonic links is characterized primarily by noise figure and dynamic range. In Chapter 3, the sources of noise that are commonly encountered in radio-frequency (RF) photonic links were discussed. This chapter provides a discussion of signal distortion beyond what was provided in Chapter 2 and also highlights some common sources of distortion. As compared to the mathematical treatment of distortion in Chapter 2, Section 4.1 follows a more intuitive approach, relying on descriptions of transfer functions and power spectra. This method is carried into Section 4.2, where nonlinearities imposed by some electrical-to-optical conversion processes are introduced as a precursor to later chapters. The distortions caused by various optical amplifiers in compression are described in Section 4.3, building on the material from Section 3.5. Finally, the nonlinearities of photodetectors typically used for microwave photonics applications are analyzed in Section 4.4.

4.1 INTRODUCTION

Much attention has been devoted to describing systems with linear system theory. A linear system can be completely described by its impulse

Fundamentals of Microwave Photonics, First Edition.
Vincent J. Urick Jr, Jason D. McKinney, and Keith J. Williams.

response, a mathematical description of how the system responds to an impulse (infinite bandwidth) stimulus. A system is said to be linear if the principle of superposition applies. If the system output for an input signal $x(t)$ is $y(t)$, superposition states that the system output for $ax_1(t) + bx_2(t)$ is $ay_1(t) + by_2(t)$ for any scalar constants a and b. When superposition applies and the impulse response is known, the output of a linear system can be described by convolving the input time domain waveform with its impulse response. If the system is nonlinear, no such general mathematical description can be formed. In practice, a small-signal approximation is used for a nonlinear system where the system is characterized about some operating point using a functional expansion such as a power series, followed by the condition that the input signals about this operating point be "small." This then allows linear system theory to be used to describe the output mathematically about the operating point so long as the deviation from the operating point is minimal.

For a single sinusoidal input signal, a nonlinear system can cause harmonic distortion in the output. The nonlinear system in Figure 4.1 shows that a single-tone input yields not only a weighted output of the fundamental frequency, ω, but also an infinite sum of harmonics at $m\omega$, where m is an integer. Consider a transfer function, $f(x)$, as shown in Figure 4.2.

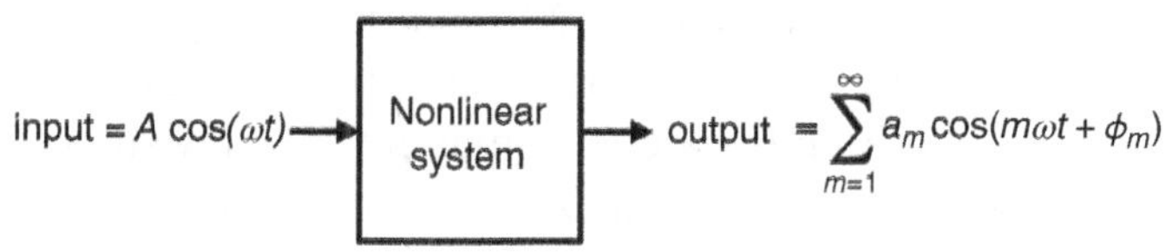

Figure 4.1. Diagram of a nonlinear system that produces an infinite sum of sinusoidal harmonics at the frequencies of $m\omega$ ($m = 1, 2, 3, \ldots$) from a single sinusoidal input stimulus at the frequency ω.

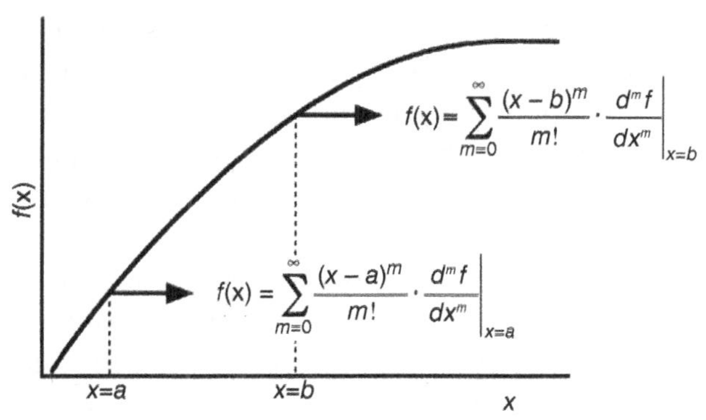

Figure 4.2. Plot of a nonlinear transfer function, $f(x)$, with Taylor series expansions around the points $x = a$ and $x = b$.

Such a curve could describe the current–power relationship in a high speed directly modulated laser diode (see also Figure 1.11). The output power, $f(x)$, can be described mathematically with a Taylor series approximation [see Equation (2.10)] expanded around a DC bias current as shown in Figure 4.2 for two DC bias currents, $x = a$ and $x = b$. An intensity modulation from the laser is created by modulating the drive current, $x = C + A\sin(\omega t)$ where C is the DC-bias current. In the case of a small-signal modulation approximation (small A), $f(x)$ can be described by a Taylor series. Note how the curvature of $f(x)$ is greater at $x = b$ than for $x = a$; thus one might assume the distortion components to be greater for the former. This can be a dangerous assumption for several reasons. Consider the case of a laser, where the transfer characteristics are often measured at DC and the stimulus is often at very high frequency. This requires the transfer characteristics, $f(x)$, to be characterized at a very high rate. When phenomena with drastically different timescales are present, for example, thermal effects in the presence of RF modulation, $f(x)$ may only be valid over a small range of stimulus frequencies. In direct-modulation lasers, power saturation (roll-over) at high bias currents may arise due to temperature effects that may be limited to microsecond or lower timescales due to the thermal mass of a typical semiconductor laser medium. Modulation at gigahertz rates may not allow sufficient time for the laser to heat up, and thus the functional dependence of $f(x)$ at high frequencies may not display such saturation and the output may be more linear. Another problem with attempting to determine the linearity of a transfer function on the basis of a cursory observation is that it is difficult to visually determine what higher order derivatives are present. The curves in Figure 4.3 will be used to demonstrate this point.

Consider the three calculated transfer functions shown in Figure 4.3. One curve is a sinusoidal transfer function, $f(x) = \sin(x)$, which is a very important function in analog photonic links and will be analyzed in further detail starting in Chapter 6. The function $\sin(x)$ is nearly linear for small x but diverges from $f(x) = x$ as x deviates from $x = 0$. A Taylor series expansion around $x = 0$ for $\sin(x)$ is an odd function of x with the coefficient of x^{2n+1} given by $(-1)^n/(2n+1)!$ for $n = 0, 1, 2 \ldots$. Therefore, the odd-order distortion terms can be easily calculated for a sinusoidal transfer function. Another function with one-half of the third-order term of $\sin(x)$ is also shown, $f(x) = \sin(x) + x^3/12$. As can be seen, this function is more linear than $\sin(x)$ but is still not perfectly linear. A third function, $f(x) = x - 5x^3/6$, is plotted and has five times

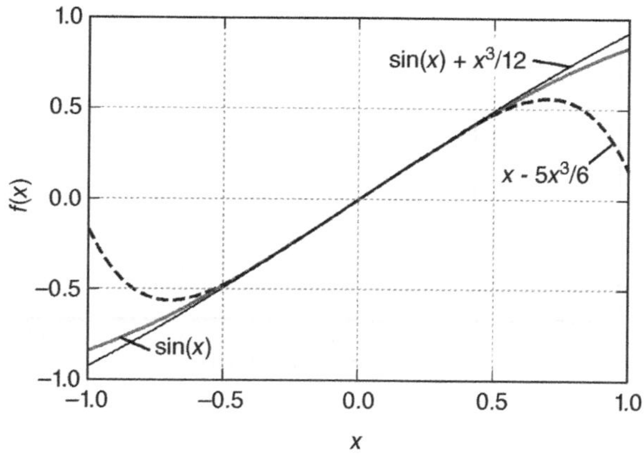

Figure 4.3. Plots of three different "transfer functions."

the third-order content of sin(x) but none of the higher order contributions. All three functions in Figure 4.3 look nearly "linear" for small x but obviously have very different harmonic content. This graphical illustration demonstrates the difficulty in determining linearity by visual inspection of transfer characteristics; thus linearity is best determined mathematically or experimentally.

Measurements of distortion can vary, and accurate characterization of nonlinearity—particularly in cascaded systems and/or in components with multiple physical sources of nonlinearity—requires careful selection of the measurement modality. The most basic (and widely accepted) characterization schemes are the single-tone and two-tone distortion measurements introduced in Chapter 2. In a single-tone distortion measurement, a single-frequency sinusoidal stimulus is presented at the input, and the powers of harmonics are measured at the output (as in Figure 4.1). In the frequency domain, Figure 4.4(a) would represent the single-sided spectrum of the distortion as integer harmonics (shown to fourth order). The power spectrum is more complicated when a two-tone test is performed. In this test, two equal-amplitude input sinusoidal signals at f_1 and f_2 are input to the device or system under test. The output spectrum may contain many mixing products including second-order terms at $(f_2 - f_1)$ and $(f_1 + f_2)$ and third-order intermodulation products at $(2f_2 - f_1)$ and $(2f_1 - f_2)$ as shown in Figure 4.4(b). For simplicity, only the lowest order distortion products are plotted in Figure 4.4(b), because the spectrum including higher order terms becomes cluttered with many intermodulation products. Harmonic distortion can often be filtered

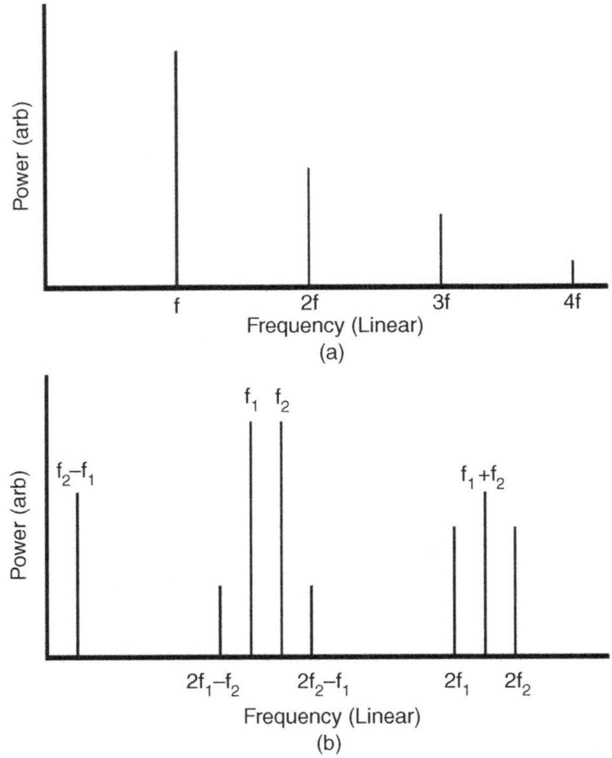

Figure 4.4. Plot of (a) the nonlinear power spectrum from a single-tone input, *f*, with the second, third, and fourth harmonics and (b) the nonlinear power spectrum from a two-tone input with associated second- and third-order products.

such as in single-octave systems where the harmonic distortion products are outside of the system bandwidth. This is not the case in multioctave systems. Third-order intermodulation terms are very important in all systems because they lie too close to the signal frequency content to be filtered.

In analog CATV systems where many subcarriers (different carrier frequencies) are present and the distortion products are many, distortion is often characterized by the composite second-order (CSO) and the composite triple-beat (CTB) [see e.g., Phillips and Darcie (1997)]. These types of distortion measurements are useful when discrete carriers are dominant with low modulation depths per channel. As the modulation depth increases, the spectrum can appear to be continuous as the number of intermodulation products grows. In the limit that the

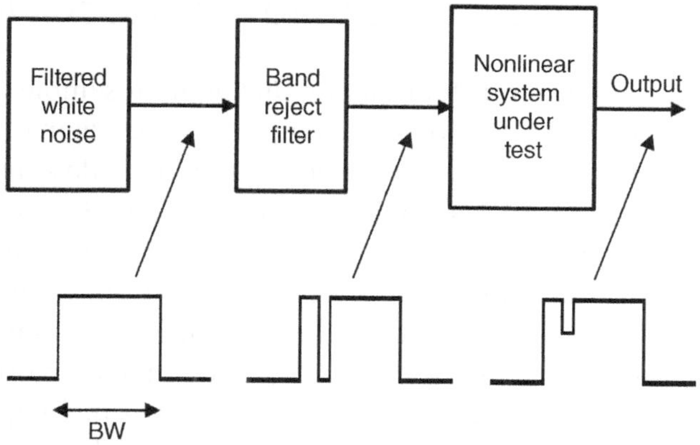

Figure 4.5. Block diagram of the white noise loading test and associated characteristic power spectra.

number of carriers is continuous, the white noise method or the white noise loading test (Yeh 1962) is sometimes employed. In this test (see Figure 4.5), a band rejection filter is placed within the bandwidth of a continuous noise signal. This yields an input test signal that has continuous frequency content over the bandwidth. This signal is then input to a nonlinear system, and the power within the notch is measured as a function of input power. This type of test gives a good picture of the distortion that remains in-band for any nonlinear system.

As mentioned in Chapter 2, the fidelity of a measurement apparatus must be considered when conducting tests such as those described previously. When beginning a system test, care should be taken to first characterize the measurement apparatus. On the stimulus side of the device or system under test, it is important to present the system with only a pure stimulus spectrum. Signal generators and signal combiners (for multitone tests) often have their own limits on linearity and can create a test spectrum exhibiting unwanted distortion. On the output side, instruments used for power measurement can also be nonlinear. It is required that the stimulus and measurement system nonlinearities be significantly less than the system under test in order to obtain accurate results. Care must also be taken when inserting filters or other components to remove or attenuate signals in the spectrum, as many filters achieve their rejection performance by reflecting energy; this reflected energy can re-enter the output of the device under test and create additional distortion.

4.2 DISTORTION IN ELECTRICAL-TO-OPTICAL CONVERSION

In the earlier section, a transfer function that might resemble direct laser intensity modulation was used as an example for the origin of distortion in an analog signal. In this section, a frequency domain description along the same lines as Figure 4.4 is employed to describe analog optical intensity modulation with an external Mach–Zehnder modulator (MZM) and pure phase modulation. An extensive description of each is provided in Chapters 6 and 7, respectively, including time domain optical fields and photocurrents that lead to RF performance metrics. Additional analog optical modulation types are covered in Chapter 8. The intention in this case is to provide a method of analysis on the basis of phasors in the optical domain that has application to analog optical modulation formats.

As introduced in Section 1.3 and detailed in Chapter 6, intensity modulation with an external MZM comprises angle (phase) modulation that is converted to intensity modulation via an integrated interferometer. The resulting transfer function is a sinusoidal optical intensity as a function of applied voltage as shown in Figures 1.13 and 6.2. When a chirp-free MZM is driven by a single tone, the output optical field can be described by an infinite sum of Bessel functions [Equation (6.15)]. The corresponding optical power spectrum contains terms at the optical carrier with dual modulation sidebands offset by the fundamental frequency and its entire set of harmonics. This is shown pictorially in Figure 4.6 as might be displayed on an optical spectrum analyzer. Equation (6.15) was employed for Figure 4.6 with a drive voltage of $0.3V_\pi$ and a quadrature bias. Detection of the field corresponding to Figure 4.6 with an ideal photodiode results in an RF spectrum devoid of even harmonics but with the odd-order terms present [see Equation (6.19)]. An illustration of the power spectrum at the output of a chirp-free, quadrature-biased MZM driven by two tones is shown in Figure 4.7 as calculated from Equation (6.39). For this calculation, a lower modulation depth than the single-tone case was employed, corresponding to a drive of $0.08V_\pi$. Similarly to the results in Figure 4.6, both even- and odd-order distortion terms are shown in Figure 4.7, but only odd-order terms are present in the photocurrent [Equation (6.40)]. Therefore, the illustrations in Figures 4.6 and 4.7 alone do not give the full picture. Rather, the relative "phases" of the sidebands must be considered, which can be represented by a phasor diagram such as shown in Figure 4.8.

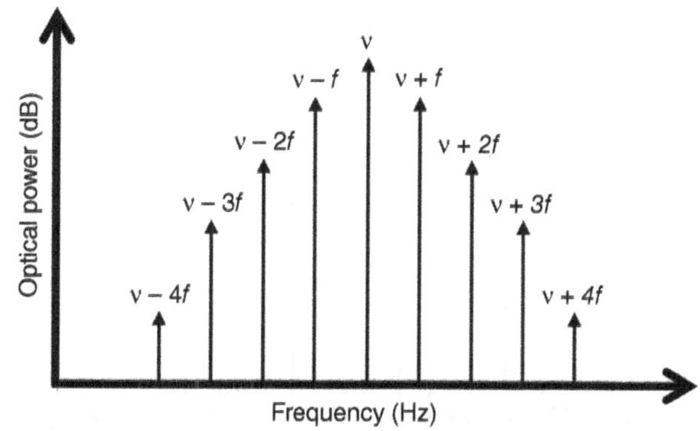

Figure 4.6. Calculated optical power spectrum [Equation (6.15)] at the output of a push-pull MZM biased precisely at quadrature and driven at by a single tone with a peak amplitude of $0.3V_\pi$. Terms up to fourth order are shown, and the relative amplitude of each frequency component is to scale.

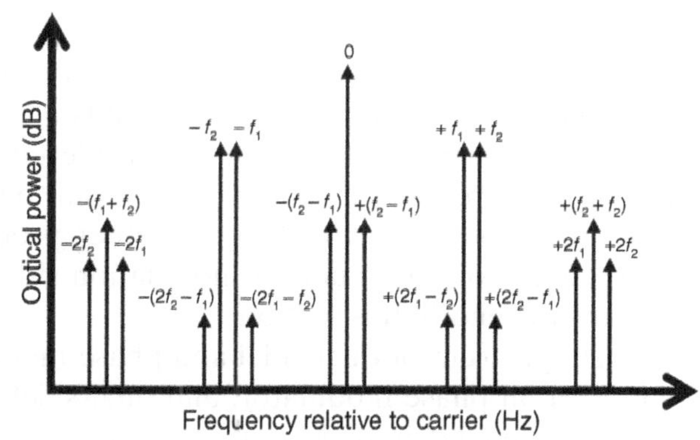

Figure 4.7. Calculated optical power spectrum [Equation (6.39)] at the output of a push-pull MZM biased precisely at quadrature and driven at by two tones, each with peak amplitude of $0.08V_\pi$. Terms up to second order are shown, and the relative amplitude of each frequency component is to scale.

Shown in Figure 4.8 is a diagram depicting phasors for the frequency components in Figure 4.6. The phasors are shown at a fixed point in time where the optical carrier can be written as $A\sin(2\pi\nu t)$. (A similar diagram can be made for the two-tone case using Equation (6.39) but for brevity is not shown in this chapter.) When this entire spectrum is demodulated with a photodiode, the even-order terms are nonexistent in the photocurrent, whereas all of the odd-order terms are present. Therefore, the relationships shown in Figure 4.8 present a special case where all mixing products corresponding to even-order terms

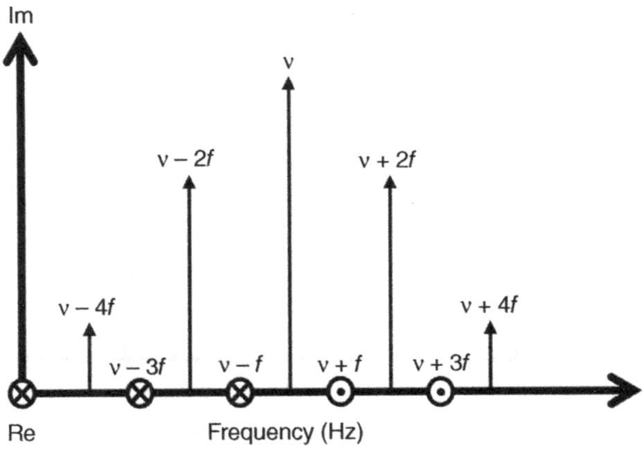

Figure 4.8. Calculated phasor diagram [Equation (6.15)] for the output of a push-pull MZM biased precisely at quadrature and driven at by a single tone with peak amplitude of $0.3V_\pi$. The real-imaginary plane is normal to the page, with the positive real axis being into the page.

cancel. Significant distortion products can be generated if the delicate amplitude and/or phase relationships of the sidebands are altered. A simple example includes carrier suppression as considered by Esman and Williams (1995) to increase the sensitivity of a link but with the consequence of induced even-order distortion. Other processes that can disturb the phase and/or amplitude of modulation sidebands will be considered toward the end of this section.

The optical power spectrum of dual-sideband phase modulation, as obtained with an external phase modulator, and dual-sideband intensity modulation, originating in an MZM as described previously, at the same modulation depth are indistinguishable. However, direct detection of such intensity modulation when quadrature biased results in an RF photocurrent with odd-order terms only, whereas direct detection of phase modulation produces no oscillatory photocurrent. [This assumes that the phase modulation is pure, containing no residual intensity modulation as observed by Tavlykaev et al. (2012)]. Shown in Figures 4.9 and 4.10 are phasor diagrams for single- and two-tone phase modulation, using the same reference phase for the optical carrier as in Figure 4.8. All of the frequency components present in the intensity modulation spectra are also present in the phase modulation spectra. However, comparing Figures 4.8 and 4.9, it can be seen that the phase relationships of the different terms vary. The specific relationships shown in Figures 4.9 and 4.10 correspond to a constant-intensity optical field that results in no RF photocurrent after direct detection. Demodulation of a phase

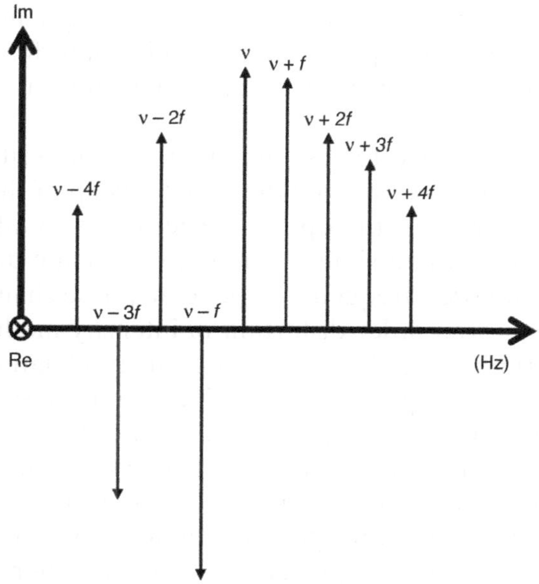

Figure 4.9. Calculated phasor diagram [Equation (7.4)] for the output of a phase modulator driven by a single tone with a peak amplitude of $0.3V_\pi$. The relative amplitudes of the components are to scale. The real-imaginary plane is normal to the page, with the positive real axis being into the page.

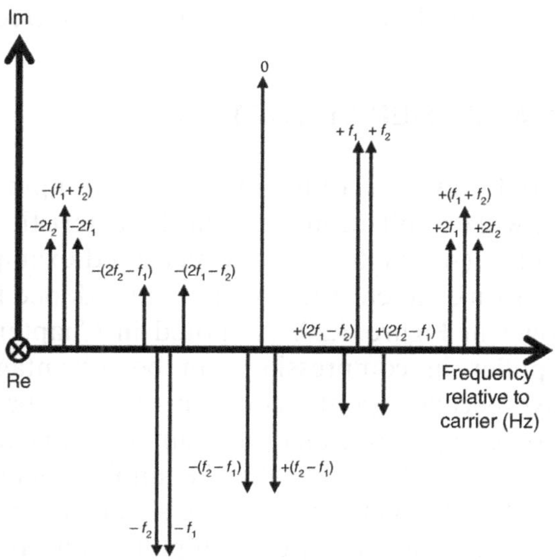

Figure 4.10. Calculated phasor diagram [Equation (7.12)] for the output of a phase modulator driven by two tones, each with a peak amplitude of $0.08V_\pi$. The relative amplitudes of the components are to scale. The real-imaginary plane is normal to the page, with the positive real axis being into the page.

modulation field can be achieved by implementing some mechanism to change these relationships, such as an optical interferometer, other types of optical filtering, or injection of an optical local oscillator (see Chapter 7).

Mechanisms that convert phase modulation to intensity modulation by altering the phase and/or amplitude relationship of the modulation sidebands with respect to the optical carrier can also affect an intensity modulated signal in a similar manner. Chromatic dispersion (see Chapter 5) can impose a frequency-dependent phase change, which can cause fading and even-order distortion in intensity-modulated signals and also convert phase modulation to intensity modulation. Optical filters can have a large dispersion slope, which can have similar effects if the optical carrier is not properly placed within the filter. The amplitude response in the optical domain is also important. For example, the cancellation of even-order terms in the intensity modulation case mentioned previously requires precise amplitude balance of the terms on both sides of the optical carrier. Optical components that have a frequency-dependent amplitude transfer functions over the range of the RF modulation can therefore disrupt the output signal fidelity. Numerous optical components can exhibit such a response, such as filters and amplifiers that may exhibit significant gain slope or ripple over their bandwidth. As described in the following section, optical amplifiers can exhibit significant gain tilt over a range of microwave frequencies.

4.3 OPTICAL AMPLIFIER DISTORTION

The optical output power as a function of optical input power for an optical amplifier will exhibit a compression characteristic similar to that for an RF amplifier (see Figure 2.5). A measured example of the top end of a compression characteristic for an erbium-doped fiber amplifier (EDFA) is shown in Figure 3.29. As noted in Chapter 3, operating an optical amplifier in compression can be advantageous from a noise perspective because spontaneous emission can be minimized in saturation. However, operating an optical amplifier in compression can distort an intensity-modulated signal if the amplifier gain medium can respond to the intensity fluctuations of the signal. Therefore, the time constant(s) associated with the optical amplification process is critical for determining the effects on an RF signal due to optical amplifier compression. An optical amplifier will exhibit a low-pass filter response to gain modulation with a characteristic cutoff frequency (f_c) that corresponds to the response time of the medium. If an optical amplifier is operated in compression, this translates to an RF response that is similar to a high-pass filter with the same f_c; RF signals below f_c experience gain compression, whereas signals well above f_c are unaffected by the optical saturation. In this section, the characteristics of some optical amplifiers in compression will be examined, demonstrating that an EDFA has numerous advantages over other optical amplifier technologies because of its relatively slow gain dynamics.

There has been considerable work on the RF compression characteristics of semiconductor optical amplifiers (SOAs) because of their potential use in integrated microwave photonic circuits. However, the carrier recombination time in an SOA is typically $\tau_c \leq 200\,\text{ps}$, which corresponds to $f_c \geq 5\,\text{GHz}$. This can inhibit their use in high fidelity analog applications, causing RF compression as described previously and also RF crosstalk in multichannel applications. Some calculated curves demonstrating the frequency response of an SOA in compression are shown in Figure 4.11. The calculations there employ closed form expressions for the normalized fundamental response and relative second- and third-harmonic distortions, as derived by Herrera et al. (2003) from solutions to the SOA rate equations. The data in Figure 4.11 are plotted over 100 MHz to 20 GHz, a range over which Herrera et al. (2003) demonstrated excellent agreement with experiment for the parameters employed in this chapter. The fundamental response can be derived from Herrera et al. (2003) as

$$|S_{21}|^2 = A \cdot \frac{x_s^2 + 2x_s + \Omega^2\tau_c^2 + 1}{e^{2g_o}x_s^2 + 2e^{g_o}x_s + \Omega^2\tau_c^2 + 1}, \tag{4.1}$$

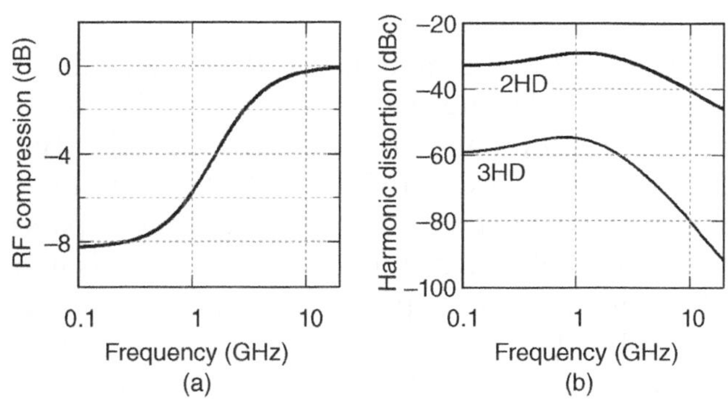

Figure 4.11. (a) Fundamental response for a compressed SOA and (b) relative second- (2HD) and third-harmonic (3HD) distortion levels for the same SOA. The theory for these plots is according to Herrera et al. (2003).

where A is the uncompressed link gain factor, $x_s = P_{sig}/P_{sat}$ is the ratio of the SOA optical input power to the saturated optical output power, Ω is the angular frequency, and g_o is the optical gain factor. The optical gain factor can be related to the small-signal optical gain factor ($g_{o,ss}$) as Herrera et al. (2003)

$$x_s(e^{g_o} - 1) = g_{o,ss} - g_o. \tag{4.2}$$

Equation (4.1) is plotted in Figure 4.11(a) for an SOA with the following parameters: $A = 1, P_{sig} = 2\,\text{mW}, P_{sat} = 9\,\text{mW}, \tau_c = 200\,\text{ps}$, and $g_{o,ss} = 4.2(6.2\,\text{dB})$. This last parameter yields $g_o = 2.3$ by way of Equation (4.2). The curve in Figure 4.11(a) exhibits a high-pass response, where frequencies $\ll 1/\tau_c = 5\,\text{GHz}$ experience gain compression and those $\gg$5 GHz do not.

The ratios of the second- and the third-harmonic current to the fundamental current are given as Herrera et al. (2003)

$$\frac{I_{2\Omega}}{I_\Omega} = -\frac{mx_s\left(e^{g_o} - 1\right)\left(1 + i2\Omega\tau_c\right)\left[1 + x_s\left(e^{g_o} + 1\right)/2 + i\Omega\tau_c\right]}{2\left(1 + x_s + i\Omega\tau_c\right)\left(1 + x_s e^{g_o} + i\Omega\tau_c\right)\left(1 + x_s e^{g_o} + i2\Omega\tau_c\right)}, \tag{4.3}$$

$$\frac{I_{3\Omega}}{I_\Omega} = \frac{m^2x_s^2\left(e^{g_o} - 1\right)\left(1 + i3\Omega\tau_c\right)}{4\left(1 + x_s e^{g_o} + i\Omega\tau_c\right)\left(1 + x_s e^{g_o} + i3\Omega\tau_c\right)}$$
$$\times\left\{\frac{e^{g_o}\left[1 + x_s\left(e^{g_o} + 1\right)/2 + i\Omega\tau_c\right]}{\left(1 + x_s e^{g_o} + i\Omega\tau_c\right)\left(1 + x_s e^{g_o} + i2\Omega\tau_c\right)} + \frac{e^{g_o} - 1}{2\left(1 + x_s + i\Omega\tau_c\right)}\right\}, \tag{4.4}$$

respectively, where m is the modulation depth. Equations (4.3) and (4.4) can be used to calculate the relative power for the second- and third-harmonics, $|I_{2\Omega}/I_\Omega|^2$ and $|I_{3\Omega}/I_\Omega|^2$, respectively. The results of such a calculation are plotted in Figure 4.11(b) using the parameters listed previously and with $m = 0.13$. As expected, a low-pass response is shown; however, the relative distortion above 5 GHz is non-negligible due to the slow roll off in the response. Some have proposed increasing the carrier recombination time in SOAs to up to 2 ns by using a type-II quantum well (Khurgin et al. 2002) to mitigate such high frequency issues in SOAs. Nonetheless, fast gain dynamics will remain a tradeoff for employing gain-compressed SOAs in microwave photonics applications.

The previous calculations for an SOA are based on a continuous-wave (CW) RF input where harmonic distortion arises. Transient effects in compressed optical amplifiers must also be considered in microwave photonics applications where pulsed RF waveforms at large modulation depth are being processed. Such time-domain effects have been studied in on-off keyed optical transmission systems, where a leading-edge overshoot followed by an undershoot can occur. The leading edge of the pulse can experience a short period of unsaturated gain, followed by gain depletion, and then a steady-state condition. The time constants associated with this process are tied to the speed of the amplifier gain dynamics. The transient effects have been studied extensively for Raman amplifiers (Chen and Wong, 2001, Gray 2002) and EDFAs (Giles et al. 1989, Sun et al. 1997). The gain saturation and recovery for the latter is relatively long, on the order of hundreds of microseconds (Giles et al. 1989).

The cutoff frequency is important not only as it relates to fundamental response and distortion, but also for noise transfer. As mentioned previously, the gain modulation for an optical amplifier will exhibit a low-pass response with a characteristic frequency f_c. Therefore, noise on the pump can be transferred to the signal through corresponding changes in the amplifier gain. For SOAs, pump noise transfer can be a problem unless low-noise electric current sources are used. On the other hand, relative intensity noise (RIN) transfer is an issue for optically pumped amplifiers. Significant pump RIN transfer has been observed at frequencies well above 1 GHz in both Raman (Fludger et al. 2001) and parametric (Marhic et al. 2005) amplifiers. The EDFA has an advantage over other optical amplifier technology in regards to RIN transfer because only low frequency noise will convert due to the relatively slow gain dynamics.

The characteristic frequency for an EDFA is only about 100 Hz, owing to the long inverted state lifetime of about 10 ms. As mentioned previously, this results in many advantages over other optical amplifiers. The gain modulation response of an EDFA exhibits a low-pass filter response that results in little crosstalk in multichannel links at frequencies above 5 kHz (Giles et al. 1989). The signal response for a compressed EDFA has a shape similar to that in Figure 4.11(a) but with f_c = 100 Hz (Freeman and Conradi 1993). The resulting distortion response has a low-pass characteristic. At frequencies above 1 kHz, even-order distortion falls off with a slope of about 20 dB/decade, whereas that for odd-order distortion has been measured at 40 dB/decade (Chen et al. 1991).

The gain flatness of an optical amplifier is an important metric when employing multiple optical channels and/or if the so-called gain tilt is significant over the RF modulation bandwidth in single-channel applications. This is not a major concern in wideband optical amplifiers such as SOAs and Raman amplifiers but may present a problem in EDFAs and Brillouin amplifiers where different modulation sidebands can experience different optical gain if the gain tilt is large. In fact, Brillouin amplifiers have often been employed to amplify individual modulation sideband components selectively (Yao 1998). Gain tilt in an EDFA can cause distortion by converting frequency or phase modulation components into intensity modulation. Such processes have been studied in intensity modulation links when there is significant residual chirp resulting from either direct laser modulation (Kikushima and Yoshinaga 1991, Clesca et al. 1993) or external modulation with an unbalanced MZM (Willems and Muys 1996). In such cases, significant second- and third-order distortion has been observed. The use of Er–Al codoping to flatten the gain (Kikushima and Yoshinaga 1991) or low-chirp modulation can be used to mitigate these issues.

4.4 PHOTODETECTOR DISTORTION

Harmonic distortion was first observed in avalanche photodiodes by Ozeki and Hara (1976). Ozeki also measured a p–i–n photodiode and noticed a second harmonic, but only when biased at low voltages, and therefore concluded that p–i–n photodiodes were reasonably linear under normal bias conditions. The first measurements of p–i–n photodiodes having significant nonlinearity were conducted by Esman and Williams (1990). This work, re-plotted in Figure 4.12, shows harmonic content up to the 40th harmonic at significant levels for photocurrents of only 1 mA. The distortions were later explained by Williams and Esman (1992) as being the result of distortion in the time-domain waveforms likely due to the high space-charge fields arising from photogenerated carriers in the intrinsic region. These observations have led to an extensive study of photodetector nonlinearities by many groups and will be discussed further in Section 4.4.2.

Distortion in photodetectors is important for high-SFDR photonic links. In Chapter 6, expressions for the link OIP2 [Equation (6.47)] and OIP3 [Equation (6.42)] are derived when sinusoidal intensity modulators are used. Reducing modulator distortion can be cumbersome in practical link designs; therefore, it is desired that the modulator be the

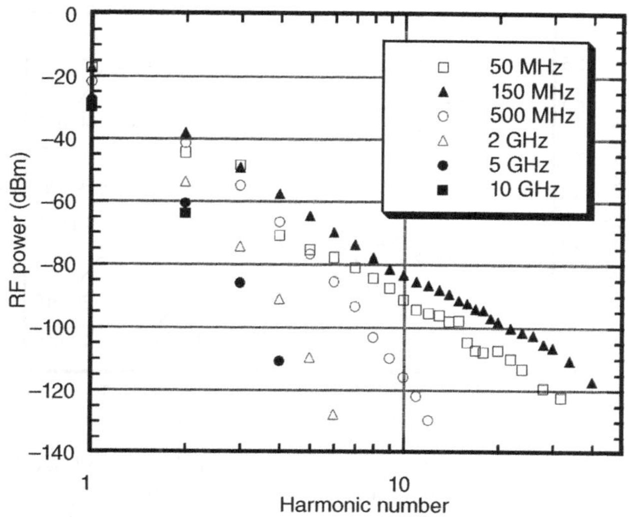

Figure 4.12. Photodetector RF power versus harmonic number for five different drive frequencies where a harmonic number of one corresponds to the fundamental (Esman and Williams 1990).

dominant source of distortion and that all other components used attain intercept points well exceeding that of the modulator. This ensures a modulator-limited link SFDR. The O/E convertor or photodetector, a basic link component, must be investigated to ensure it meets this criterion. To explain the importance of high linearity photodetectors in photonic links, Williams et al. (1998) performed link distortion measurements using high current balanced photodetectors and further characterized the high current photodetectors separately without a modulator present.

The measured OIP2 and OIP3 for photodetectors as a function of fundamental power (modulation depth) are shown in Figure 4.13. Over the range of average photocurrents and modulation depths measured, the OIP3 of the photodetectors straddled the OIP3 of the modulator-limited link. For photocurrents of 30 mA per photodetector, the link-limited OIP3 of 28.5 dBm [see Equation (6.42)] is below the photodetector OIP3 of greater than 30 dBm at all measured modulation depths. For photocurrents of 50 mA per photodetector, the link-limited OIP3 of 33 dBm is just above the photodetector OIP3 of greater than 30 dBm at low modulation depths, decreasing to 25 dBm at high modulation depths. Measurements of the corresponding link distortion are shown in Figure 4.14 demonstrating the impact of photodetector distortion on overall link performance. At 30 mA per

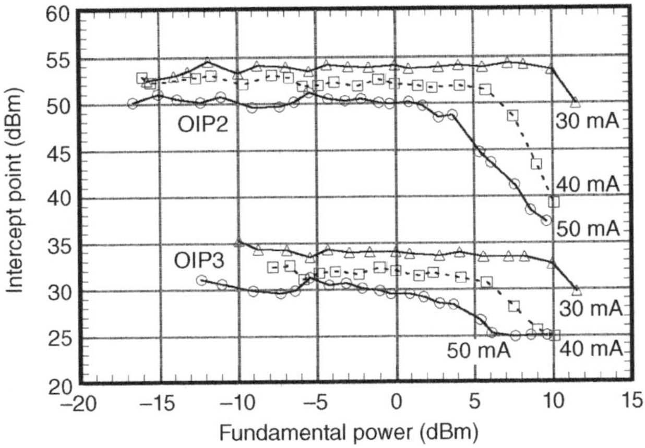

Figure 4.13. Measured small-signal intercept points of the photodiode from Williams et al. (1998). The triangles, squares, and circles correspond to average photocurrents of 30, 40, and 50 mA, respectively. The fundamental frequency is 500 MHz.

photodetector, the link SFDR of 119 dB · $Hz^{2/3}$ is in agreement with the modulator-limited value, and the distortion data at all modulation depths appears to follow a constant slope of 30 dB/decade of input power. At 50 mA per photodetector, the link SFDR of 119.5 dB · $Hz^{2/3}$ is 1.0 dB below the expected modulator-limited value, and the distortion data deviates from a constant slope of 30 dB/decade of input power as the modulation depth increases due to the reduction in the photodiode OIP3 as measured separately in Figure 4.13. This data highlights the importance of high linearity photodetectors in photonic links, especially those that operate at high average photocurrents. In addition, links that employ modulator linearization, achieving higher modulator-limited performance, will place additional demands on photodetector linearity.

The measured link OIP2 performance in Figure 4.14 was substantially better than the photodetector OIP2 measured alone. This was suggested as resulting from either balancing between the two photodetectors or a cancellation effect between the photodetector- and modulator-induced distortions. In later work by Hastings et al. (2008), it was demonstrated that it was possible to cancel second-order photodetector distortion within a balanced-output link biased at quadrature. It was also shown by Urick et al. (2013) that modulator second-order distortion could be used to cancel photodetector second-order distortion in a single-output link scenario. Therefore, both hypotheses have been proven to be possible to reduce overall

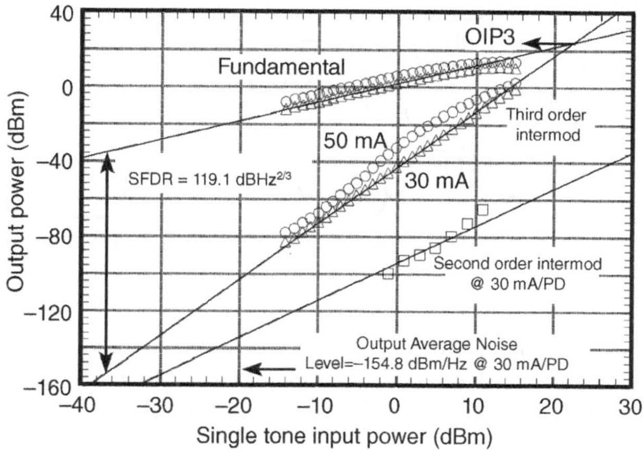

Figure 4.14. Measured two-tone intermodulation distortion for a photonic link (Williams et al. 1998). The RF frequencies were 1.0 and 1.1 GHz. The received average photocurrents were 30 mA (triangles) and 50 mA (circles) per photodetector. The second order distortion (squares) was measured at 30 mA per photodetector.

even-order link distortion. Such techniques should be very useful for designing links to attain high OIP2, an important system metric for many wideband RF systems.

4.4.1 Photodetector Distortion Measurement Systems

The sources of photodetector nonlinearity are dramatically more complicated than the relatively straightforward analysis of many modulation-induced distortions. Closed form mathematical solutions for the former have not been derived owing to the complexity of the nonlinear mechanisms and the number of possible sources of nonlinearity (see Section 4.4.2). Therefore, the measurement of photodetector nonlinearities becomes an important tool in quantifying photodetector nonlinearity. Earlier in this chapter, it was highlighted that any signal source used to test for linearity should be ideally linear so as not to corrupt the nonlinearity measurements. For photodetectors, such a system was developed by Williams et al. (1989) on the basis of the mixing of two single frequency lasers. This system is shown in Figure 4.15 where two single frequency lasers are phase locked to provide a stable RF beat tone at a frequency of $\nu_2 - \nu_1$. The system can generate a pure sinusoid devoid of any harmonic content, because there are no laser spectral components at harmonic integers of the generated RF frequency. This is true so long as the lasers are well isolated such that any reflections from one laser into another do not

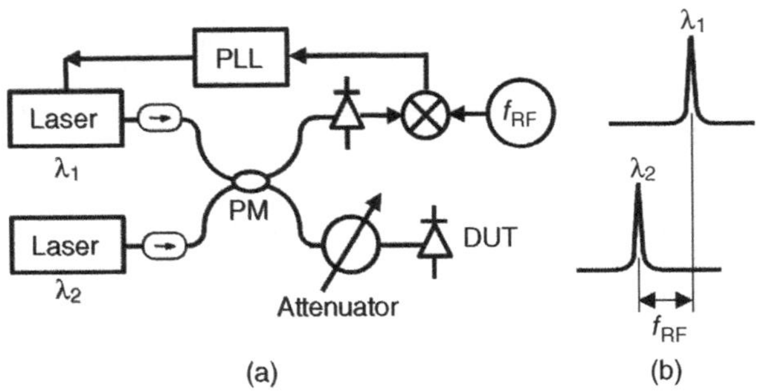

Figure 4.15. Two-laser heterodyne setup for measuring photodetector nonlinearity where two single frequency lasers are offset phase locked to generate a pure RF sinusoid. PM: polarization-maintaining, PLL: phase-locked loop, DUT: device under test.

generate other spurious laser modes. This can be an issue when the second laser is near a secondary mode of the first laser and the isolation is inadequate. Such instances will only occur over a narrow bandwidth and a small number of well-known frequencies that can be avoided.

A third laser can be added to the phase-locked pair to allow the average power, and thus the modulation depth, to be adjusted independently of the sinusoidal stimulus amplitude (Davis et al. 1996). This arrangement can mimic a MZM-based IMDD link biased at quadrature where a large DC photocurrent with variable modulation depth modulation impinges on the photodetector. The wavelength of the third laser should be offset far enough such that the mixing products are well above the photodetector bandwidth. Separations from 500 GHz to 1 THz (>5 nm) should be sufficient for gigahertz-bandwidth photodiodes. As demonstrated by Williams and Esman (1998), misleading results can arise if the wavelength separation is too great (e.g., 1.32-µm phase-locked lasers combined with a 1.55-µm DC laser). This is due to the exponential carrier absorption profile and the variation in absorption coefficients for widely separated wavelengths [see Humphreys et al. (1985)].

To conduct an intermodulation product measurement via a two-tone test, it is necessary to generate two sinusoids and provide for an independent DC photocurrent level. Thus, four lasers are required to generate the two tones (Scott et al. 2000) with a fifth laser to control the DC photocurrent. Such an apparatus is shown in Figure 4.16, where two RF tones are generated at $|\nu_2 - \nu_1|$ and $|\nu_4 - \nu_3|$. In this case, it is again necessary to separate the wavelengths properly so as to not generate in-band mixing products, but close enough to avoid the problems associated with a varying absorption coefficient. Although a complicated

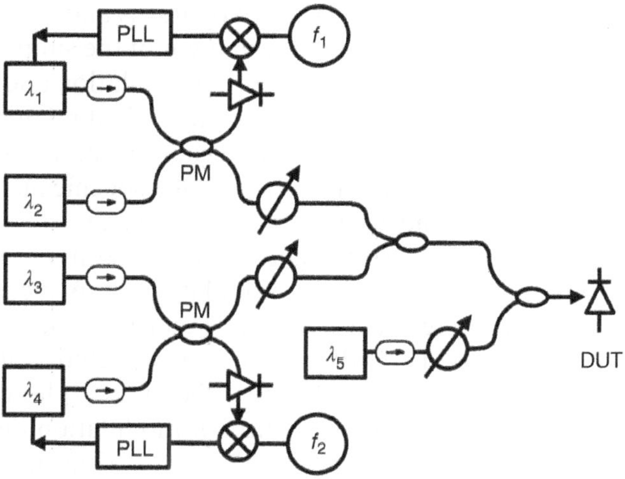

Figure 4.16. Five-laser two-tone heterodyne setup. PM: polarization-maintaining, PLL: phase-locked loop, DUT: device under test.

arrangement, the five-laser two-tone heterodyne apparatus has been used extensively due to the signal purity it can achieve.

In some cases, it may be more convenient to construct the distortion stimulus apparatus first used by Ozeki and Hara (1976). In this approach, three lasers are externally modulated at three different frequencies. This setup is shown in Figure 4.17 with the addition of a fourth laser for the independent adjustment of DC power. Each modulator input is a sinusoid at a different frequency yielding second-order products at the sums/differences between two of the three frequencies and third-order products at algebraic sums/differences of all three frequencies. Ramaswamy et al. (2010) conducted a study of a

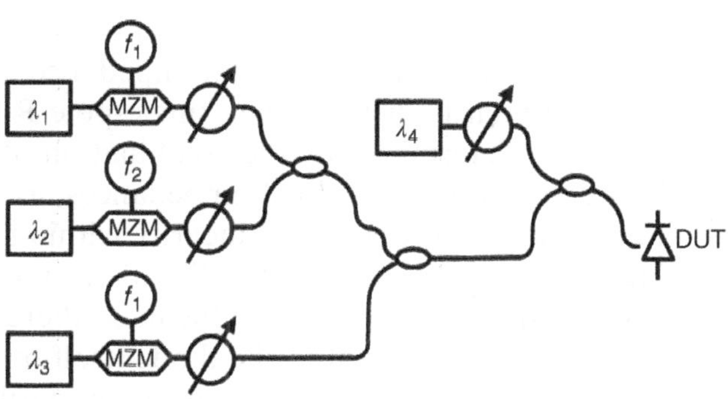

Figure 4.17. Four-laser three-tone MZM measurement setup.

two-modulator versus a three-modulator apparatus and outlined when it is necessary to use the third modulator in favor of the simpler, but less accurate, two-modulator approach. The intermodulation and harmonic products are mathematically related to the individual mixing products between the three modulated lasers when using the stimulus approach in Figure 4.17. However, this requires the use of a specific mathematical model for the behavior of the nonlinearity. Draa et al. (2011) conducted a detailed analysis between the two measurement setups in Figures 4.16 and 4.17, concluding that the three-modulator setup measurements agreed well with the more-ideal five-laser heterodyne measurements so long as the distortion data adhered to the expected mathematical slopes (30 dB/decade for the third-order). For many photodetectors, this occurs only over a very narrow region of modulation depths or in cases where a single nonlinear process dominates. Thus, it is generally more desirable to use the five-laser heterodyne measurement system when characterizing photodetectors for two-tone intermodulation distortion, as it generates a more accurate characteristic input spectrum. However, this highlights the weakness of the individual types of nonlinearity measurements. The two-tone measurement is more widely accepted, but the three-tone or even an *n*-tone measurement may be more appropriate for a particular application (such as legacy CATV, for example). Therefore, it may sometimes be more appropriate to characterize photodetectors with the white-noise loading test shown in Figure 4.5.

4.4.2 Photodetector Nonlinear Mechanisms

The most prevalent photodetector used in photonic links is the p–i–n junction photodiode as discussed in Chapter 1 and depicted in Figure 1.17. The basic device consists of an intrinsic absorbing semiconductor sandwiched between highly doped p-type and n-type materials. Electrically, the device behaves similarly to a current source in parallel with the junction capacitance and a high shunt impedance. The device frequency response is primarily dictated by a single-pole RC low-pass filter, with the resistance determined by the combination of the series impedance with the matching and load impedances as shown in Figure 1.17. There are many sources of nonlinearity within this type of photodiode. Table 4.1 shows an abbreviated list of six nonlinearities that have received attention in the literature. Included there is a listing of whether the nonlinearity is internal to the diode itself or a function of the external circuit, whether the nonlinearity is primarily affected by the optical power density within the photodiode, and whether the

TABLE 4.1 Several Photodiode Nonlinearities

Nonlinearity	Internal effect	Load dependent (external)	Power-density dependent	Frequency dependence
Responsivity (V)	–	Yes	No	Independent
Responsivity (I)	Yes	–	No	Independent
Capacitance (I)	Yes	–	Possibly	Increases at high f
Capacitance (V)	–	Yes	No	Increases at high f
Electron velocity (E)	Yes	Yes	Internal: Yes External: No	Increases at high f
Hole velocity (E)	Yes	Yes	Internal: Yes External: No	Increases at high f

nonlinearity has any frequency dependence. This section will cover some aspects of these nonlinearities in detail.

The first photodiode harmonic distortions observed by Esman and Williams (1990) were later explained by Williams and Esman (1992) to be caused by a transit-time nonlinearity. The carrier velocities were observed to decrease as the optical intensity increased. Shown in Figure 4.18 is the measured sinusoidal stimulus response for a photodiode using the heterodyne stimulus setup from Figure 4.15 without the DC laser (Williams and Esman 1992). The low-power 225-μW response appears to be a pure sinusoid. As the power was increased to 1.6 mW, the signal peak was lower than expected with an apparent

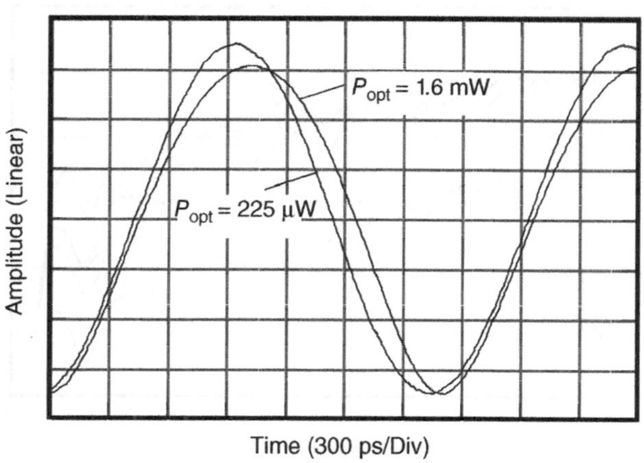

Figure 4.18. Oscilloscope trace of a 500-MHz photodetector signal at low (225 μW) and high (1.6 mW) incident optical powers showing waveform distortion (Williams and Esman 1992).

delay as the signal level increased, with a subsequent faster recovery as the power returned to a low level. Thus, the time-domain response of the photodiode is a nonlinear function of the current level, leading to distortion that appears as harmonics of the fundamental frequency. In addition to the observed harmonic distortion, such transit-time nonlinearities caused the RF response of the photodiodes to compress because the power in the fundamental frequency must decrease if power is being converted into harmonics. Response compression was also observed by Williams and Esman (1992) where the frequency response of the photodiode was characterized as a function of applied optical power, repeated in this chapter in Figure 4.19. Notice how the normalized RF response decreases as the incident optical power increases. Such characteristics are important to measure in photodiodes optimized to linearly detect much higher photocurrents as needed to improve link performance (see Chapter 9).

Another experimental effort by Kuhl et al. (1992) studied the effects on the impulse response of a photodetector directly as a function of optical power or pulse energy. In these detailed sets of measurement data, pulse width narrowing with a corresponding increase in the pulse "tail" was observed at higher pulse energies consistent with a carrier transit-time nonlinearity. This behavior is consistent with the study by Williams et al. (1994), where the cause of transit-time nonlinearities was shown to be a buildup of space charge within the depletion region causing the carrier velocities to be a function of photocurrent. Eventually,

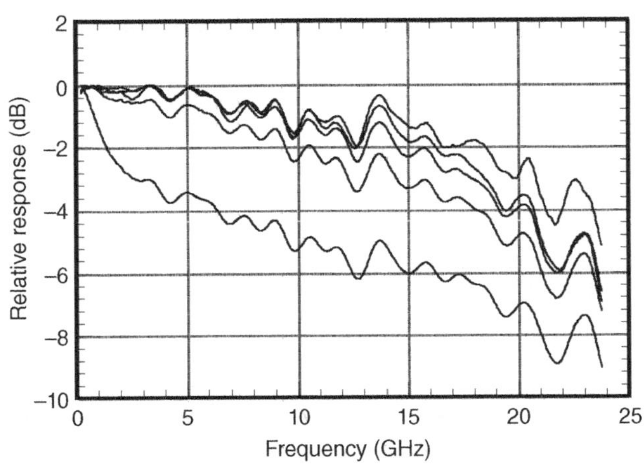

Figure 4.19. Normalized frequency response against incident optical power for P_{opt} = 100, 200, 400, 800, and 1600 μW, ordered from top to bottom curves, respectively (Williams and Esman 1992).

the space charge fields build to a point where the electric field over portions of the depletion region collapses, resulting in significant signal response compression.

To aid in the understanding of these photodiode nonlinearities, detailed numerical modeling must be employed, as the device physics are too complex for closed-form mathematical solutions. The first one-dimensional model reported on p–i–n photodetectors to calculate nonlinearities was reported by Dentan and de Cremoux (1990), although they did not report any corresponding experimental data. The one-dimensional model of Williams et al. (1996) was used successfully to calculate time domain distortions and response compressions, accurately predicting harmonic distortion levels over a wide range of operating conditions.

Models to calculate carrier transport are in a general class of drift-diffusion models employed to simulate a wide variety of semiconductor devices. The modeling is based on the solution to the following set of three coupled equations. The first is Poisson's equation:

$$\nabla \cdot E = \frac{q}{\varepsilon}(p - n + N_{\mathrm{d}} - N_{\mathrm{a}}), \tag{4.5}$$

which determines the electric field, E, everywhere within the device as a function of the location of the charge, both fixed-charge densities (dopants), N_{d} and N_{a}, and free carrier charge, p and n. In Equation (4.5), q is the electronic charge constant (taken to be positive) and ε is the permittivity of the device material. The carrier continuity equations are also used, one for holes and one for electrons:

$$\frac{\partial p}{\partial t} = G - R - \frac{1}{q}\nabla \cdot (J_{p\text{-drift}} + J_{p\text{-diff}}), \tag{4.6}$$

$$\frac{\partial n}{\partial t} = G - R + \frac{1}{q}\nabla \cdot (J_{n\text{-drift}} + J_{n\text{-diff}}), \tag{4.7}$$

respectively, where G is the generation rate, R is the recombination rate, and J are the electron and hole drift and diffusion current densities, with p denoting holes and n denoting electrons. The algorithm for the drift-diffusion model is executed as follows. Given an initial distribution of charges, the electric field according to Poisson's equation is calculated. Then, all diffusion and drift currents are calculated accounting for generation and recombination, all as a function of position. Time is then incremented and the charge is allowed to move according to the continuity equations. Since the charge distribution has changed, the electric field is then recalculated via Poisson's equation and the process is iterated. This algorithm, used by Williams et al. (1996), has proven to predict the nonlinear behavior in many devices accurately, although it was implemented in only one dimension (assuming a uniform intensity distribution radially away from the axial direction).

The numerical model by Williams et al. (1996) is useful in calculating nonlinear effects including distortion levels attributable to many nonlinearities such as transit-time effects. Transit-time nonlinearities are the result of the carrier velocities in the depletion region (for both electrons and holes) being a function of the instantaneous current level. In a depletion region where local electric fields can change dramatically, the carrier velocities are appreciable functions of the instantaneous current level. Functional dependencies for the carrier velocities are reviewed by Williams et al. (1996), and empirical expressions are plotted in Figure 4.20 for InGaAs with experimental data for electrons by Windhorn et al. (1982) and for holes by Hill et al. (1987). With typical depletion region lengths of 1 μm (for 25 GHz devices) at an applied voltage of between 5 and 10 V, electric field strengths vary between 50 and 100 kV/cm, allowing carrier velocities to vary by 10% or more (Figure 4.20). A positional dependence of the carrier velocities throughout the depletion region by itself does not create nonlinearity if the electric field as a function of position is current independent. However, because of current flow and the change in photocurrent due to modulation, the spatial distribution of charge is modified, thus changing the electric field via Poisson's equation.

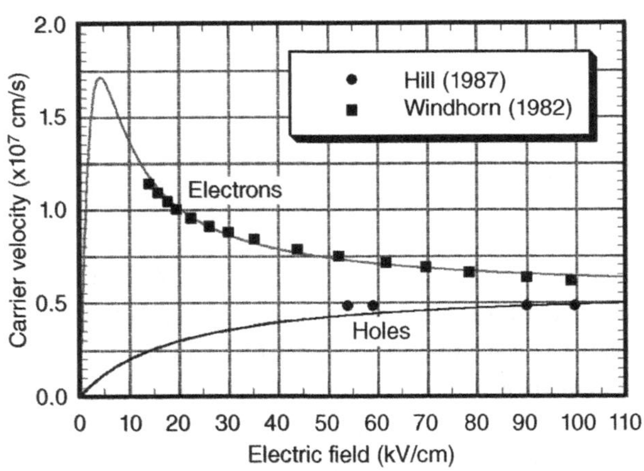

Figure 4.20. Empirical plots for the electron (top curve) and hole (bottom curve) velocities in InGaAs versus electric field. Experimental data are shown with symbols for electrons (Windhorn et al. 1982) and holes (Hill et al. 1987).

An estimate of the change in depletion region electric field can be made if the problem is simplified as in Kato (1999), a useful endeavor to gain insight into the magnitude of the space charge field. For such simplifying assumptions, firstly, the background dopant densities are assumed to be zero. This makes the dark electric field strength equal to the depletion region width divided by the applied voltage. Secondly, the carrier velocities are assumed to be fixed at the level of the dark electric field strength. For a 1-µm long depletion region at 5 V reverse bias, the dark electric field is 50 kV/cm yielding electron and hole velocities of 0.73×10^7 cm/s and 0.42×10^7 cm/s, respectively. The third assumption is that the carriers are generated uniformly throughout the depletion layer (as opposed to exponentially decaying absorption in the direction of propagation) and that the recombination rate is zero. With a uniform generation rate and carriers at constant velocity, the steady state carrier densities within the depletion region are linear functions of position as shown in Figure 4.21(a). The maximum hole (p_{max}) and electron (n_{max}) densities are inversely proportional to their velocity (assumed to be constant in this example), so it is evident that the total number of holes is substantially greater in the depletion layer than the number of electrons. The electric field [Equation (4.5)] can be directly integrated with $p(x)$ and $n(x)$ given by Figure 4.21(a) and the additional boundary condition that the total integral of the electric field be equal to the applied photodiode terminal voltage. The electric field will thus take a parabolic shape, given linear functions for the carrier densities. If no series impedance is assumed, then the electric field can

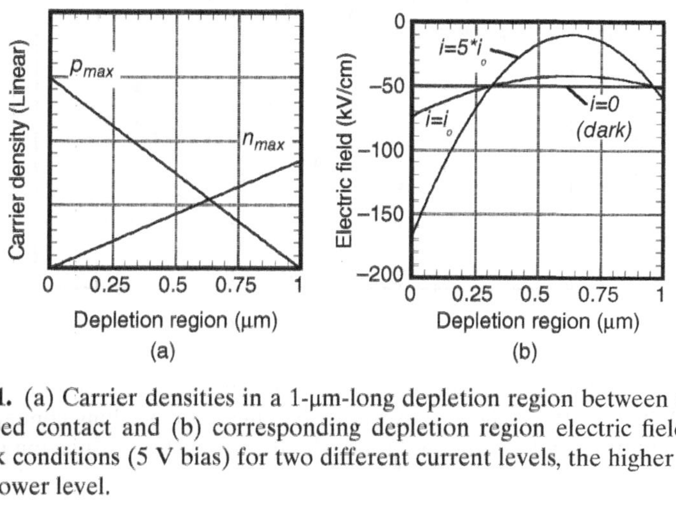

Figure 4.21. (a) Carrier densities in a 1-µm-long depletion region between a p-doped and n-doped contact and (b) corresponding depletion region electric field strength under dark conditions (5 V bias) for two different current levels, the higher being five times the lower level.

be calculated for several current densities as shown in Figure 4.21(b). Note how the large number of depletion region holes cause the electric field near the *p*-side of the depletion region to increase dramatically, which could result in breakdown. On the *n*-side of the depletion region, the electric field tends toward zero as current increases. If the current continues to increase such that the field in this region collapses to near zero, both holes and electrons will build up in this region (they no longer have sufficient velocity to move quickly), and the response will compress as observed in Figure 4.19. The exact current level at which this happens is a function of photodiode design, geometry, and illumination conditions.

When a full one-dimensional model is implemented for a particular photodiode epitaxial layer design and the field-dependent velocities are considered, good agreement can be obtained with the measured nonlinear distortion. Simulations of time domain distortions similar to the data in Figure 4.18 were performed by Williams et al. (1996) and the associated response compressions in Figure 4.19 calculated by Williams (1994a), both showing excellent agreement with experimental data. Simulations by Williams (1994a) of harmonic distortion as a function of current also yielded excellent agreement with experimental data, and a typical result is shown in Figure 4.22. Such one-dimensional simulations are certainly an approximation to the actual geometry as the intensity profile of real devices has a Gaussian distribution with

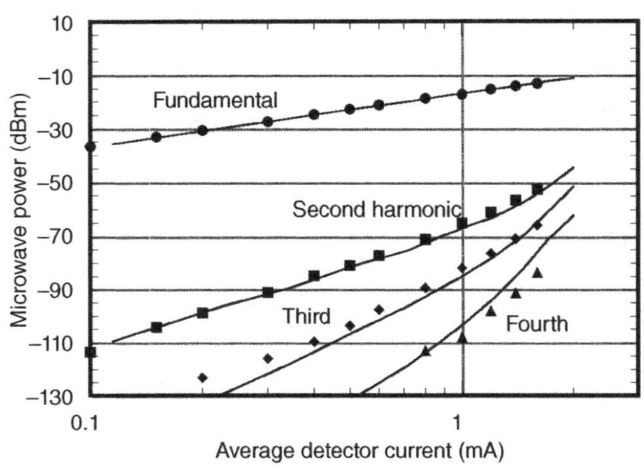

Figure 4.22. One-dimensional simulations and experimental data of harmonic distortion from Williams (1994a) showing excellent agreement in describing nonlinear effects for fundamental, second-, third-, and fourth-harmonic levels as a function of average photocurrent with a 7.5-μm spot size.

radial symmetry. However, one-dimensional drift-diffusion models have been tested against more accurate, but computationally intense, Monte-Carlo methods by Wilson et al. (1997) and have been found to perform exceptionally well. With the improvement in computing power, incorporation of additional physical effects and advanced numerical techniques, the one-dimensional model has been expanded to two-dimensions (radial symmetry requires only two dimensions) by Hu et al. (2014), with preliminary results showing the ability to more accurately predict other subtle features in the observed nonlinear behavior.

Another photodetector nonlinearity studied and found to be important is in the general class of quantum efficiency or responsivity nonlinearities. The first responsivity nonlinearity observed experimentally was by Williams (1994a), although in this work, the nonlinearity was attributed to absorption in an undepleted p-type absorber adjacent to the depletion region. It was not until the later work of Williams and Esman (1998) that the nonlinearity was identified as a responsivity that was a function of the photodetector instantaneous current. Qualitatively, responsivity nonlinearities can be understood by observation of Figure 4.23. Plotted there is a calculated photocurrent for an ideal photodetector driven with a sinusoid (dashed line) against that of a photodetector having a responsivity that is a function of photocurrent (solid line). The responsivity for the latter is a linear function of the

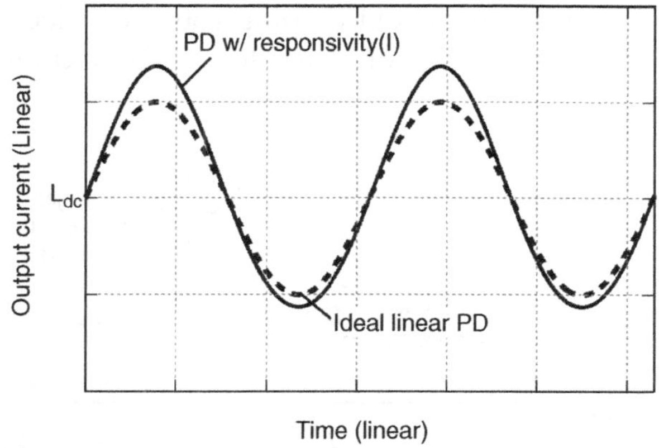

Figure 4.23. Calculated output photocurrents for two photodiodes. The dashed line represents an ideal photodiode where the responsivity is constant. The solid line represents a photodiode with a responsivity that is a function of the instantaneous current. The DC photocurrent for the linear photodiode is represented by I_{dc}, at which point the responsivity for both are equal.

instantaneous photocurrent, with a responsivity equal to that for the ideal case at its average current (I_{dc}). If the nonlinearity is fast, in that the responsivity can change during the evolution of the sinewave, then as the current increases, the current will be more than expected if the responsivity were equal to that which generated I_{dc}. As the current decreases below I_{dc}, the responsivity is lower, leaving the actual current lower than the ideal response.

A responsivity that increases with photocurrent was identified by Williams and Esman (1998) to be a source of photodetector nonlinearity at high electric fields. The diagram in Figure 4.24 is helpful in understanding the physical origin of this responsivity nonlinearity. In a single heterostructure photodiode, absorption occurs throughout the depletion region and the p-doped InGaAs contact. In the undepleted p-region, absorption creates a hole–electron pair. The minority carrier, the electron, is generally assumed to move only by diffusion in this region; however, because there is a majority carrier hole drift current, there is a small proportional p-region electric field (this region acts similar to a conductor with majority carrier holes). In Figure 4.24, the scenario at the top assumes a 1-mA detector current yielding a small electric field (3 V/cm). The minority electron will travel until it recombines, an average distance of about 2.4 nm for E ~3 V/cm and τ ~100 ps recombination time. If the electron is generated further than 2.4 nm (on average) from the depletion region edge, it will recombine before entering the depletion region and will not contribute to the current in the external circuit. The scenario at the bottom of Figure 4.24 assumes a 10-mA detector current yielding a 30-V/cm electric field. The electron will travel an average of 24 nm for E ~30 V/cm in τ ~100 ps.

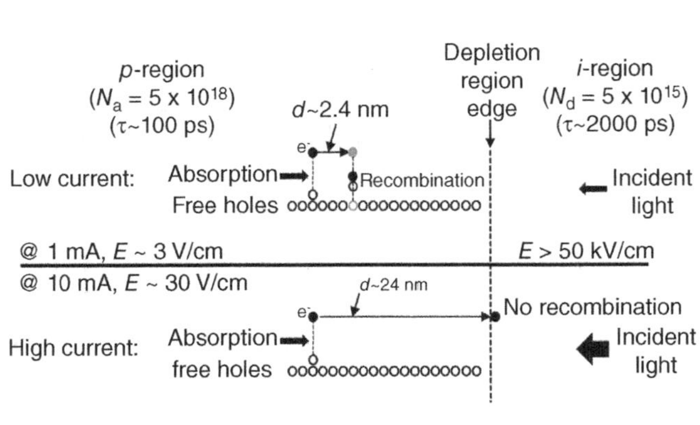

Figure 4.24. An illustration of recombination near the depletion region edge (p–i interface) at low (upper) and high (lower) photocurrents (Williams and Esman 1998).

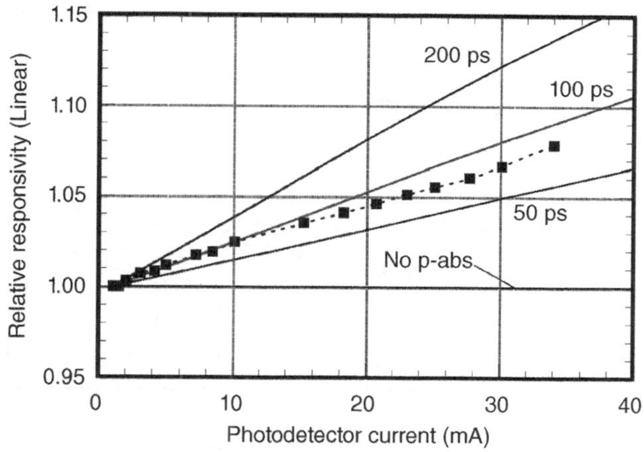

Figure 4.25. Measured (symbols) and simulated (solid lines) responsivity normalized to the responsivity at low currents. The simulation is shown for recombination times of 50, 100, and 200 ps (Williams and Esman 1998).

In this case, more electrons will make it into the depletion region, avoid recombination, and contribute to the total current at 10 mA as compared to 1 mA, thus resulting in a power-density-dependent photocurrent and a source of nonlinear distortion.

The measured responsivity of a photodetector is plotted in Figure 4.25 as a function of photocurrent (Williams and Esman 1998) where it can clearly be seen that the responsivity is indeed a function of current. Also plotted is a simulation of the photodiode with various p-region recombination times, illustrating the impact of recombination time on the responsivity nonlinearity. Williams (1994a) also noted that at high intrinsic-region electric fields where space-charge nonlinearities are less impactful, the nonlinear distortion became independent of the stimulus frequency. Figure 4.26 shows a measurement of harmonic distortion against simulations with and without p-region absorption (Williams 1994a). The harmonic distortion model more closely agrees with measurement when the p-region absorption is included (responsivity nonlinearity included). In another work (Williams 1994b), the 100 ps simulation result that more accurately predicts the responsivity nonlinearity (Figure 4.25) was also close to the value needed to model the harmonic distortion accurately. A responsivity measurement as a function of current must be measured with care unless the effect is substantial as it is in Figure 4.25. Other heating-related effects can cause a current-dependent responsivity such as band shifting as observed by Stievater and Williams (2004), where temperature can

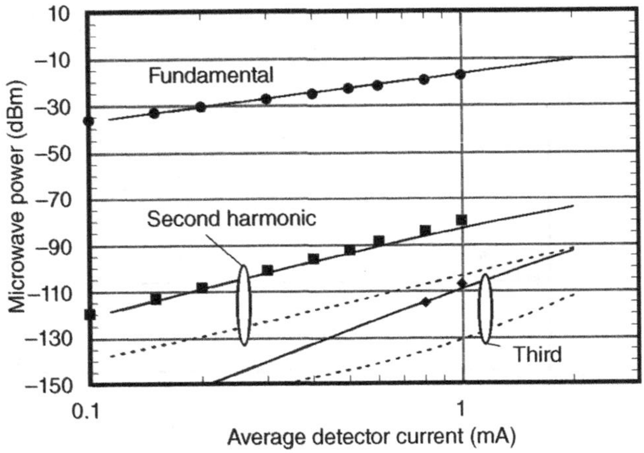

Figure 4.26. Measured and simulated photodetector fundamental, second, and third harmonics. Simulations are performed with (solid) and without (dashed) p-contact absorption (Williams 1994a).

shift the band edge and cause the absorption coefficient to change slightly. Temperature-related effects should have a time constant in the microsecond range, so they should not create significant nonlinear distortion at gigahertz frequencies as was discussed earlier in this chapter.

Measurements of the current-dependent responsivity can be used to estimate the nonlinear distortion via a Taylor series expansion, if it is measured in a timeframe commensurate with the inverse of the frequency range over which the nonlinearity is being modeled. Expressions for the harmonic distortion can then be formed on the basis of the measurements of the responsivity as was performed by Hastings et al. (2011). The responsivity can be described by a Taylor series about some average photocurrent as

$$\Re(I) = \Re(I_{\text{avg}}) + \Re'(I_{\text{avg}})(I - I_{\text{avg}}) + \frac{1}{2}\Re''(I_{\text{avg}})(I - I_{\text{avg}})^2, \tag{4.8}$$

where $\Re'(I_{\text{avg}})$ and $\Re''(I_{\text{avg}})$ denote the first and second derivatives of the responsivity with respect to current evaluated at the average current, I_{avg}. Expressions for the ratio of the harmonic to fundamental powers can then be approximated using methods described by Hastings et al. (2011) as

$$\frac{P_{2\text{f}}}{P_{\text{f}}} = \left(\frac{I_{2\text{f}}}{I_{\text{f}}}\right)^2 \approx \left\{\frac{mP_{\text{avg}}}{2}\left[I_{\text{avg}}\Re''(I_{\text{avg}}) - \Re'(I_{\text{avg}}) - \frac{3}{2}P_{\text{avg}}\Re(I_{\text{avg}})\,\Re''(I_{\text{avg}})\right]\right\}^2 \tag{4.9}$$

$$\frac{P_{3f}}{P_f}=\left(\frac{I_{3f}}{I_f}\right)^2\approx\left[\frac{1}{8}m^2P_{avg}^2\Re(I_{avg})\,\Re''(I_{avg})\right]^2 \qquad (4.10)$$

where m is the modulation depth, and P_{avg} is the average optical power. The main assumptions in moving from Equation (4.8) to Equations (4.9) and (4.10) are $I \approx \Re(I_{avg})P_o$, the input optical power is $P_o = P_{avg} + mP_{avg}\sin(\Omega t)$ and $I_f \approx m\Re(I_{avg})P_{avg}$. As a simple example, Equation (4.9) can be used with the data from Figure 4.25 to estimate the second harmonic measured in Figure 4.26. An estimate of the first derivative of Figure 4.25 is (0.002/mW) over the range of 0–23 mA. With $m = 1$ and $P_{avg} = 1$ mW, this gives a second harmonic relative to the fundamental of −60 dBc if the second derivative is neglected, close to the −65 dBc at 1 mA from Figure 4.26. Similar agreements were obtained by Hastings et al. (2011) with another style photodetector for both second- and third-harmonic distortion. This simple technique can be useful in determining if the nonlinearities measured could be the result of nonlinear responsivity or whether another nonlinear mechanism is needed to explain an observed distortion level.

The current-dependent responsivity nonlinearity in Table 4.1 is listed as being independent of stimulus frequency. For the example cited previously, this is basically true at low frequency, but it will not hold as the frequency increases to the point where the minority electron transit time in the p-doped absorber approaches the scale of the oscillation period. As for the other listed characteristics, the current-dependent responsivity should be independent of optical spot size (optical intensity), because the majority carrier hole density that determines the p-region electric field is high and largely unaffected by absorption. Finally, this type of current-dependent responsivity is listed as an internal effect because all the variables that determine the nonlinearity are internal to the photodiode to first order.

One type of responsivity nonlinearity that is a function of optical spot size is a bleaching effect. Bleaching in semiconductors is well known. Bleaching or absorption saturation occurs when intense optical fields either deplete the number of available final energy states or depopulate the number of initial states. In either case, the absorption coefficient is reduced as the number of available absorption sites decreases. Bleaching is usually observed in short-pulse applications where optical intensities can be high. Bleaching has been suggested by Juodawlkis et al. (2002) to cause sufficient responsivity reduction in pulsed optical links for photonic analog-to-digital convertors (PADC) resulting in nonlinear distortion large enough to limit the PADC performance. Because bleaching itself is power density dependent, it follows that the associated photodiode nonlinearity would also be power density dependent. To estimate if this nonlinearity could result in nonlinear distortion under CW illumination, a simple calculation of the first derivative of the responsivity with respect to current can be performed. The worse-case photodiode studied by Juodawlkis et al. (2002) exhibited a 10% change in responsivity for a 50-pJ energy change in a 30 ps pulse. This yields an estimate for the first derivative of about 6×10^{-5}/mW or about 100 times smaller than the current-dependent responsivity responsible for the distortion in Figure 4.26. Although this is a significantly smaller second-order nonlinearity in this case, it is not generally insignificant, as the current-dependent responsivity in the device for Figures 4.25 and 4.26 is substantially larger than most other photodetectors. Thus, further study is needed to quantify the nonlinearity imposed by bleaching for CW-illuminated photodiodes.

The responsivity nonlinearities discussed to this point are a function of only internal device characteristics. A set of responsivity nonlinearities that are affected by external factors are those nonlinearities that are a function of the photodetector terminal voltage. For example, current flow in the load impedance could reduce the terminal voltage and cause a transit-time nonlinearity. In many photodiodes, the responsivity can also be a function of the photodetector terminal voltage. This type of nonlinearity is labeled a voltage-dependent responsivity in Table 4.1.

Voltage-dependent responsivity nonlinearities were tied directly to nonlinear distortion by Beling et al. (2008) where a 4% change in responsivity was observed for a 7 V change in the bias voltage. It was suggested that the origin of this was impact ionization, whereby the acceleration of holes and/or electrons in a high field region can ionize additional carriers leading to increased current or, equivalently, responsivity gain. This effect was also observed and attributed to impact ionization under high illumination conditions in p–i–n diodes by Wilson and Walker (1998). To quantify the specifics of voltage-dependent responsivity nonlinearities, Hastings et al. (2009) showed experimental harmonic distortion data, measurements of the voltage-dependent responsivity, and a brief numerical treatment. The measurement of the responsivity as a function of applied voltage from Hastings et al. (2009) is shown in Figure 4.27 along with a calculation for the impact-ionization responsivity increase based on the model from McIntyre (1966). As can be seen, impact ionization accounts for the major feature of increasing responsivity as a function of depletion region electric field. The origin of the other oscillations is unknown;

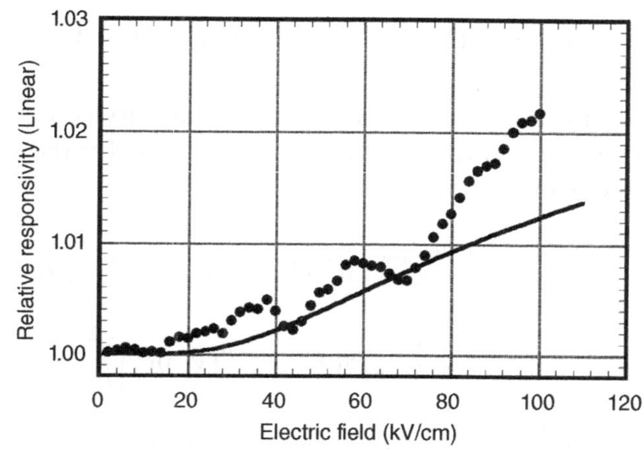

Figure 4.27. Measured relative responsivity as a function of intrinsic region electric field (applied voltage/depletion region width) and an ionization calculation based on the model of MacIntyre (1966).

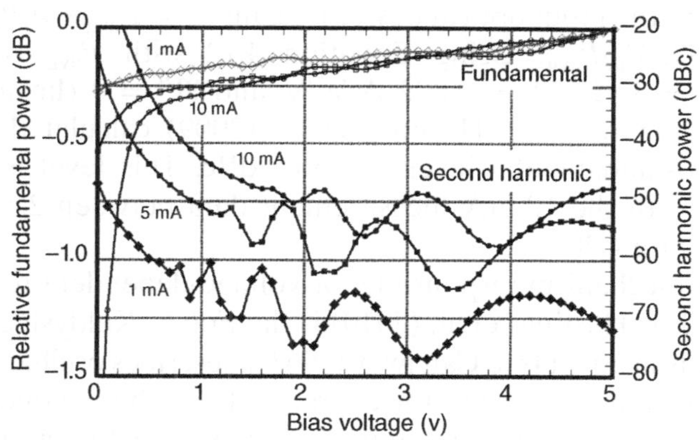

Figure 4.28. Measured fundamental and second harmonic for the photodiode measured in Figure 4.27 at 1, 5, and 10 mA (Hastings et al. 2009).

however, it does appear from the data in Figure 4.28 that the oscillations in the second harmonic amplitudes have similar periodicity to the oscillations in the measured voltage-dependent responsivity.

A Taylor series expansion can also be used for the voltage-dependent responsivity to derive expressions for estimation of the nonlinear distortion on the basis of the effect. A Taylor series for the voltage-dependent responsivity centered about the applied voltage, V_a, was given by Hastings et al. (2010) as

$$\Re(V) = \Re(V_a) + \Re'(V_a)(V - V_a) + \frac{1}{2}\Re''(V_a)(V - V_a)^2, \tag{4.11}$$

where $\Re^n(V_a)$ are the derivatives of the responsivity centered around V_a. The corresponding ratios of harmonic distortion for the second and third harmonics can be written as

$$\frac{P_{2f}}{P_f} = \left(\frac{I_{2f}}{I_f}\right)^2 \approx \left\{\frac{mP_{avg}}{2}\left[\Re''(V_a)V_aR_L - \Re'(V_a)R_L - \frac{3}{2}P_{avg}\Re(V_a)\Re''(V_a)R_L^2\right]\right\}^2 \tag{4.12}$$

$$\frac{P_{3f}}{P_f} = \left(\frac{I_{3f}}{I_f}\right)^2 \approx \left[\frac{1}{8}m^2R_L^2P_{avg}^2\Re(V_a)\Re''(V_a)\right]^2 \tag{4.13}$$

where m is the modulation depth, R_L is the load resistance, and P_{avg} is the average power. The main assumptions in moving from Equation (4.11) to Equations (4.12) and (4.13) are $V \approx \Re(V_a)P_oR_L$, $I \approx \Re(V_a)P_o$, the input optical power is again $P_o = P_{avg} + mP_{avg}\sin(\Omega t)$ and $I_f \approx m\Re(V_a)P_{avg}$. Using Equation 4.12 with P_{avg} = 30 mW, m = 1, R_L = 50 Ω, $\Re'$ = 0.002 A/W/V, and $\Re''$ = 0 (linear approximation of Figure 4.27), Hastings et al. (2009) calculated a second harmonic-to-fundamental level of −56.5 dBc. This level is in rough agreement with the 10-mA measurement data between 2.5- and 5-V biases in Figure 4.28.

Another mechanism responsible for voltage-dependent responsivity investigated by Hastings et al. (2010) is the Franz–Keldysh effect. The Franz–Keldysh effect [see Callaway (1963)] describes oscillations in the carrier transition probability for energies above the conduction band edge and electric-field-assisted tunneling probability for energies below the band edge in the presence of an applied electric field. The net effect is a responsivity change as a function of wavelength especially near the band edge. Shown in Figure 4.29 is the measured responsivity of a photodiode around the band edge for applied bias voltages from 3 to 15 V in 2-V increments. Note how the band edge appears to shift to shorter wavelengths with higher applied voltage. The result is a net responsivity decrease at wavelengths beyond 1575 nm and a net responsivity increase for wavelengths below 1525 nm. Also contained within the data in Figure 4.29 are the effects of impact ionization. At 1450 nm, far away from the band edge where the Franz–Keldysh effect is negligible, the photodiode experiences a 4.5% increase in responsivity attributed to impact ionization that is in agreement with the theory of McIntyre for

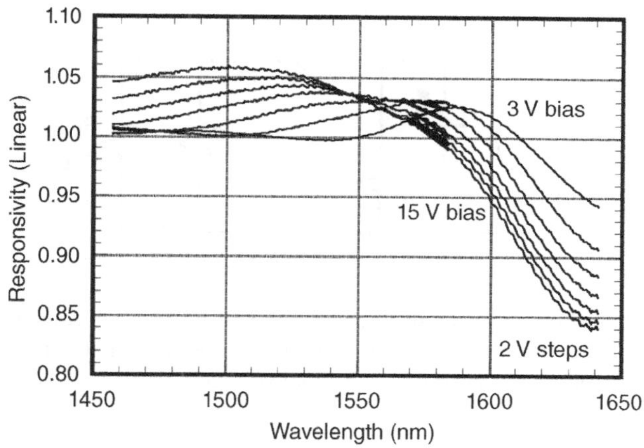

Figure 4.29. Measured responsivity as a function of wavelength for various applied voltages demonstrating the Franz–Keldysh effect. All curves are normalized to the responsivity at 3 V and 1550 nm.

this photodiode structure. As the responsivity increase due to impact ionization can be offset near the band edge with a decrease in responsivity due to the Franz–Keldysh effect, these two nonlinearities can be used to compensate each other. Hastings et al. (2010) demonstrated that the harmonic distortion in both the second and third harmonics have distinct local minima when measured as a function of wavelength. This compensation method reduced the measured second and third harmonics by up to 15 and 10 dB, respectively, near 1545–1550 nm. Similarly to the current-dependent responsivity nonlinearity associated with nonlinear recombination, both of these voltage-dependent responsivity nonlinearities (impact ionization and the Franz–Keldysh effect) are independent of optical power density, and both are independent of frequency in the low gigahertz ranges.

The last sources of photodetector nonlinearities listed in Table 4.1 are capacitive effects, where the device capacitance is a function of voltage or current. This can be represented within the equivalent circuit in Figure 4.30, allowing the junction capacitance, shunt and/or series resistances to be dependent on the photodetector current or the photodetector terminal voltage. Current dependencies in the junction capacitance and shunt resistance were first suggested by Jiang and Yu (1998). In this work, the current dependencies were modeled and numerically fit to RF reflection coefficient data to explain the nonlinear distortion present in the photodetector output with a simple circuit representation of the nonlinear output current. A later article by Williams and Goetz (2000)

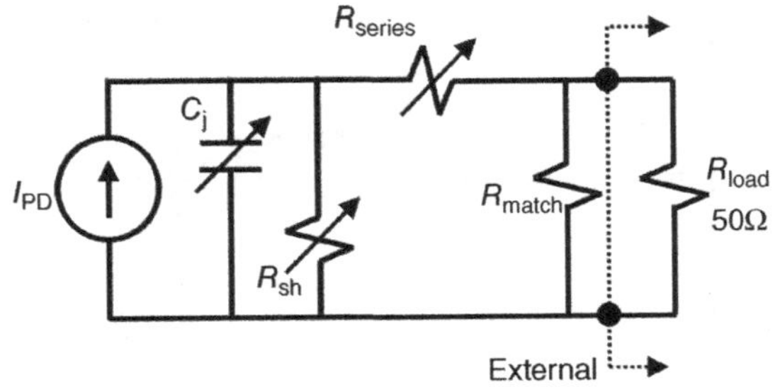

Figure 4.30. Photodiode equivalent circuit similar to Figure 1.18 considering that the junction capacitance, the shunt resistance, and the series resistance can be variable and a function of photodetector current or terminal voltage.

used model results from a drift-diffusion simulation to show that the current dependence in the capacitance value could be due to the internal redistribution of charge in the depletion region. In an article by Hollis (2011), the existence of space-charge impedance was also identified as a possible contributor to the current dependence in the equivalent circuit elements.

Whatever the cause of a current dependence of the equivalent circuit values of a photodetector, a nonlinearity will result. This can be understood with the help of Figure 4.31 where a simple single-pole R–C amplitude response is plotted as a function of the normalized frequency. Also shown are two additional amplitude response curves where the capacitance has been slightly raised and lowered. This could represent

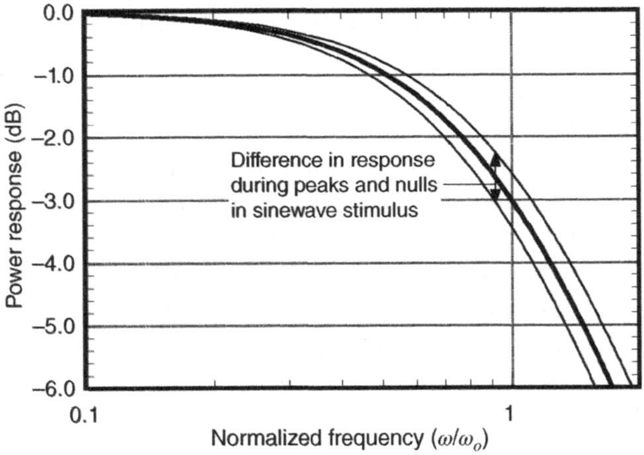

Figure 4.31. Normalized amplitude response of a single-pole R–C filter (center, bold) plotted against that for slightly higher (bottom curve) and lower capacitances (top curve).

the equivalent capacitance values of the response at the peaks and nulls of a sinewave. Therefore, similarly to the arguments used to describe Figure 4.23, as the photodetector current oscillates around its average or DC value, the R–C roll off experienced by the peaks of the sinusoid will be different from that experienced by the nulls of the sinusoid. Depending on the sign of the change, this will result in additional current at the peak/null and suppression of the other. This will happen so long as the change in capacitance can occur faster than the current in the photodiode changes. Current-dependent impedances should be treated as internal effects (not load impedance dependent), and more work is needed to determine their ultimate performance limitations on high linearity devices.

Another important capacitive effect is the voltage-dependent capacitance. In the same manner as a voltage-dependent responsivity effect, if the capacitance of a photodiode is a function of applied bias (almost always the case), then any voltage transients due the load impedance will cause the photodetector terminal voltage to change dynamically during modulation. The same argument associated with Figure 4.31 can be made to explain how this effect can lead to harmonic distortion. This effect has been studied by Pan et al. (2009) and compared against measurement data, appearing to be an important photodetector nonlinearity.

Photodiode nonlinearity is a more complicated problem than the analysis of many optical modulation schemes covered in this book. While the latter is typically expressible by closed-form analytic equations, the majority of the work to this point on photodiode distortion is empirical and numerical. While the sources of photodiode-induced distortion are not easily expressed, the associated impact on photonic link performance can be easily cast in terms of measured levels. That is, the degradation in link performance can be quantified in terms of photodiode OIP2 and OIP3, which can be measured using the methods described in Section 4.4.1. An explicit example of this process is given in Section 6.4 as it applies to IMDD links using an external MZM.

REFERENCES

Beling, A., Pan, H., Chen, H., Campbell, J. C., Hastings, A. S., Tulchinsky, D. A., and Williams, K. J., "Impact of voltage-dependent responsivity on photodiode non-linearity," in *21st Annual Meeting of the IEEE Lasers and Electrooptics Society (LEOS)*, 157–158, (2008).

Callaway, J., "Optical absorption in an electric field," *Physical Review*, 130(2), 549–553 (1963).

Chen, C. J. and Wong, W. S., "Transient effects in saturated Raman amplifiers," *Electronics Letters*, 37(6), 371–373 (2001).

Chen, L. K., Lau, K. Y., and Huber, D. R., "Fundamental distortion characteristics of the erbium fiber amplifier," in *Optical Fiber Communications Digest*, paper WL6, p. 112 (1991).

Clesca, B., Bousselet, P., and Hamon, L., "Second-order distortion improvements or degradations brought by erbium-doped fiber amplifiers in analog links using directly modulated lasers," *IEEE Photonics Technology Letters*, 5(9), 1029–1031 (1993).

Davis, G. A., Weiss, R. E., LaRue, R. A., Williams, K. J., and Esman, R. D., "A 920-1650-nm high-current photodetector," *IEEE Photonics Technology Letters*, 8(10), 1373–1375 (1996).

Dentan, Denton M. and de Cremoux, B., "Numerical simulation of the nonlinear response of a p-i-n photodiode under high illumination," *Journal of Lightwave Technology*, 8(8), 1137–1144 (1990).

Draa, M. N., Hastings, A. S., and Williams, K. J., "Comparison of photodiode nonlinearity measurement systems," *Optics Express*, 19(13), 12635–12645 (2011).

Esman, R. D. and Williams, K. J., "Measurement of harmonic distortion in microwave photodetectors," *IEEE Photonics Technology Letters*, 2(7), 502–504 (1990).

Esman, R. D. and Williams, K. J., "Wideband efficiency improvement of fiber optic systems by carrier subtraction," *IEEE Photonics Technology Letters*, 7(2), 218–220 (1995).

Fludger, C. R. S., Handerek, V., and Mears, R. J., "Pump to signal RIN transfer in Raman fiber amplifiers," *Journal of Lightwave Technology*, 19(8), 1140–1148 (2001).

Freeman, J. and Conradi, J., "Gain modulation response of erbium-doped fiber amplifiers," *IEEE Photonics Technology Letters*, 5(2), 224–226 (1993).

Giles, C. R., Desurvire, E., and Simpson, J. R., "Transient gain and cross talk in erbium-doped fiber amplifiers," *Optics Letters*, 14(16), 880–882 (1989).

Gray, S., "Transient gain dynamics in wide bandwidth discrete Raman amplifiers," in *Optical Fiber Communications Digest*, paper ThR2, p. 512–513 (2002).

Hastings, A. S., Urick, V. J., Sunderman, C., Diehl, J. F., McKinney, J. D., Tulchinsky, D. A., Devgan, P. S., and Williams, K. J., "Suppression of even-order photodiode nonlinearities in multioctave photonic links," *Journal of Lightwave Technology*, 26(15), 2557–2562 (2008).

Hastings, A. S., Tulchinsky, D. A., and Williams, K. J., "Photodetector nonlinearities due to voltage-dependent responsivity," *IEEE Photonics Technology Letters*, 21(21), 1642–1644 (2009).

Hastings, A. S., Tulchinsky, D. A., Williams, K. J., Pan, H., Beling, A., and Campbell, J. C., "Minimizing photodiode nonlinearities by compensating voltage-dependent responsivity effects," *Journal of Lightwave Technology*, 28(22), 3329–3333 (2010).

Hastings, A. S., Draa, M., and Williams, K., "Quantifying current-dependent quantum efficiency photodiode nonlinearities," in *IEEE Topical Meeting on Microwave Photonics Digest*, 413–416 (2011).

Herrera, J., Ramos, F., and Marti, J., "Nonlinear distortion generated by semiconductor optical amplifier boosters in analog optical systems," *Optics Letters*, 28(13), 1102–1104 (2003).

Hill, P., Schlafer, J., Powazinik, W., Urban, M., Eichen, E., and Olshansky, R., "Measurement of hole velocity in n-type InGaAs," *Applied Physics Letters*, 50(18), 1260–1262 (1987).

Hollis, M. A., "Space-charge impedance in photodiodes," *IEEE Photonics Technology Letters*, 23(18), 1307–1309 (2011).

Hu, Y., Menyuk, C. R., Urick, V. J., and Williams, K. J., "Modeling sources of nonlinearity in a simple p-i-n photodetector," *Journal of Lightwave Technology*, 32(20), 3710–3720 (2014).

Humphreys, D. A., King, R. J., Jenkins, D., and Moseley, A. J., "Measurement of absorption coefficients of $Ga_{0.47}In_{0.53}As$ over the wavelength range 1.0-1.7 μm," *Electronics Letters*, 21(25/26), 1187–1189 (1985).

Jiang, H., Yu, P. K. L., "Equivalent circuit analysis of harmonic distortions in photodiode," *IEEE Photonics Technology Letters*, 10(11), 1608–1610 (1998).

Juodawlkis, P. W., O'Donnell, F. J., Hargreaves, J. J., Oakley, D. C., Napoleone, A., Groves, S. H., Mahoney, L. J., Molvar, K. M., Missaggia, L. J., Donnelly, J. P., Williamson, R. C., and Twichell, J. C., "Absorption saturation nonlinearity in InGaAs/InP p-i-n photodiodes," in *15th Annual Meeting of the IEEE Lasers and Electrooptics Society (LEOS)*, 2, 426–427 (2002).

Kato, K., "Ultrawide-band/high-frequency photodetectors," *IEEE Transactions on Microwave Theory and Techniques*, 47(7), 1265–1281 (1999).

Khurgin, J. B., Vurgraftman, I., Meyer, J. R., Xu, S., and Kang, J. U., "Reduced crosstalk semiconductor optical amplifiers based on type-II quantum wells," *IEEE Photonics Technology Letters*, 14(3), 278–280 (2002).

Kikushima, K. and Yoshinaga, H., "Distortion due to gain tilt of erbium-doped fiber amplifiers," *IEEE Photonics Technology Letters*, 3(10), 945–947 (1991).

Kuhl, D., Hieronymi, F., Bottcher, E. H., Wolf, T., Bimberg, D., Kuhl, J., and Klingenstein, M., "Influence of space charges on the impulse response of InGaAs metal–semiconductor-metal photodetectors," *Journal of Lightwave Technology*, 10(6), 753–759 (1992).

Marhic, M. E., Kalogerakis, G., Wong, K. K. Y., and Kazovsky, L., "Pump-to-signal transfer of low-frequency intensity modulation in fiber optical parametric amplifiers," *Journal of Lightwave Technology*, 23(3), 1049–1055 (2005).

McIntyre R. J., "Multiplication noise in uniform avalanche diodes," *IEEE Trans. on Electron Devices*, 13(1), 164–168 (1966).

Ozeki T. and Hara, E. H., "Measurements of nonlinear distortion in photodiodes," *Electronics Letters*, 12(3), 80–83 (1976).

Pan, H. Beling, A., Chen, H., Campbell, J. C., and Yoder, P. D., "A High-linearity modified uni-traveling carrier photodiode with offset effects of nonlinear capacitance," *Journal of Lightwave Technology*, 27(20), 4435–4439 (2009).

Phillips, M. R., and Darcie, T. E., "Lightwave analog video transmission," in *Optical Fiber and Telecommunications IIIA*, Kaminov, I. P. and Koch, T. L. (editors), Academic Press (1997).

Ramaswamy, A., Nunoya, N., Williams, K. J., Klamkin, J., Piels, M., Johansson, L. A., Hastings, A. S., Coldren, L. A., and Bowers, J. E., "Measurement of intermodulation distortion in high-linearity photodiodes," *Optics Express*, 18(3), 2317–2324 (2010).

Scott, D. C., Vang, T. A., Elliott, J., Forbes, D., Lacey, J., Everett, K., Alvarez, F., Johnson, R., Krispin, A., Brock, J., Lembo, L., Jiang, H., Shin, D. S., Zhu, J. T., and Yu, P. K. L., "Measurement of IP3 in p-i-n photodetectors and proposed performance requirements for RF fiber-optic links," *IEEE Photonics Technology Letters*, 12(4), 422–424 (2000).

Stievater T. H. and Williams, K. J., "Thermally induced nonlinearities in high-speed p-i-n photodetectors," *IEEE Photonics Technology Letters*, 16(1), 239–241 (2004).

Sun, Y., Zyskind, J. L., and Srivastava, A. K., "Average inversion level, modeling, and physics of erbium-doped fiber amplifiers," *IEEE Journal of Selected Topics in Quantum Electronics*, 3(4), 991–1007 (1997).

Tavlykaev, R., Wiberg, A. O. J., Gopalakrishnan, G. K., and Radic, S., "Nonlinear distortions in EO phase modulators," *IEEE Photonics Technology Letters*, 24(19), 1778–1780 (2012).

Urick, V. J., Hutchinson, M. N., Singley, J. M., McKinney, J. D., and Williams, K. J., "Suppression of even-order photodiode Distortions via predistortion linearization with a bias-shifted Mach-Zehnder modulator," *Optics Express*, 21(12), 14368–14376 (2013).

Willems, F. W. and Muys, W., "Distortion caused by the interaction of residual chirp in Mach-Zehnder intensity modulators with optical cavities, fiber dispersion, and EDFA gain-slope in analog CATV lightwave systems," *Optical Fiber Technology*, 2, 63–68 (1996).

Williams, K. J., "Nonlinear mechanisms in microwave photodetectors operated with high intrinsic region electric fields," *Applied Physics Letters*, 65(10), 1219–1221 (1994a).

Williams, K. J., "*Microwave Nonlinearities in Photodiodes*," Doctoral Dissertation, UMI (1994b).

Williams, K. J. and Esman, R. D., "Observation of photodetector nonlinearities," *Electronics Letters*, 28(8), 731–732 (1992).

Williams, K. J. and Esman, R. D., "Photodiode DC and microwave nonlinearity at high current due to carrier recombination nonlinearities," *IEEE Photonics Technology Letters*, 10(7), 1015–1017 (1998).

Williams, K. J., and Goetz, P. G., "Photodiode compression due to current-dependent capacitance," in *IEEE Int. Topical Meeting on Microwave Photonics*, 221–224 (2000).

Williams, K. J., Goldberg, L., Esman, R. D., Dagenais, M., and Weller, J. F., "6-34 GHz offset phase locking of Nd:YAG 1319 nm nonplanar ring lasers," *Electronics Letters*, 25(18), 1242–1243 (1989).

Williams, K. J., Esman, R. D., and Dagenais, M., "Effects of high space-charge fields on the response of microwave photodetectors," *IEEE Photonics Technology Letters*, 6(5), 639–641 (1994).

Williams, K. J., Esman, R. D., and Dagenais, M., "Nonlinearities in p-i-n microwave photodetectors," *Journal of Lightwave Technology*, 14(1), 84–96 (1996).

Williams, K. J., Nichols, L. T., and Esman, R.D., "Photodetector nonlinearity Limitations on a high-dynamic range 3 GHz fiber optic link," *Journal of Lightwave. Technology*, 16(2), 192–199 (1998).

Wilson, S. P. and Walker, A. B., "Mechanism of impact ionization enhancement in GaAs p-i-n diodes under high illumination conditions," *Semiconductor Science and Technology*, 13(2), 190–193 (1998).

Wilson, S. P., Woods, S. J., and Walker, A. B., "Modeling of p-i-n photodiodes under high illumination conditions," *International Journal of Numerical Modeling: Electronic Networks, Devices and Fields*, 10(3), 139–151 (1997).

Windhorn, T. H., Cook, L. W., and Stillman, G. E., , "The electron velocity-field characteristic for n-InGaAs at 300 K," *IEEE Electron Device Letters*, 3(1), 18–20, (1982).

Yao, X. S., "Brillouin selective sideband amplification of microwave photonic signals," *IEEE Photonics Technology Letters*, 10(1), 138–140 (1998).

Yeh, L. P., "Noise loading test of a complete frequency-division multiples voice point-to-point communication system," *Transactions of the American Institute of Electrical Engineering*, 81(1), 40–43 (1962).

CHAPTER 5

PROPAGATION EFFECTS

5.1 INTRODUCTION

This chapter provides a detailed introduction to the impact of propagation effects on an analog signal impressed on an optical carrier in fiber. Modulation-specific propagation effects are covered in the three chapters that follow, which build on the principles in this chapter. Optical fiber exhibits numerous troublesome yet interesting nonlinearities, much more so than microwave transmission media. The fiber effects most prevalent in microwave photonics are covered in this chapter.

The effects studied in this chapter range from linear scattering mechanisms to relatively complicated nonlinear processes. Elastic Rayleigh scattering is the dominant loss mechanism for links at 1550 nm and therefore is pertinent to the radio-frequency (RF) loss in analog optical links. Linear Rayleigh scattering is analyzed in Section 5.2, particularly in the context of double Rayleigh scattering (DRS) events that can cause multiple-path interference (MPI) in fiber links. The RF phase stability can be important in microwave photonic links, particularly in multichannel applications intended to maintain the relative RF phase of each channel, and is discussed in Section 5.3. Two effects related to the index of refraction are treated in Sections 5.4 and 5.9. The frequency

Fundamentals of Microwave Photonics, First Edition.
Vincent J. Urick Jr, Jason D. McKinney, and Keith J. Williams.

dependence of the index of refraction in fiber gives rise to chromatic dispersion (CD), the topic of Section 5.4. Because real-world optical fibers are not perfectly cylindrical, two orthogonal polarization modes experience different indices of refraction. The difference between the two modal refractive indices—the birefringence—changes randomly in the fiber leading to polarization mode dispersion (PMD), which is covered in Section 5.9. Two nonlinear and inelastic scattering processes, stimulated Brillouin scattering (SBS) and stimulated Raman scattering (SRS), are the topics of Sections 5.5 and 5.6, respectively. Both SBS and SRS were introduced in Section 3.5.2 in the context of optical amplification; in this chapter, propagation impairments imposed by both processes are described.

Electromagnetic fields in an optical fiber will induce a polarization in the dielectric medium. This induced polarization **P** is a nonlinear function of the applied electric field **E** and can be written as a power series expansion (Agrawal 2013):

$$\mathbf{P} = \varepsilon_0(\chi^{(1)} \cdot \mathbf{E} + \chi^{(2)} \cdot \mathbf{EE} + \chi^{(3)} \cdot \mathbf{EEE} + \cdots), \tag{5.1}$$

where ε_0 is permittivity of vacuum and $\chi^{(n)}$ is the nth order susceptibility. The first-order term including $\chi^{(1)}$ describes linear propagation in fiber. All other higher order terms in Equation (5.1) involve products of electric fields and therefore produce nonlinearities. For example, the second-order susceptibility can lead to second-harmonic generation. However, symmetric molecules such as SiO_2 do not exhibit second-order susceptibility, and $\chi^{(2)}$ is negligible for typical optical fibers (Agrawal 2013). The most important nonlinear term for microwave photonics is the third-order susceptibility, $\chi^{(3)}$. The gain coefficients associated with Brillouin and Raman scattering are related to the imaginary part of $\chi^{(3)}$. The third-order susceptibility causes parametric four-wave mixing (FWM), a process mathematically similar to third-order intermodulation distortion described in Chapter 2. The nonlinear refractive index also originates from $\chi^{(3)}$, which can cause self-phase modulation (SPM) and cross-phase modulation (XPM). The polarization dependence of XPM can result in a nonlinear birefringence that causes cross-polarization modulation (XPolM). These important nonlinearities associated with $\chi^{(3)}$ are covered in Sections 5.7–5.9. Finally, many of the propagation effects listed in this chapter can act in concert to cause complicated detriments in microwave photonic links, and sample cases are provided throughout.

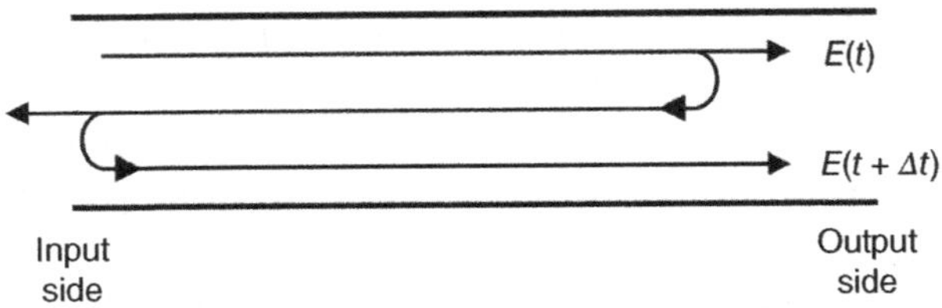

Figure 5.1. Conceptual diagram of multiple-path interference (MPI) in a fiber optic link.

5.2 DOUBLE RAYLEIGH SCATTERING

Multiple reflection points in a fiber optic link can cause interference and noise. In particular, MPI can convert phase noise to intensity noise. Consider the cartoon shown in Figure 5.1. A portion of a forward-traveling wave can be reflected to propagate in the reverse direction. A small amount of this backscattered wave can then be reflected to co-propagate with the signal. In this case, the field is a superposition of the original laser field and a delayed copy of itself, thus allowing laser phase fluctuations to be converted into intensity fluctuations. If the output of such a link is passed to a photodetector, laser phase fluctuations will manifest as baseband intensity noise in the electrical domain. This situation is not unlike the delayed self-homodyne technique to measure laser linewidth as described in Section 3.4. The level of the delayed signal will typically be much smaller than the undelayed version. For example, discrete low level reflections in the fiber link can be caused by bad splices or connectors (Gimlett and Cheung 1989). These types of MPI can be managed with proper link construction and maintenance. However, a distributed reflection due to Rayleigh scattering occurs in the fiber, which is a process that is not as easily mitigated.

Rayleigh scattering in fiber is the primary source of attenuation near 1.55 μm, dominating other sources such as absorption and radiative losses. Rayleigh scattering arises from small inhomogeneities in the fiber and results in light scattering in all directions. Rayleigh scattering in fiber is usually described as being elastic and will be treated as such in this section. However, Okusaga et al. (2012) have observed an approximately 50-kHz frequency shift attributed to inelastic spontaneous Rayleigh scattering in fiber (Okusaga et al. 2013). Moreover, the discussion in this chapter is limited to spontaneous Rayleigh scattering, although stimulated Rayleigh scattering in fiber has been studied (Zhu et al. 2010). DRS is an important source of MPI in long fiber optic links. The DRS process provides a mechanism for a nearly continuous set of reflections of the type shown in Figure 5.1. The relative intensity noise (RIN) at a link output due to DRS can be derived using a theory similar to that given by Wan and Conradi (1996) as

$$\mathrm{RIN}_{\mathrm{DRS}} = 2R_{\mathrm{bs}}^2(2\alpha L + e^{-2\alpha L} - 1)\frac{\Delta\nu}{\pi[f^2 + (\Delta\nu)^2]}, \tag{5.2}$$

where R_{bs} is the Rayleigh backscattering reflectance for the fiber, α is the attenuation coefficient, L is the fiber length, $\Delta\nu$ is the FWHM Lorentzian linewidth, and f is the electronic frequency. Equation (5.2) is valid only for a laser exhibiting a Lorentzian lineshape and for L much larger than the laser coherence length.

Shown in Figure 5.2 are two RIN spectra measured at the output of 6- and 19-km spans of standard single-mode fiber both with an average photocurrent of $I_{\mathrm{dc}} = 0.92$ mA. An apparatus such as that depicted in Figure 3.10 was employed for the measurements. A low noise RF amplifier with a noise figure of $\mathrm{NF}_{\mathrm{LNA}} = 2$ dB was used to amplify the signals above the spectrum analyzer noise floor. A semiconductor laser was used as the optical input, having $\Delta\nu = 481$ kHz as measured with the setup shown in Figure 3.19. The attenuation coefficient was measured as $\alpha = 0.18$ dB/km for each span. The R_{bs} in Equation (5.2) was used as a fit parameter to the 6-km spectrum, resulting in $R_{\mathrm{bs}} = -30.5$ dB. This same value was then used in the calculation for $L = 19$ km. Neither shot noise nor amplifier noise in the amount $F_{\mathrm{LNA}}k_{\mathrm{B}}T$ [see Equation (3.31)]

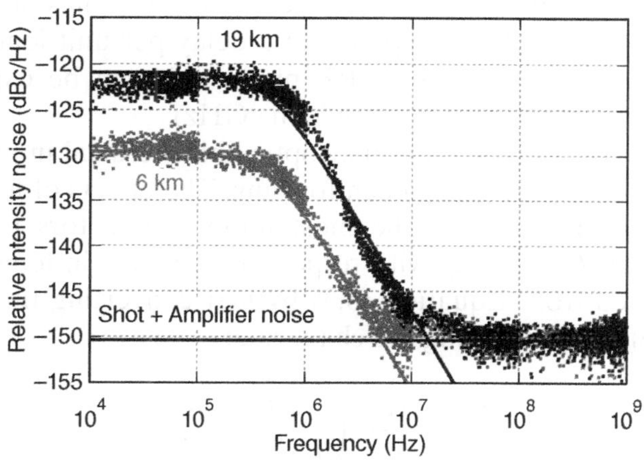

Figure 5.2. Measured (symbols) and calculated DRS-induced (lines) relative intensity noise at the output of 6 km (gray) and 19 km (black) of standard single-mode optical fiber. Also shown is the calculated level due to shot noise and RF amplifier noise from the measurement apparatus.

was subtracted from the measured results. The resulting noise floor set by these two sources is shown in Figure 5.2 at a level of –150.3 dBc/Hz.

The measured data and theoretical results in Figure 5.2 agree well, justifying the use of Equation (5.2) for link design and analysis. Again, Equation (5.2) was derived for a laser exhibiting a Lorentzian lineshape and for MPI where the differential path delays are outside the coherence time of the laser [$\tau_c = 1/(\pi \cdot \Delta\nu)$ for a Lorentzian]. A more general expression can be obtained by replacing the Lorentzian lineshape function in Equation (5.2) with the Fourier transform of the magnitude squared of the autocorrelation function for the optical field (Wan and Conradi 1996).

5.3 RF PHASE IN FIBER OPTIC LINKS

The finite speed of light results in a propagation delay for a signal transmitted through a fiber optic link, which can cause detrimental latency in some systems. However, low loss fiber can be very advantageous when used as a storage medium for buffering "bottlenecks" in a system or for delay-line signal processing. The time delay for traversing a fiber span of length L is

$$t = \frac{nL}{c}, \tag{5.3}$$

where n is the effective group index of refraction and c is the speed of light in vacuum. Standard single-mode fiber has an $n = 1.468$ (Corning 2011), which corresponds to a delay per unit length of 4.90 ns/m = 4.90 μs/km. In terms of RF phase, this can be expressed as $1.76 \times 10^{-6}\,\mathrm{deg}/(\mathrm{m} \cdot \mathrm{Hz}) = 17.6\,\mathrm{deg}/(\mathrm{cm} \cdot \mathrm{GHz})$.

The RF phase stability of a fiber optic link is crucial in many applications such as phased-array beamforming. The effects of temperature and stress differentials are the most important factors to consider. The change in RF phase ϕ with respect to a change in temperature T can be derived from Equation (5.3) by first converting time delay to RF phase and then applying the chain rule:

$$\frac{d\phi}{dT} = \frac{2\pi Lf}{c}\left(\frac{n}{L}\frac{dL}{dT} + \frac{dn}{dT}\right), \tag{5.4}$$

where f is the frequency. The term $(1/L) \cdot dL/dT$ is the thermal expansion coefficient and has a value of $(1/L) \cdot dL/dT = 5.6 \times 10^{-7}/°\mathrm{C}$ for typical silica fiber (Campillo et al. 2004). The differential index change

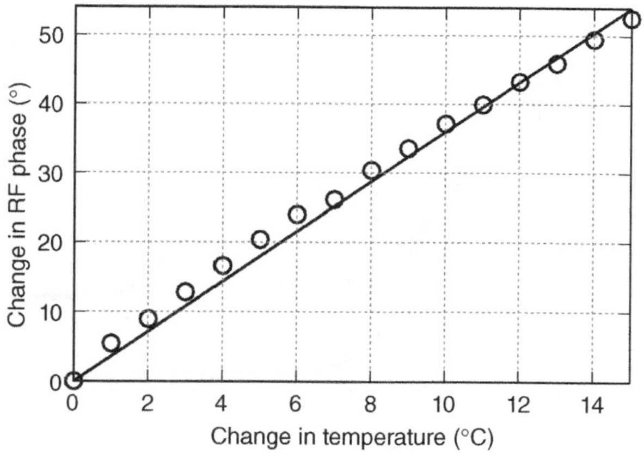

Figure 5.3. Measured phase change for a 12-GHz signal propagated through 20 m of fiber as a function of temperature change. The solid line shows theory, and the circles are experimental data.

with respect to temperature is $dn/dT = 1.2 \times 10^{-5}/°\text{C}$ for typical silica fiber at 1550 nm (Campillo et al. 2004). Therefore, the index change with temperature is the more important of the two effects. Inserting $n = 1.468$ into Equation (5.4), the phase change can be written as $d\phi/dT = 0.27\,\text{rad}/(\text{km}\cdot\text{GHz}\cdot°\text{C}) = 15\,\text{deg}/(\text{km}\cdot\text{GHz}\cdot°\text{C})$. Shown in Figure 5.3 are measured data plotted with a calculation of Equation (5.4) for $n = 1.468$ (Urick et al. 2008). For this experiment, 20 m of standard single mode fiber carrying a 12-GHz signal was placed in a temperature chamber. The relative phase change was then measured as a function of temperature change. As can be seen, the theoretical and experimental results agree nicely.

For the results shown in Figure 5.3, the fiber being tested was not coiled on a spool but rather laid loosely in the temperature chamber. Therefore, the effects of stress could be neglected. This may not be the case in practical applications. The effects of stress can be determined by differentiating Equation (5.3) with respect to a longitudinally applied stress (Sun et al. 1995):

$$\frac{d\phi}{d\sigma} = \frac{2\pi Lf}{c}\left(\frac{n}{L}\frac{dL}{d\sigma} + \frac{dn}{d\sigma}\right). \tag{5.5}$$

In Equation (5.4) $dn/d\sigma = -3.4 \times 10^{-12}/\text{Pa}$ and $[(1/L)\cdot dL/d\sigma]^{-1} = 7.2 \times 10^{10}$ Pa is Young's modulus. This differential can be written as $d\phi/d\sigma = 1.3\,\text{deg}/(\text{km}\cdot\text{GHz}\cdot\text{kPa})$. Equation (5.5) can be important for installed systems, particularly those employing bundled fiber optic

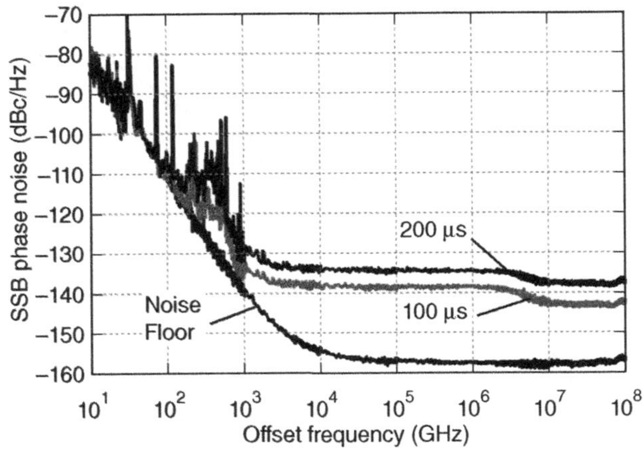

Figure 5.4. Measured single-sideband (SSB) phase noise for a 10.24-GHz signal at the output of a 20-km (100-μs) and 40-km (200-μs) fiber optic link. The measurement noise floor is also shown.

cables, which have been studied for commercial (Serizawa et al. 1997) and military (Roman et al. 1998a) applications.

In addition to the phase stability discussed previously in this section, the single-sideband (SSB) phase noise close to the RF carrier at the output of a link is also important. The SSB phase noise imposed by various noise sources in a fiber link was treated in Chapter 3. The fundamental limits due to thermal fluctuations in a fiber have been treated by Foster et al. (2007) and the references therein. It has long been known that the SSB phase noise performance in fiber links can be acceptable for numerous applications, such as UHF shipboard antenna remoting (Sun et al. 1995) and X-Band radar applications (Roman et al. 1998b). Analog fiber optic links are also acceptable for distribution of high quality frequency standards, where active cancellation techniques are sometimes employed (Ye et al. 2003). Shown in Figure 5.4 are measured examples of the achievable performance in passive fiber optic delay lines (Urick et al. 2012). The SSB phase noise for a 10.24-GHz signal at level of −15 dBm after passing through a 20- and 40-km link is shown in Figure 5.4. Both links employ erbium-doped fiber amplifiers (EDFAs) to maintain an adequate signal level, and the noise at offset frequencies above 10 kHz is dominated by optical-amplifier noise. The noise at lower offset frequencies is near the measurement noise floor, save some spikes in the link data. These spikes are attributed to noise transferred from the EDFA pump lasers, as they occur at frequencies below the typical cutoff for erbium ions in fiber (see Section 4.3).

5.4 CHROMATIC DISPERSION

The frequency dependence of the index of refraction, or CD, of optical fiber may significantly impact the RF performance of long optical links. This impact may be advantageous when the dispersion is introduced intentionally, for example, in dispersion-based optical filtering geometries (Supradeepa et al. 2012) or certain types of photonic architectures for ultrabroadband electrical waveform generation (Wang and Yao 2010, McKinney and Weiner 2013). In general, however, the impact is deleterious for most links where dispersion-induced fading, phase-to-intensity conversion, and distortion degrade the link performance. In this section, a general treatment of dispersion with respect to a linear link response is presented. In particular, the effect of CD on the RF transmission of purely intensity- or phase-modulated links is considered and the increased RF noise floor that may result from conversion of laser phase noise to measurable electrical noise is addressed.

The analysis in this case begins with the simple picture of a link employing pure intensity or phase modulation, without assuming a particular implementation of the modulator. This simple case allows for prediction of the fundamental response for actual dispersive microwave photonic links with great accuracy. For an optical link operating in the small-signal regime, the modulated optical field may be written as

$$e(t) = Ee^{i\omega_0 t}[1 + m(t)], \tag{5.6}$$

where ω_0 is the source laser frequency and $m(t)$ is the applied modulation that satisfies the condition $|m(t)| \ll 1$. In this case, the optical modulation is taken to be conventional continuous-wave (CW) double-sideband modulation such as would arise using a simple optical intensity or phase modulator—more complex modulation techniques will be considered in Chapter 8. The optical intensity (which results in measurable photocurrent variations) is proportional to $|e(t)|^2$, which yields

$$p(t) \propto E^2[1 + 2\mathrm{Re}\{m(t)\}]. \tag{5.7}$$

Note that because small-signal operation was assumed, the term proportional to $|m(t)|^2$ is vanishingly small and has been excluded in Equation (5.7). It is seen that the modulated optical intensity is then proportional to the real component of the modulated electric field. In the case of intensity modulation [$m(t)$ real], the optical intensity variations are directly proportional to the applied modulation as expected. For pure optical phase modulation [$m(t)$ imaginary] the intensity is constant, which is intuitively satisfying.

For the RF signals of interest in most analog optical systems, the impact of dispersion is most readily analyzed in the frequency domain. To that end, it is useful to consider the complex modulation spectrum given by the Fourier transform of the time-domain modulation:

$$M(\omega) = \int_{-\infty}^{\infty} m(t)e^{-i\omega t}dt. \tag{5.8}$$

In the case of pure intensity modulation, the modulation signal $m(t)$ (real) results in a Hermitian complex spectrum (the real portion of the spectrum is an even function of frequency, and the imaginary portion is an odd function of frequency) (Bracewell 2000). For the case of optical phase modulation where the modulation signal $m(t)$ is purely imaginary, the resulting complex spectrum is anti-Hermitian (even- and odd-nature of the real and imaginary portions of the spectrum are reversed). Determination of the complex photocurrent spectrum at the terminus of an analog link then amounts to finding the Hermitian portion of the modulated optical field.

The propagation constant is introduced to describe CD mathematically (Agrawal 2013):

$$\beta(\omega) = \frac{\omega}{c}n(\omega), \tag{5.9}$$

where ω represents optical frequency, c is the speed of light in vacuum, and $n(\omega)$ is the frequency-dependent refractive index. For the bandwidths of interest in the vast majority of link applications, the frequency dependence of $\beta(\omega)$ is well described by a three-term Taylor expansion about the optical carrier frequency:

$$\beta(\omega) = \beta_0 + \beta_1(\omega - \omega_o) + \frac{1}{2}\beta_2(\omega - \omega_o)^2, \tag{5.10}$$

where

$$\beta_n = \left.\frac{d^n\beta(\omega)}{d\omega^n}\right|_{\omega=\omega_o}. \tag{5.11}$$

In this case, β_0 yields the absolute optical phase and is safely ignored. The second term (β_1) is the group delay, which defines the finite propagation speed of signals in the fiber as discussed in the preceding section. Although the group delay is integral to the design of optical delay lines and ensures causality in optical filtering geometries, this term is generally not included in the mathematical analysis of analog optical links. The CD of the fiber is described by β_2, which characterizes the frequency dependence of the group delay. Generally, there are both material and waveguide contributions to the CD, although for typical single-mode fibers away from the zero-dispersion wavelength (where β_2 vanishes), the material contribution to the dispersion dominates (Agrawal 2013). Specialty dispersion-shifted and dispersion-compensating fibers—where the material and modal indices are tailored to produce a desired dispersion—do exist and find application in dispersion-managed links as well as in nonlinear fiber optics applications such as parametric amplification (Agrawal 2013). In the fiber optics community, the dispersion parameter D (ps/nm/km) is frequently used in lieu of β_2 (ps^2/km)—the two are related through the relation

$$D = \frac{d\beta_1}{d\lambda} = -\frac{2\pi c}{\lambda^2}\beta_2. \tag{5.12}$$

For standard single-mode fiber, $\beta_2 \approx -22\,\text{ps}^2/\text{km}$ ($D \approx 17$ ps/nm/km) at a center wavelength of $\lambda = 1550$ nm ($\omega/2\pi = 193.55$ THz) (Corning 2011). There are numerous fiber-dependent empirical expressions for $D(\lambda)$ written in terms of the zero-dispersion wavelength, λ_0, and the zero-dispersion slope, S_0 (Corning 2001); values for standard single-mode fiber are $\lambda_0 \approx 1310$ nm and $S_0 \approx 0.092$ ps/nm^2/km (Corning 2011).

To include the effects of CD in an optical link, the complex optical spectrum is multiplied by a phase exponential representing propagation through the dispersive medium. Given the complex spectrum of the modulated input field [calculated via the Fourier transform of Equation (5.6)], the field after propagation through a length L of dispersive fiber is expressed as

$$E(\widetilde{\omega}) = [E_{\text{in}}(\widetilde{\omega}) * M(\widetilde{\omega})] \times Q(\widetilde{\omega}), \tag{5.13}$$

where $*$ denotes convolution, the offset from the laser center frequency is defined to be $\widetilde{\omega} = \omega - \omega_o$, and $Q(\widetilde{\omega})$ is given by

$$Q(\widetilde{\omega}) = \exp\left(-i\frac{1}{2}\beta_2 L\widetilde{\omega}^2\right). \tag{5.14}$$

As is evident from Equations (5.13) and (5.14), multiplication of the modulation spectrum with the complex exponential representing dispersion will induce a phase change between the upper and lower optical sidebands of the modulated field incident on the dispersive optical fiber.

This alteration of the spectral symmetry will convert intensity modulation to phase modulation and vice versa leading to frequency-dependent structure not present in the original modulation signal (see also the phasor diagrams in Chapter 4). This may be mitigated using a variety of techniques such as optical SSB modulation as discussed in Chapter 8.

Although a full treatment of dispersion in analog links (including dispersion-induced distortion) requires explicit evaluation of Equation (5.13), insight into the first-order effects of dispersion-induced fading in intensity-modulated links and phase-to-intensity conversion may be gained by viewing the effects of dispersion as a filter (Wang and Petermann 1992, McKinney and Diehl 2007). This is accomplished by first expanding the exponential in Equation (5.14) using Euler's identity,

$$\exp\left(-i\frac{1}{2}\beta_2 L\tilde{\omega}^2\right) = \cos\left(\frac{1}{2}\beta_2 L\tilde{\omega}^2\right) - i\sin\left(\frac{1}{2}\beta_2 L\tilde{\omega}^2\right). \tag{5.15}$$

Note that the first term in Equation (5.15) is Hermitian while the second is anti-Hermitian. Since the product of two Hermitian functions—or two anti-Hermitian functions—is Hermitian, the response of a dispersive optical link in the case of pure intensity modulation and pure phase modulation may be determined by inspection. For the case of intensity modulation where the modulation spectrum $M(\omega)$ is Hermitian, the complex spectrum of the photocurrent [determined from the Fourier transform of Equation (5.7)] at the link output will be proportional to

$$I_{\text{out}}(\omega) \propto M(\omega)\cos\left(\frac{1}{2}\beta_2 L\tilde{\omega}^2\right). \tag{5.16}$$

In the case of pure phase modulation where $M(\omega)$ is anti-Hermitian, the complex photocurrent spectrum is then

$$I_{\text{out}}(\omega) \propto M(\omega)\sin\left(\frac{1}{2}\beta_2 L\tilde{\omega}^2\right). \tag{5.17}$$

To illustrate the impact of CD on the response of an analog link, the normalized RF power responses of dispersive intensity- and phase-modulated approximately 50-km links are shown in Figure 5.5(a) and (b), respectively. In this case, the circles represent measured data and the solid lines are calculations on the basis of Equations (5.16) and (5.17), with $\beta_2 = -21.47\,\text{ps}^2/\text{km}$. For an ideal intensity-modulated

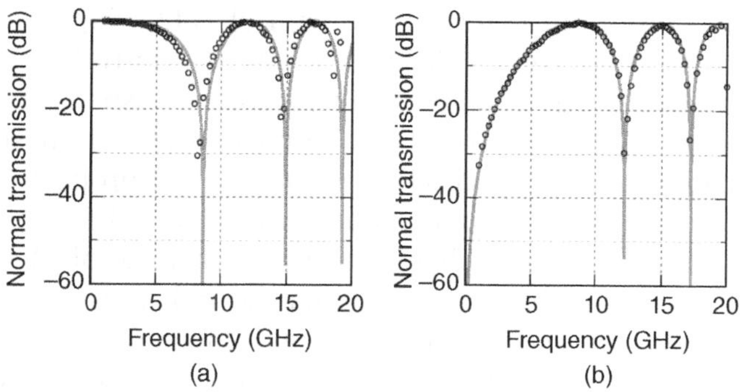

Figure 5.5. Radio-frequency response for a 50-km link employing (a) optical intensity modulation and (b) optical phase modulation. In both plots, the circles are measured data and the lines are theoretical calculations based on Equations (5.16) and (5.17) with $\beta_2 = -21.47\,\text{ps}^2/\text{km}$.

analog optical link, the RF transmission would be constant as a function of frequency (or, in practice, show structure arising only from the frequency response of the optical modulator and photodiode). Considering Figure 5.5(a), it can be seen that dispersion introduces fading of the RF signal as evidenced by the strong nulls in the RF response. At the frequencies where the RF response vanishes, the applied intensity modulation is fully converted to optical phase modulation, that is, the optical intensity becomes constant, which gives no measurable RF response at the photodiode. Conversely, for the case of optical phase modulation shown in Figure 5.5(b), the RF transmission of the link should ideally be zero (in linear units) neglecting any imperfections in the optical modulator (residual intensity modulation due to reflections for example) or other fiber nonidealities such as DRS (see Section 5.2). However, dispersion periodically converts the applied phase modulation to near perfect intensity modulation as shown by the peaks in the RF response. It should be noted that the intensity- and phase-modulation responses are complimentary, that is, maximum intensity-to-phase conversion (nulls) in the intensity-modulated link response correspond to maximum phase-to-intensity conversion (peaks) in the phase-modulated link response. The slight offset of the null locations between the measured data and theoretical calculation in the intensity-modulated link response arises from a slight residual chirp (optical phase modulation) in the intensity modulator used for the measurement (Koyama and Iga 1988).

For modulation signals with non-negligible bandwidth, CD may be used as a tool for generation of microwave waveforms as well as optical signal processing (Azana and Muriel 2000, Supradeepa et al. 2012, McKinney and Weiner 2013) but may also introduce additional deleterious effects. For example, in addition to dispersive fading or phase-to-intensity conversion of the applied modulation signal, CD may also lead to an increased RF noise floor if not properly managed. Particularly, in links using semiconductor distributed feedback lasers, conversion of laser optical phase noise into measurable RF noise at the link output can substantially impact the link noise figure. In the following section, the basic process for analyzing a dispersive link with finite-bandwidth modulation signals is described and a detailed discussion of link behavior with stochastic fields is provided. Experimental data illustrating the impact of laser phase noise in a dispersive link is considered as an example.

In many physical situations, it is insightful to consider the effect of CD directly in the time domain, as opposed to using the frequency-domain picture using a multiplicative quadratic phase exponential. This first requires calculation of the output optical field from the fiber from which other desired quantities (photocurrent, RF power spectrum, etc.) may be readily determined. The output field after dispersion may be expressed as

$$e_{\text{out}}(t) = e_{\text{in}}(t) * q(t), \tag{5.18}$$

where $q(t)$ is the time-domain impulse response of the fiber given by the inverse Fourier transform of $Q(\widetilde{\omega})$ as given in Equation (5.14):

$$q(t) = \frac{1}{2\pi}\int_{-\infty}^{\infty} e^{-i\beta_2 L\widetilde{\omega}^2/2} e^{i\widetilde{\omega}t}\,dt = \left(\frac{1}{i2\pi\beta_2 L}\right)^{-1/2} e^{it^2/(2\beta_2 L)}. \tag{5.19}$$

To analyze the effects of the convolution in Equation (5.18), it is useful to consider the time-domain behavior of $q(t)$ [an excellent discussion of quadratic phase filters is given in (Papoulis 1994)]. The normalized real (solid black) and imaginary (solid gray) parts of $q(t)$ are shown in Figure 5.6 for a 50-km length of SMF-28 ($\beta_2 L \approx 1100\,\text{ps}^2$). For comparison, a notional input field envelope $e_{\text{in}}(t)$ is shown by the dashed black line. Considering either the real or the imaginary part of $q(t)$, it is seen that the nonzero contribution to the convolution integral occurs over a time of approximately $t_q = 2\sqrt{\pi\beta_2 L}$. Beyond this point, the oscillations between positive and negative values do not contribute to the value of the integral [easily seen by comparing the time scale of the oscillations with variations in the complex field envelope (dashed black line) in Figure 5.6]. Thus, if the temporal variations in the input field envelope

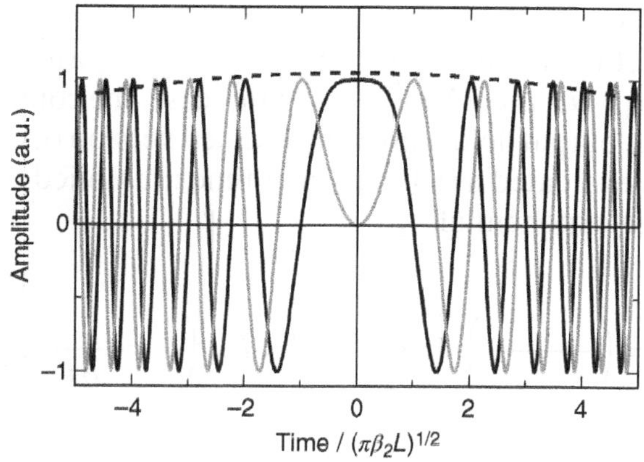

Figure 5.6. Time-domain representation of the real (solid black line) and imaginary (gray line) parts of the fiber impulse response $q(t)$ in Equation (5.18). For comparison purposes, a notional input complex field envelope $e_{\text{in}}(t)$ is shown by the dashed black line.

are much longer than t_q or $\beta_2 L \to 0$, $q(t)$ may be treated as a Dirac delta function and the output field is equivalent to the input field (from the sifting property of the delta function):

$$e_{\text{out}}(t) = e_{\text{in}}(t) * \delta(t) = e_{\text{in}}(t). \tag{5.20}$$

In the opposite regime where there is significant dispersion and t_{q} is much greater than the time scale of fluctuations in the input field envelope, it is well known that the output field shows a strong frequency-dependent delay. To see the effects of this delay, it is useful to write out the convolution in Equation (5.20) explicitly

$$e_{\text{out}}(t) = \left(\frac{1}{i2\pi\beta_2 L}\right)^{-1/2} e^{i\frac{1}{2\beta_2 L}t^2} \int e_{\text{in}}(t') e^{i\frac{1}{2\beta_2 L}t'^2} e^{-i\frac{1}{\beta_2 L}t't} dt'. \tag{5.21}$$

Defining the instantaneous frequency to be

$$\omega = \frac{1}{\beta_2 L} t, \tag{5.22}$$

it is readily seen that the complex output time-domain field envelope is given by

$$e_{\text{out}}(t) = \left(\frac{1}{i2\pi\beta_2 L}\right)^{-1/2} e^{it^2/(2\beta_2 L)} \times [E_{\text{in}}(\omega) * e^{-i\beta_2 L\omega^2/2}]. \tag{5.23}$$

When the complex spectrum of the input field, $E_{\text{in}}(\omega)$, is band limited with a bandwidth that is much greater than $f_Q = 1/\sqrt{\pi\beta_2 L}$, the complex exponential $\exp(-i\beta_2 L\omega^2/2) \to \delta(\omega)$. In this case, the complex temporal envelope of the output field is given by a scaled version of the input complex spectrum evaluated as a function of time multiplied by a quadratic temporal phase exponential:

$$e_{\text{out}}(t) = \left(\frac{1}{i2\pi\beta_2 L}\right)^{-1/2} e^{it^2/(2\beta_2 L)} \times E_{\text{in}}\left(\omega = \frac{1}{\beta_2 L}t\right). \tag{5.24}$$

In this case, the instantaneous frequency ω was replaced with the expression of Equation (5.22). Equation (5.24) clearly shows the optical frequency-to-time conversion used to describe pulse broadening in the ultrafast optics community (Weiner 2009) and used extensively in optical synthesis of broadband electrical waveforms (Wang and Yao 2010, McKinney and Weiner 2013), optical techniques for microwave signal processing (Supradeepa et al. 2012), and optical/microwave measurements (Azana and Muriel 2000, Dorrer 2004), when $E_{\text{in}}(\omega)$ is both time and band limited.

For input fields that are band limited, but exhibit stochastic complex field envelopes [e.g., amplified spontaneous emission (Desurvire 2002) or laser phase noise (Armstrong 1966)], the procedure outlined previously loses its intuitive character, in that, the output field envelope $e_{\text{out}}(t)$ represents one sample of a random process determined by the random nature of the phase of the input electric field. In microwave photonics applications, the effect of random optical input signals (i.e., noise) on system performance is to increase the RF noise floor—an effect that is most clearly seen by determining the average RF power spectrum as opposed to the time-domain envelope of the optical field (or intensity). This path of analysis will therefore be followed. Others have described the impact of laser phase noise in interferometric (Tkach and Chraplyvy 1986) and dispersive systems (Yamamoto et al. 1990, Wang and Petermann 1992, Marshall et al. 2000). The end result presented in this case is identical to that shown by Marshall—the derivation, however, differs in approach to provide the reader with physical insight into the role dispersion plays with respect to random signals.

The aim in this case is to determine the output RF noise power spectral density (PSD) arising from phase-to-intensity conversion of the source laser phase noise from which the associated RIN can be determined. From the Wiener–Khintchine Theorem, the power spectrum of the photocurrent is given by

$$S_{\mathrm{rf}}(\widetilde{\omega}) = R_{\mathrm{o}} \int_{-\infty}^{\infty} \langle I(t)I(t+\tau)\rangle e^{-i\widetilde{\omega}\tau}\, d\tau \tag{5.25}$$

provided that the photocurrent is a wide-sense stationary random process (Papoulis 1991). In this case, $\langle\rangle$ denotes the ensemble average and R_{o} is the load resistance seen by the photodiode. The photocurrent, in turn, is proportional to the magnitude squared of the output optical field,

$$I(t) = \Re |e_{\mathrm{in}}(t) * q(t)|^2, \tag{5.26}$$

where $e_{\mathrm{in}}(t)$ is again the complex time-domain envelope of the optical field incident on the dispersive link (to be discussed further in the following section) and $\Re$ is the average responsivity (A/W) of the photodiode used to perform the measurement. Writing out the integral expression of Equation (5.26) results in

$$I(t) = \frac{\Re}{2\pi\beta_2 L} \int_{-\infty}^{\infty} \int_{-\infty}^{\infty} e_{\mathrm{in}}(t+t_1)e_{\mathrm{in}}^*(t+t_2)e^{i(t_1^2-t_2^2)/(2\beta_2 L)}\, dt_2 dt_1, \tag{5.27}$$

where $*$ denotes complex conjugation. Making the substitutions $t + t_1 = u_1$ and $t + t_2 = u_1 + u_2$ and subsequently defining the instantaneous frequency to be $\widetilde{\omega} = u_2/(\beta_2 L)$, the photocurrent expression may be rewritten as

$$I(t) = \frac{\Re}{2\pi} \int_{-\infty}^{\infty} e^{-i\beta_2 L\widetilde{\omega}^2/2} e^{i\widetilde{\omega}t}\, d\widetilde{\omega} \int_{-\infty}^{\infty} e_{\mathrm{in}}(u_1)e_{\mathrm{in}}^*(u_1 + \beta_2 L\widetilde{\omega})e^{-i\widetilde{\omega}u_1}\, du_1. \tag{5.28}$$

Clearly, the u_1 integral is in the form of a Fourier transform. Realizing that the integrand in Equation (5.28) possess a random temporal (and, hence, spectral) phase, the multiplicative quadratic phase exponential can be ignored, as it will not contribute to the central moments of the random process $I(t)$ (Papoulis 1991). In this case, the double integration corresponds to back-to-back Fourier and inverse-Fourier transformations. The photocurrent is then given by

$$I(t) = \Re e_{\mathrm{in}}(t)e_{\mathrm{in}}^*(t + \beta_2 L\widetilde{\omega}). \tag{5.29}$$

The RF power spectrum is then found from Equation (5.25) to be (setting $t = 0$ since the choice of time origin is arbitrary for a stationary random process)

$$S_{\mathrm{rf}}(\widetilde{\omega}) = \Re^2 R_{\mathrm{o}} \int_{-\infty}^{\infty} \langle e_{\mathrm{in}}(0)e_{\mathrm{in}}^*(\beta_2 L\widetilde{\omega})e_{\mathrm{in}}(\tau)e_{\mathrm{in}}^*(\tau + \beta_2 L\widetilde{\omega})\rangle e^{-i\widetilde{\omega}\tau}\, d\tau. \tag{5.30}$$

Note that Equation (5.30) is in exact agreement with the result shown by Marshall et al. (2000). The details of calculating $S_{\mathrm{rf}}(\omega)$ will then depend on the underlying statistics of the phase of the input electric field envelope.

Consider the input electric field of the form $e_{\mathrm{in}}(t) = \sqrt{P_{\mathrm{o}}}\exp[\phi(t)]$, which maps fluctuations in the laser phase to a stationary random process. In this case, P_{o} is the average laser power, and intensity fluctuations of the laser are ignored to describe the noise contribution arising from the laser phase noise explicitly. Inserting this expression into Equation (5.30) yields

$$S_{\mathrm{rf}}(\widetilde{\omega}) = \Re^2 P_{\mathrm{o}}^2 R_{\mathrm{o}} \int_{-\infty}^{\infty} \langle e^{i[\phi(0)-\phi(\beta_2 L\widetilde{\omega})]} e^{i[\phi(\tau)-\phi(\tau+\beta_2 L\widetilde{\omega})]}\rangle e^{-i\widetilde{\omega}\tau}\, d\tau. \qquad (5.31)$$

Taking the phase difference to be a Gaussian random process (Papoulis 1991) allows Equation (5.31) to be rewritten as

$$S_{\mathrm{rf}}(\widetilde{\omega}) = \Re^2 P_{\mathrm{o}}^2 R_{\mathrm{o}} \int_{-\infty}^{\infty} e^{-\langle[\phi(0)-\phi(\beta_2 L\widetilde{\omega})+\phi(\tau)-\phi(\tau+\beta_2 L\widetilde{\omega})]^2\rangle/2} e^{-i\widetilde{\omega}\tau}\, d\tau. \qquad (5.32)$$

Defining the phase difference to be $\psi_{t_1,t_2} = \phi(t_1) - \phi(t_2)$, algebraic manipulation of the argument of the exponential in Equation (5.32) gives

$$\begin{aligned}(\psi_{0,\beta_2 L\widetilde{\omega}} + \psi_{\tau,\tau+\beta_2 L\widetilde{\omega}})^2 &= \psi^2_{0,\beta_2 L\widetilde{\omega}} + \psi^2_{\tau,\tau+\beta_2 L\widetilde{\omega}} \\ &\quad + \psi^2_{\tau+\beta_2 L\widetilde{\omega},\beta_2 L\widetilde{\omega}} + \psi^2_{\tau,0} \\ &\quad - \psi^2_{\tau,\beta_2 L\widetilde{\omega}} - \psi^2_{\tau+\beta_2 L\widetilde{\omega},0}.\end{aligned} \qquad (5.33)$$

If the phase noise PSD is approximately white, ψ_{t_1,t_2} is described by free Brownian motion (Lax 1967) and

$$\langle \psi^2_{t_1,t_2} \rangle = \Delta\omega |t_1 - t_2|, \qquad (5.34)$$

where $2\Delta\omega$ is the full-width-half-maximum (FWHM) Lorentzian linewidth. Note that $\Delta\omega$ is the laser FWHM linewidth in the optical domain as discussed in Section 3.4. Using Equations (5.33) and (5.34) in Equation (5.32) and performing the integration yield the noise PSD

$$\begin{aligned}S_{\mathrm{rf}}(\widetilde{\omega}) &= \Re^2 P_{\mathrm{o}}^2 R_{\mathrm{o}} \delta(\omega) + \Re^2 P_{\mathrm{o}}^2 R_{\mathrm{o}} \frac{2\Delta\omega}{\Delta\omega^2 + \widetilde{\omega}^2} \\ &\quad \left\{1 - e^{-\Delta\omega|\beta_2 L\widetilde{\omega}|}\left[\cos\left(\beta_2 L\widetilde{\omega}^2\right) + \frac{\Delta\omega}{\widetilde{\omega}}\sin(\beta_2 L\widetilde{\omega}^2)\right]\right\}.\end{aligned} \qquad (5.35)$$

Subtracting the delta-function at zero frequency [equivalent to subtracting the average photocurrent squared in Equation (5.25) and treating the RIN as the second central moment of $I(t)$], normalizing to the average DC power dissipated by the photodiode ($\Re^2 P_o^2 R_o = I_{dc}^2 R_o$), and multiplying by a factor of two yield the single-sided RIN spectrum arising from phase-to-intensity conversion of the laser phase noise

$$\mathrm{RIN}(\widetilde{\omega}) = \frac{4\Delta\omega}{\Delta\omega^2 + \widetilde{\omega}^2} \times \left\{1 - e^{-\Delta\omega|\beta_2 L\widetilde{\omega}|}\left[\cos\left(\beta_2 L\widetilde{\omega}^2\right) + \frac{\Delta\omega}{\widetilde{\omega}}\sin(\beta_2 L\widetilde{\omega}^2)\right]\right\}, \tag{5.36}$$

in agreement with previous work (Marshall et al. 2000). Two items are of particular interest with respect to the RIN spectrum arising from the laser phase noise. Firstly, the RIN spectrum is intimately tied to the laser lineshape as evidenced by the Lorentzian function in Equation (5.36). Therefore, the RIN level decreases with frequency more rapidly as the laser linewidth decreases. Secondly, when the dispersion is sufficiently large that the frequency-dependent delay over the bandwidth of interest is much larger than the coherence length of the laser ($|\beta_2 L\omega_{\mathrm{max}}| \gg 2\pi/\Delta\omega$), the measured RIN spectrum is equivalent to the lineshape measurement performed using the delayed self-heterodyne measurement discussed in Section 3.4. It should be noted, however, that this limit is unrealistic in a practical sense. The required fiber length to achieve this large-dispersion limit will be extremely long. For example, consider a laser with a FWHM Lorentzian linewidth of $\Delta\omega/2\pi = 1$ MHz (coherence time ~1 μs). To measure the corresponding lineshape over 10-MHz bandwidth using standard single-mode fiber ($|\beta_2| \approx 22\,\mathrm{ps}^2/\mathrm{km}$ and taking $\gg$ as one order-of-magnitude) would require a length $L \approx 7.2 \times 10^9$ km (about 1.8×10^5 times the circumference of the Earth).

To illustrate the impact of laser phase noise and dispersion on link noise performance, consider Figure 5.7, which shows the measured RIN spectrum for a 50-km single-mode fiber link ($\beta_2 L \approx -1145\,\mathrm{ps}^2$) employing a DFB laser with FWHM linewidth $\Delta\omega/2\pi = 1$ MHz (Urick et al. 2011). The measured RIN spectrum is shown by the solid black line, and the calculated RIN value from Equation (5.36) is shown by the dark gray line. When the back-to-back noise without the fiber in place of ~ −143 dBm/Hz across the band of interest (dominated by ASE from an EDFA at the link output) is incoherently added with the calculated noise value arising from the laser phase noise, the resulting RIN (light gray line) agrees quite well with the measured RIN spectrum.

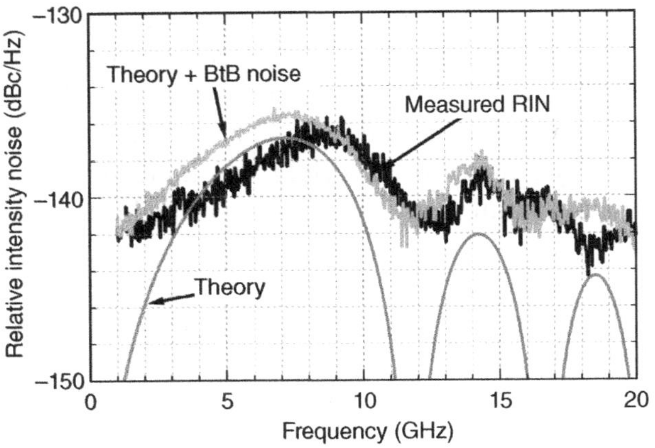

Figure 5.7. Measured RIN spectrum (black) for a 50-km intensity-modulated direct-detection link $\Delta\omega/2\pi = 1$ MHz, total dispersion $\beta_2 L \approx -1145\,\text{ps}^2$ and center wavelength of $\lambda = 1558$ nm. The incoherent addition of the calculated noise arising from only phase-to-intensity conversion of the laser phase noise (dark gray) and the back-to-back noise without the fiber span in place (~ -143 dBm/Hz across the frequency range above) is shown by the light gray curve and agrees well with the measured RIN (Urick et al. 2011).

As detailed in this section, CD can affect the fundamental RF response and provide a mechanism for conversion of optical noise between quadratures. The specific effects of CD on intensity- and phase-modulated analog links are revisited in Chapters 6 and 7, respectively. Even-order distortion induced by CD in intensity-modulated links employing an MZM is covered in Chapter 6. A further discussion on the complimentary responses to CD of intensity and phase modulation is provided in Chapter 7, including the mention of some mitigation techniques on the basis of in-phase and quadrature modulation. A thorough summary of CD mitigation methods is given by Agrawal (1997). Notable mitigation techniques employed in the microwave photonics community include SSB modulation (see Section 8.3) and construction of a fiber span with low net dispersion ($D \times L$). This latter method can be implemented with concatenated fiber spans having opposite D, a technique that can be simultaneously used to combat other fiber issues such as SBS.

5.5 STIMULATED BRILLOUIN SCATTERING

SBS is a low threshold nonlinearity and therefore very important for microwave photonics applications employing any appreciable amount

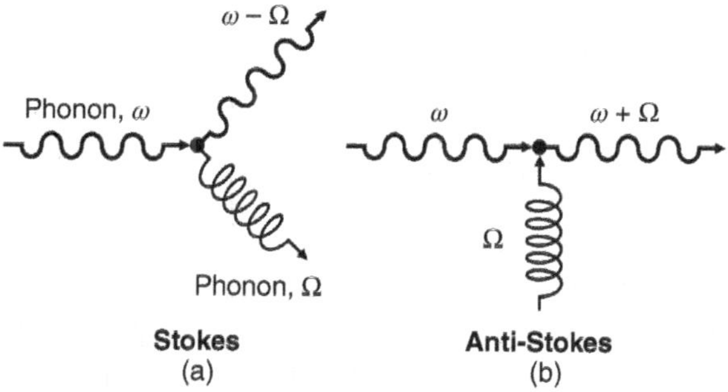

Figure 5.8. Schematic of the Brillouin (and Raman) scattering process.

of fiber. The SBS process was introduced in Section 3.5.3 in the context of fiber amplifiers but will be discussed more thoroughly in this section. From a classical point of view, Brillouin scattering is a process in which light propagating in a material is scattered by acoustic vibrations in the material lattice. The light scattered by the acoustic wave is Doppler shifted, resulting in a new optical frequency component. The Bragg condition determines the Brillouin scattering angle. A quantum-mechanical schematic of the Brillouin scattering process is shown in Figure 5.8.[1] The scattering is inelastic and can occur in two possible ways. In the Stokes scattering, a photon is annihilated with the subsequent creation of a lower frequency photon and a phonon. Anti-Stokes scattering involves the annihilation of a photon and a phonon followed by the emission of a higher frequency photon. These scattering processes occur when light propagates in standard fiber, where the Stokes process is much more probable, in a direction counter-propagating with respect to the input signal. The frequency of the phonon involved in Brillouin scattering is near 11 GHz for standard fiber, and the backscattered light is shifted by this amount. At low signal levels, the backscattered light from spontaneous Brillouin scattering is typically below the Rayleigh backscattered light. However, the Brillouin scattering can be stimulated at input powers above a certain point, a level typically referred to as the SBS threshold.

A complete derivation of the SBS threshold involves solving coupled differential equations that describe the pump and probe evolution along the fiber length. These methods are detailed elsewhere (Agrawal 2013),

[1]The diagrams in Figure 5.8 are not proper Feynman diagrams, which would require more interaction vertices such as shown in Balkanski and Wallis (2000).

and the main results are presented in this chapter. A useful approximation for the SBS threshold power is (Smith 1972)

$$P_{\mathrm{SBS}} \approx \frac{21A_{\mathrm{eff}}}{g_{\mathrm{B}}(f_{\mathrm{B}}) \cdot L_{\mathrm{eff}}}, \tag{5.37}$$

where A_{eff} is the effective area for the fiber, g_{B} is the peak Brillouin gain factor at the Brillouin center frequency f_{B} (typically near 11 GHz), and L_{eff} is the effective length for the fiber. A fiber parameter that is often specified is the mode-field diameter (MFD), which, assuming a Gaussian beam approximation to the power in the waveguide, can be related to A_{eff} by

$$A_{\mathrm{eff}} = \frac{\pi(\mathrm{MFD})^2}{4}. \tag{5.38}$$

The MFD for a step-index fiber can also be calculated using an empirical approximation (Marcuse 1977):

$$\mathrm{MFD} = d_1\left(0.65 + \frac{1.619}{V^{3/2}} + \frac{2.879}{V^6}\right), \tag{5.39}$$

where d_1 is the core diameter and V is given by Equation (1.1). The MFD $\approx 10.4\ \mu\mathrm{m}$ for standard single-mode fiber resulting in $A_{\mathrm{eff}} \approx 85\ \mu\mathrm{m}^2$ by way of Equation (5.38). The Brillouin gain spectrum has a Lorentzian shape that can be written as (Agrawal 2013)

$$g_{\mathrm{B}}(f) = \frac{g_{\mathrm{B}}(f_{\mathrm{B}}) \cdot (\Delta f_{\mathrm{B}})^2}{4(f - f_{\mathrm{B}})^2 + (\Delta f_{\mathrm{B}})^2}, \tag{5.40}$$

where Δf_{B} is the FWHM Brillouin linewidth with typical values of Δf_{B} ≈ 16 MHz. The peak Brillouin gain is (Agrawal 2013)

$$g_{\mathrm{B}}(f_{\mathrm{B}}) = \frac{2\pi\gamma_{\mathrm{e}}^2 X}{c n_{\mathrm{p}} \lambda_{\mathrm{p}}^2 \rho_0 v_{\mathrm{a}}(\Delta f_{\mathrm{B}})}, \tag{5.41}$$

where γ_{e} is the electrostrictive constant for the core, X is a factor accounting for overlap of the acoustic and optical modes, c is the speed of light in vacuum, n_{p} is the index of refraction for the pump, λ_{p} is the wavelength of the pump, ρ_0 is the density, and v_{a} is the acoustic velocity. For typical silica fibers, $g_{\mathrm{B}}(f_{\mathrm{B}}) = 3 - 5 \times 10^{-11}$ m/W (Agrawal 2013).

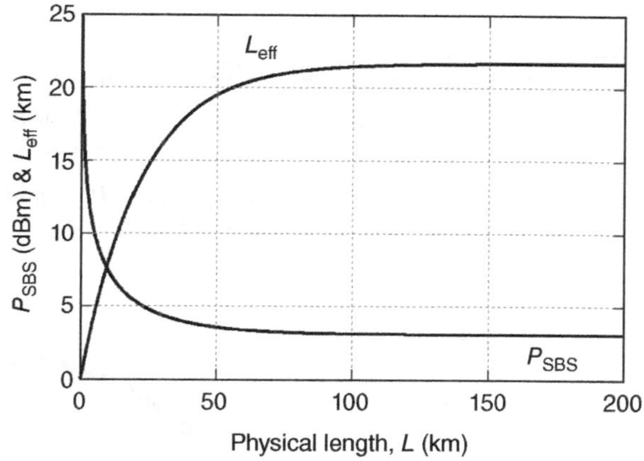

Figure 5.9. Calculated SBS threshold [Equation (5.37)] and effective length [Equation (5.42)] for $A_{eff} = 85\ \mu m^2$, $g_B(f_B) = 4 \times 10^{-11}$ m/W, and $\alpha = 0.2$ dB/km.

The final parameter needed to calculate Equation (5.37) is the effective length, which can be related to the physical length (L) by

$$L_{eff} = \frac{1 - e^{-\alpha L}}{\alpha}, \tag{5.42}$$

where α is the attenuation constant for the source laser [see also Equation (3.75)].

Equation (5.37) and (5.42) are plotted in Figure 5.9 as a function of physical length for parameters typical of standard single-mode fiber: $A_{eff} = 85\,\mu m^2$, $g_B(f_B) = 4 \times 10^{-11}$ m/W, and $\alpha = 0.2$ dB/km. As shown, the effective length levels out to about 21.7 km at about a physical length of 150 km. (In general, when αL is very large, $L_{eff} \rightarrow 1/\alpha$.) Therefore, for $L > 125$ km, the SBS threshold is relatively constant at 3.1 dBm (2.0 mW). Calculations such as those shown in Figure 5.9 are useful when the fiber parameters are well known. However, it is best practice to characterize the fiber being used (if possible) when SBS is a concern.

An architecture to measure the SBS threshold is shown in Figure 5.10. A laser is launched through a well-calibrated optical coupler into the fiber span under test. An optical power meter can be employed to measure the output power and the power at two ports on the coupler. The coupler outputs are used to determine the launch power into the span and the backscattered power. A data set taken with such an apparatus for 50 km of standard single-mode fiber is shown in Figure 5.11. In the linear response regime, the output power is

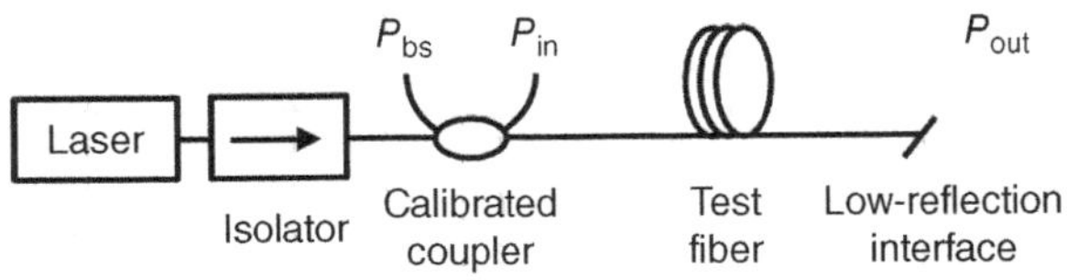

Figure 5.10. Apparatus for measuring optical output power (P_{out}) and backscattered power (P_{bs}) for a fiber span as a function of optical input power (P_{in}).

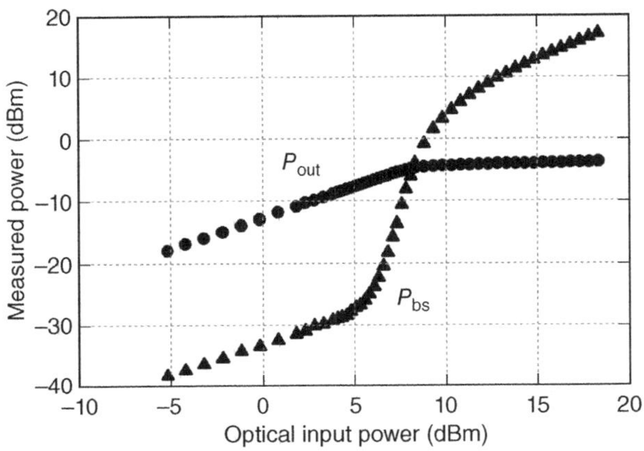

Figure 5.11. Measured optical output power (P_{out}) and backscattered power (P_{bs}) as a function of input power for a 50-km span of standard single-mode fiber.

about 12.5 dB lower than the input power, which corresponds to an attenuation coefficient of 0.25 dB/km. The backscattered power in this region is about 33 dB below the launch power, a reasonable level for Rayleigh scattering. The backscattered power begins to increase rapidly just below 5 dBm as the SBS threshold is achieved. Inserting $A_{eff} = 85\ \mu m^2$, $g_B(f_B) = 4 \times 10^{-11}$ m/W, and $\alpha = 0.25$ dB/km into Equation (5.37) yields $P_{SBS} = 4.3$ dBm, which is in good agreement with the experimental result. Increasing the input power beyond the SBS threshold results in minimal change to the output power. Such saturation of the power out of a fiber span can impact photonic links, where the received optical power will determine the RF performance.

Although some microwave photonic links have been operated beyond the SBS threshold (Devgan et al. 2008), it is generally good practice to operate links in the regime where Brillouin scattering is negligible compared to Rayleigh scattering. Depletion of the laser signal is the most obvious detriment imposed by SBS, which in turn will limit the achievable RF performance of a link. However, there are

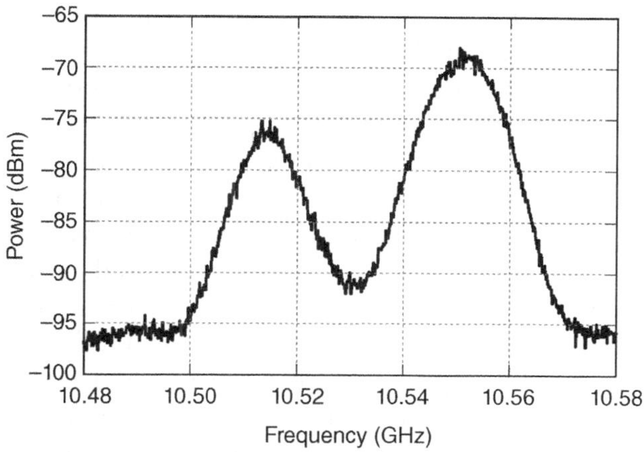

Figure 5.12. Measured electrical spectrum at the output of a link employing two fiber types with different SBS frequencies (Urick et al. 2005).

other issues with operating a microwave photonic link near or beyond the SBS threshold. Microwave photonic links operating in X-band can be seriously degraded by SBS because of significant noise at the Brillouin frequency (~11 GHz). The backscattered Brillouin wave can be reflected to co-propagate with the signal through elastic Rayleigh scattering. The consequence at the link output is a beat note between the signal and the frequency-shifted Brillouin wave. An example of this noise was shown in Figure 3.35 for 20 km of standard single mode-fiber, where a single noise peak near 10.8 GHz was observed. Another spectrum showing two Brillouin peaks is shown in Figure 5.12. Depicted are two noise peaks near 10.52 and 10.55 GHz measured at the output of a custom 20-km fiber delay line constructed from two different fiber types (Urick et al. 2006). Other more-complicated detriments can arise from SBS, including SBS acting in concert with DRS (Fotiadi and Kiyan 1998) and phase-to-intensity-noise conversion arising from SBS (Zhang and Phillips 2005). Since the performance of photonic links scales with output power, mitigation techniques for SBS must be considered for most long-haul links.

Employing alternating sections of fiber with different SBS frequencies can significantly improve the SBS threshold as compared to a continuous span of one fiber type (Mao et al. 1992). If the SBS frequencies are largely nonoverlapping, such as shown in Figure 5.12, then the backscatter from the different fiber types are decoupled. For example, the 20-km link that produced the data in Figure 5.12 exhibited a 13-dBm SBS threshold (Urick et al. 2006), about 8-dB higher than a 20-km span of standard single-mode fiber. This concatenated fiber technique is useful for analog fiber optic delay lines where a spool of fiber is co-located with the modulation and demodulation stages but can be quite impractical for long-distance point-to-point links. Using spans of fiber with SBS frequency distributions has also been demonstrated (Shiraki et al. 1996a). Fiber spans with nonuniform fiber dopants (Shiraki et al. 1996b) and unique refractive index profiles (Ruffin et al. 2005) have likewise shown increased power handling. Applying a temperature (Hansryd et al. 2001) or strain (Boggio et al. 2005) distribution to the fiber has also been shown to increase the SBS threshold but is again impractical for point-to-point links. Mitigation techniques on the modulation-side of the link include polarization scrambling (Deventer and Boot 1994) and broadening of the input signal (Wilson et al. 1996). Such techniques can be useful in digital systems but can be problematic for high fidelity analog applications. For example, when a phase dither is applied to broaden the input signal, the link will be more susceptible to phase-to-intensity noise conversion. Suppressed-carrier modulation formats are very attractive for circumventing SBS in microwave photonic links as will be discussed in Sections 8.2 and 8.4.

5.6 STIMULATED RAMAN SCATTERING

SRS was introduced in Section 3.5.2 in the context of fiber amplifiers. Raman scattering is similar to Brillouin scattering in that Figure 5.8 can also be employed to describe the process. However, there are numerous significant differences between Raman and Brillouin scattering. Where Brillouin scattering involves an acoustic phonon, optical phonons interact with the photons in Raman scattering. Thus, the associated frequency shift of the light is much larger for Raman scattering, about 13 THz for optical fibers. The gain spectrum is much broader for Raman scattering, as shown in Figure 3.34, and can span tens of terahertz. While Brillouin scattering is dependent on the incident angle and occurs only in a direction opposite to the pump signal, Raman scattered light can counter- or co-propagate with the pump. The SRS process occurs at a much larger threshold than SBS. Therefore, while SBS is a major concern for single-channel photonic links, SRS does not pose a significant detriment to links using a single optical carrier. However, SRS can cause serious problems for multichannel links in the form of interchannel crosstalk.

For a single-channel link, where the signal acts as the pump for the spontaneous Raman process, the threshold at which Raman scattering becomes stimulated can be calculated. A simple approximation for the SRS threshold power can be used (Smith 1972)

$$P_{SRS} \approx \frac{16A_{eff}}{g_R L_{eff}}, \tag{5.43}$$

where g_R is Raman gain coefficient. The Raman gain coefficient can be estimated using previously published data (Stolen and Ippen 1973) with a triangular fit having a slope of 5.0×10^{-15} m/W/THz (Phillips and Ott 1999). For 1550-nm light in standard optical fiber, the peak Raman gain occurs at a frequency shift of about 13 THz, where $g_R = 6.5 \times 10^{-14}$ m/W. Inserting this value along with the same parameters for calculating the minimum SBS threshold power ($A_{eff} = 85\ \mu m^2$, $\alpha = 0.2$ dB/km, and $L_{eff} = 21.7$ km) into Equation (5.43) yields 30 dBm (1 W) for the minimum SRS threshold. The calculated SBS threshold using the same parameters is 3 dBm (2 mW). Clearly, SBS will inhibit performance at much lower power levels than SRS. However, the effects of SRS in terms of RF crosstalk are important as detailed in the following section.

The effects of SRS on the RF crosstalk in a multichannel link involve solving coupled differential equations that describe the evolution of the different wavelengths propagating in the fiber. A two-channel system is assumed in this treatment where the equations are (Phillips and Ott 1999)

$$\frac{\partial I_1}{\partial L} + \frac{1}{v_1}\frac{\partial I_1}{\partial t} = I_1(g_R I_2 - \alpha_1). \tag{5.44a}$$

$$\frac{\partial I_2}{\partial L} + \frac{1}{v_2}\frac{\partial I_2}{\partial t} = I_2(-g_R I_1 - \alpha_2). \tag{5.44b}$$

In Equations (5.44), the subscripts correspond to the signals at wavelengths λ_1 and λ_2, I is the optical intensity, v is the group velocity, and α is the fiber loss. The Raman gain coefficient is taken to be positive when $\lambda_1 > \lambda_2$ and negative when $\lambda_1 < \lambda_2$ and can be approximated as (Campillo et al. 2004)

$$g_R = \left(5.0 \times 10^{-15} \frac{\text{m}}{\text{W} \cdot \text{THz}}\right)\left(\frac{1}{\lambda_2} - \frac{1}{\lambda_1}\right)c. \tag{5.45}$$

Assuming that λ_1 is an unmodulated CW signal and that λ_2 is modulated by an RF signal, the RF crosstalk (Xtalk) from λ_2 onto λ_1 is defined as

$$\text{Xtalk} = \frac{P_{\text{RF},1}(\Omega)}{P_{\text{RF},2}(\Omega)}, \tag{5.46}$$

which is the ratio of the RF power on channel 1 (CW) at the link output to that on channel 2 (modulated). Equations (5.44) can be solved for the amplitude ($\text{Xtalk}_{\text{SRS}}$) and RF phase ($\Theta_{\text{SRS}}$) of the crosstalk due to SRS yielding the following equations (Phillips and Ott 1999)

$$\text{Xtalk}_{\text{SRS}} = \left(\frac{\rho_{\text{SRS}} g_{\text{R}} P_2}{A_{\text{eff}}}\right)^2 \frac{1 + e^{-2\alpha L} - 2e^{-\alpha L}\cos(\Omega d_{12} L)}{\alpha^2 + \Omega^2 d_{12}^2}$$

$$\left(1 + \frac{\rho_{\text{SRS}} g_{\text{R}} P_2 L_{\text{eff}}}{A_{\text{eff}}}\right)^{-2}, \tag{5.47}$$

$$\Theta_{\text{SRS}} = \tan^{-1}\left(\frac{-\Omega d_{12}}{-\alpha}\right) + \tan^{-1}\left[\frac{e^{-\alpha L}\sin(\Omega d_{12} L)}{e^{-\alpha L}\cos(\Omega d_{12} L) - 1}\right], \tag{5.48}$$

where ρ_{SRS} is the polarization overlap factor that is one when the two beams are in the same polarization and nearly zero when they are in orthogonal polarizations, P_2 is the average optical power in the modulated channel at the fiber input (assumed to be amplitude modulation), Ω is the angular drive frequency, $\alpha_1 = \alpha_2 = \alpha$ is assumed, and d_{12} is the walk-off parameter. This last parameter describes the effect of CD and is defined as (Agrawal 2013)

$$d_{12} = \beta_1(\lambda_1) - \beta_1(\lambda_2) = \frac{1}{v_1} - \frac{1}{v_2} \approx D(\lambda_1 - \lambda_2), \tag{5.49}$$

where β_1 and D are as defined in Section 5.4. The RF phase of the SRS-induced crosstalk [Equation (5.48)] is important when multiple sources of crosstalk are considered, which need to be combined with their relative phases preserved. The arguments of the inverse tangent functions in Equation (5.48) are written to be compatible with an "atan2" function, which is commonly used in many computer languages.

Equations (5.47) and (5.48) have been validated previously (Phillips and Ott 1999) and are compared to measured data in the following section. Equation (5.47) can be employed to predict the level of

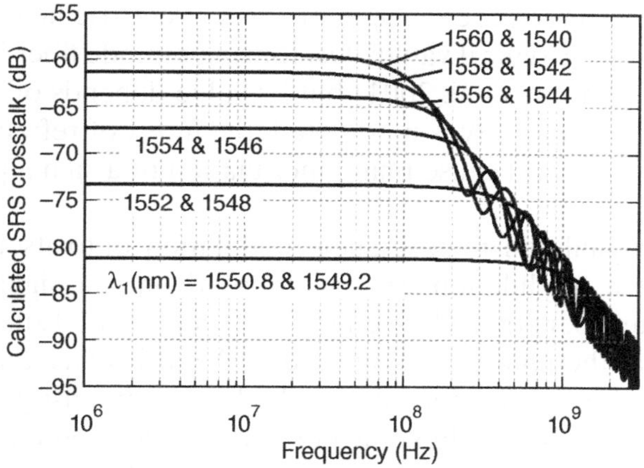

Figure 5.13. Calculated crosstalk due to stimulated Raman scattering in a 25-km link with a modulated channel at 1550 nm. The CW wavelengths are designated in nanometers.

crosstalk due to SRS as shown in Figure 5.13, where the SRS-induced crosstalk from $\lambda_2 = 1550$ nm onto 12 values of λ_1 are shown. The following inputs are used for the calculation: $\rho_{\text{SRS}} = 1$, $\alpha = 0.2$ dB/km, $L = 25$ km, $P_2 = 0$ dBm, $A_{\text{eff}} = 85\ \mu\text{m}^2$, and $D = 16.5$ ps/nm/km. As given by Equation (5.47), the amplitude of the SRS crosstalk is the same for equal separation above or below the modulated channel. The effect of SRS worsens with larger channel spacing, with the peak Raman gain occurring near 1450 nm for a 1550-nm signal. As shown in Figure 5.13, the SRS crosstalk is larger and more sensitive to channel spacing at low modulation frequencies. This point is important to consider for HF and VHF implementations of analog photonics such as for antenna remoting applications.

5.7 CROSS-PHASE MODULATION

The linear part of the refractive index (n) has been treated previously in terms of RF propagation delay and CD. The total refractive index (n_{T}) will depend on the optical intensity in the fiber due to higher order terms in the susceptibility:

$$n_T = n + n_2 |E|^2, \tag{5.50}$$

where

$$n_2 = \frac{3}{8} n \cdot \text{Re}[\chi^{(3)}] \tag{5.51}$$

is the nonlinear index coefficient and E is the electric field in the fiber (Agrawal 2013). A typical value of n_2 for single-mode fiber is $n_2 = 2.6 \times 10^{-20}$ m^2/W (Antona et al. 2001) but its value depends on the specific fiber type. If the intensity is modulated, the index of refraction will be modulated, which will cause phase modulation of a signal propagating in the fiber.

A single signal can impose such modulation onto itself, the SPM process. Phase modulation can be imposed onto a signal by other signals in the fiber, the XPM effect. An analysis of Equation (5.50) can show that the phase shift imposed by XPM is twice that due SPM for two equal-intensity optical signals. For an intensity modulation direct-detection (IMDD) link, the nonlinear phase modulation by itself may not be a problem because a photodiode alone is insensitive to optical phase fluctuations. However, SPM and XPM can be converted to intensity modulation by CD. Likewise, a constant intensity modulation, such as phase modulation, will not cause SPM or XPM. However, if CD converts the phase modulation to intensity modulation, the intensity modulation can cause SPM and XPM, which can then distort the original phase-modulated signal. The effects of SPM and XPM are therefore strongly dependent on the dispersion map for a link, not simply the net dispersion.

Some measured data exemplifying the interplay between XPM and CD are shown in Figure 5.14 (Rogge et al. 2005). For these experiments, an optical carrier near 1550 nm was modulated in phase or intensity by an RF signal and the crosstalk at +50-GHz channel spacing was measured. While XPM dominates the crosstalk in Figure 5.14, other mechanisms do contribute such as FWM (Rogge et al. 2005), the topic of the following section. Two fiber configurations were used to probe the effect of dispersion map. The first configuration consisted of 21 km of standard single-mode fiber (SMF) followed by 12.5 km of managed low dispersion fiber (MLDF); the second configuration simply swaps the order of these two spans. The SMF had a dispersion parameter of $D = 16.5$ ps/nm/km resulting in a net dispersion of 347 ps/nm. The MLDF span consisted of four alternating sections of fiber with $D = \pm$ 4 ps/nm/km, thus having low dispersion and nearly zero net dispersion. The dispersion map strongly influences the crosstalk as shown in Figure 5.14. In the intensity-modulated link, the signal creates XPM, which is converted to intensity-modulated crosstalk via CD. This process is more efficient with the MLDF span first. The phase modulation is constant intensity and therefore does not induce XPM. Therefore, the crosstalk mechanism for the phase-modulated link is a

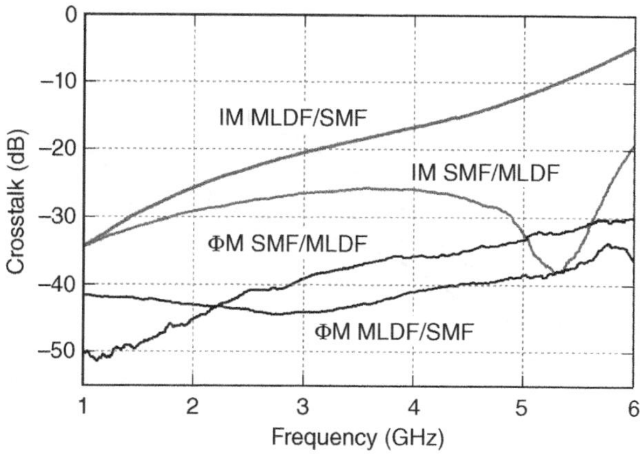

Figure 5.14. Measured crosstalk for a 33.5-km link employing intensity modulation (IM) and phase modulation (ΦM). Two fiber configurations were employed for each format: single-mode fiber followed by managed low dispersion fiber (SMF/MLDF) and vice versa (MLDF/SMF) (Rogge et al. 2005).

three-step process. Firstly, dispersion transforms the phase-modulation into intensity modulation. This intensity modulation then causes XPM, which is converted to intensity-modulated crosstalk via dispersion or the phase-sensitive receiver. In contrast to the case of intensity modulation, this process is more efficient with the higher dispersion fiber first for most of the frequency range shown.

The calculation of XPM-induced crosstalk in a two-channel intensity-modulated link involves many of the same parameters used in the previous section to describe SRS crosstalk. For two wavelengths λ_1 (unmodulated CW) and λ_2 (intensity modulated), the crosstalk can be derived using the wave-envelope perturbation analysis technique (Phillips et al. 1991). The results in terms of the amplitude ($\text{Xtalk}_{\text{XPM}}$) and RF phase ($\Theta_{\text{XPM}}$) of the crosstalk due to XPM are (Phillips and Ott 1999)

$$\text{Xtalk}_{\text{XPM}} = \left(\frac{2n_2\lambda_1 DP_2\Omega^2\rho_{\text{XPM}}}{A_{\text{eff}}c}\right) \times \frac{1+e^{-2\alpha L}-2e^{-\alpha L}(1-\alpha L)\cos(d_{12}\Omega L)-2L\left[\alpha+d_{12}\Omega e^{-\alpha L}\sin(d_{12}\Omega L)\right]+(\alpha^2+d_{12}^2\Omega^2)L^2}{(\alpha^2+d_{12}^2\Omega^2)^2}, \tag{5.52}$$

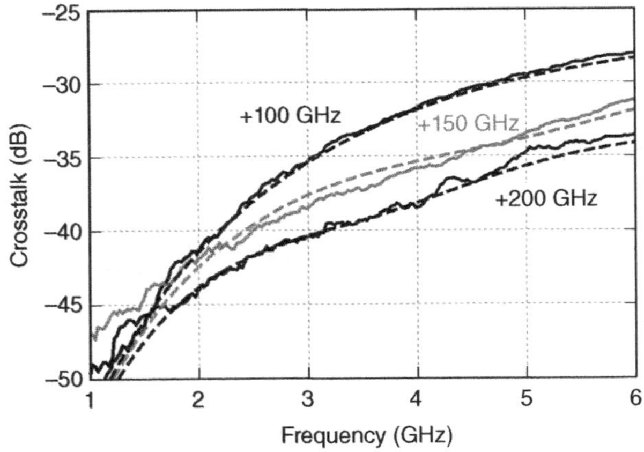

Figure 5.15. Measured crosstalk for a 25-km link (Bucholtz et al. 2004) plotted with theory (dashed lines) from Equation (5.51).

$$\Theta_{XPM} = \tan^{-1}\left(\frac{2\Omega d_{12}\alpha}{\alpha^2 - \Omega^2 d_{12}^2}\right) + \tan^{-1}\left[\frac{e^{-\alpha L}\sin(\Omega d_{12}L) - \Omega d_{12}L}{e^{-\alpha L}\cos(\Omega d_{12}L) - 1 + \alpha L}\right], \tag{5.53}$$

where all parameters are as defined previously, and ρ_{XPM} is the polarization overlap factor for the XPM process. Neglecting the effects of PMD, ρ_{XPM} ranges from one-third for perpendicular polarizations to one for parallel polarizations. Equation (5.52) is plotted against measured data for an intensity-modulated link using 25 km of Corning MetroCor fiber in Figure 5.15. The parameters for Figure 5.15 are $\lambda_2 = 1550.16$ nm, $\rho_{XPM} = 1$, $n_2 = 2.7 \times 10^{-20}\ \text{m}^2/\text{W}$, $\alpha = 0.2$ dB/km, $P_2 = 8.45$ dBm, $A_{\text{eff}} = 58\ \mu\text{m}^2$, and $D = -8$ ps/nm/km. The data and theory agree quite well for the three channel spacings shown: +100 GHz, +150 GHz, and +200 GHz.

Some trends for XPM-induced crosstalk are shown in Figure 5.16. Depicted are plots of Equation (5.52) for $\lambda_2 = 1550$ nm, $\rho_{XPM} = 1$, $n_2 = 2.6 \times 10^{-20}\ \text{m}^2/\text{W}$, $\alpha = 0.2$ dB/km, $P_2 = 0$ dBm, $A_{\text{eff}} = 85\ \mu\text{m}^2$, and $D = 16.5$ ps/nm/km, as a function of λ_1 and Ω. Similarly to the SRS case, the crosstalk due to XPM is symmetric about the modulation channel for the wavelengths shown. At low modulation frequencies, the crosstalk is very low as governed by the first term in Equation (5.52). However, the crosstalk due to XPM increases rapidly with frequency as shown in Figure 5.16. The effect of channel spacing for XPM is opposite

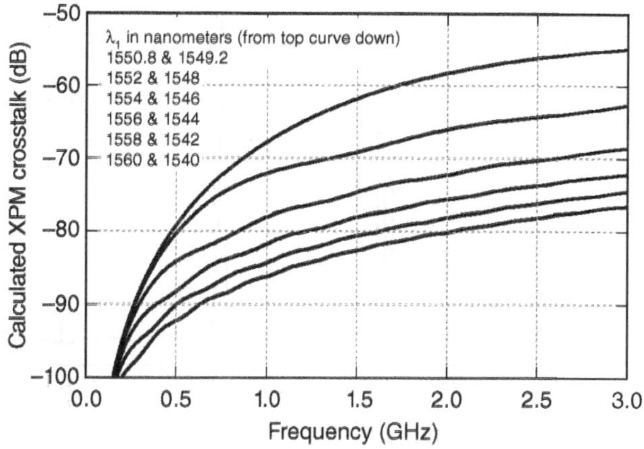

Figure 5.16. Calculated crosstalk due to cross-phase modulation in a 25-km link with a modulated channel at 1550 nm.

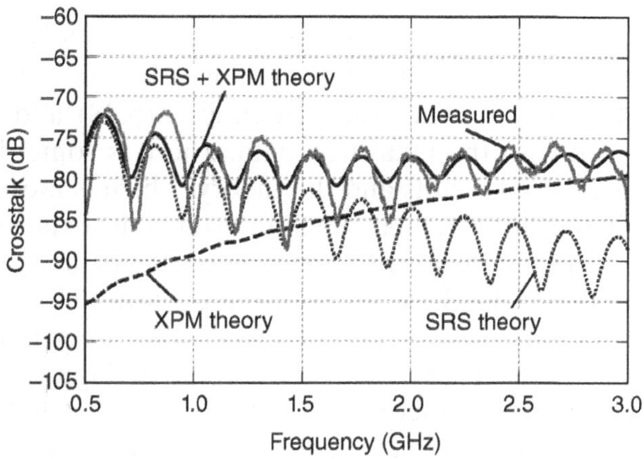

Figure 5.17. Measured crosstalk for a 19.1-km link compared against theory for SRS- and XPM-induced crosstalk (Sunderman and Urick 2014).

to that for SRS; the efficiency of XPM-induced crosstalk is higher at narrower channel spacing.

As mentioned previously, RF phase must be included when analyzing crosstalk arising from more than one physical mechanism. An example of this is shown in Figure 5.17, where crosstalk with contributions from both SRS and XPM is plotted (Sunderman and Urick 2014). The parameters for Figure 5.18 are $\lambda_1 = 1565.00$ nm, $\lambda_2 = 1551.57$ nm, $\rho_{SRS} = 0.9$, $\rho_{XPM} = 0.8$, $n_2 = 2.7 \times 10^{-20}$ m^2/W, $\alpha = 0.2$ dB/km, $L = 19.1$ km, $P_2 = 1.7$ dBm, $A_{eff} = 85$ μm^2, and $D = 16.5$ ps/nm/km. The SRS and

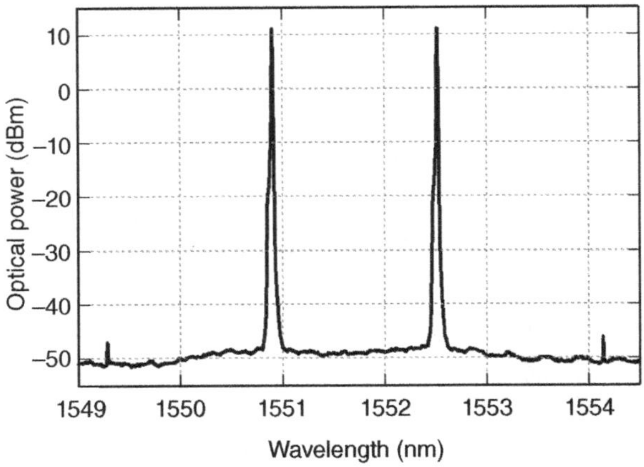

Figure 5.18. Measured spectrum at the output of a span of dispersion shifted fiber demonstrating two FWM products resulting from the input waves at 1552.6 and 1551.0 nm (Sunderman and Urick 2014).

XPM theory, Equations (5.47) and (5.52), respectively, do not alone predict the measured crosstalk. The coherent addition of the SRS and XPM crosstalk was obtained using Equations (5.48) and (5.53) and agrees fairly well with the measured data. There is some discrepancy between the measured results and the theory, which is most likely due to dispersion in the 200-GHz flat-top filter used in the experiment. Such filter dispersion has been shown to be a major factor in analog optical crosstalk (Campillo and Bucholtz 2006) and should be kept in mind for multichannel microwave photonic links. Other fiber nonlinearities could have also caused lower level crosstalk in Figure 5.17 that is not accounted for in the theory. Such mechanisms include FWM or polarization effects, which are the subjects of the following two sections.

5.8 FOUR-WAVE MIXING

FWM, or four-photon mixing, is a nonlinear fiber effect originating from the third-order susceptibility in Equation (5.1). Shown in Figure 5.18 are measured data demonstrating an effect of FWM, where two optical frequencies at ν_1 and ν_2 interact via $\chi^{(3)}$ to generate new frequencies at $2\nu_1 - \nu_2$ and $2\nu_2 - \nu_1$. This process can be viewed as a type of third-order intermodulation distortion in the optical domain, similar to that described in Section 2.3.2. Indeed, FWM can cause third-harmonic generation as well. A full analysis of FWM requires a vector-field treatment owing to the polarization dependence of the effects (Agrawal 2013). Furthermore, FWM is most efficient when the propagation constants of the interacting waves are matched (sometimes referred to as phase matching); thus frequency spacing and dispersion play a strong role. The spectrum shown in Figure 5.18 was measured at the output of a span of dispersion-shifted fiber with a zero-dispersion wavelength near 1.5 μm. Two lasers at 193.10 and 193.30 THz (1552.6 and 1551.0 nm, respectively) were combined with a polarization-maintaining coupler and launched into the fiber. The FWM mixing products at 192.90 THz (1554.2 nm) and 193.50 THz (1549.5 nm) can be clearly seen. In this case, polarization matching at the input and low CD allowed for fairly efficient FWM observable with an optical spectrum analyzer.

Similarly to other fiber nonlinearities, FWM can be described by a gain parameter. The parametric gain coefficient for standard fiber at 1550 nm is about $g_P = 1.1 \times 10^{-13}$ m/W (Agrawal 2013). Note that this value is larger than that for SRS, $g_R = 6.5 \times 10^{-14}$ m/W, and smaller than the SBS, $g_B = 4 \times 10^{-11}$ m/W. However, these scattering processes are self-phase matching, whereas FWM requires phase matching between the pump signals (Stolen 1975). Therefore, the FWM process can be very inefficient in long fiber lengths and/or large channel spacing where the phase matching conditions are not satisfied (Shibata et al. 1987). While SRS and SBS are generally more problematic than FWM for analog systems, there are some deleterious effects of FWM in microwave photonics applications.

Although clearly demonstrating the FWM effect, the associated RF issues may not be clear from the data in Figure 5.18. That is, the two mixing products at $2\nu_1 - \nu_2$ and $2\nu_2 - \nu_1$ are typically at frequency offsets much larger than the RF channel bandwidth around either of the carries. However, significant crosstalk, second-order distortion, and third-order distortion can result from FWM in a two-wavelength system due to second-order FWM (Phillips et al. 2008). FWM is often described by the case where nine new frequencies are generated by the nonlinear interaction of three input frequencies. For inputs at ν_1, ν_2, and ν_3, the third-order terms occur at $\nu_{ijk} = \nu_i + \nu_j - \nu_k$, where i, j, and k have values of 1, 2, or 3 (Chraplyvy 1990). A classic example of FWM-induced interchannel crosstalk occurs when the three signal are equally spaced, that is, $\nu_2 - \nu_1 = \nu_3 - \nu_2$. An analog optical signal, either phase or amplitude modulated, comprises a carrier and two sidebands. Thus, such a signal presents the classical crosstalk scenario mentioned previously. Therefore, FWM can potentially play a negative role even in single-channel systems particularly when the modulation depth is large.

5.9 POLARIZATION EFFECTS

The purpose of this section is to provide a largely qualitative overview of important polarization effects in fiber optic systems, such as polarization-dependent loss (PDL), PMD, and XPolM. The elements of Jones calculus and Mueller calculus must be introduced to that end, but a full discussion of either is too lengthy for this work. Kliger et al. (1990) provides a good treatment of these two representations of polarization in optics and references the original sources therein.

Light propagating in the z-direction can be represented as a sum of two components aligned to the x- and y-axis. Such an electromagnetic wave can be written in vector form as

$$\mathbf{E} = \begin{bmatrix} E_x e^{i\phi_x} \\ E_y e^{i\phi_y} \end{bmatrix} e^{i(\omega t - \beta z)}, \tag{5.54}$$

where E_x and E_y are the amplitudes of the two components, ϕ_x and ϕ_y are their phases, ω is the angular frequency, and β is the propagation constant. The 2×1 vector in (5.54) is known as a Jones vector and contains the polarization information of the light. In Jones calculus, a 2×2 matrix is employed to describe an optical component, which can be multiplied by the input Jones vector to determine the output Jones vector. Jones calculus is physically intuitive because it is directly tied to Maxwell's equations. However, the parameters of a Jones vector are not directly measureable, and the vector is, in general, imaginary.

Mueller calculus uses 4×1 vectors, Stokes vectors, to represent the polarization of a lightwave, and 4×4 matrices for optical components. The Stokes parameters are real and constructed to be related to measureable optical intensities. Let I_x be the intensity in the x-component, I_y the intensity in the y-component, I_{45} the intensity in the linearly polarized component at 45°, I_{-45} the intensity in the linearly polarized component at −45°, I_R the intensity in the right-circular component, and I_L the intensity in the left-circular component. These intensities can be measured via six experiments using a combination of linear polarizers and quarter waveplates (Damask 2005). The Stokes parameters are related to these intensities and to Equation (5.54) by (Damask 2005)

$$S_0 = I_x + I_y = \frac{1}{2}(E_x^2 + E_y^2).$$

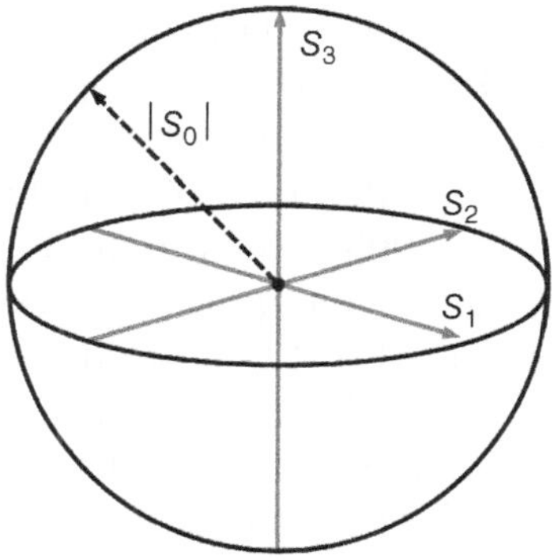

Figure 5.19. The Poincaré sphere showing the associated Stokes parameters.

$$S_1 = I_x - I_y = \frac{1}{2}(E_x^2 - E_y^2).$$
$$S_2 = I_{45} - I_{-45} = E_x E_y \cos(\phi_y - \phi_x).$$
$$S_3 = I_R - I_L = E_x E_y \sin(\phi_y - \phi_x). \qquad (5.55)$$

Finally, the Stokes vector is composed of these four components as

$$\mathbf{S} = \begin{bmatrix} S_0 \\ S_1 \\ S_2 \\ S_3 \end{bmatrix}. \qquad (5.56)$$

The common representation of Stokes space is the Poincaré sphere, which is a unit sphere normalized to the total optical intensity (S_0) with the three axes shown in Figure 5.19. Note that Equation (5.56) is sometimes referred to as a Mueller vector, reserving Stokes vector to designate a 3×1 vector on the Poincaré sphere. The equator of the Poincaré sphere corresponds to linear polarization states, with one full trip around the equator amounting to a 180° polarization rotation. The north and south poles represent right- and left-circular polarization, respectively. Generally, traversing half the sphere circumference along a geodesic will rotate the polarization to an orthogonal state.

The previous two sections provide enough background to introduce the reader to PDL and PMD. However, it is worth pointing to important works relating the Jones and Mueller calculus for fiber optic analysis. It can be shown (Frigo 1986) that Jones and Stokes vectors are elegantly related by Pauli spin matrices.[2] This "spin-vector calculus of polarization" (Damask 2005) was first applied to guided waves by Frigo (1986), elegantly reviewed in context of PMD (Gordon and Kogelnik 2000) and later extended to include optical propagation phase (Frigo and Bucholtz 2009). Such analyses are well beyond the scope of this book but have been applied to RF analog optical phase modulated links (Frigo et al. 2013) that are the topic of Chapter 7 and remain an active area of research for optical communications (Karlsson 2014). The advanced reader is encouraged to consult the associated rich body of literature for descriptions of these and other treatments of polarization effects in fiber optic transmission media.

The PDL of an optical component refers to transmission loss that is a function of the input polarization. There can be numerous underling physical mechanisms that give rise to PDL, such as dichroism. Regardless of the source, PDL is defined as (Damask 2005)

$$\text{PDL [dB]} \equiv 10 \log \left(\frac{P_{\text{max}}}{P_{\text{min}}} \right), \tag{5.57}$$

where P_{max} and P_{min} are maximum and minimum output powers of an optical component, measured at a fixed input power varied over all polarization states. For example, a perfect polarizer would completely attenuate one polarization state, in which case $P_{\text{min}} = 0$ and PDL $\rightarrow \infty$. PDL can be represented as a vector in Jones or Stokes space in the direction of the polarization state that relates to P_{max}. The polarization associated with P_{min} is always orthogonal to the PDL vector. The PDL of modern fiber is very low, but other fiber optic components can exhibit significant PDL. PDL will convert polarization fluctuations to intensity fluctuations, and this can be problematic for IMDD links. Polarization fluctuations can come about in numerous ways, such as PMD that is described shortly. The PDL of fiber-to-air interfaces used for coupling to photodiodes causes RF gain variations correlated with polarization fluctuations at a link output. Such instabilities created a problem for a sensitive analog photonic link employed in radio astronomy, which required an automatic gain control system to be implemented (White 2000). Polarization-dependent gain (PDG) can occur in optical amplifiers, such as EDFAs (Mazurczyk and Zyskind 1994), which can cause intensity fluctuations in links when the polarization is not fixed.

[2]Frigo's (1986) formalism related relates a 2×1 Jones vector,$|s\rangle$, to a 3×1 vector in Stokes space,$\hat{s}$. The relationship can be written in Dirac notation as $\hat{s} = \langle s|\vec{\sigma}\,|s\rangle$, where $\vec{\sigma}$is the Pauli spin vector (Damask 2005).

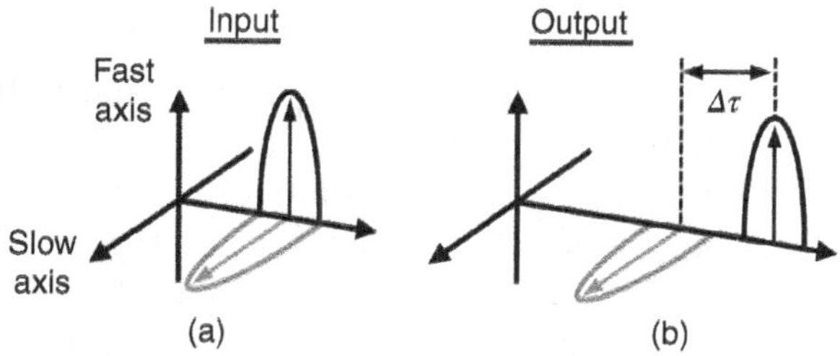

Figure 5.20. Pictorial representation of PMD. Signals in the two eigenstates for the material remain in the same polarizations after propagation but are dispersed in time an amount $\Delta\tau$.

Polarization-modulated links are susceptible to PDL (Campillo 2006), which can cause fading and distortion. As alluded to the previous section, PDL can act in concert with other propagation effects (Rossi et al. 2014), such as PMD, to create further impairments in photonic links.

PMD (Poole and Wagner 1986) arises from random fluctuations in the linear birefringence (the polarization dependence of the refractive index) in fiber. Analysis of PMD can become quite complicated. Early work on the statistics of PMD was completed by Foschini and Poole (1991), and later textbook treatments of PMD include those by Kogelnik and Jopson (2002) and Damask (2005). Gordon and Kogelnik (2000) provided an elegant summary of the prevailing PMD models and their relationships. In the principle states model, there are two orthogonal polarization eigenstates whose polarizations do not change after propagating the fiber span under analysis. The difference in propagation time for these two eigenstates, the differential group delay (DGD), is maximum for these two states. As shown in Figure 5.20, these eigenstates are commonly designated by a slow and fast axis, not necessarily aligned with the x and y directions. Analysis of PMD is conducted using a PMD vector ($\vec{\tau}$) whose magnitude ($\Delta\tau$) is the DGD and whose direction is along the slow axis in Stokes space. A good approximation of for long fibers ($L \gg 10\,\text{m}$) is (Agrawal 2013)

$$\Delta\tau \approx D_{\text{p}}\sqrt{L}, \tag{5.58}$$

where D_{p} is the mean PMD parameter.

The PMD of fiber has been studied extensively (Brodsky et al. 2006), and the PMD for different fiber types can vary significantly. For example, the PMD of standard telecommunications fiber is on the order of $0.1\,\text{ps}/\sqrt{\text{km}}$ (Corning 2011) but that of erbium-doped fiber can be as high as $7.5\,\text{ps}/\sqrt{\text{km}}$ (Desurvire 2002). Signal fading due to fiber PMD is a problem for digital systems and had been likened to multipath effects (Poole et al. 1991). Relatively less work has been performed to describe the effects of PMD on analog systems, where signal fading and distortion can be induced by PMD (Poole and Darcie 1993). As mentioned previously, the effects of PMD can be complex. Other considerations include the frequency dependence of PMD ($d\vec{\tau}/d\omega$), termed second-order PMD, which can give rise to issues such as polarization-dependent CD, PCD (Kogelnik and Jopson 2002). In addition, PMD can interact with other nonlinearities (Menyuk and Marks 2006) and propagation effects. For example, an optical phase modulation can be converted to a polarization modulation by PMD, which can then be converted to intensity modulation via PDL (Huttner et al. 2000).

The PDL and PMD are rooted in the linear birefringence of fiber. Nonlinear birefringence can cause further problems, most notably XPolM. The nonlinear index introduced in Section 5.7 assumes a fixed polarization. More generally, the nonlinear part of Equation (5.50) can be broken into two polarization components (Agrawal 2013)

$$\Delta n_x = n_2\left(|E_x|^2 + \frac{2}{3}|E_y|^2\right), \quad \Delta n_y = n_2\left(\left|E_y\right|^2 + \frac{2}{3}|E_x|^2\right). \tag{5.59}$$

Equation (5.59) results in a nonlinear modal birefringence, $|n_x - n_y|$. In a multiwavelength system, intensity modulation on one channel can impart polarization modulation on other channels via the nonlinear birefringence. Numerous terms have been synonymously employed to describe this effect; those listed by Phillips et al. (2006) include XPolM, the optical Kerr effect (OKE), and XPM-induced depolarization. The analytical details of this effect beyond Equation (5.59) can become involved. There has been a lot of research focused on OKE for high-speed telecommunications applications (Collings and Boivin 2000, Winter et al. 2009). The effects of XPolM for microwave photonics applications have been studied by Campillo (2006) and Phillips et al. (2006). Study of PMD became important in digital systems as the data rate (frequency content) increased and polarization multiplexing techniques were employed. The same will hold for analog fiber optic links, where PMD effects will be impactful in very high frequency links and/or links employing polarization encoding.

REFERENCES

Agrawal, G. P., *Fiber-Optic Communications Systems*, 2nd edition, Wiley, New York (1997).

Agrawal, G., *Nonlinear Fiber Optics*, 5th edition, Academic Press (2013).

Antona, J. C., Bigo, S., and Kosmalski, S., "Nonlinear index measurements of various fibre types over C+L bands using four-wave mixing," in *Proceedings of the 27th European Conference on Optical Communications*, 270–271 (2001).

Armstrong, J. A., "Theory of interferometric analysis of laser phase noise," *Journal of Optical Society of America*, 56(8), 1024–1031 (1966).

Azana, J. and Muriel, M. A., "Real-time optical spectrum analysis based on time-space duality in chirped fiber gratings," *IEEE Journal of Quantum Electronics*, 36(5), 517–526 (2000).

Balkanski, M. and Wallis, R. F., *Semiconductor Physics and Applications*, Oxford (2000).

Boggio, M. C., Marconi, J. D., and Fragnito, H. L., "Experimental and numerical investigation of the SBS-threshold increase in an optical fiber by applying strain distributions," *Journal of Lightwave Technology*, 23(11), 3808–3814 (2005).

Bracewell, R. N., *The Fourier Transform and its Applications*, McGraw-Hill (2000).

Brodsky, M., Frigo, N. J., Boroditsky, M., Tur, M., "Polarization mode dispersion of installed fibers," *Journal of Lightwave Technology*, 24(12), 4584–4599 (2006).

Bucholtz, F., Urick, V. J., and Campillo, A. L., "Comparison of crosstalk for amplitude and phase modulation in an analog fiber optic link," in *Proceedings of the IEEE Topical Meeting on Microwave Photonics*, 66–69 (2004).

Campillo, A. L., "Interchannel nonlinear crosstalk in analog polarization modulated WDM systems," *Journal of Lightwave Technology*, 24(3), 1186–1193 (2006).

Campillo, A. L. and Bucholtz, F., "Increased nonlinear crosstalk in analog wavelength-division-multiplexed links due to the chromatic dispersion of optical filters," *IEEE Photonics Technology Letters*, 18(19), 2014–2016 (2006).

Campillo, A. L., Funk, E. E., Tulchinsky, D. A., Dexter, J. L., and Williams, K. J., "Phase performance of an eight-channel wavelength-division-multiplexed analog-delay line," *Journal of Lightwave Technology*, 22(2), 440–447 (2004).

Chraplyvy, A. R., "Limitations on lightwave communications imposed by optical-fiber nonlinearities," *Journal of Lightwave Technology*, 8(10), 1548–1557 (1990).

Collings, B. C. and Boivin, L., "Nonlinear polarization evolution induced b cross-phase modulation and its impact on transmission systems," *IEEE Photonics Technology Letters*, 12(11), 1582–1584 (2000).

Corning, Inc., "Single-Mode Dispersion Measurement Method," Accessed 21 January 2014: http://www.corning.com/WorkArea/downloadasset.aspx?id=7929 (2001).

Corning, Inc., "Corning SMF-28e+ Optical Fiber Product Information," Accessed 2 March 2013: http://www.corning.com/opticalfiber/products/SMF-28e+_fiber.aspx (2011).

Damask, J. N., *Polarization Optics in Telecommunications*, Springer (2005).

Desurvire, E., *Erbium-Doped Fiber Amplifiers Principles and Applications*, Wiley (2002).

Deventer, M. and Boot, A. J., "Polarization properties of stimulated Brillouin scattering in single-mode fibers," *Journal of Lightwave Technology*, 12(4), 585–590 (1994).

Devgan, P. S., Diehl, J. F., Urick, V. J., and Williams, K. J., "Long-haul microwave analog link with shot-noise-limited performance above the stimulated brillouin scattering threshold threshold," in *Proceedings of the IEEE Topical Meeting on Microwave Photonics*, 326–329 (2008).

Dorrer, C., "Temporal van Cittert-Zernike theorem and its application to the measurement of chromatic dispersion," *Journal of Optical Society of America B*, 21(8), 1417–1423 (2004).

Foschini, G. J. and Poole, C. D., "Statistical theory of polarization dispersion in single mode fibers," *Journal of Lightwave Technology*, 9(11), 1439–1456 (1991).

Foster, S., Tikhomirov, A., and Milnes, M., "Fundamental thermal noise in distributed feedback fiber lasers," *IEEE Journal of Quantum Electronics*, 43(5), 378–384 (2007).

Fotiadi, A. A. and Kiyan, R. V., "Cooperative stimulated Brillouin and Rayleigh backscattering process in optical fiber," *Optics Letters*, 23(23), 1805–1807 (1998).

Frigo, N. J., "A generalized geometrical representation of coupled mode theory," *IEEE Journal of Quantum Electronics*, 22(11), 2131–2140 (1986).

Frigo, N. J. and Bucholtz, F., "Geometric representation of optical propagation phase," *Journal of Lightwave Technology*, 27(15), 3283–3293 (2009).

Frigo, N. J., Bucholtz, F., and McLaughlin, C. V., "Polarization in phase modulated optical links: Jones- and generalized Stokes-space analysis," *Journal of Lightwave Technology*, 31(9), 1503–1511 (2013).

Gimlett, J. L. and Cheung, N. K., "Effects of phase-to-intensity noise conversion by multiple reflections on Gigabit-per-second DFB laser transmission systems," *Journal of Lightwave Technology*, 7(6), 888–895 (1989).

Gordon, J. P. and Kogelnik, H., "PMD fundamentals: polarization mode dispersion in optical fibers," *Proceedings of National Academy of Sciences*, 97(9), 4541–4550 (2000).

Hansryd, J., Dross, F., Westlund, M., Andrekson, P. A., and Knudsen, S. N., "Increase of the SBS threshold in a short highly nonlinear fiber by applying a temperature distribution," *Journal of Lightwave Technology*, 19(11), 1691–1697 (2001).

Huttner, B., Geiser, C., and Gisin, N., "Polarization-induced distortions in optical fiber networks with polarization-mode dispersion and polarization-dependent losses," *IEEE Journal of Selected Topics in Quantum Electronics*, 6(2), 317–329 (2000).

Karlsson, M., "Four-dimensional rotations in coherent optical communications," *Journal of Lightwave Technology*, 32(6), 1246–1257 (2014).

Kliger, D. S., Lewis, J. W., and Randall, C. E., *Polarized Light in Optics and Spectroscopy*, Academic (1990).

Kogelnik, H. and Jopson, R. M., "Polarization-mode dispersion," in *Optical Fiber Telecommunications IVB*, I. Kaminow and T. Li (editors), Academic (2002).

Koyama, F. and Iga, K., "Frequency chirping in external modulators," *Journal of Lightwave Technology*, 6(1), 87–93 (1988).

Lax, M., "Classical noise V: noise is self-sustained oscillators," *Physical Review*, 160(2), 290–300 (1967).

Mao, X. P., Tkach, R. W., Chraplyvy, A. R., Jopson, R. M., and Derosier, R. M., "Stimulated Brillouin threshold dependence on fiber type and uniformity," *IEEE Photonics Technology Letters*, 4(1), 66–69 (1992).

Marcuse, D., "Loss analysis of single-mode fiber splices," *The Bell System Technical Journal*, 56(5), 703–718 (1977).

Marshall, W. K., Crosignani, B., and Yariv, A., "Laser phase noise to intensity noise conversion by lowest-order group-velocity dispersion in optical fiber: exact theory," *Optics Letters*, 25(3), 165–167 (2000).

Mazurczyk, V. J. and Zyskind, J. L., "Polarization dependent gain in erbium doped-fiber amplifiers," *IEEE Photonics Technology Letters*, 6(5), 615–618 (1994).

McKinney, J. D. and Diehl, J. F., "Measurement of chromatic dispersion using the baseband radio-frequency response of a phase-modulated analog optical link employing a reference fiber," NRL Memorandum Report, NRL/MR/5652-07-9072 (2007).

McKinney, J. D. and Weiner, A. M., "Photonic synthesis of ultrabroadband arbitrary electromagnetic waveforms," in *Microwave Photonics*, C. H. Lee (editor), CRC Press, 243–306 (2013).

Menyuk, C. R. and Marks, B. S., "Interaction of polarization mode dispersion and nonlinearity in optical fiber transmission systems," *Journal of Lightwave Technology*, 24(7), 2806–2826 (2006).

Okusaga, O., Cahill, J., Docherty, A., Zhou, W., and Menuk, C. R., "Guided entropy mode Rayleigh scattering in optical fibers," *Optics Letters*, 37(4), 683–685 (2012).

Okusaga, O., Cahill, J., Docherty, A., Menuk, C. R., and Zhou, W., "Spontaneous inelastic Rayleigh scattering in optical fibers," *Optics Letters*, 38(4), 549–551 (2013).

Papoulis, A., *Probability, Random Variables, and Stochastic Processes*, McGraw-Hill (1991).

Papoulis, A., "Pulse compression, random variables, and diffraction: a unified approach," *Journal of Optical Society of America A*, 11(1), 3–13 (1994).

Phillips, M. R. and Ott, D. M., "Crosstalk due to optical fiber nonlinearities in WDM CATV lightwave systems," *Journal of Lightwave Technology*, 17(10), 1782–1792 (1999).

Phillips, M. R., Darcie, T. E., Marcuse, D., Bodeep, G. E., and Frigo, N. J., "Nonlinear distortion generated by dispersive transmission of chirped intensity-modulated signals," *IEEE Photonics Technology Letters*, 3, 481–483 (1991).

Phillips, M. R., Woodwad, S. L., and Smith, R. L., "Cross-polarization modulation: theory and measurement in subcarrier-modulated WDM systems," *Journal of Lightwave Technology*, 24(11), 4089–4099 (2006).

Phillips, M. R., Wu, K. Y., and Villarruel, F. X., "Four-wave-mixing-induced crosstalk and distortion in subcarrier-multiplexed lightwave links: theory and measurement," *Journal of Lightwave Technology*, 26(15), 2647–2652 (2008).

Poole, C. D. and Darcie, T. E., "Distortion related to polarization-mode dispersion in analog lightwave systems," *Journal of Lightwave Technology*, 11(11), 1749–1759 (1993).

Poole, C. D. and Wagner, R. E., "Phenomenological approach to polarization dispersion in long single-mode fibers," *Electronics Letters*, 22, 1029–1030 (1986).

Poole, C. D., Tkach, R. W., Chraplyvy, A. R., and Fishman, D. A., "Fading in lightwave systems due to polarization-mode dispersion," *IEEE Photonics Technology Letters*, 3(1), 68–70 (1991).

Rogge, M. S., Urick, V. J., Bucholtz, F., Williams, K. J., and Knapp, P., "Comparison of amplitude and phase modulation crosstalk in hyperfine WDM fiber optic links," in *CLEO Technical Digest*, paper CMH2 (2005).

Roman, J. E., Frankel, M. Y., Williams, K. J., and Esman, R. D., "Optical fiber cables for synchronous remoting of numerous transmitters/receivers," *IEEE Photonics Technology Letters*, 10(4), 591–593 (1998a).

Roman, J. E., Nichols, L. T., Williams, K. J., Esman, R. D., Tavik, G. C., Livingston, M., and Parent, M. G., "Fiber-optic remoting of ultrahigh dynamic range radar," *IEEE Transactions on Microwave Theory and Techniques*, 46(12), 2317–2323 (1998b).

Rossi, N., Serena, P., and Bononi, A., "Polarization-dependent loss impact on coherent optical systems in presence of fiber nonlinearity," *IEEE Photonics Technology Letters*, 26(4), 334–337 (2014).

Ruffin, A. B., Li, M. J., Chen, X., Kobyakov, A., and Annunziata, F., "Brillouin gain analysis for fibers with different refractive indices," *Optics Letters*, 32(23) 3123–3125 (2005).

Serizawa, Y., Myoujin, M., Miyazaki, S., and Kitamura, K., "Transmission delya variations in OPGW and overhead fiber-optic cable links," *IEEE Transactions on Power Delivery*, 12(4), 1415–1421 (1997).

Shibata, N., Braun, R. P., and Waarts, R. G., "Phase-mismatch dependence of efficiency of wave generation through four-wave mixing in a single-mode optical fiber," *IEEE Journal of Quantum Electronics*, 23(7), 1205–1210 (1987).

Shiraki, K., Ohashi, M., and Tateda, M., "SBS threshold of a fiber with aBrillouin frequency shift distribution," *Journal of Lightwave Technology*, 14(1), 50–57 (1996a).

Shiraki, K., Ohashi, M., and Tateda, M., "Performance of strain-free stimulated Brillouin scattering suppression fiber," *Journal of Lightwave Technology*, 14(4), 549–554 (1996b).

Smith, R. G., "Optical power handling capacity of low loss optical fibers as determined by stimulated Raman and Brillouin scattering," *Applied Optics*, 11(11), 2489–2494 (1972).

Stolen, R. H., "Phase-matched-stimulated four-photon mixing in silica-fiber waveguides," *IEEE Journal of Quantum Electronics*, 11(3), 100–103 (1975).

Stolen, R. H. and Ippen, E. P., "Raman gain in glass optical waveguides," *Applied Physics Letters*, 22(6), 276–278 (1973).

Sun, C. K., Anderson, G. W., Orazi, R. J., Berry, M. H., Pappert, S. A., and Shadaram, M., "Phase and amplitude stability of broadband fiber optic links," in *Proceedings of SPIE*, 2560, 50–56 (1995).

Sunderman, C. E. and Urick, V. J., "Fiber-optic propagation effects in long-haul HF/VHF/UHF analog photonic links," NRL Memorandum Report, NRL/MR/5650-14-9537 (2014).

Supradeepa, V. R., Long, C. M., Wu, R., Ferdous, F., Hamidi, E., Leaird, D. E., and Weiner, A. M., "Comb-based radiofrequency photonic filters with rapid tunability and high selectivity," *Nature Photonics*, 6(2), 186–194 (2012).

Tkach, R. W. and Chraplyvy, A. R.,"Phase noise and linewidth in an InGaAsP DFB laser," *Journal of Lightwave Technology*, 4(11), 1711–1716 (1986).

Urick, V. J., Swingen, L., Rogge, M. S., Campillo, A. L., Bucholtz, F., Dexter, J. L., and Knapp, P. F., "Design and characterization of long-haul single-channel intensity-modulated analog fiber-optic links," NRL Memorandum Report, NRL/MR/5650-05-8904 (2005).

Urick, V. J., Rogge, M. S., Knapp, P. F., Swingen, L., and Bucholtz, F., "Wideband predistortion linearization for externally-modulated long-haul analog fiber optic-links," *IEEE Transactions on Microwave Theory and Techniques*, 54(4), 1458–1463 (2006).

Urick, V. J., Diehl, J., Hastings, A., Sunderman, C., McKinney, J. D., Devgan, P. S., Dexter, J. L., and Williams, K. J., "Analysis of fiber-optic links for HF antenna remoting," NRL Memorandum Report, NRL/MR/5650-08-9101 (2008).

Urick, V. J., Bucholtz, F., McKinney, J. D., Devgan, P. S., Campillo, A. L., Dexter, J. L., and Williams, K. J., "Long-haul analog photonics," *Journal of Lightwave Technology*, 29(8), 1182–1205 (2011).

Urick, V. J., Singley, J. M., Sunderman, C. E., Diehl, J. F., and Williams, K. J., "Design and performance of Ka-band fiber-optic delay lines," Naval Research Laboratory Memorandum Report, NRL/MR/5650-12-9456 (2012).

Wan, P. and Conradi, J., "Impact of double Rayleigh backscatter noise on digital and analog fiber systems," *Journal of Lightwave Technology*, 14(3), 288–297 (1996).

Wang, J. and Petermann, K., "Small signal analysis for dispersive optical fiber communication systems," *Journal of Lightwave Technology*, 10(1), 96–100 (1992).

Wang, J. and Yao, J., "Large time-bandwidth product microwave arbitrary waveform generation using a spatially discrete chirped fiber Bragg grating," *Journal of Lightwave Technology*, 28(11), 1652–1660 (2010).

Weiner, A. M., *Ultrafast Optics*, John Wiley & Sons (2009).

White, S. D., "Implementation of a photonic automatic gain control system for correcting gain variations in the Green Bank Telescope fiber optic system," *Review of Scientific Instruments*, 71(8), 3196–3199 (2000).

Wilson, G. C., Wood, T. H., Zyskind, J. L., Sulhoff, J. W., Johnson, J. E., Tanbun-Ek, T., and Morton, P. A., "SBS and MPI suppression in analogue systems with integrated electroabsorption modulator/DFB laser transmitters," *Electronics Letters*, 32(16), 1502–1504 (1996).

Winter, M., Bunge, C. A., Setti, D., and Petermann, K., "A statistical treatment of cross-polarization modulation in DWDM systems," *Journal of Lightwave Technology*, 27(17), 3739–3751 (2009).

Yamamoto, S., Edagawa, N., Taga, H., Yoshida, Y., and Wakabayashi, H., "Analysis of laser phase noise to intensity noise conversion by chromatic dispersion in intensity modulation and direct detection optical-fiber transmission," *Journal of Lightwave Technology*, 8(11), 1716–1722 (1990).

Ye, J., Peng, J. L., Jones, R. J., Holman, K. W., Hall, J. L., Jones, D. J., Diddams, S. A., Kitching, J., Bize, S., Bergquist, J. C., Hollberg, L. W., Robertson, L., and Ma, L .S., "Delivery of high-stability optical and microwave frequency standards over an optical fiber network," *Journal of Optical Society of America B*, 20(7), 1459–1467 (2003).

Zhang, J. and Phillips, M. R., "Modeling intensity noise caused by stimulated Brillouin scattering in optical fibers," in *Proceedings of the Conference on Lasers and Electro-Optics Digest*, 140–142 (2005).

Zhu, T., Bao, X., Chen, L., Liang, H., and Dong, Y., "Experimental study on stimulated Rayleigh scattering in optical fibers," *Optics Express*, 18(22), 22958–22963 (2010).

CHAPTER 6

EXTERNAL INTENSITY MODULATION WITH DIRECT DETECTION

An analysis of an external intensity modulation direct-detection (IMDD) link employing a Mach–Zehnder modulator (MZM) is provided in this chapter. The IMDD format with an external MZM is arguably the most prevalent architecture presently in use and, therefore, receives a thorough treatment in this chapter. Section 6.1 provides a largely qualitative description of these links with the intention to provide some intuition into their operation. A thorough quantitative analysis of IMDD links with MZMs is given in Sections 6.2 and 6.3. The analysis focuses on a mathematical description of the relevant link performance metrics, and supporting experimental data are shown throughout. Section 6.3.4 introduces some of the design trade space for these links. Equations are presented in linear and decibel form, where the latter facilitates compact equations for link design and analysis. Photodetector effects that can limit the intrinsic performance are the topic of Sections 6.4. Schemes to linearize the modulation transfer function are reviewed in Section 6.5. The chapter concludes with an analysis of propagation effects in Section 6.6 that builds on the material in Chapter 5.

Fundamentals of Microwave Photonics, First Edition.
Vincent J. Urick Jr, Jason D. McKinney, and Keith J. Williams.

6.1 CONCEPT AND LINK ARCHITECTURES

The three core components to the IMDD links under consideration are a laser, a MZM, and a photodetector. All of these have been introduced in previous chapters. The MZM modulates the radio-frequency (RF) information onto the laser intensity, and the photodetector converts the signal back into the electrical domain. Consider firstly the two basic MZM structures shown in Figure 6.1. Light from the laser is coupled into an input waveguide that is split into two paths and then recombined at the output. Employing a Y-branch coupler on the output provides a single output [Figure 6.1(a)]. As shown in Figure 6.1(b), a directional coupler can be used to provide two outputs from an MZM. The most prevalent material for MZMs is the electro-optic $LiNbO_3$ crystal. However, MZMs have been constructed using a wide range of materials such as GaAs, Si, and polymers. The MZM structures shown in Figure 6.1 are simple Mach–Zehnder interferometers, as their name indicates. The interference between the two light beams at the output coupler depends on their relative phase at the recombination point. If the MZM is constructed using an electro-optic material, such as $LiNbO_3$, this relative phase can be controlled by applying a voltage across one or both arms of the interferometer. An applied voltage will change the index of refraction that will in turn change the phase. Photodiodes alone cannot demodulate phase modulation, so the MZM structure is used to convert electro-optic phase modulation into intensity modulation.

A typical transfer characteristic for an MZM is shown in Figure 6.2 (see also Figure 1.3). Shown in Figure 6.2 is the normalized optical transmission for a continuous-wave (CW) optical input as a function of applied electrical voltage for a two-output MZM. Full constructive interference occurs when the signals combine in phase and peak

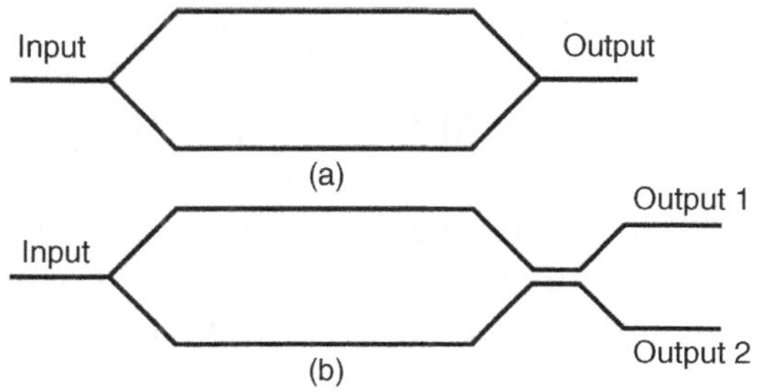

Figure 6.1. Two structures for a Mach–Zehnder modulator (MZM) using (a) a Y-branch output coupler and (b) a directional coupler on the output.

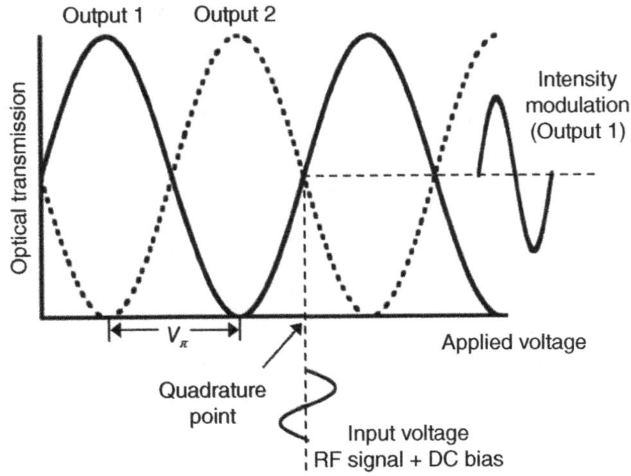

Figure 6.2. Typical MZM transfer curves.

transmission occurs at these points. Peak transmission on one MZM output corresponds to null transmission on the other output, that is, the full destructive interference points. At midpoints between the peak and null transmission, the quadrature points, a fairly linear response is obtained. A DC bias voltage can be applied to set the MZM at quadrature, and a reasonably linear optical intensity modulation will result when an RF signal is applied. The intensity modulation at the MZM output can then be easily converted back to an electrical signal via demodulation with a photodetector. The two MZM outputs carry intensity modulations that have a relative phase of 180°, which can be exploited in balanced-detection architectures.

Two methods of applying the RF signal to an MZM are shown in Figure 6.3. Both MZMs shown in Figure 6.3 are dual-output, but the methods apply to a single-output MZM as well. Shown in Figure 6.3(a) is a single-arm-drive MZM where the input RF signal is applied to only one arm of the MZM. A dual-drive or push–pull MZM is shown in Figure 6.3(b). For a dual-drive MZM, the input RF signal is split equally between the two arms, but the signals applied to each arm have a relative phase of 180°. The electric fields at the output of a single-arm and dual-drive MZM are compared in Section 6.2. Although the electrical fields exhibit distinct differences, it is shown that the RF photocurrents are of the same form neglecting any propagation effects.

Three types of IMDD links employing an external MZM are shown in Figures 6.4–6.6. In all of the links, a CW laser is connected to the input

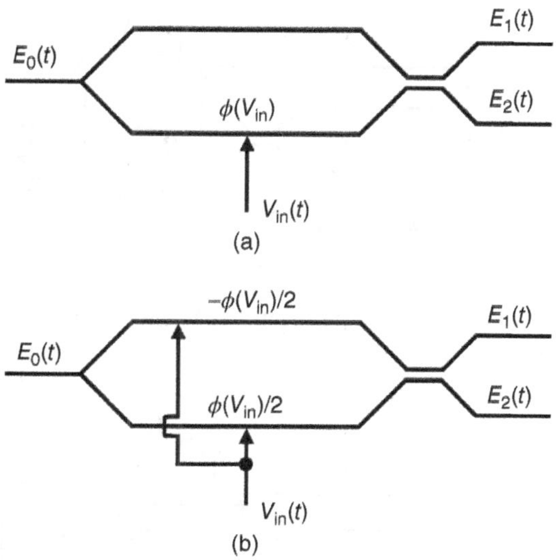

Figure 6.3. Schematics for (a) a single-arm-drive MZM and (b) a dual-drive or push–pull MZM.

of an MZM, typically via a polarization-maintaining (PM) fiber. The PM fiber is necessary to align the laser to the proper axis of the $LiNbO_3$ crystal. The electrical signal comprises a DC bias and an RF signal; these two voltages are often applied on separate electrodes, but a single electrical input can be employed with a bias tee. Shown in Figure 6.4 is a single-output link that is demodulated with a single photodiode. One output of a dual-output MZM can also be used to implement this type of link. Both outputs of a dual-output MZM can be used to construct a balanced links such as shown in Figures 6.5 and 6.6. As described throughout this chapter, balanced links can result in better performance than single-output links at the cost of requiring two RF-phase-matched fiber runs. The balanced link shown in Figure 6.5 employs photodiodes

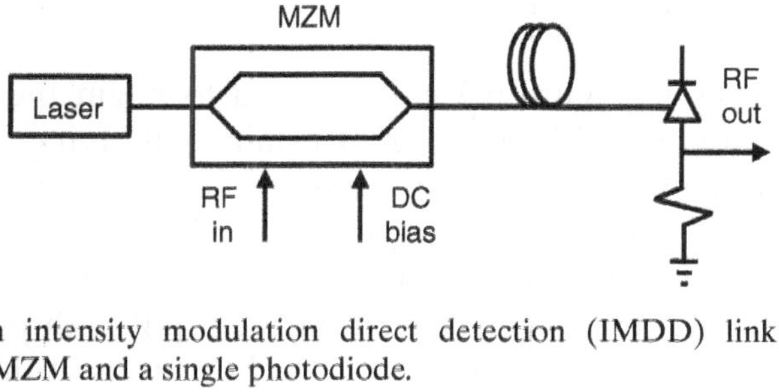

Figure 6.4. An intensity modulation direct detection (IMDD) link employing a single-output MZM and a single photodiode.

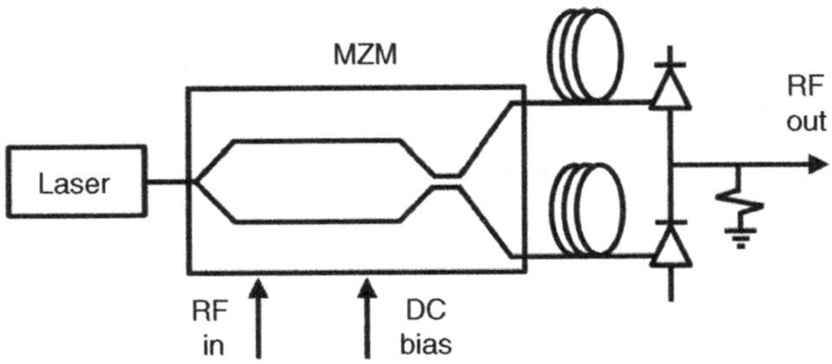

Figure 6.5. An IMDD link with a dual-output MZM and balanced detection via direct photocurrent subtraction.

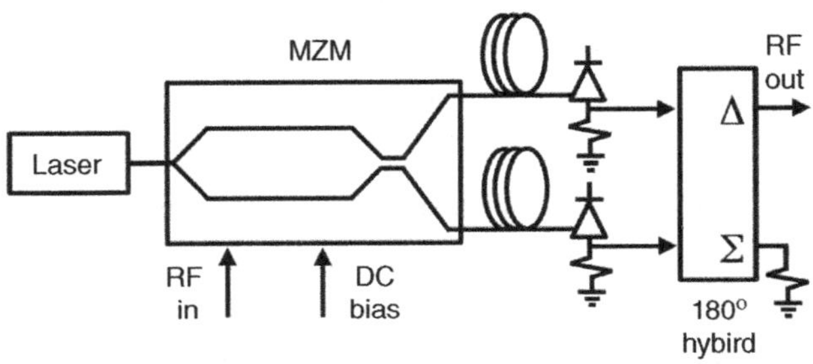

Figure 6.6. An IMDD link with a dual-output MZM and balanced detection using an RF hybrid coupler.

that are wired to subtract their two photocurrents directly before delivering the signal to a load. An RF hybrid coupler can also be used to subtract the photocurrents as depicted in Figure 6.6. The main differences between direct current subtraction and an RF hybrid coupler are that the latter will typically exhibit less bandwidth, is bulkier, and results in 3-dB more intrinsic loss. A quantitative analysis of these architectures is the focus of the remainder of this chapter.

6.2 SIGNAL TRANSFER AND GAIN

The architectures in Figure 6.3 will be used to begin the analysis of an externally modulated IMDD link. Certain simplifying assumptions are made throughout. Firstly, it is assumed that a scalar field analysis can be employed, that is, optical polarization effects can be ignored. A $LiNbO_3$ MZM is highly polarization dependent, but it is assumed that polarization-maintaining fiber is used on the input with proper alignment. Secondly, it is assumed that the coupling ratios at the input and output of the MZM are precisely 50%. For typical commercial MZMs, this is a safe assumption in terms of the RF metrics derived in this case. Nonideal coupling can be treated with a transfer matrix-formalism as used in the following section and has been studied elsewhere (Betts 2002). The two arms of the MZM are assumed to have zero relative time delay unless a voltage is applied; in practice, this may not be the case. However, this would result in a small static phase shift, which is trivial to include in the differential phase shift and nonconsequential in terms of the analysis given in the following section. For a push–pull MZM, it is assumed that the voltages applied to each arm are of equal amplitude and have a relative phase of precisely 180°.

Transfer matrices will be employed for the analysis of a MZM. A short general description of this technique is described in this section before applying the simplifying assumptions summarized previously. Fiber optic couplers are often described with a transfer matrix. A general matrix for a 2 × 2 coupler can be written as (Priest 1982)

$$\begin{bmatrix} E_{\text{out},1} \\ E_{\text{out},2} \end{bmatrix} = \begin{bmatrix} A & B \\ C & D \end{bmatrix} \begin{bmatrix} E_{\text{in},1} \\ E_{\text{in},2} \end{bmatrix}, \tag{6.1}$$

where $E_{\text{in},1}$ and $E_{\text{in},2}$ are the fields at the input of the coupler, $E_{\text{out},1}$ and $E_{\text{out},2}$ are the fields at the output of the coupler, and A, B, C, and D can be complex. If back reflections can be neglected and the coupler is symmetric, Equation (6.1) can be rewritten as a product of an overall transmission factor, $\sqrt{a}$, and a unitary matrix (Moslehi et al. 1984):

$$\begin{bmatrix} E_{\text{out},1} \\ E_{\text{out},2} \end{bmatrix} = \sqrt{a} \begin{bmatrix} \sqrt{1-K} & i\sqrt{K} \\ i\sqrt{K} & \sqrt{1-K} \end{bmatrix} \begin{bmatrix} E_{\text{in},1} \\ E_{\text{in},2} \end{bmatrix}, \tag{6.2}$$

where K is the coupling coefficient. In the case of an ideal, lossless 50/50 coupler, $a = 1$ and $K = 1/2$. The analysis that follows assumes such couplers on the input and output of the MZM. The phase shifts in the two arms of the MZM can be described by a phase matrix (Weihs et al. 1996)

$$\begin{bmatrix} \sqrt{a_1}F(\phi_1) & 0 \\ 0 & \sqrt{a_2}F(\phi_2) \end{bmatrix}, \tag{6.3}$$

where a_1 and a_2 are the loss factors for each arm, and $F(\phi_1)$ and $F(\phi_2)$ are functions that describe the phase evolution through the arms. Only relative phase shifts are carried through in the following analysis, and it is assumed that $a_1 = a_2$. Furthermore, the phase matrix for an ideal push–pull MZM used in the following section exhibits $F(\phi_1) = F^*(\phi_2)$ when the F's describe relative phase differences.

The initial input voltage for the analysis is a single RF tone with a DC bias

$$V_{\text{in}}(t) = V_{\text{dc}} + V_{\text{rf}} \sin(\Omega t), \tag{6.4}$$

where V_{dc} is the bias voltage, and Ω is the angular radio frequency of the signal with peak voltage V_{rf}. The field in the fiber at the optical input to the MZM is taken to be CW and is written as

$$E_0(t) = \sqrt{\frac{2P_{\text{o}}}{A}} \left(\frac{\mu}{\varepsilon}\right)^{\frac{1}{4}} e^{i\omega t}, \tag{6.5}$$

where A is the cross-sectional area of the fiber core, ε is the permittivity of the fiber core, μ is the permeability of the fiber core, P_{o} is the average optical power, and ω is angular optical frequency. Equation (6.5) is written in a such a way that

$$P_{\text{o}} = \frac{A}{2} \left(\frac{\varepsilon}{\mu}\right)^{\frac{1}{2}} E_0^* E_0, \tag{6.6}$$

which is how optical power will be calculated from complex fields in the following section.

The first step in the analysis is to consider the electric fields at the outputs of the MZMs shown in Figure 6.3. Both electric fields are calculated for a dual-output MZM, noting that one of the outputs can describe a single-output MZM. Given the aforementioned assumptions, the transfer matrix for the MZM shown in Figure 6.3(a) can be written as

$$\begin{bmatrix} E_1(t) \\ E_2(t) \end{bmatrix} = \frac{\sqrt{l_{\text{mzm}}}}{2} \begin{bmatrix} 1 & i \\ i & 1 \end{bmatrix} \begin{bmatrix} e^{i\phi(t)} & 0 \\ 0 & 1 \end{bmatrix} \begin{bmatrix} 1 & i \\ i & 1 \end{bmatrix} \begin{bmatrix} E_0(t) \\ 0 \end{bmatrix}, \tag{6.7}$$

where l_{mzm} is the optical power loss factor for the MZM, and ϕ is the relative phase shift between the two arms of the MZM. The phase shift is

$$\phi(t) = \phi_{\text{dc}} + \phi_{\text{rf}} \sin(\Omega t), \tag{6.8}$$

where $\phi_{\text{dc}} = \pi V_{\text{dc}}/V_\pi(0)$ and $\phi_{\text{rf}} = \pi V_{\text{rf}}/V_\pi(\Omega)$, with V_π being the frequency-dependent voltage required to produce π radians relative phase shift in the MZM as shown in Figure 6.2. Inserting Equations (6.5) and (6.8) into Equation (6.7) yields

$$\begin{bmatrix} E_1(t) \\ E_2(t) \end{bmatrix} = \frac{\gamma}{2} e^{i\omega t} \begin{bmatrix} e^{i\phi(t)} - 1 \\ ie^{i\phi(t)} + i \end{bmatrix}, \tag{6.9}$$

where constants have been collected into γ,

$$\gamma = \sqrt{\frac{2l_{\rm mzm}P_{\rm o}}{A}}\left(\frac{\mu}{\varepsilon}\right)^{\frac{1}{4}}. \tag{6.10}$$

The fields in Equation (6.9) can be expanded in terms of Bessel functions (with the aid of Appendices IV and VI) to yield

$$\begin{aligned}\mathrm{Re}\,(E_1) = & -\frac{\gamma}{2}\cos(\omega t) + \frac{\gamma}{2}\cos(\phi_{\rm dc})\sum_{k=-\infty}^{\infty} J_k(\phi_{\rm rf})\cos[(\omega + k\Omega)t] \\ & -\frac{\gamma}{2}\sin(\phi_{\rm dc})\sum_{m=-\infty}^{\infty} J_m(\phi_{\rm rf})\sin[(\omega + m\Omega)t],\end{aligned} \tag{6.11}$$

$$\begin{aligned}\mathrm{Re}\,(E_2) = & -\frac{\gamma}{2}\sin(\omega t) - \frac{\gamma}{2}\cos(\phi_{\rm dc})\sum_{k=-\infty}^{\infty} J_k(\phi_{\rm rf})\sin[(\omega + k\Omega)t] \\ & -\frac{\gamma}{2}\sin(\phi_{\rm dc})\sum_{m=-\infty}^{\infty} J_m(\phi_{\rm rf})\cos[(\omega + m\Omega)t],\end{aligned} \tag{6.12}$$

where J_n are nth-order Bessel functions of the first kind (Appendix VI). Equations (6.11) and (6.12) describe a dual-sideband modulation with an infinite number of sidebands spaced symmetrically around the optical carrier. The spacing is dictated by the RF modulation frequency, and the amplitude of each term is determined by the RF drive voltage in the argument of the Bessel functions. An important aspect of these fields as compared to a push–pull MZM is their dependence on the DC bias. In Equations (6.11) and (6.12), modulation sidebands at the same frequency appear in both summations. The total power in any given modulation sideband is proportional to $(\sin^2\phi_{\rm dc} + \cos^2\phi_{\rm dc})$ and, therefore, cannot be adjusted by changing the MZM bias. However, the power in the optical carrier can be adjusted, as there are three terms that contribute. For example, the optical carrier can be extinguished in Equation (6.11) when $\phi_{\rm dc} = 2n\pi$, with n being an integer; the zeroing condition for Equation (6.12) is $\phi_{\rm dc} = (2n + 1)\pi$. This characteristic of Equations (6.11) and (6.12) can be explained qualitatively, in that the optical carrier is present in both arms of the interferometer, whereas the modulation sidebands are generated only in one arm. Therefore, there can be no interference at the modulation sideband frequencies.

The push–pull MZM shown in Fig. 6.2(b) does exhibit modulation in both arms of the interferometer and can be described by the following transfer matrix

$$\begin{bmatrix} E_1(t) \\ E_2\ (t) \end{bmatrix} = \frac{\sqrt{l_{\mathrm{mzm}}}}{2} \begin{bmatrix} 1 & i \\ i & 1 \end{bmatrix} \begin{bmatrix} e^{i\phi(t)/2} & 0 \\ 0 & e^{-i\phi(t)/2} \end{bmatrix} \begin{bmatrix} 1 & i \\ i & 1 \end{bmatrix} \begin{bmatrix} E_0(t) \\ 0 \end{bmatrix}. \tag{6.13}$$

Carrying out Equation (6.13) yields

$$\begin{bmatrix} E_1(t) \\ E_2(t) \end{bmatrix} = \frac{\gamma}{2} e^{i\omega t} \begin{bmatrix} e^{i\phi(t)/2} - e^{-i\phi(t)/2} \\ ie^{i\phi(t)/2} + ie^{-i\phi(t)/2} \end{bmatrix}. \tag{6.14}$$

The resulting electric fields can be written as

$$\begin{aligned} \mathrm{Re}(E_1) = & -\gamma \sin\left(\frac{\phi_{\mathrm{dc}}}{2}\right) \sum_{k=-\infty}^{\infty} J_{2k}\left(\frac{\phi_{\mathrm{rf}}}{2}\right) \sin\left[(\omega + 2k\Omega)t\right] \\ & + \gamma \cos\left(\frac{\phi_{\mathrm{dc}}}{2}\right) \sum_{m=-\infty}^{\infty} J_{2m+1}\left(\frac{\phi_{\mathrm{rf}}}{2}\right) \cos\left[(\omega + (2m+1)\Omega)t\right]. \end{aligned} \tag{6.15}$$

$$\begin{aligned} \mathrm{Re}(E_2) = & -\gamma \cos\left(\frac{\phi_{\mathrm{dc}}}{2}\right) \sum_{k=-\infty}^{\infty} J_{2k}\left(\frac{\phi_{\mathrm{rf}}}{2}\right) \sin\left[(\omega + 2k\Omega)t\right] \\ & + \gamma \sin\left(\frac{\phi_{\mathrm{dc}}}{2}\right) \sum_{m=-\infty}^{\infty} J_{2m+1}\left(\frac{\phi_{\mathrm{rf}}}{2}\right) \cos\left[(\omega + (2m+1)\Omega)t\right]. \end{aligned} \tag{6.16}$$

In Equations (6.15) and (6.16), the optical carrier and the even-order modulation sidebands have the same dependence on ϕ_{dc}, which is different for the odd-order sidebands. Therefore, all frequency components depend on the DC bias in a push–pull MZM. A diagram related to Equation (6.15) is plotted in Figure 4.8, showing the relative positions of phasors for each frequency component at quadrature.

In the absence of any propagation effects, it will be shown that the photocurrents from a single-arm and push–pull MZM are of the same form. The photocurrent will be calculated as $I(t) = \Re P_{\mathrm{o}}$ where $\Re$ is the photodiode responsivity and P_{o} is given by Equation (6.6). As mentioned in Chapter 3, the photodiode responsivity is often defined as $\Re = \eta q/(h\nu)$, where η is the photodiode quantum efficiency, q is the electron charge constant, h is Planck's constant, and ν is the optical frequency. Strictly speaking, this definition of $\Re$ is correct only when all of the optical power in the fiber core is delivered to the photodiode. If power is lost between the fiber and the photodiode, an effective responsivity can be employed, $\mathfrak{R}_{\text{eff}} = \alpha_{\text{pd}}\eta q/(h\nu)$, where α_{pd} would account for the power loss ($\alpha_{\text{pd}} < 1$). Applying these definitions to Equations (6.9) and (6.14) and inserting into Equation (6.6) yield the photocurrents

$$I_{1,2}(t) = \frac{\mathfrak{R} g_{\text{o}} l_{\text{mzm}} P_{\text{o}}}{2}[1 \mp \cos\phi(t)], \qquad (6.17)$$

where the subscripts "1" and "2" correspond to the "−" and "+" signs, respectively. In this case, g_{o} was added to account for net optical gain (or loss) between the MZM and photodiode. Note, E_1 in Equation (6.9) and E_1 in Equation (6.14) produce the same I_1 in Equation (6.17). The same is true for E_2. Furthermore, either E_1 or E_2 can be chosen as the output field for the single-output MZM case. Therefore, the analysis will continue in terms of the RF metrics for a dual output MZM using Equation (6.17) as the current sourced from the photodiode.

Equation (6.17) can be expanded by inserting Equation (6.8) and using Anger's series (Appendix VI):

$$\begin{aligned} I_{1,2}(t) = {} & I_{\text{dc,q}} \mp \cos(\phi_{\text{dc}}) I_{\text{dc,q}} J_0(\phi_{\text{rf}}) \\ & \mp 2\cos(\phi_{\text{dc}}) I_{\text{dc,q}} \sum_{k=1}^{\infty} J_{2k}(\phi_{\text{rf}}) \cos(2k\Omega t) \\ & \pm 2\sin(\phi_{\text{dc}}) I_{\text{dc,q}} \sum_{m=0}^{\infty} J_{2m+1}(\phi_{\text{rf}}) \sin[(2m+1)\Omega t], \end{aligned} \qquad (6.18)$$

where $I_{\text{dc,q}} = \mathfrak{R} g_{\text{o}} l_{\text{mzm}} P_{\text{o}}/2$ is the DC photocurrent at quadrature bias, defined as $\phi_{\text{dc}} = (2n+1)\pi$, with n being an integer. The performance as a function of DC bias is written explicitly in Equation (6.18) as it will be used in this chapter to described even-order distortions as small deviations from quadrature. Equation (6.18) will also be revisited in Section 8.2, where the performance of an MZM will be analyzed at large deviations from quadrature. Applying the quadrature bias condition to Equation (6.18) yields

$$I_{1,2}(t) = I_{\text{dc}} \pm 2I_{\text{dc}} \sum_{m=0}^{\infty} J_{2m+1}(\phi_{\text{rf}}) \sin[(2m+1)\Omega t], \qquad (6.19)$$

where $I_{\text{dc,q}}$ was replaced with I_{dc}.

The photocurrent term at the fundamental drive frequency is of importance for the analysis, which can be written as

$$I_{\Omega,\text{single}} = 2I_{\text{dc}}J_1(\phi_{\text{rf}})\sin(\Omega t) \tag{6.20}$$

for a single photodiode. In Equation (6.20), the positive current was chosen for convenience. For balanced photodiodes, the photocurrent is

$$I_{\text{bal}}(t) = I_2(t) - I_1(t) = 4I_{\text{dc}}\sum_{m=0}^{\infty} J_{2m+1}(\phi_{\text{rf}})\sin[(2m+1)\Omega t], \tag{6.21}$$

noting that $I_1 - I_2$ would result in a negative current of the same magnitude. The fundamental term in Equation (6.21) is

$$I_{\Omega,\text{bal}} = 4I_{\text{dc}}J_1(\phi_{\text{rf}})\sin(\Omega t). \tag{6.22}$$

Equations (6.20) and (6.22) become equivalent if I_{dc} in Equation (6.20) is replaced with $(2I_{\text{dc}})$. That is, the equations are equivalent if the DC current per photodiode in Equation (6.22) is replaced with the sum of the DC photocurrents from each photodiode. Unless otherwise noted, I_{dc} will be the DC photocurrent for the single photodiode case or the sum of the two DC photocurrents for balanced photodiodes. This convention allows for application of the equations to either MZM architecture.

The average RF output power at the fundamental frequency is calculated as

$$P_{\Omega} = \langle I_{\Omega}^2 \rangle R_{\text{o}} |H_{\text{pd}}|^2, \tag{6.23}$$

where R_{o} is the output resistance, $H_{\text{pd}}(f)$ is the photodiode circuit filter function, and $\langle\rangle$ denotes a time average. Introduced in Chapter 3, H_{pd} is intended to capture the frequency-dependent response of the photodiode itself and the circuit between the p–i–n junction and the load. Ideally, $H_{\text{pd}} = 1$ but this is not realized in practice at microwave frequencies. Wideband photodiodes typically employ an impedance-matching circuit that results a current division between the matching circuit and the load. Each of the IMDD links shown in Figures 6.4–6.6 employs such a circuit. Neglecting the component frequency responses, the links in Figures 6.4 and 6.5 exhibit $H_{\text{pd}} = 1/2$ due to current splitting in the matching circuit. This current split will result in a 6-dB RF power loss given that $P_{\Omega} \propto I_{\Omega}^2$. The balanced link in Figure 6.6 that uses a 180° hybrid results in $H_{\text{pd}} = 1/(2\sqrt{2})$, a factor of 1/2 for the impedance-matching circuits and an additional

$1/\sqrt{2}$ to account for intrinsic loss in the 180° hybrid coupler. Inserting Equation (6.20) into Equation (6.23) gives

$$P_\Omega = 2I_{dc}^2 J_1^2(\phi_{rf}) R_o |H_{pd}|^2 \tag{6.24}$$

as the output power for an externally modulated IMDD link. The response given by Equation (6.24) is dictated by a first-order Bessel function, which is analyzed in the following section.

The small-signal or linearized response is obtained from Equation (6.24) by applying the small-signal approximation for the Bessel function, $J_n(\phi_{rf}) \approx \phi_{rf}^n/(2^n n!)$ for $\phi_{rf} \ll 1$, which translates to $V_{rf} \ll V_\pi/\pi$ in this case. The resulting small-signal output power at the fundamental is

$$P_{\Omega\,ss} = \frac{I_{dc}^2 \phi_{rf}^2 R_o |H_{pd}|^2}{2}. \tag{6.25}$$

The small-signal RF gain factor [Equation (2.6)] for the link can be obtained from Equation (6.25) by inserting $\phi_{rf}^2 = 2\pi^2 P_{in} R_i / V_\pi^2$, where P_{in} is the average RF input power and R_i is the input resistance,

$$g = \frac{I_{dc}^2}{V_\pi^2} \pi^2 R_i R_o |H_{pd}|^2. \tag{6.26}$$

For a given value of V_π, the IMDD link can act as an RF amplifier if enough optical power is injected into the link, that is, if I_{dc} is large enough. A decibel form of Equation (6.26) is very useful for link design:

$$G[\mathrm{dB}] = -22.1 + 20\log\left(\frac{I_{dc}\,[\mathrm{mA}]}{V_\pi[\mathrm{V}]}\right), \tag{6.27}$$

where $R_i = R_o = 50\ \Omega$ and $H_{pd} = 1/2$ were used. Equation (6.27) would apply to the links in Figures 6.4 and 6.5, if the sum of the two photocurrents was employed for the latter. Equation (6.27) is plotted in Figure 6.7, demonstrating the achievable gain values for certain I_{dc} and V_π.

Shown in Figure 6.8 is a measured data set demonstrating the change in G for a balanced link compared to a single-output link. These data were obtained at the intermediate frequency (IF) from a 12-GHz photonic down-converting link in single-output (Figure 6.4) and balanced (Figure 6.5) configurations (Williams and Esman 1996). Each output arm was operated at a DC photocurrent of 30 mA. Therefore, $I_{dc} = 30$ mA for the single-ended configurations and $I_{dc} = 60$ mA for

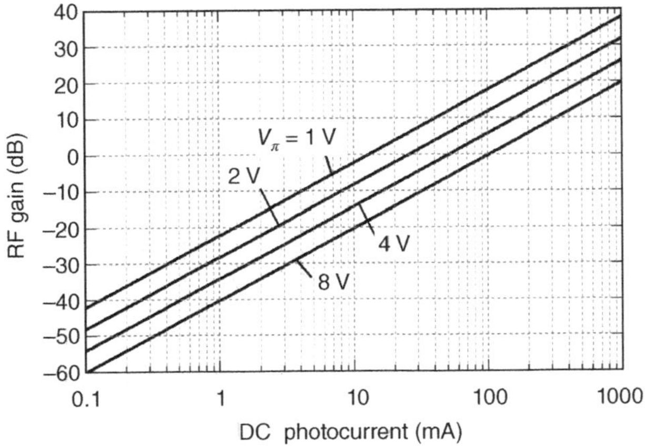

Figure 6.7. Calculated small-signal gain for an IMDD link as given by Equation (6.26) with $R_i = R_o = 50\ \Omega$ and $H_{pd} = 1/2$.

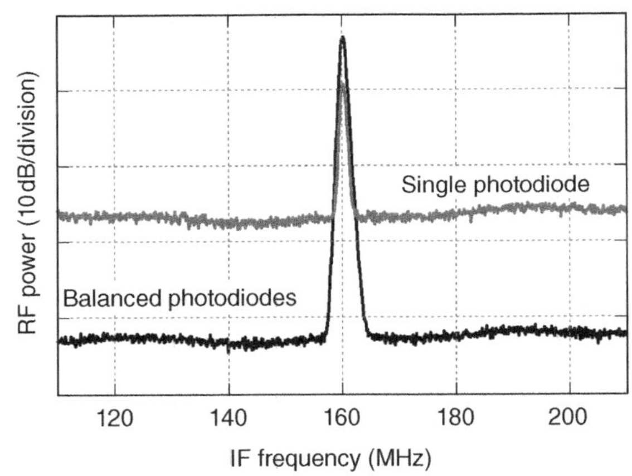

Figure 6.8. Measured outputs of a 12-GHz down-converting link using a single photodiode (gray) and balanced photodiodes (black) (Williams and Esman 1996).

the balanced link. As shown in Figure 6.8, the balanced link exhibits about 6-dB more gain than the single-ended link employing either output. Equation (6.27) agrees with this observation. This increase in gain comes at the costs of a more-complicated receiver and the need to phase match the two transmission spans. A balanced configuration does have other advantages though; also shown in Figure 6.8 is a significant (about 15.5 dB) change in the noise floor for the balanced link. This is because balanced link rejects "common-mode" noise, a concept that will be further described in Section 6.3.1.

Other qualities of Equation (6.24) that are important from a system point of view are the maximum output power and the output power at 1-dB compression. The maximum output is obtained at the maximum of the first-order Bessel function,

$$P_{\text{out,max}} = 2I_{\text{dc}}^2 J_1^2(\phi_{\max}) R_{\text{o}} |H_{\text{pd}}|^2. \tag{6.28}$$

Equation (6.28) can be evaluated using the values in Appendix VI for J_1, $J_1(\phi_{\max}) = J_1(1.8412) = 0.5819$. For $R_{\text{o}} = 50\ \Omega$ and $H_{\text{pd}} = 1/2$, Equation (6.28) can be written as

$$P_{\text{out,max}}\,[\text{dBm}] = -20.7 + 20 \log\,(I_{\text{dc}}[\text{mA}]). \tag{6.29}$$

The input power to achieve $P_{\text{out,max}}$ can be determined from $\phi_{\max} = V_{\text{in,max}}\,\pi/V_\pi$, where $V_{\text{in,max}}$ is the voltage to achieve $P_{\text{out,max}}$,

$$P_{\text{in,max}} = \frac{V_\pi^2 \phi_{\max}^2}{2\pi^2 R_{\text{i}}}. \tag{6.30}$$

With $\phi_{\max} = 1.8412$ and $R_{\text{i}} = 50\ \Omega$, Equation (6.30) can be written as

$$P_{\text{in,max}}\,[\text{dBm}] = 5.4 + 20 \log\,(V_\pi[\text{V}]). \tag{6.31}$$

Specifying the compression dynamic range (CDR) will require an expression for the output power at 1-dB compression, which can be written as

$$P_{\text{out,1dB}} = 2I_{\text{dc}}^2 J_1^2(\phi_{\text{1dB}}) R_{\text{o}} |H_{\text{pd}}|^2, \tag{6.32}$$

where ϕ_{1dB} is the peak phase shift at 1-dB compression. This quantity can be determined by setting the ratio of the linearized response [Equation (6.25)] to the actual response [Equation (6.24)] to 1 dB:

$$\frac{P_{\Omega,\text{ss}}}{P_\Omega} = 10^{0.1} \Rightarrow \frac{\phi_{\text{1dB}}^2}{J_1^2(\phi_{\text{1dB}})} = 4 \times 10^{0.1}. \tag{6.33}$$

Equation (6.33) can be solved numerically to obtain $\phi_{\text{1dB}} = 0.9504$, which can then be used to evaluate $J_1(0.9504) = 0.4235$. Inserting this value along with $R_{\text{o}} = 50\ \Omega$ and $H_{\text{pd}} = 1/2$ into Equation (6.32) gives a decibel form of the output power at 1-dB compression:

$$P_{\text{out,1dB}}[\text{dBm}] = -23.5 + 20 \log\,(I_{\text{dc}}[\text{mA}]). \tag{6.34}$$

The input compression point ($P_{in,1dB}$) is also an important metric and can be obtained from the definition of $\phi_{1dB} = V_{in,1dB}\pi/V_\pi$, where $V_{in,1dB}$ is the voltage to achieve 1-dB compression on the output. This expression can be rewritten as

$$P_{in,1dB} = \frac{V_\pi^2 \phi_{1dB}^2}{2\pi^2 R_i}. \tag{6.35}$$

Inserting $\phi_{1dB} = 0.9504$ and $R_i = 50\ \Omega$ into Equation (6.35) results in

$$P_{in,1dB}\ [\text{dBm}] = -0.4 + 20 \log\left(V_\pi[\text{V}]\right). \tag{6.36}$$

The aforementioned equations for compression and maximum powers assume that the MZM is the limiting factor. In practice, other limiting components may present themselves, particularly the photodetector.

Shown in Figure 6.9 is the measured RF output power as a function of RF input power for an IMDD link (Urick et al. 2011), illustrating some of the signal properties discussed to this point. A link such as that shown in Figure 6.4 was employed to generate these data. The average current was $I_{dc} = 2.4$ mA, and the modulation frequency was 6 GHz where $V_\pi = 2.6$ V. The output was matched to $R_o = 50\ \Omega$ and $H_{pd} = 1/2$. The shape of the curve in Figure 6.4 is governed by the first-order Bessel function as described by Equation (6.24). Equation (6.24) is plotted with the measured data demonstrating good agreement. The link compression is apparent in Figure 6.4, but the cyclic nature of the Bessel function is more clearly shown in Figure 2.5. The normalized data in Figure 2.5 were

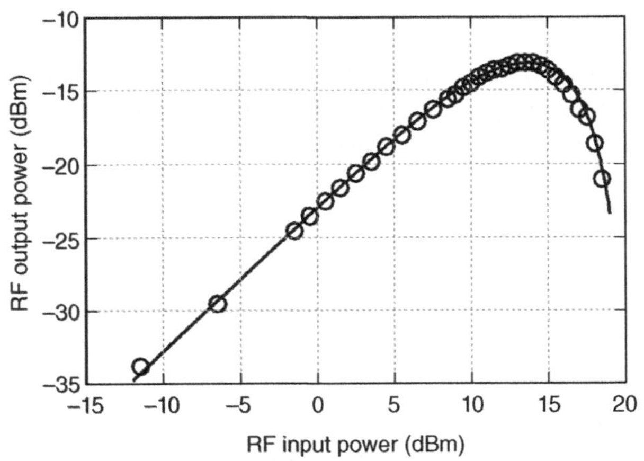

Figure 6.9. Measured (symbols) and calculated (line) RF transfer curve for an IMDD link (Urick et al. 2011).

collected with an IMDD link employing an external MZM and therefore also exhibit a response governed by Equation (6.24). Figure 6.9 demonstrates the power transfer measures discussed to this point. In the linear regime, the measured RF gain is −23 dB, which agrees nicely with Equation (6.27) where $G = -22.8$ dB. The maximum output power for the link is −13 dBm at an input power of 13.5 dBm. These results are in line with $P_{\text{out,max}} = -13.1$ dBm and $P_{\text{in,max}} = 13.7$ dBm as predicted by Equations (6.29) and (6.31), respectively. The link compresses 1 dB with an output power of −16 dBm occurring at an input power of 7.5 dBm. Equations (6.34) and (6.36) predict $P_{\text{out,1dB}} = -15.9$ dBm and $P_{\text{in,1dB}} = 7.9$ dBm, respectively.

The aforementioned single-tone analysis will allow for specifying the RF gain, noise figure, and CDR for an IMDD link. As described in Chapter 2, the linearity of an RF system is often characterized by a two-tone test. Therefore, the analysis in this section continues by considering an input signal comprising two tones:

$$V_{\text{in}}(t) = V_{\text{dc}} + V_1 \sin(\Omega_1 t) + V_2 \sin(\Omega_2 t). \tag{6.37}$$

The phase shift corresponding to Equation (6.37) is

$$\phi(t) = \phi_{\text{dc}} + \phi_1 \sin(\Omega_1 t) + \phi_2 \sin(\Omega_2 t). \tag{6.38}$$

The fields for a two-tone modulation can be obtained by inserting Equation (6.38) into Equations (6.9) or (6.14). The field expansions for a two-tone drive results in lengthy expressions, and only one example is included in this case. The field, E_1 in Equation (6.14), for a push–pull MZM can be written as

$$\begin{aligned}
\text{Re}(E_1) = &-\gamma \sin\left(\frac{\phi_{\text{dc}}}{2}\right) J_0\left(\frac{\phi_1}{2}\right) J_0\left(\frac{\phi_2}{2}\right) \sin(\omega t) \\
&+ \gamma \cos\left(\frac{\phi_{\text{dc}}}{2}\right) J_0\left(\frac{\phi_2}{2}\right) \sum_{m=-\infty}^{\infty} J_{2m+1}\left(\frac{\phi_1}{2}\right) \\
&\cos[\omega t + (2m+1)\Omega_1 t] \\
&+ \gamma \cos\left(\frac{\phi_{\text{dc}}}{2}\right) J_0\left(\frac{\phi_1}{2}\right) \sum_{u=-\infty}^{\infty} J_{2u+1}\left(\frac{\phi_2}{2}\right) \\
&\cos[\omega t + (2u+1)\Omega_2 t] \\
&- \gamma \sin\left(\frac{\phi_{\text{dc}}}{2}\right) J_0\left(\frac{\phi_2}{2}\right) \sum_{\substack{k=-\infty \\ k\neq 0}}^{\infty} J_{2k}\left(\frac{\phi_1}{2}\right) \sin(\omega t + 2k\Omega_1 t)
\end{aligned}$$

$$
\begin{aligned}
&-\gamma \sin\left(\frac{\phi_{dc}}{2}\right) J_0\left(\frac{\phi_1}{2}\right) \sum_{\substack{j=-\infty \\ j\neq 0}}^{\infty} J_{2j}\left(\frac{\phi_2}{2}\right) \sin(\omega t + 2j\Omega_2 t) \\
&\pm \gamma \sin\left(\frac{\phi_{dc}}{2}\right) \sum_{r=-\infty}^{\infty} \sum_{s=-\infty}^{\infty} J_{2r+1}\left(\frac{\phi_1}{2}\right) J_{2s+1}\left(\frac{\phi_2}{2}\right) \\
&\qquad \sin\left[\omega t + ((2r+1)\Omega_1 t \mp (2s+1)\Omega_2 t)\right] \\
&-\gamma \sin\left(\frac{\phi_{dc}}{2}\right) \sum_{\substack{j=-\infty \\ j\neq 0}}^{\infty} \sum_{\substack{k=-\infty \\ k\neq 0}}^{\infty} J_{2k}\left(\frac{\phi_1}{2}\right) J_{2j}\left(\frac{\phi_2}{2}\right) \\
&\qquad \sin\left[\omega t + (2k\Omega_1 t \pm 2j\Omega_2 t)\right] \\
&\pm \gamma \cos\left(\frac{\phi_{dc}}{2}\right) \sum_{\substack{q=-\infty \\ q\neq 0}}^{\infty} \sum_{u=-\infty}^{\infty} J_{2q}\left(\frac{\phi_1}{2}\right) J_{2u+1}\left(\frac{\phi_2}{2}\right) \\
&\qquad \cos\left[\omega t + (2q\Omega_1 t \pm (2u+1)\Omega_2 t)\right] \\
&\pm \gamma \cos\left(\frac{\phi_{dc}}{2}\right) \sum_{m=-\infty}^{\infty} \sum_{\substack{n=-\infty \\ n\neq 0}}^{\infty} J_{2m+1}\left(\frac{\phi_1}{2}\right) J_{2n}\left(\frac{\phi_2}{2}\right) \\
&\qquad \cos\left[\omega t + (2n\Omega_2 t \pm (2m+1)\Omega_1 t)\right]. \qquad (6.39)
\end{aligned}
$$

The first line in Equation (6.39) is the optical carrier, whereas the second and third lines are the upper and lower sidebands due to the odd-order fundamentals and harmonics at f_1 and f_2, respectively. The fourth and fifth lines account for the even-order harmonics. The sixth and seventh lines correspond to even-order IMD; odd-order IMD terms are contained in the eighth and ninth lines. Note that the IMD terms, lines six through nine, contain two terms each as indicated by the ± inside the trigonometric arguments and at the beginning of the lines. The two-tone fields for a single-arm-drive and a push–pull MZM follow the same trends as for the single-tone case in the small-signal limit. This point is illustrated in Figure 6.10 where measured spectra are shown for the two types of MZMs (Rogge et al. 2007). For all data in Figure 6.10, two equal-amplitude tones at 9 and 12 GHz were applied to the MZM, and the spectrum was measured with an optical spectrum analyzer. Shown in Figure 6.10(a) are the measured spectra at the output of a single-arm-drive MZM for quadrature and null bias. At quadrature, the modulation sidebands at 9- and 12-GHz offsets

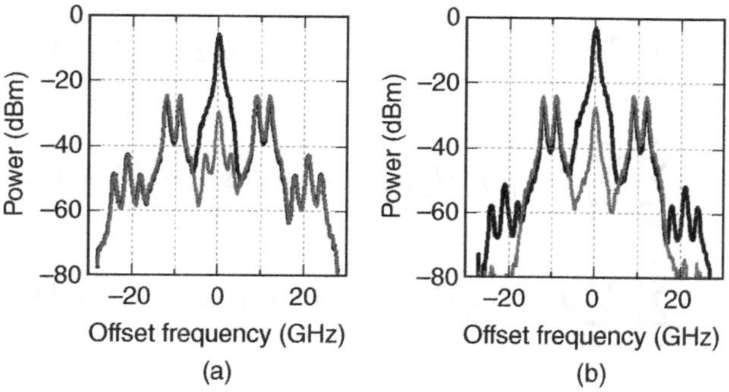

Figure 6.10. Measured optical spectra for IMDD links driven with two tones at 9 and 12 GHz. Shown are operation at quadrature (black) and null bias (gray) using (a) a single-arm-drive MZM and (b) a push–pull MZM (Rogge et al. 2007).

on either side of the carrier are clearly resolved. The second-order harmonics at ±18 and ±24 GHz are also shown along with the IMD2 at ±21 GHz (see also Figure 4.7). At null bias, about 24-dB suppression of the carrier frequency is observed with negligible change of the modulation terms, as predicted by Equations (6.11) and (6.12). The other IMD2 terms at ±3 GHz are also unveiled at null bias, as they were masked by the carrier at quadrature bias. Two spectra for a push–pull MZM are shown in Figure 6.10(b). At quadrature bias, the same frequency components observed in Figure 6.10(a) are present. However, the null-biased spectrum is much different. With a carrier suppression of about 25 dB, the even-order terms are no longer present, and the fundamental sidebands grow by about 3 dB. These characteristics are all predicted by Equation (6.39).

Equation (6.38) can be inserted into Equation (6.17) and expanded to calculate the photocurrents. The expansion is a bit more involved than the single-tone case and results in an expression that can be separated into three parts:

$$I_{1,2}^{\text{dc}} = I_{\text{dc,q}} \mp I_{\text{dc,q}} J_0(\phi_1) J_0(\phi_2) \cos(\phi_{\text{dc}}). \tag{6.40a}$$

$$I_{1,2}^{\text{odd}}(t) = \pm 2 \sin(\phi_{\text{dc}}) I_{\text{dc,q}} \times \left\{ J_0(\phi_2) \sum_{j=0}^{\infty} J_{2j+1}(\phi_1) \sin[(2j+1)\Omega_1 t] \right.$$
$$\left. + J_0(\phi_1) \sum_{k=0}^{\infty} J_{2k+1}(\phi_2) \sin[(2k+1)\Omega_2 t] \right.$$

$$
\begin{aligned}
&-\sum_{j=0}^{\infty}\sum_{m=1}^{\infty} J_{2j+1}(\phi_1)J_{2m}(\phi_2)\sin\left[(2m\Omega_2-(2j+1)\Omega_1)t\right]\\
&-\sum_{k=0}^{\infty}\sum_{h=1}^{\infty} J_{2k+1}(\phi_2)J_{2h}(\phi_1)\sin\left[(2h\Omega_1-(2k+1)\Omega_2)t\right]\\
&+\sum_{j=0}^{\infty}\sum_{m=1}^{\infty} J_{2j+1}(\phi_1)J_{2m}(\phi_2)\sin\left[(2m\Omega_2+(2j+1)\Omega_1)t\right]\\
&\left.+\sum_{k=0}^{\infty}\sum_{h=1}^{\infty} J_{2k+1}(\phi_2)J_{2h}(\phi_1)\sin\left[(2h\Omega_1+(2k+1)\Omega_2)t\right]\right\}.
\end{aligned}
\tag{6.40b}
$$

$$
\begin{aligned}
I_{1,2}^{\text{even}}(t) = \mp 2\cos(\phi_{\text{dc}})I_{\text{dc,q}} \times &\left\{ J_0(\phi_2)\sum_{k=1}^{\infty} J_{2k}(\phi_1)\cos(2k\Omega_1 t)\right.\\
&+J_0(\phi_1)\sum_{m=1}^{\infty} J_{2m}(\phi_2)\cos(2m\Omega_2 t)\\
&-\sum_{n=0}^{\infty}\sum_{p=0}^{\infty} J_{2n+1}(\phi_1)J_{2p+1}(\phi_2)\cos\left[((2p+1)\Omega_2-(2n+1)\Omega_1)t\right]\\
&+\sum_{n=0}^{\infty}\sum_{p=0}^{\infty} J_{2n+1}(\phi_1)J_{2p+1}(\phi_2)\cos\left[((2p+1)\Omega_2+(2n+1)\Omega_1)t\right]\\
&+\sum_{k=1}^{\infty}\sum_{m=1}^{\infty} J_{2k}(\phi_1)J_{2m}(\phi_2)\cos\left[2(m\Omega_2-k\Omega_1)t\right]\\
&\left.+\sum_{k=1}^{\infty}\sum_{m=1}^{\infty} J_{2k}(\phi_1)J_{2m}(\phi_2)\cos\left[2(m\Omega_2+k\Omega_1)t\right]\right\}.
\end{aligned}
\tag{6.40c}
$$

Equation (6.40a) is the average current for each arm, where the term $I_{\text{dc,q}}$ is again used to identify the DC current at quadrature bias [$\phi_{\text{dc}} = (2n+1)\pi$ with n being an integer]. The odd-order terms including the two fundamentals are collected in Equation (6.40b), and the even-order distortion is given by Equation (6.40c). The single summations include the harmonic distortion, whereas the double summations represent intermodulation distortion (IMD). The even order terms are zero at precisely quadrature. However, the bias dependence is included in Equation 6.40a–6.40c to analyze the performance as a function of small deviations from quadrature.

The largest small-signal distortion terms at quadrature are third-order IMD (IMD3) products at frequencies $(2f_2 - f_1)$ and $(2f_1 - f_2)$ given by the third and fourth lines of Equation (6.40b), respectively. The average power at small-signal for these terms can be obtained by applying the small-angle Bessel approximation:

$$P_{\mathrm{IMD3,ss}} = \frac{I_{\mathrm{dc}}^2 \phi_{\mathrm{rf}}^6 R_{\mathrm{o}} |H_{\mathrm{pd}}|^2}{128}. \tag{6.41}$$

The small-signal fundamental response for the two-tone drive is equivalent to the single-tone drive as given by Equation (6.25). The OIP3 [Equation (2.18)] for the distortion in Equation (6.41) can be obtained by equating Equation (6.25) to Equation (6.41) and solving for ϕ_{rf}, which results in $\phi_{\mathrm{rf}} = 2\sqrt{2}$. Inserting this value for ϕ_{rf} into Equation (6.25) or (6.41) yields

$$\mathrm{OIP3} = 4I_{\mathrm{dc}}^2 R_{\mathrm{o}} |H_{\mathrm{pd}}|^2. \tag{6.42}$$

Sometimes, third-harmonic distortion is employed to characterize the linearity of an IMDD link. An analysis similar to that mentioned previously can be used to show that the third-harmonic OIP3 is a factor of three larger than Equation (6.42). Equation (6.42) can be put into decibel form with $R_{\mathrm{o}} = 50\ \Omega$ and $H_{\mathrm{pd}} = 1/2$:

$$\mathrm{OIP3\ [dBm]} = -13.0 + 20\log\left(I_{\mathrm{dc}}[\mathrm{mA}]\right). \tag{6.43}$$

The IIP3 [Equation (2.20)] is obtained by dividing Equation (6.42) by Equation (6.26):

$$\mathrm{IIP3} = \frac{4V_{\pi}^2}{\pi^2 R_{\mathrm{i}}}, \tag{6.44}$$

which can be converted to decibel form for $R_{\mathrm{i}} = 50\ \Omega$:

$$\mathrm{IIP3\ [dBm]} = 9.1 + 20\log\left(V_{\pi}[\mathrm{V}]\right). \tag{6.45}$$

Measured data used to determine the OPI3 for an IMDD link are shown in Figure 6.11 (Williams et al. 1998). The link employed a dual-output MZM and balanced photodiodes with $H_{\mathrm{pd}} = 1/2$ at 1 GHz to match the output to 50 Ω. Each photodiode was operated at 30 mA average current. Two input signals of equal amplitude at 1.0 and 1.1 GHz were employed for the stimulus. The results in Figure 6.11 show the measured responses for the fundamental and IMD3 obtained

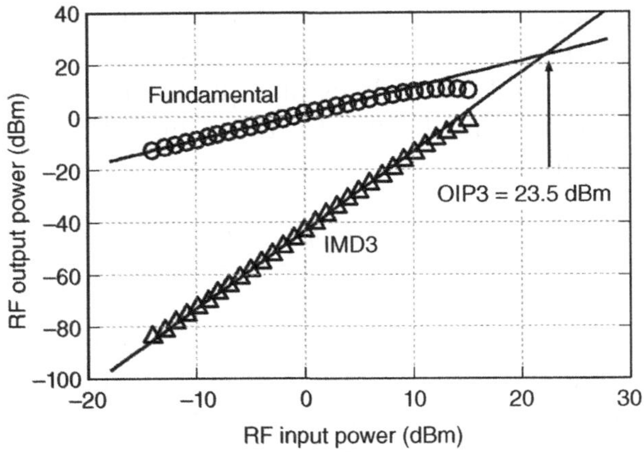

Figure 6.11. Measured OIP3 for an IMDD link (Williams et al. 1998).

with an electrical spectrum analyzer. The measured OIP3 = 23.5 dBm, as compared to OIP3 = 22.6 dBm predicted by Equation (6.43) with $I_{\mathrm{dc}} = 60$ mA. The 0.9-dB discrepancy between measurement and theory is within the experimental uncertainty.

An exact quadrature bias is difficult to obtain in practice, and it is therefore useful to analyze the even-order distortion as a function of bias phase. The largest even-order distortion is the second-order IMD (IMD2) at frequencies $|f_2 - f_1|$ and $(f_2 + f_1)$ given by the fifth and sixth lines of Equation (6.40c), respectively. The average small-signal power for these terms is

$$P_{\mathrm{IMD2,ss}} = \frac{\cos^2(\phi_{\mathrm{dc}}) I_{\mathrm{dc,q}}^2 \phi_{\mathrm{rf}}^4 R_{\mathrm{o}} |H_{\mathrm{pd}}|^2}{8}. \tag{6.46}$$

As given by Equation (6.40b), the fundamental power will scale as $\sin^2(\phi_{\mathrm{dc}})$; multiplying Equation (6.25) with $\sin^2(\phi_{\mathrm{dc}})$ and inserting $I_{\mathrm{dc}} = I_{\mathrm{dc,q}}$ give the fundamental output power as a function of ϕ_{dc}. Equating this expression to Equation (6.46) yields $\phi_{\mathrm{rf}} = 2\tan(\phi_{\mathrm{dc}})$. Inserting this value for ϕ_{rf} into Equation (6.46) gives the OIP2 as a function of ϕ_{dc}

$$\mathrm{OIP2}(\phi_{\mathrm{dc}}) = \frac{2\sin^4(\phi_{\mathrm{dc}}) I_{\mathrm{dc,q}}^2 R_{\mathrm{o}} |H_{\mathrm{pd}}|^2}{\cos^2(\phi_{\mathrm{dc}})}. \tag{6.47}$$

It can be shown that the second-harmonic OIP2 is four times larger than Equation (6.47), a result that was also seen in Chapter 2 for the Taylor series expansion of a two-tone signal. Similarly to the fundamental

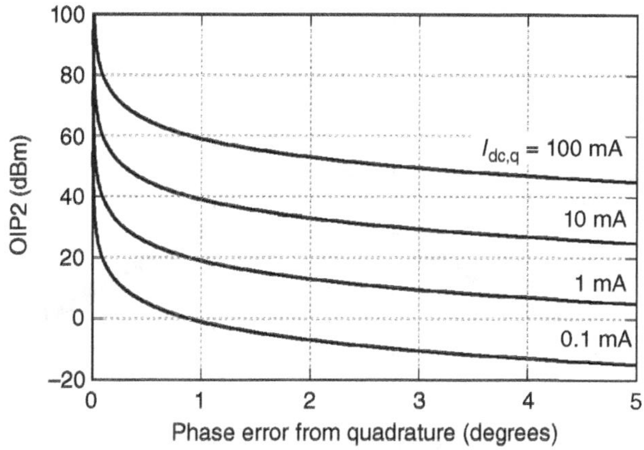

Figure 6.12. Calculated OIP2 for an IMDD link as given by Equation (6.47) with R_o = 50 Ω and H_{pd} = 1/2.

power, the RF gain as a function of ϕ_{dc} will also scale as $\sin^2(\phi_{dc})$. Dividing Equation (6.47) by Equation (6.26) multiplied by $\sin^2(\phi_{dc})$ gives the IIP2 as

$$\text{IIP2}(\phi_{dc}) = \frac{2\tan^2(\phi_{dc})V_\pi^2}{\pi^2 R_i}. \tag{6.48}$$

Equation (6.47) is plotted in Figure 6.12 for R_o = 50 Ω and H_{pd} = 1/2. The horizontal axis is the magnitude of the phase error from quadrature. The curve is symmetric about quadrature, and the OIP2 decreases significantly for even modest bias phase error. In multioctave systems, bias errors can degrade the overall spurious-free dynamic range (SFDR) as compared to the intrinsic third-order SFDR limited by the MZM. The total SFDR for an IMDD link will be analyzed in the following section. The theory given by Equation (6.47) is compared to measured results in Figure 6.13 (Urick et al. 2012). The experimental parameters for these data are $I_{dc,q}$ = 5.2 mA, R_o = 50 Ω, and H_{pd} = 1/2, which are also used for the calculation. The agreement between the measurement and the theory is excellent. The maximum measured OIP2 level of about 35 dBm is limited by photodiode distortion.

6.3 NOISE AND PERFORMANCE METRICS

The most important signal characteristics of the IMDD link have been derived starting from transfer matrices in the preceding section. Performance metrics such as noise figure, signal-to-noise ratio (SNR), and

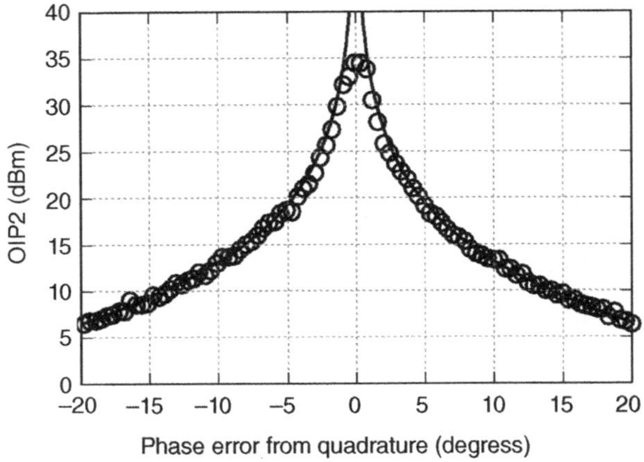

Figure 6.13. Measured (symbols) and calculated (line) OIP2 as a function of MZM bias error (Urick et al. 2012).

dynamic range depend on those properties. However, the noise in an IMDD link is also critical to such metrics. Therefore, the treatment now turns to an analysis of noise leveraging the material in Chapter 3 when necessary.

6.3.1 General Equations

An expression for the output noise power spectral density for an IMDD link using an external quadrature-biased MZM can we written as

$$N_{\mathrm{out}} = \frac{I_{\mathrm{dc}}^2}{V_\pi^2}\pi^2 R_{\mathrm{i}} R_{\mathrm{o}} |H_{\mathrm{pd}}|^2 k_{\mathrm{B}} T_{\mathrm{s}} + k_{\mathrm{B}} T_{\mathrm{s}} + 2qI_{\mathrm{dc}} R_{\mathrm{o}} |H_{\mathrm{pd}}|^2$$

$$+ \mathrm{RIN} \cdot I_{\mathrm{dc}}^2 R_{\mathrm{o}} |H_{\mathrm{pd}}|^2. \tag{6.49}$$

As in the previous section, I_{dc} is the total detected DC current allowing Equation (6.49) to be applied to any of the links in Figures 6.4–6.6. The first term in Equation (6.49) is due to thermal noise (Section 3.2) at the input of the link multiplied by the small-signal gain [Equation (6.26)] in order to translate it to the link output. The second term accounts for thermal noise at the link output. Both thermal noise expressions are given in terms of the standard noise temperature such that Equation (6.49) can be used to calculate the standard noise figure. There are some situations where additional terms may be needed to describe thermal noise. For example, lossy electrodes or thermal noise originating in the termination resistor can result in additional additive terms (Cox and Ackerman 2013). The third term in Equation (6.49) quantifies the contribution from shot noise (Section 3.3). The final term captures optical noise in excess of the fundamental thermal and shot noise sources as captured by the relative intensity noise (RIN) metric. Equation (3.4) defines the RIN, which can include contributions from laser noise (Section 3.4), optical amplification (Section 3.5), or propagation effects (Chapter 5). The varied sources of noise manifest differently in an IMDD link, and some noise sources can be mitigated with a dual-output MZM.

Shown in Figure 6.14 is a data set demonstrating a major advantage of the balanced IMDD architecture employing a dual-output MZM. Three RIN curves are shown in Figure 6.14—two for either arm alone in an IMDD link employing a dual-output MZM and one in the balanced configuration (Urick 2007). The link employed a semiconductor laser amplified by an erbium-doped fiber amplifier (EDFA) before the MZM. Either single-arm configuration operated at $I_{dc} = 50$ mA, while the balanced configuration operated at 100 mA total average detected current. Therefore, the RIN data is normalized to four times more average power for the balanced link as compared to the single-arm links. This is appropriate because the RF gain for a balanced link is larger by the same factor. The total link length was 309 m, and the two paths were well matched with a relative time delay of 30 ps. There are numerous sources of noise in this particular link; the dominant ones are laser

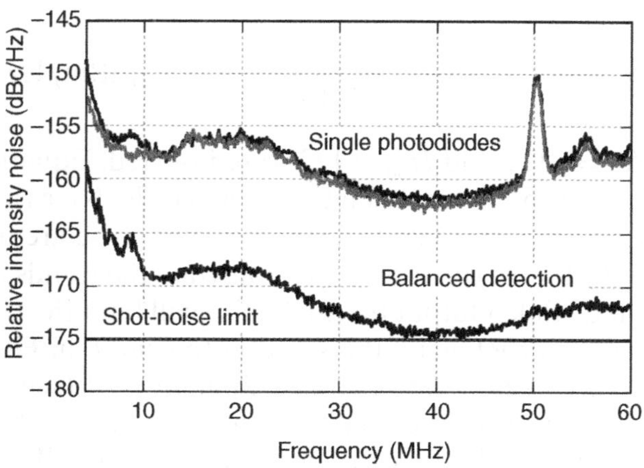

Figure 6.14. Measured relative intensity noise (RIN) for an IMDD link measured with both single outputs (top) and with balanced detection (bottom). The calculated shot noise limit is also shown (Urick 2007).

noise, noise from the EDFA, electrical noise, and shot noise. Because the laser noise and the noise from the EDFA are introduced before the modulation stage, they are referred to as common-mode noise. The electrical noise and shot noise are not common mode. A balanced link can cancel common-mode noise if the amplitude and phase of each arm are well matched. The peak near 50 MHz in the single-arm links was determined to originate from the laser and is rejected by about 25 dB in the balanced link. The remainder of the spectrum for the balanced link is about 12 dB below the single-arm spectra. This represents rejection of the white noise arising from the EDFA. The RIN level for the shot noise at 100 mA as given by Equation (3.27) is −175 dBc/Hz as marked in Figure 6.14. A balanced link cannot achieve better than shot-noise-limited performance. Finally, the remaining noise above the shot noise level for the balanced configuration is electrical noise input to the MZM that is not common mode and therefore cannot be removed by balanced detection. The metrics for single-arm and balanced links take the same form, but the terms contained in N_{out} may differ depending on the architectures under consideration.

The noise factor for an IMDD link can be obtained from the standard definition [Equation (2.6)]:

$$F = \frac{V_\pi^2 N_{\text{out}}}{I_{\text{dc}}^2 \pi^2 R_{\text{i}} R_{\text{o}} |H_{\text{pd}}|^2 k_{\text{B}} T_{\text{s}}}, \tag{6.50}$$

where N_{out} is given by Equation (6.49) and Equation (6.26) was inserted for the gain factor. Equation (6.50) can be rewritten in decibel form as

$$\text{NF[dB]} = 196.1 - 20 \log \left(\frac{I_{\text{dc}}\,[\text{mA}]}{V_\pi [\text{V}]} \right) + N_{\text{out}}\ [\text{dBm/Hz}], \tag{6.51}$$

where $R_{\text{i}} = R_{\text{o}} = 50\ \Omega$, and $H_{\text{pd}} = 1/2$ were inserted. Equation (6.51) is plotted in Figure 6.15. Shown in Figure 6.15(a) is the NF as a function of I_{dc} for various V_π values. Only thermal and shot noise are included for Figure 6.15(a) (RIN = 0) in order to demonstrate the trends in a "fundamentally limited" link. Output thermal noise will limit the performance only at low values of RF gain; the slope of the curve in Figure 6.15(a) for this limit is $m = -2$. In the shot noise limit, $m = -1$ (see also Section 6.3.2). If the RF gain is large enough to achieve the input-thermal-noise limit, then $m = 0$ and NF → 0 dB. The limit of NF → 0 dB is due to the definition of noise figure and the expression for input thermal noise in Equation (6.49). Other sources of thermal noise such as described by Cox and Ackerman (2013) may limit the minimum NF to a value larger

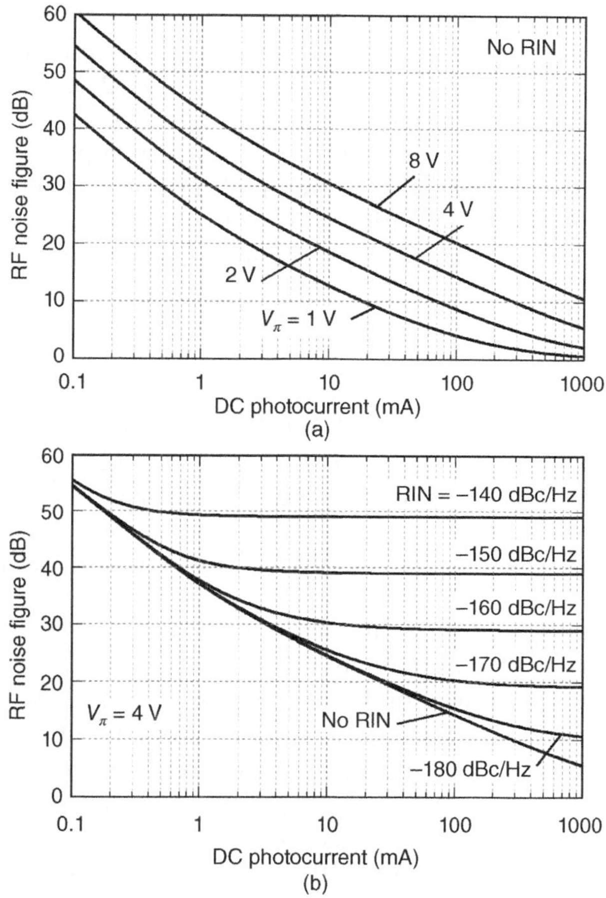

Figure 6.15. Calculated noise figure as given by Equation (6.51). Shown are (a) the effect of V_π with RIN = 0 and (b) the effect of RIN with V_π = 4 V.

than 0 dB. Shown in Figure 6.15(b) are curves for a fixed $V_\pi = 4$ V and different RIN values. As can be seen in this case, even modest RIN can seriously degrade the fundamental NF particularly at large photocurrents.

SNR was introduced in the context of noise figure in Equation (2.8). The SNR at the output of an RF component or system is

$$\mathrm{SNR_{out}} = \frac{P_{\mathrm{out}}}{N_{\mathrm{out}} B}. \tag{6.52}$$

The equations in Section 6.2 allow for four SNR values to be calculated for an IMDD link using Equation (6.52) with Equation (6.49) inserted for N_{out}. Inserting Equation (6.24) gives a general expression for the SNR, using Equation (6.25) gives the SNR in the linear regime, the maximum SNR is determined by inserting Equation (6.28), and the SNR at 1-dB compression is obtained with Equation (6.32).

The dynamic range for an IMDD link is a very important metric for many analog applications. The CDR and SFDR were defined generally in Section 2.3. Inserting Equation (6.32) into the definition of 1-dB CDR [Equation (2.14)] and calculating the numerical constants yield

$$\mathrm{CDR}_{1\mathrm{dB}} = \frac{(0.4516)I_{\mathrm{dc}}^2 R_{\mathrm{o}} |H_{\mathrm{pd}}|^2}{N_{\mathrm{out}} B} \tag{6.53}$$

for the 1-dB CDR of an IMDD link, where N_{out} is given by Equation (6.49). Using $R_o = 50\ \Omega$, $H_{pd} = 1/2$, and $B = 1$ Hz, Equation (6.53) can be converted to decibel form as

$$\mathrm{CDR}_{1\mathrm{dB}}[\mathrm{dB}\cdot\mathrm{Hz}] = -22.5 + 20\log\left(I_{\mathrm{dc}}[\mathrm{mA}]\right) - N_{\mathrm{out}}[\mathrm{dBm/Hz}]. \tag{6.54}$$

Equation (6.54) is plotted in Figure 6.16 for $R_i = 50\ \Omega$. Shown in Figure 6.16(a) is the fundamentally limited $\mathrm{CDR}_{1\mathrm{dB}}$ as a function of I_{dc} for different values of V_π. The $\mathrm{CDR}_{1\mathrm{dB}}$ increases with I_{dc} and is independent of V_π until contributions from input thermal noise become appreciable. As shown in Figure 6.16(a), the V_π value will set the maximum attainable $\mathrm{CDR}_{1\mathrm{dB}}$, which will be achieved at a lower I_{dc} for a lower V_π. The maximum $\mathrm{CDR}_{1\mathrm{dB}}$ is determined by the input-thermal-noise limit. When thermal noise originating at the link input is the dominant noise source at the output, the dynamic range essentially becomes "clamped" because the noise floor and output power at 1-dB compression [Equation (6.32)] both scale as I_{dc}^2. This limit can be expressed mathematically by inserting the first term in Equation (6.49), which is the output noise power spectral density due to input thermal noise, into Equation (6.53):

$$\mathrm{CDR}_{1\mathrm{dB,max}} = \frac{(0.4516)V_\pi^2}{\pi^2 R_{\mathrm{i}} k_{\mathrm{B}} T_{\mathrm{s}} B}. \tag{6.55}$$

As described previously, the NF is optimized by minimizing V_π, so there are important tradeoffs between dynamic range and NF. The RIN level can also cut into the intrinsic $\mathrm{CDR}_{1\mathrm{dB}}$ for an IMDD link. The effect of RIN on $\mathrm{CDR}_{1\mathrm{dB}}$ is shown in Figure 6.16(b) for a fixed $V_\pi = 4$ V.

For many applications, the SFDR is more important and typically smaller than the $\mathrm{CDR}_{1\mathrm{dB}}$. The SFDR for an ideal IMDD link is

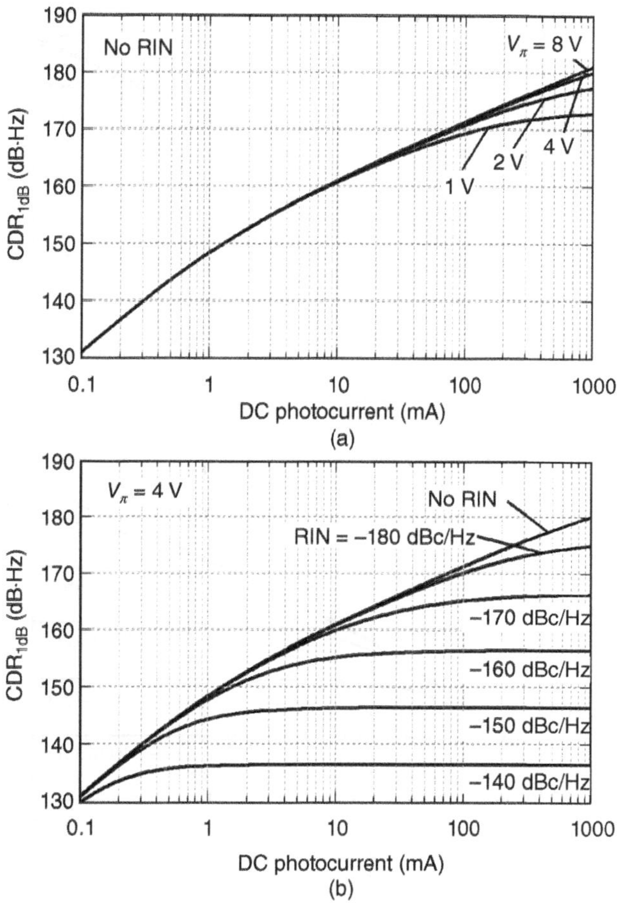

Figure 6.16. Calculated compression dynamic range as given by Equation (6.54) with $R_i = 50\ \Omega$. Shown are (a) the effect of V_π with RIN = 0 and (b) the effect of RIN with $V_\pi = 4$ V.

third-order limited and can be obtained by inserting Equation (6.42) into Equation (2.16):

$$\mathrm{SFDR}_3 = \left(\frac{4I_{dc}^2 R_o \left| H_{pd} \right|^2}{N_{out} B} \right)^{2/3}, \qquad (6.56)$$

where N_{out} is given by Equation (6.49). A useful decibel version for this expression can be written as

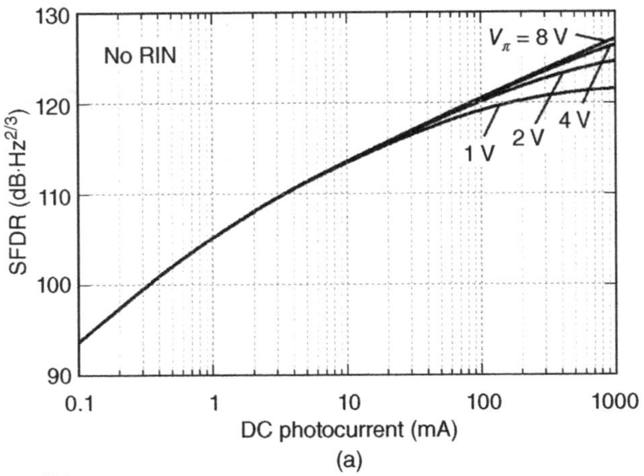

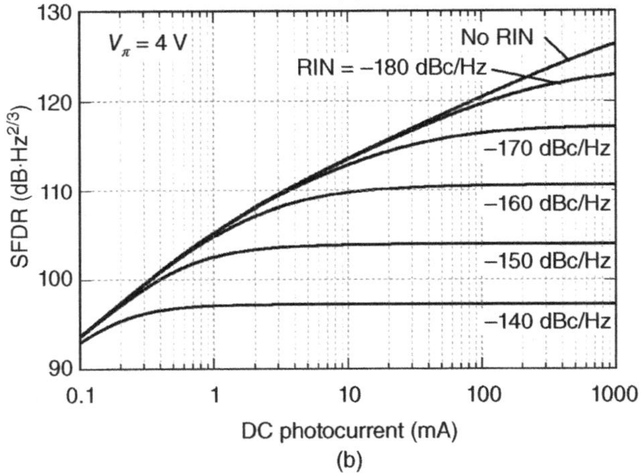

Figure 6.17. Calculated spurious-free dynamic range as given by Equation (6.57) with $R_i = 50\ \Omega$. Shown are (a) the effect of V_π with RIN = 0 and (b) the effect of RIN with $V_\pi = 4$ V.

$$\mathrm{SFDR}_3\ [\mathrm{dB}\cdot\mathrm{Hz}^{2/3}] = -8.7 + \frac{40}{3}\log\left(I_{dc}[\mathrm{mA}]\right) - \frac{2}{3}N_{out}\ [\mathrm{dBm/Hz}], \tag{6.57}$$

where the values $R_o = 50\ \Omega$, $H_{pd} = 1/2$, and $B = 1$ Hz were employed. Shown in Figure 6.17 are various plots of Equation (6.57) using $R_i = 50\ \Omega$. Shown in Figure 6.17(a) is the fundamentally limited SFDR as a function of I_{dc} with four V_π values. Trends similar to those noted previously for CDR_{1dB} are apparent. When output thermal noise limits the SFDR, the slope of the plots in Figure 6.17(a) is $m = I_{dc}^{4/3}$. The slope is $m = I_{dc}^{2/3}$ in the shot noise limit (Section 6.3.2). As with CDR, the SFDR becomes independent of I_{dc} when input thermal noise dominates. An expression for the maximum SFDR is obtained by inserting the first term in Equation (6.49) into Equation (6.56):

$$\mathrm{SFDR}_{\max} = \left(\frac{4V_\pi^2}{\pi^2 R_i k_B T_s B}\right)^{2/3}. \tag{6.58}$$

The SFDR for an ideal IMDD link biased precisely at quadrature will be third-order limited as described by Equations (6.56)–(6.58). However, as given by Equation (6.47), second-order distortion can be an issue for a link with bias drift away from quadrature. Equation (6.47) can be inserted into the SFDR definition [Equation (2.16)] to yield

$$\mathrm{SFDR}_2(\phi_{dc}) = I_{dc,q}\frac{\sin^2(\phi_{dc})}{|\cos(\phi_{dc})|}\left(\frac{2R_o\left|H_{pd}\right|^2}{N_{out}(\phi_{dc})B}\right)^{1/2} \tag{6.59}$$

as the second-order SFDR for an IMDD link as a function of the DC bias phase. In order to evaluate Equation (6.59), an expression for $N_{out}(\phi_{dc})$ is required. Equation (6.49), N_{out} at quadrature bias, can be used as a starting point to this end. Of the four terms in Equation (6.49), only the $k_B T_s$ term due to output thermal noise is independent of ϕ_{dc}. Using Equation (6.40b), it can be shown that the first term in Equation (6.49), which accounts for input thermal noise, can be written as a function of ϕ_{dc} by making the substitution $I_{dc} = I_{dc,q}$ and multiplying by $\sin^2(\phi_{dc})$. The third and fourth terms in Equation (6.49) are expressions for the output shot noise and noise associated with RIN, respectively. The level of these two terms depends on the average photocurrent, which is given as a function of ϕ_{dc} by Equation (6.40a) with the small-signal approximation for the Bessel functions. The resulting N_{out} as a function of ϕ_{dc} is

$$\begin{aligned} N_{out}(\phi_{dc}) = {} & \left[\frac{I_{dc,q}^2}{V_\pi^2}\sin^2(\phi_{dc})\,\pi^2 R_i R_o |H_{pd}|^2 + 1\right] k_B T_s \\ & + 2qI_{dc,q}R_o|H_{pd}|^2[1-\cos(\phi_{dc})] \\ & + \mathrm{RIN}\cdot I_{dc,q}^2 R_o|H_{pd}|^2[1-2\cos(\phi_{dc})+\cos^2(\phi_{dc})]. \end{aligned} \tag{6.60}$$

An assumption in Equation (6.60) is that the RIN level itself is not a function of ϕ_{dc}. This assumption holds for RIN that originates before the MZM, such as laser RIN or RIN from an optical preamplifier. However, post-MZM optical amplification would have a RIN level that is a function of the injected power, in which case, Equation (6.60) would need to include the RIN(ϕ_{dc}). Further discussion on this point is taken up in Section 8.2.

The limiting SFDR as a function of ϕ_{dc} is a very important system metric. To obtain this value, SFDR$_2$(ϕ_{dc}) and SFDR$_3$(ϕ_{dc}) must be compared; the limiting SFDR is the smaller of the two. The SFDR$_2$(ϕ_{dc}) is given by Equations (6.59) and (6.60). The OIP3(ϕ_{dc}) can be derived from Equation (6.40b) as

$$\mathrm{OIP3}(\phi_{dc}) = 4\sin^2(\phi_{dc}) I_{dc,q}^2 R_o |H_{pd}|^2. \tag{6.61}$$

The desired expression for SFDR$_3$(ϕ_{dc}) is then obtained by inserting Equations (6.60) and (6.61) into Equation (2.16). There are many parameters that determine the limiting SFDR. Shown in Figure 6.18 are plots of SFDR(ϕ_{dc}) under various conditions. The limiting SFDR(ϕ_{dc}) is determined by calculating both SFDR$_2$(ϕ_{dc}) and SFDR$_3$(ϕ_{dc}) using the aforementioned equations and taking the smaller one at each ϕ_{dc} as the SFDR. For all curves in Figure 6.18, $I_{dc,q} = 10$ mA, $V_\pi = 10$ V, $R_i = R_o = 50\ \Omega$, $H_{pd} = 1/2$, and $\phi_{dc} = \pi/2$ at quadrature. Shown in Figure 6.18(a) is a plot of SFDR(ϕ_{dc}) for RIN = −250 dBc/Hz (negligible) and for different values of B. Near quadrature, the curves appear flat and are limited by SFDR$_3$ in this region. [There is actually a small slope in these regions determined by the form of SFDR$_3$(ϕ_{dc}).] The region over which the link is limited by SFDR$_3$ is smaller for smaller B and this is a general trend. The more-curved regions of Figure 6.18(a) are where the SFDR$_2$ limits. As shown in this case, even modest deviations from quadrature can significantly degrade the intrinsic SFDR$_3$. Shown in Figure 6.18(b) is the SFDR(ϕ_{dc}) for various RIN values with $B = 1$ Hz. Similar trends as those for Figure 6.18(b) can be seen. The requirement on the bias accuracy to remain limited by SFDR$_3$ is lessened as the RIN increases. Similarly to the dependence on B, this trend is due to the different slopes of the second- and third-order distortion and the level of the total output noise power.

6.3.2 Shot-Noise-Limited Equations

The analysis in this section concentrates on certain metrics in the limit that shot noise dominates the output noise power spectral density. The shot noise limit is an important operation condition for an IMDD link employing an external MZM. As will be shown in the following

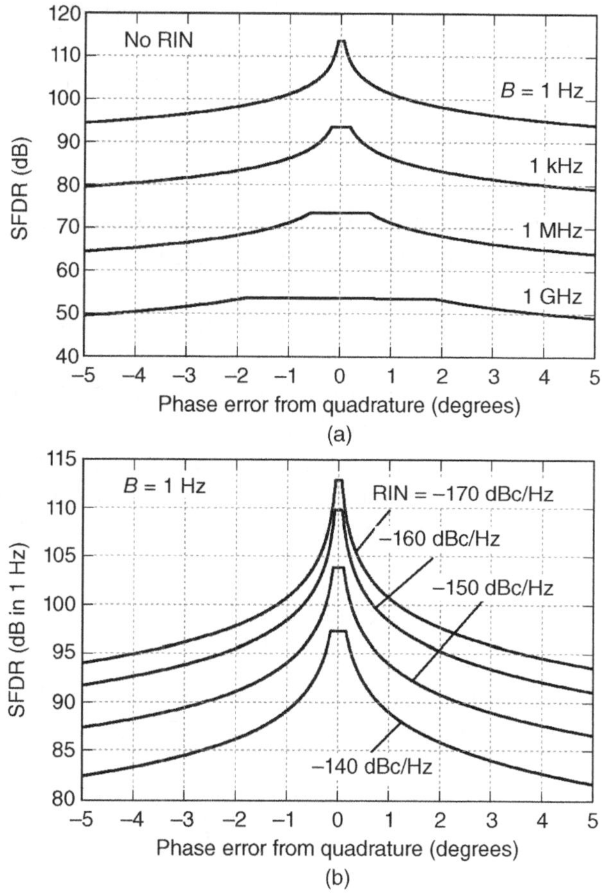

Figure 6.18. Calculated spurious-free dynamic range for an IMDD link as a function of MZM bias error with $I_{dc,q} = 10$ mA, $V_\pi = 10$ V, $R_i = R_o = 50\ \Omega$, and $H_{pd} = 1/2$. Shown are the effects of (a) electrical bandwidth (B) for RIN = 0 and (b) excess RIN with $B = 1$ Hz.

section, the RF performance for such a link improves as a function of average photocurrent. This improvement is due to the fact that the RF output power scales as I_{dc}^2, whereas the shot noise at the output scales as I_{dc}. All of the following expressions assume a precise quadrature bias. The following equations can be applied to a single-output link (Figure 6.4) or a balanced link (Figures 6.5 and 6.6), where the DC photocurrent is the sum of the two average photocurrents for the latter. When appropriate, decibel-scale equations are provided as references for link design and analysis. The decibel equations use the following parameter values: $R_i = R_o = 50\,\Omega$, $H_{pd} = 1/2$, and $B = 1$ Hz, which apply to the links shown in Figures 6.4 and 6.5.

The output noise power spectral density from Equation (6.49) is rewritten as

$$N_{sh} = 2qI_{dc}R_o|H_{pd}|^2 \tag{6.62}$$

to describe the shot noise limit. Inserting $R_o = 50\,\Omega$ and $H_{pd} = 1/2$ into Equation (6.62) and converting to decibel yield

$$N_{sh}\ [\mathrm{dBm/Hz}] = -174.0 + 10\log\,(I_{dc}\ [\mathrm{mA}]). \tag{6.63}$$

Equations (6.62) and (6.63) can then be inserted into the equations in Section 6.3.1 to derive shot-noise-limited equations.

Four SNR equations can be derived for the shot noise limit by using Equation (6.52). Inserting Equations (6.24) and (6.62) into Equation (6.52) gives a general expression for the SNR in the shot noise limit:

$$\mathrm{SNR}_{sh} = \frac{I_{dc}J_1^2(\phi_{rf})}{qB}. \tag{6.64}$$

The small-signal shot-noise-limited SNR is obtained by inserting Equations (6.26) and (6.62) into Equation (6.52) as

$$\mathrm{SNR}_{sh,ss} = \frac{I_{dc}\pi^2 R_i P_{in}}{2qV_\pi^2 B}. \tag{6.65}$$

Equation (6.65) can be normalized to $B = 1$ Hz and converted to decibel form

$$\mathrm{SNR}_{sh,ss}\ [\mathrm{dB \cdot Hz}] = 151.9 + P_{in}\ [\mathrm{dBm}] + 10\log\,(I_{dc}\ [\mathrm{mA}])$$

$$- 20\log\,(V_\pi[\mathrm{V}]), \tag{6.66}$$

where $R_i = 50\,\Omega$ was inserted. The maximum SNR in the shot noise limit can be written as

$$\mathrm{SNR}_{sh,max} = \frac{(0.3386)I_{dc}}{qB}, \tag{6.67}$$

which results from inserting Equations (6.28) and (6.62) into Equation (6.52). Equation (6.67) can be rewritten as

$$\mathrm{SNR}_{sh,max}\ [\mathrm{dB \cdot Hz}] = 153.3 + 10\log\,(I_{dc}\ [\mathrm{mA}]). \tag{6.68}$$

The SNR at 1-dB compression is 2.8 dB lower than the maximum SNR, with the former written in linear and decibel form as

$$\mathrm{SNR}_{\mathrm{sh,1dB}} = \frac{(0.1794)I_{\mathrm{dc}}}{qB}. \tag{6.69}$$

$$\mathrm{SNR}_{\mathrm{sh,1dB}}\ [\mathrm{dB}\cdot\mathrm{Hz}] = 150.5 + 10\log\left(I_{\mathrm{dc}}\ [\mathrm{mA}]\right). \tag{6.70}$$

Equations (6.69) and (6.70) were obtained by inserting Equations (6.32) and (6.62) into Equation (6.52).

When superposed with an RF signal, shot noise will contribute equally to the single-sideband (SSB) amplitude and phase noise in an IMDD link. The SSB amplitude or phase noise can be related to the SNR aforementioned equations as

$$\mathrm{SSB\ Noise}_{\mathrm{sh}}\ [\mathrm{dBc/Hz}] = -\mathrm{SNR}_{\mathrm{sh}}\ [\mathrm{dB}\cdot\mathrm{Hz}] - 3. \tag{6.71}$$

In Equation (6.71), $\mathrm{SSB\,Noise}_{\mathrm{sh}}$ refers to either SSB amplitude or phase noise in the shot noise limit, and $\mathrm{SNR}_{\mathrm{sh}}$ can be given by Equations (6.64)–(6.70). The data in Figure 3.9 are for an IMDD link employing an external MZM, showing the SSB phase noise as a function of frequency offset from a 10.24-GHz carrier. Over much of the measurement range, shot noise limits the performance and the data agree with theory.

The RF noise factor in the shot noise limit is obtained by inserting Equation (6.62) into Equation (6.50):

$$F_{\mathrm{sh}} = \frac{2qV_{\pi}^{2}}{I_{\mathrm{dc}}\pi^{2}R_{\mathrm{i}}k_{\mathrm{B}}T_{\mathrm{s}}}. \tag{6.72}$$

For $R_{\mathrm{i}} = 50\ \Omega$, Equation (6.72) can be written as

$$\mathrm{NF}_{\mathrm{sh}}\ [\mathrm{dB}] = 22.1 - 10\log\left(I_{\mathrm{dc}}\ [\mathrm{mA}]\right) + 20\log\left(V_{\pi}[\mathrm{V}]\right). \tag{6.73}$$

Equation (6.73) predicts that I_{dc} and V_{π} determine the NF in the shot noise limit. Shown in Figure 6.19 are measured V_{π} data for a MZM as a function of frequency (McKinney et al. 2007). This particular MZM was employed in a low-noise-figure IMDD link. The NF for the link is plotted in Figure 6.20 for three configurations (McKinney et al. 2007). The NF for each of the single outputs is shown, each running at about 18.5 mA average current. The NF for the balanced configuration is considerably lower, owing to the rejection of common-mode noise. Equation (6.73) is plotted along with the NF data in Figure 6.20.

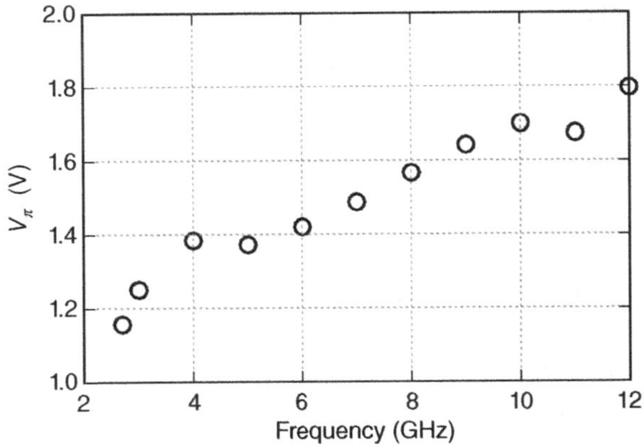

Figure 6.19. Measured V_π as a function of frequency for a high efficiency MZM (McKinney et al. 2007).

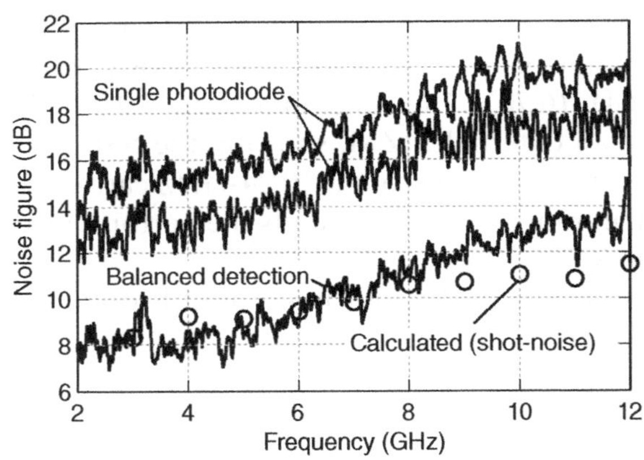

Figure 6.20. Measured noise figure for a high performance IMDD link measured for each single-photodiode output and with balanced detection. The symbols designate the calculated noise figure in the shot noise limit using the V_π values from Figure 6.19.

For the calculation, $I_{dc} = 37$ mA, and the data in Figure 6.19 were used for V_π. The data and theory agree quite well except at higher frequencies, where the contribution of output thermal noise raises the NF above the shot noise limit.

The CDR in a shot-noise-limited IMDD link is given by Equation (6.53) with Equation (6.62) inserted for N_{out}:

$$\mathrm{CDR}_{1\mathrm{dB,sh}} = \frac{(0.2258)I_{dc}}{qB}. \tag{6.74}$$

Equation (6.74) can be normalized to $B = 1$ Hz and expressed in decibel form as

$$\mathrm{CDR}_{1\mathrm{dB,sh}}\ [\mathrm{dB} \cdot \mathrm{Hz}] = 151.5 + 10 \log\left(I_{\mathrm{dc}}\ [\mathrm{mA}]\right). \tag{6.75}$$

Similarly to the shot-noise-limited CDR, the shot-noise-limited SFDR for an IMDD link is a function of I_{dc} and B only. Inserting Equation (6.62) into Equation (6.56) yields

$$\mathrm{SFDR}_{\mathrm{sh}} = \left(\frac{2I_{\mathrm{dc}}}{qB}\right)^{2/3}. \tag{6.76}$$

A decibelform of the this expression can be written as

$$\mathrm{SFDR}_{\mathrm{sh}}\ [\mathrm{dB} \cdot \mathrm{Hz}^{2/3}] = 107.3 + \frac{20}{3} \log\left(I_{\mathrm{dc}}\ [\mathrm{mA}]\right) \tag{6.77}$$

for $B = 1$ Hz.

It is important to note that all of Equations (6.64) through (6.77) are independent of the photodiode frequency response and post-photodiode circuit as described by H_{pd}. This is because the signal and shot noise will experience the same filter function at the link output. Therefore, metrics such as SNR, F, CDR, and SFDR, which involve ratios of the signal and shot noise, will be independent of H_{pd}. This would also be the case for other sources of noise that are present in the optical domain such as input thermal noise and excess RIN.

6.3.3 RIN-Limited Equations

The equations in this section quantify the performance of an IMDD link employing an external MZM in the limit that RIN is the dominant noise source. A RIN measure is typically used to describe excess noise in a photonic link, such as from lasers (Section 3.4) or optical amplifiers (Section 3.5). However, the usage of RIN can be generalized to include any noise source (Urick et al. 2011). In this case, the equations in this section can be widely applied as a useful and compact formulation of the link metrics. For example, Equation (3.21) can be employed with the following equations to determine the performance when thermal noise is the limiting factor. Likewise, the shot-noise RIN as given by Equation (3.26) can be inserted into the equations of this section to derive the equations in Section 6.3.2. It is assumed in the following analysis that the RIN is normalized to 1 Hz bandwidth and therefore has units of

$(\text{Hz})^{-1}$ (or dBc/Hz in a dB scale). As in the previous section, decibel equations are provided with the following assumptions: $R_\text{i} = R_\text{o} = 50\ \Omega$, $H_\text{pd} = 1/2$, $B = 1$ Hz, and $T_\text{s} = 290$ K, which apply to the links shown in Figures 6.4 and 6.5.

For this analysis, it is assumed that the output noise power spectral density is given by

$$N_\text{RIN} = \text{RIN} \cdot I_\text{dc}^2 R_\text{o} |H_\text{pd}|^2 \tag{6.78}$$

for $R_\text{o} = 50\ \Omega$ and $H_\text{pd} = 1/2$, Equation (6.78) can be rewritten in decibel form as

$$N_\text{RIN}[\text{dBm/Hz}] = -19.0 + \text{RIN}[\text{dBc/Hz}] + 20 \log (I_\text{dc}[\text{mA}]). \tag{6.79}$$

The SNR in the RIN limit is obtained by inserting Equations (6.24) and (6.78) into Equation (6.51):

$$\text{SNR}_\text{RIN} = \frac{2J_1^2(\phi_\text{rf})}{\text{RIN} \cdot B}. \tag{6.80}$$

There are three other useful SNR expressions in the RIN limit. The small-signal SNR when RIN dominates is

$$\text{SNR}_\text{RIN,ss} = \frac{\pi^2 R_\text{i} P_\text{in}}{\text{RIN} \cdot V_\pi^2 B}, \tag{6.81}$$

which can be written in decibel form as

$$\text{SNR}_\text{RIN,ss}[\text{dB} \cdot \text{Hz}] = -3.1 + P_\text{in}[\text{dBm}] - \text{RIN}[\text{dBc/Hz}] - 20 \log (V_\pi[\text{V}]) \tag{6.82}$$

for $R_\text{i} = 50\ \Omega$ and $B = 1$ Hz. Equation (6.81) results from inserting Equations (6.25) and (6.78) into Equation (6.52). Equations (6.28), (6.78), and (6.53) can be combined to yield the SNR in the RIN limit as

$$\text{SNR}_\text{RIN,max} = \frac{0.6772}{\text{RIN} \cdot B}. \tag{6.83}$$

Normalizing Equation (6.83) a 1-Hz bandwidth and converting to decibel form result in

$$\text{SNR}_\text{RIN,max}\ [\text{dB} \cdot \text{Hz}] = -1.7 - \text{RIN}\ [\text{dBc/Hz}]. \tag{6.84}$$

Finally, the SNR at 1-dB compression is obtained by inserting Equations (6.32) and (6.78) into Equation (6.52):

$$\text{SNR}_{\text{RIN,1dB}} = \frac{0.3587}{\text{RIN} \cdot B}, \tag{6.85}$$

which can be converted to a decibel scale as

$$\text{SNR}_{\text{RIN,1dB}}[\text{dB} \cdot \text{Hz}] = -4.5 - \text{RIN}[\text{dBc/Hz}] \tag{6.86}$$

for $B = 1$ Hz.

The RF noise figure, CDR, and SFDRs for an IMDD link can be written in a compact form as a function of RIN. The noise factor is obtained by inserting Equation (6.78) into Equation (6.50):

$$F_{\text{RIN}} = \frac{\text{RIN} \cdot V_{\pi}^2}{\pi^2 R_{\text{i}} k_{\text{B}} T_{\text{s}}}. \tag{6.87}$$

The noise figure is the decibel version of Equation (6.87),

$$\text{NF}_{\text{RIN}}[\text{dB}] = 177.0 + \text{RIN}[\text{dBc/Hz}] + 20 \log (V_{\pi}[\text{V}]), \tag{6.88}$$

where $R_{\text{i}} = 50\ \Omega$ was used. Inserting Equation (6.78) into Equation (6.53) gives the CDR in the RIN limit as

$$\text{CDR}_{\text{RIN,1dB}} = \frac{0.4516}{\text{RIN} \cdot B}. \tag{6.89}$$

A logarithmic version of Equation (6.89) for $B = 1$ Hz is

$$\text{CDR}_{\text{RIN,1dB}}[\text{dB} \cdot \text{Hz}] = -3.5 - \text{RIN}[\text{dBc/Hz}]. \tag{6.90}$$

Finally, the SFDR in the RIN limit can be written as

$$\text{SFDR}_{\text{RIN}} = \left(\frac{4}{\text{RIN} \cdot B}\right)^{2/3}, \tag{6.91}$$

which was obtained by inserting Equation (6.78) for N_{out} in Equation (6.56). Equation (6.91) can be normalized to $B = 1$ Hz resulting in

$$\text{SFDR}_{\text{RIN}}[\text{dB} \cdot \text{Hz}^{2/3}] = 4.0 - \frac{2}{3}\text{RIN}[\text{dBc/Hz}]. \tag{6.92}$$

The aforementioned equations demonstrate the utility of a RIN formalism for describing the performance of an IMDD link. The maximum SNR, SNR at 1-dB compression, CDR at 1-dB compression, and SFDR are functions of RIN only. The noise figure depends only on RIN and the MZM half-wave voltage, V_π. Many sources of RIN, such as laser RIN, are independent of the average photocurrent. Therefore, specification of the minimum RIN for a link will determine many of the maximum attainable performance levels.

6.3.4 Trade Space Analysis

There are numerous system design trades in an external IMDD link that may not be at first obvious given the aforementioned equations. The gain, noise figure, and SFDR are three metrics that will typically be encountered in the early stages of a link design. As given by Equation (6.26), the gain factor becomes larger as I_{dc} is increased or as V_π is decreased. Inserting Equation (6.49) into Equation (6.50) can provide more insight into the noise factor trends:

$$F = 1 + \frac{V_\pi^2}{\pi^2 R_i}\left(\frac{1}{I_{dc}^2 R_o \left|H_{pd}\right|^2} + \frac{2q}{I_{dc} k_B T_s} + \frac{\mathrm{RIN}}{k_B T_s}\right). \tag{6.93}$$

The first through fourth terms in Equation (6.93) correspond to input thermal noise, output thermal noise, shot noise, and RIN due to excess optical noise, respectively. Similarly to gain factor, noise factor improves (decreases) as I_{dc} is increased or as V_π is decreased. Decreasing the RIN also improves the noise factor. The effects of these parameters on the SFDR can be seen by inserting Equation (6.49) into Equation (6.56):

$$\mathrm{SFDR}_3 = \left(\frac{\pi^2 R_i k_B T_s B}{4V_\pi^2} + \frac{k_B T_s B}{4 I_{dc}^2 R_o \left|H_{pd}\right|^2} + \frac{qB}{2I_{dc}} + \frac{\mathrm{RIN}\cdot B}{4}\right)^{-2/3}. \tag{6.94}$$

As with Equation (6.93), the first through fourth terms in Equation (6.94) correspond to input thermal noise, output thermal noise, shot noise, and RIN due to excess optical noise, respectively. Increasing I_{dc} or decreasing RIN will improve the SFDR. However, the SFDR is independent of V_π except in cases when input thermal noise limits the link. When this is the case, a larger V_π will raise the ceiling on SFDR as

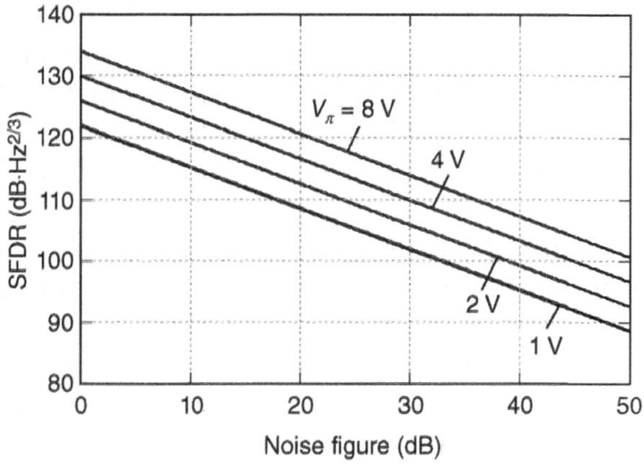

Figure 6.21. Calculated third-order spurious-free dynamic range as a function of noise figure using Equation (6.95) with $R_i = 50\ \Omega$ and $B = 1$ Hz.

described by Equation (6.58). Therefore, the design trades depend on the region of operation of the link.

The tradeoffs between F and SFDR can be investigated further as follows. The SFDR can be written as a function of F by first multiplying Equations (6.50) and (6.56) and then solving for SFDR:

$$\mathrm{SFDR}_3 = \left(\frac{4V_\pi^2}{\pi^2 R_i k_B T_s B}\right)^{2/3} \left(\frac{1}{F}\right)^{2/3}. \tag{6.95}$$

Plots of Equation (6.95) are shown in Figure 6.21 for various values of V_π with $R_i = 50\ \Omega$ and $B = 1$ Hz. Some simple conclusions can be drawn from Figure 6.21. Firstly, for a given noise figure value, the SFDR is larger with larger V_π. However, by Equation (6.91), a large V_π will require large I_{dc} and low RIN to obtain a small noise figure. Conversely, for a given SFDR level, the noise figure is improved as V_π is decreased. In this case, the input thermal noise given by the first term in Equation (6.94) can increase the restrictions on I_{dc} and RIN. A method developed by Bucholtz et al. (2008) delves further into these tradeoffs and uses a graphical technique to convey the results.

6.4 PHOTODETECTOR ISSUES AND SOLUTIONS

The treatment of IMDD to this point has assumed that the photodiode is precisely linear. As described in Chapter 4, the photodiodes encountered in microwave photonics can exhibit a host of nonlinearities.

In this section, the impairments imposed by photodiode distortion on an IMDD link are quantified assuming that the second- and third-order photodiode nonlinearities can be described by an OIP2 and OIP3, respectively. In the later part of this section, some mitigation techniques for photodiode-induced even-order distortion in IMDD links are detailed.

The impact of photodiode nonlinearities on the single-octave SFDR for an IMDD link is conducted using a cascade analysis as described in Chapter 2. The assumption is that an IMDD link employing an ideal photodiode is the first stage followed by a nonlinear photodiode having unity gain and a third-order output intercept point $\mathrm{OIP3_{pd}}$. The cascaded OIP3 of these two stages is given by Equation (2.31) with Equation (6.42) inserted for the link OIP3:

$$\mathrm{OIP3} = \left(\frac{1}{4I_{\mathrm{dc}}^2 R} + \frac{1}{\mathrm{OIP3_{pd}}} \right)^{-1}. \tag{6.96}$$

In Equation (6.96), H_{pd} was set to unity such as to avoid any confusion as to where the $\mathrm{OIP3_{pd}}$ is specified. Inserting Equation (6.96) into Equation (2.16) yields the cascaded $\mathrm{SFDR_3}$. The ratio of this cascaded $\mathrm{SFDR_3}$ to the $\mathrm{SFDR_3}$ of an IMDD link with an ideal photodiode [Equation (6.56)] results in the penalty imposed by the photodiode (Urick et al. 2008):

$$\Delta\mathrm{SFDR_3} = \left(1 + \frac{4I_{\mathrm{dc}}^2 R_{\mathrm{o}}}{\mathrm{OIP3_{pd}}} \right)^{-2/3}. \tag{6.97}$$

Equation (6.97) is independent of N_{out} and B. Calculations of Equation (6.97) are plotted in Figure 6.22 for $R_{\mathrm{o}} = 50\ \Omega$, demonstrating the intuitive result that the photodiode OIP3 must be larger at higher average photocurrents in order to maintain the intrinsic link linearity.

Under ideal conditions, an IMDD link with an MZM biased precisely at quadrature does not produce even-order distortion. For multioctave applications, the required OIP2 of the photodiode ($\mathrm{OIP2_{pd}}$) to maintain the third-order dynamic range of the link is given by Equation (2.25) with Equation (6.42) inserted for OIP3:

$$\mathrm{OIP2_{pd}} \geq (4I_{\mathrm{dc}}^2 R_{\mathrm{o}})^{4/3} (N_{\mathrm{out}} B)^{-1/3} \tag{6.98}$$

for $H_{\mathrm{pd}} = 1$. The equality in Equation (6.98) is plotted in Figure 6.23 with Equation (6.49) used for N_{out}, with $R_{\mathrm{i}} = R_{\mathrm{o}} = 50\ \Omega$, $V_{\pi} = 4$ V, and

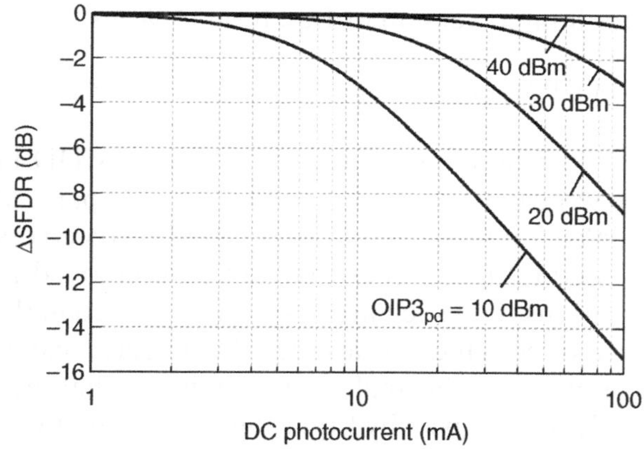

Figure 6.22. Calculated change in third-order spurious-free dynamic range [Equation (6.97)] relative to an ideal IMDD link due to photodiode distortion, which is characterized by an output intercept point ($OIP3_{pd}$).

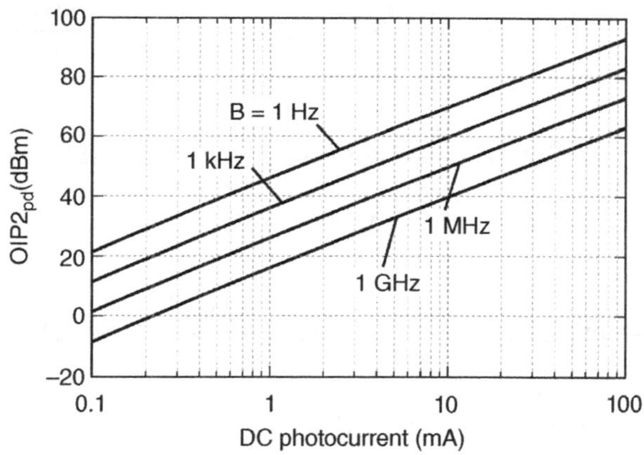

Figure 6.23. Calculated photodiode OIP2 required to maintain the intrinsic third-order SFDR of an IMDD link as given by Equation (6.98) with $R_i = R_o = 50\ \Omega$, $V_\pi = 4$ V, and RIN = 0.

RIN = 0. The effect of electrical bandwidth is shown in Figure 6.23, where larger bandwidth relaxes the requirement on $OIP2_{pd}$. The effects of RIN in Equation (6.98) have been studied by Urick et al. (2012). Equation (6.98) can be written for the shot noise limit as (Urick et al. 2008)

$$OIP2_{pd,shot} \geq (2I_{dc})^{7/3}(qB)^{-1/3}R_o. \tag{6.99}$$

Similarly, a RIN-limited version of Equation (6.98) is

$$\mathrm{OIP2}_{\mathrm{pd,RIN}} \geq 4(2)^{2/3} I_{\mathrm{dc}}^2 (\mathrm{RIN} \cdot B)^{-1/3} R_{\mathrm{o}}. \tag{6.100}$$

The photodiode requirements in a multioctave IMDD link as given by Equation (6.98) can be quite difficult to achieve, particularly in low noise links or links that feed narrowband (small B) receivers. However, there are architectural techniques that can be employed to overcome the device limitations. Two such methods are discussed in the following. The explanations of both techniques use a Taylor series expansion of the photodiode response. For the analysis, it is assumed that the drive signal constitutes a DC portion given by Equation (6.40a) and two small-signal equal-amplitude ($\phi_1 = \phi_2 = \phi \ll 1$) sinusoidal tones given by the first two lines of Equation (6.40b) with $j = k = 1$. Using this as the input to the photodiode and conducting a Taylor series expansion [Equation (2.10)] about I_{dc} give an output current of

$$\begin{aligned} I_{\mathrm{pd}} &= (a_0 + a_2 I^2) + \left(a_1 I + \frac{9a_3 I^3}{4}\right) \sin(\Omega_1 t) + \left(a_1 I + \frac{9a_3 I^3}{4}\right) \sin(\Omega_2 t) \\ &\quad - \frac{a_2 I^2}{2} \cos(2\Omega_1 t) - \frac{a_2 I^2}{2} \cos(2\Omega_2 t) + a_2 I^2 \cos[(\Omega_1 - \Omega_2)t] \\ &\qquad - a_2 I^2 \cos[(\Omega_1 + \Omega_2)t] \\ &\quad - \frac{a_3 I^3}{4} \sin(3\Omega_1 t) - \frac{a_3 I^3}{4} \sin(3\Omega_2 t) + \frac{3a_3 I^3}{4} \sin[(2\Omega_1 - \Omega_2)t] \\ &\qquad + \frac{3a_3 I^3}{4} \sin[(2\Omega_2 - \Omega_1)t] \\ &\quad - \frac{3a_3 I^3}{4} \sin[(2\Omega_1 + \Omega_2)t] - \frac{3a_3 I^3}{4} \sin[(2\Omega_2 + \Omega_1)t] + \cdots, \end{aligned} \tag{6.101}$$

where $I = \phi I_{\mathrm{dc,q}} \sin(\phi_{\mathrm{dc}})$ and I_1 in Equation (6.40) was used for the input.

The first technique takes advantage of the complementary nature of the two MZM outputs. Consider a Taylor series expansion given by Equation (6.101) when calculated for I_2 in Equation (6.40). The result is an output current just as Equation (6.101) except the odd-order terms are 180° out of phase and the even-order terms have the same phase. Therefore, if two identical photodiodes (having the same Taylor

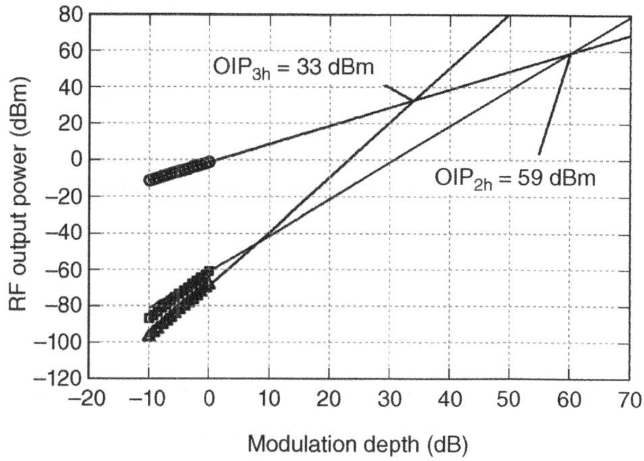

Figure 6.24. Measured fundamental, second-harmonic and third-harmonic response for a photodiode driven with a pure 5-MHz tone. The harmonic output intercept points are shown (Hastings et al. 2008).

coefficients) are employed in an ideal balanced configuration (same amplitude and true 180° phase shift), then the even-order terms should be rejected. This technique was first demonstrated by Hastings et al. (2008) with the data in Figures 6.24–6.26. Shown in Figure 6.24 are the measured output intercept points due to second- and third-order harmonic distortion, OIP_{2h} and OIP_{3h}, respectively, for a photodiode driven by a pure 5-MHz tone with an average current of $I_{dc} = 6.5$ mA. A common technique to increase the effective linearity of a nonlinear device is to array a number of such devices. If all the N devices in the array exhibit identical nonlinearity and their output currents are then combined linearly without loss or phase shifts, an increase by a factor of N^2 is expected in the output intercept point as compared to the intercept point of any individual device. An array of four photodiodes was constructed using devices very similar to the one that produced the data in Figure 6.24. Two photodiodes were attached to each output of a quadrature-biased MZM. The currents from the two photodiodes on each MZM arm were added, and the composite current from each MZM arm was subtracted. The total photocurrent for the link was $I_{dc} = 26$ mA (6.5 mA per photodiode). The OIP_{2h} and OIP_{3h} for this link are shown in Figure 6.25. The "array gain" for the receiver should have been 12 dB, resulting in $OIP_{2h} = 71$ dBm and $OIP_{3h} = 45$ dBm if the photodiodes were the only factor. The measured $OIP_{3h} = 27$ dBm is limited by the MZM-generated distortion. However, the measured $OIP_{2h} = 105$ dBm is attributed to the predicted cancellation effect.

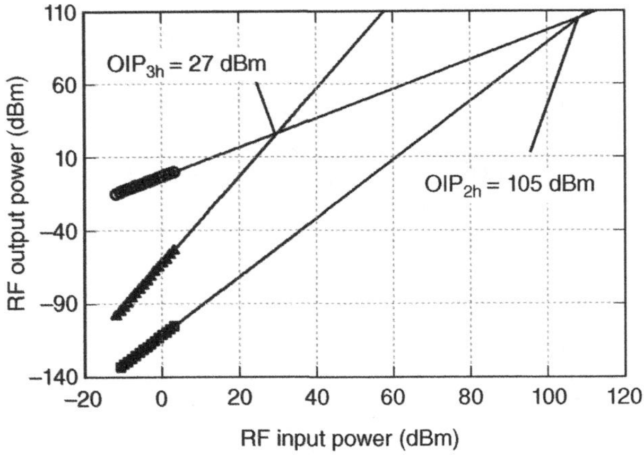

Figure 6.25. Measured fundamental (5 MHz), second- and third-harmonic response for an IMDD link. The harmonic output intercept points are shown (Hastings et al. 2008).

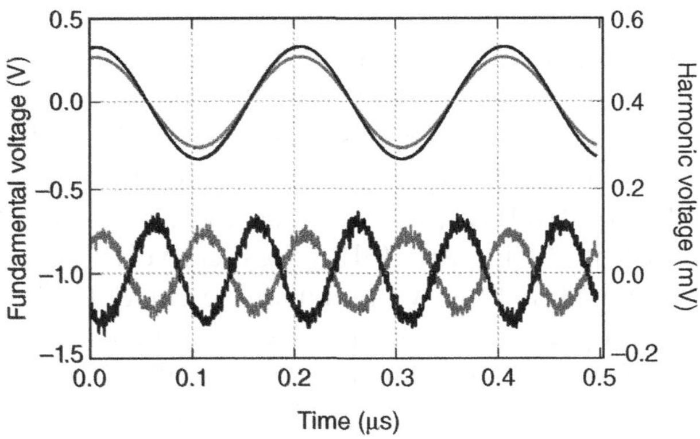

Figure 6.26. Measured waveforms at the output of a balanced IMDD link showing (top, left axis) the fundamental tones from each arm that have been shifted to be in phase after the subtraction and (bottom, right axis) the second harmonics that are 180° out of phase (Hastings et al. 2008).

The reason for cancellation can be seen more clearly in Figure 6.26, where the waveforms for the fundamentals and second harmonics from each MZM are shown as measured at the receiver output. Clearly, the fundamentals are in phase, and the second harmonics are out of phase.

There are some drawbacks of the balanced-receiver technique for suppressing photodiode even-order distortion. The method requires two phase-matched fibers, which can be impractical for long point-to-point links and/or high modulation frequencies. The technique also relies on a well-matched set of photodiodes to perform the cancellation. A technique developed by Urick et al. (2013) eliminates these issues by canceling even-order distortion of a photodiode with a single-output MZM. The method entails shifting the MZM bias to intentionally generate even-order distortion matching the amplitude of the photodiode-induced distortion; the equations in this chapter predict that the two sources can produce even-order distortions that are out of phase. A brief summary of the theory and some experimental data for this technique are provided in this section.

Consider a single-output MZM driving a single photodiode as shown in Figure 6.4. The IMD2 for such a link can be modeled as the composite response of the IMD2 in Equations (6.40c) and (6.101). The resulting equation is

$$I_{\mathrm{imd2}} = \pm\phi^2 I_{\mathrm{dc,q}} \left[\frac{\cos\left(\phi_{\mathrm{dc}}\right)}{2} + a_2 I_{\mathrm{dc,q}} \sin^2(\phi_{\mathrm{dc}}) \right] \cos\left[(\Omega_2 \mp \Omega_1)t\right], \tag{6.102}$$

where the first term in the bracket is due to the MZM and the second term originates from the photodiode. The phase of the MZM term is different by 180° on either side of a quadrature point. However, the MZM-generated fundamental has the same phase about a quadrature point [by Equation (6.40b)], which implies that a_2 should have the same sign about a quadrature point as well [by Equation (6.101)]. The magnitude of a_2 can be determined from a photodiode OIP2 measurement, and the other parameters in Equation (6.101) are easily obtained. Therefore, cancellation can be achieved at some bias phases, which are given by Equation (6.102).

Shown in Figure 6.27 is the measured OIP2 for an Applied Optoelectronics PD3000 photodiode reverse biased at 1 V with $I_{\mathrm{dc}} = 3.0$ mA. These data were obtained with a two-tone setup employing two MZMs, one driven at 0.9 GHz and the other at 1.1 GHz, designed to isolate the photodiode-generated IMD2 at 2.0 GHz (Urick et al. 2013). The measured OIP2 = 13.5 dBm due to this distortion. A two-tone test was applied to an IMDD link employing this photodiode, the results of which are shown in Figure 6.28. The IMD2 at 2.0 GHz resulted in an OIP2 = 12.5 dBm when the MZM was biased at quadrature with $I_{\mathrm{dc}} = 3.0$ mA. This level is very close to that measured for the photodiode alone, indicating that the photodiode is limiting the IMD2. The MZM bias was adjusted to determine the minimum IMD2, which was observed well away from quadrature at an average current of $I_{\mathrm{dc}} = 2.5$ mA. The fundamental output powers at quadrature and at

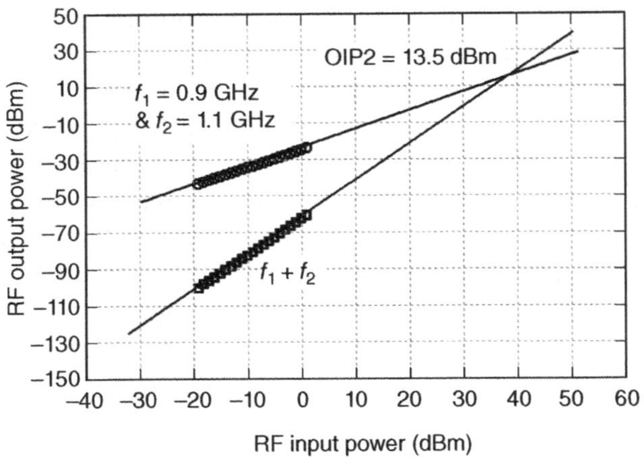

Figure 6.27. Measured OIP2 for the photodiode used to demonstrate a novel linearization scheme. (Urick et al. 2013).

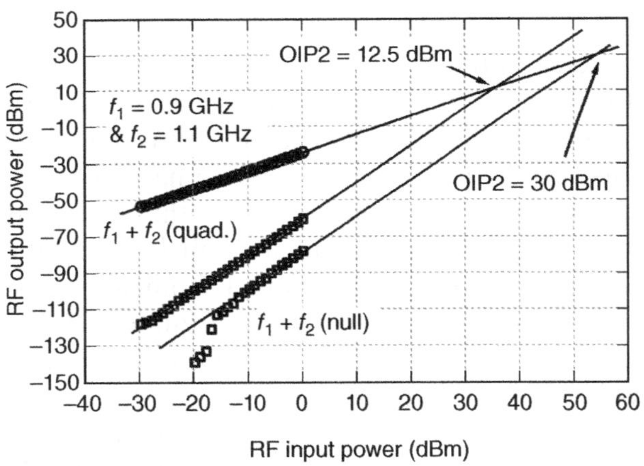

Figure 6.28. Measured OIP2 for an IMDD link employing a single-output MZM at quadrature bias and at the nonquadrature bias point where IMD2 from the photodiode is suppressed (Urick et al. 2013).

the cancellation condition are nearly identical, another advantage of this method. That is, some linearization techniques achieve reduced distortion at the cost of decreased output power at the fundamental. The IMD2 at the cancellation point is also plotted in Figure 6.28. Very strong suppression was measured at input powers below −15 dBm; the suppression is good but much less at higher input powers. A second-order function (slope of two in logarithmic scale) is fit to the data at higher input powers resulting in OIP2 = 30 dBm. This limiting

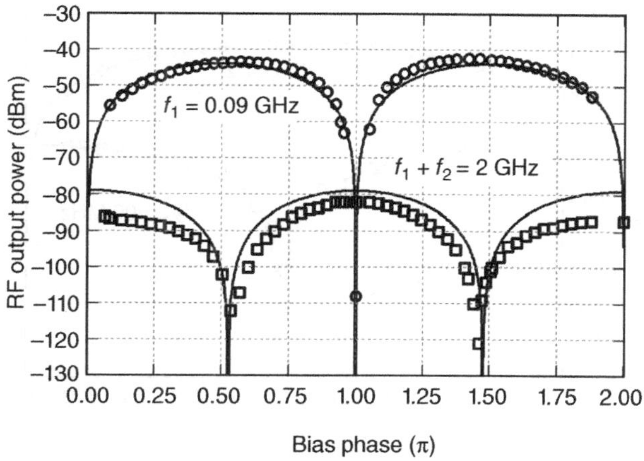

Figure 6.29. Measured (symbols) and calculated (lines) fundamental power and IMD2 power for an IMDD link as a function of MZM bias phase (Urick et al. 2013).

OIP2 is attributed to reflections in the apparatus but is still 16.5 dB better than the OIP2 of the photodiode. The carrier-to-intermodulation ratio (CIR) is much better for the cancellation condition. At high input powers, the CIR is 17.5 dB higher at 2.5 mA than that at 3.0 mA (MZM quadrature). In the strong cancellation region, the CIR difference is upward of 40 dB.

The theoretical and measured fundamental and IMD2 for the link as a function of MZM bias are plotted in Figure 6.29. As shown in this case, the calculated fundamental power [from Equation (6.101)] and the IMD2 power [Equation (6.102)] follow the experimental results. The calculated and experimental fundamental powers agree well at quadrature ($I_{dc} = 3.0$ mA). However, the measured values diverge above the calculation below 3.0 mA and go below the calculation at photocurrents above 3.0 mA. The IMD2 curves follow the same trend with a larger divergence but do agree quite well at the cancellation condition. The reason for the divergences is attributed to photodiode compression, which is worse at higher photocurrents and at higher frequencies. The photodiode compression is an artifact of the experimental preparation, where the photodiode was intentionally operated at a low bias in order to move the cancellation far enough away from the quadrature to demonstrate the technique definitively. As described by Hutchinson et al. (2014), including more terms from the Taylor series [Equation (6.101)] in the calculation can account for photodetector compression.

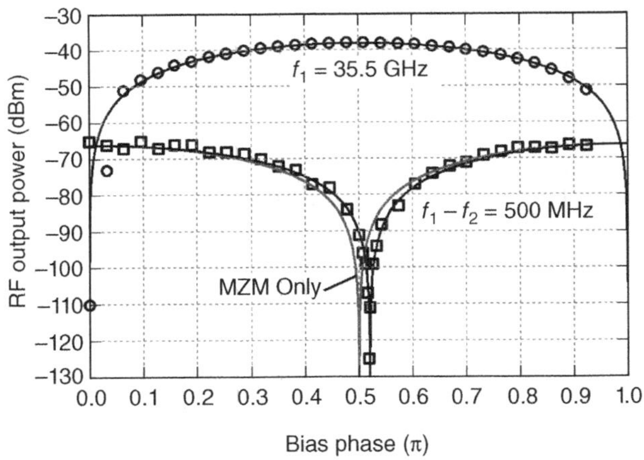

Figure 6.30. Measured (symbols) and calculated (lines) fundamental power and IMD2 power for a Ka-Band IMDD link as a function of MZM bias phase (Hutchinson et al. 2014).

A similar data set is shown in Figure 6.30, where a high frequency photodiode (U^2T 75-GHz) is employed in an IMDD link with a single-output MZM (Hutchinson et al. 2014). The photocurrent at quadrature for this link was 2.5 mA, and the two tones were at 35.0 and 35.5 GHz. The IMD2 at 500 MHz and one of the fundamentals are plotted as a function of MZM bias in Figure 6.30. As can be seen in this case, the theory and measured data agree precisely for an uncompressed photodiode. Also shown in Figure 6.30 is the calculated IMD2 response for the link parameters with an ideal photodiode as given by Equation (6.40). This plot further emphasizes the impact of the photodiode IMD2 outside the full cancellation regime.

6.5 LINEARIZATION TECHNIQUES

An ideal MZM biased precisely at quadrature exhibits no even-order distortion, making the IMDD link an ideal candidate for multioctave applications. There have been numerous demonstrations of techniques to linearize the intrinsic modulation transfer function of an MZM to reduce the third-order distortion that can inhibit single-octave implementations. Shown in Figure 6.31 is an early method where two MZMs are optically connected in series and driven by the same RF signal. There are two main controls in the structure shown in Figure 6.31, the coupling ratio for the RF signal split between the two MZMs and the

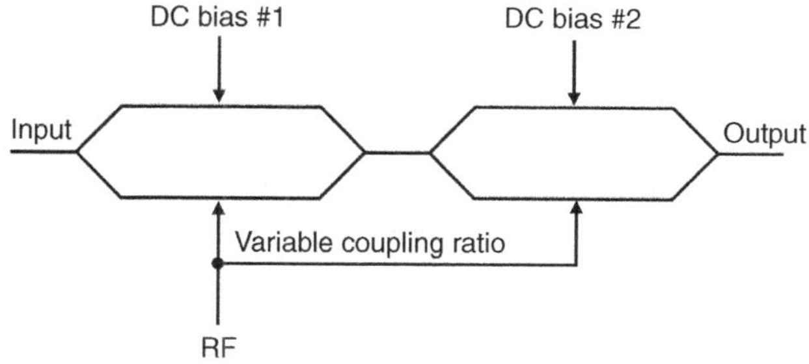

Figure 6.31. Diagram of dual series Mach–Zehnder modulators.

DC bias on each MZM. When implemented as shown in Figure 6.31, with single-output MZMs having 50/50 optical coupling ratios, these controls can be optimized to achieve third-order linearization at the cost of even-order distortions (Betts 1994). Additional controls can be achieved in a similar structure by using variable optical power coupling or dual-output and dual-input MZMs (Skeie and Johnson 1991). These structures allow for simultaneous second- and third-order linearization at the cost of increased control complexity. The transfer-matrix formalism used in Section 6.2 can be applied to the analysis of series MZMs in order to derive the linearization conditions.

Two MZMs in parallel have also been employed to linearize analog optical links (Korotky and de Ridder, 1990). The basic structure for such a dual parallel MZM (DPMZM) is shown in Figure 6.32, where two MZMs are nested into each arm of a third MZM. The resulting structure affords four possible parallel optical paths, the relative phases of which are controlled via three DC electrodes. The RF signal is applied to each of the two nested MZMs. By varying the relative amplitude and phase of the modulation on each MZM through the numerous controls, the distortion from one MZM can be arranged to cancel that from the other. Since the early work by Korotky and de Ridder (1990), there has been resurgence in the use of DPMZMs for analog applications (Kawanishi et al. 2007) owing to their widespread commercial availability. Integrated DPMZMs have been demonstrated useful for numerous telecommunication formats, such as optical quadrature amplitude modulation (QAM) (Winzer et al. 2010), where digital signals are applied to the nested MZMs. Some interesting analog results have been recently demonstrated using DPMZMs such as linearized single-sideband modulation (Kawanishi and Izutsu 2004), linearized double-sideband suppressed-carrier modulation (Zhu et al. 2009), and wideband RF phase shifting (Chan et al. 2012). Again,

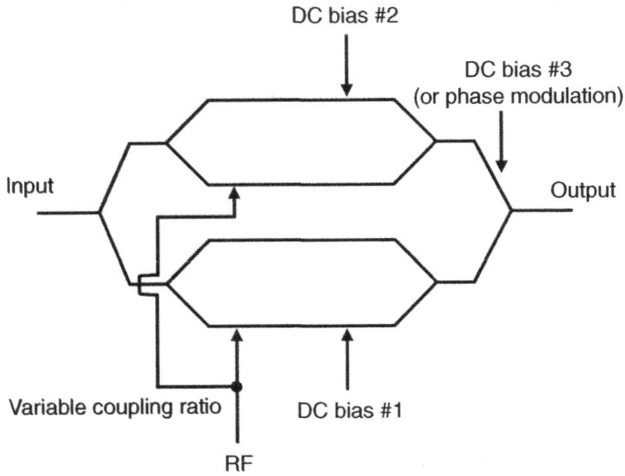

Figure 6.32. Diagram of dual parallel Mach-Zehnder modulators (DPMZM).

the matrix approach introduced in Section 6.2 can be applied to the DPMZM structure, although the analysis is a bit more complicated than for the dual series MZMs because description of a DPMZM requires 4×4 matrices.

Although the dual-series MZM and DPMZM structures shown previously can be fabricated on a single chip, the linearization of a single MZM can be desirable. Optical techniques can be employed to this end, such as dual-polarization and dual-wavelength structures. In a dual-polarization scheme (Johnson and Roussell 1988), the polarization of the light input into an MZM is intentionally misaligned. The linearization condition is then obtained by decomposing the input polarization into two orthogonal states (TE and TM), which exhibit very different electro-optic phase shifts. The distortion on one polarization can then be arranged to cancel that on the other. A dual-wavelength technique (Ackerman 1999) is similar, where two wavelengths are input into a single MZM. The distortion from one wavelength can suppress the other's through control of the MZM bias and optical power in each wavelength.

The linearization schemes discussed to this point can be viewed as "imbedded," where the linearization mechanism is part of the link itself. External circuits can also be used to linearize an IMDD link, including digital or analog electronic circuits, electro-optic "circuits," or hybrid approaches. Linearization methods for photonic links can use technology developed for electronic amplifier linearization including feed-forward, feed-back, and predistortion concepts (Katz 2001).

External predistortion linearization and feed-forward linearization techniques were widely employed in legacy cable television (CATV) photonic links (Nazarathy et al. 1993). To implement predistortion linearization, the nonlinear behavior of a link is characterized a priori and a circuit is placed before the link to generate frequency components to cancel the link's distortion. A good example of an analog electronic predistortion technique is given by MacDonald et al. (2005), where a 4–12-GHz predistortion linearizer reduced the IMD3 of an IMDD link in excess of 20 dB. For feed-forward linearization, a link's distortion is measured in an external circuit that generates a correction signal. This correction signal is then combined with the distorted link signal somewhere forward in the transmission line to cancel the link distortion. A 2–18-GHz feed-forward-linearized photonic link has demonstrated upward of 25-dB distortion suppression using a combination of electro-optics, microwave circuitry, and digital signal processing (O'Connor et al. 2008). A single metric does not typically give the entire picture for a linearization scheme, and it is of utmost importance for these and other linearization schemes to quantify the performance clearly.

Numerous metrics have been used to quantify the performance of a linearization scheme. As described in Chapter 2, measures such as output intercept point must be used cautiously. Data from a linearized 6–12-GHz IMDD link exemplifying this point are shown in Figure 6.33. Plotted are the measured fundamental and IMD3 for an IMDD link with and without a predistortion linearizer in dBm against the input power also in dBm (Urick et al. 2006). The stimulus for both links comprised two tones at 9.000 and 9.011 GHz. Lines with slopes of one and three fit the unlinearized link data well over most of the data set, with some compression in both curves at the highest input powers. The linearized link, however, exhibits an IMD3 that does not exhibit a slope of three over much of the measurement range. In fact, the distortion at input power levels above −35 dBm rises with a slope greater than three. Therefore, when specifying the performance of each link in terms of OIP3 = 5.7 dBm (unlinearized) and OIP3 = 14.6 dBm (linearized), it must be clearly identified over which input powers these metrics are valid. In this case, both links have the same $N_{out} = -148$ dBm/Hz. A better comparison of the two links is obtained by plotting the SFDR in decibels as a function of output bandwidth (B) as shown in Figure 6.34. As shown, the linearized link enjoys a 6-dB SFDR advantage at $B < 10$ MHz, but both links have about the same SFDR at $B > 100$ MHz. This is a nontrival point for a 6–12-GHz link and

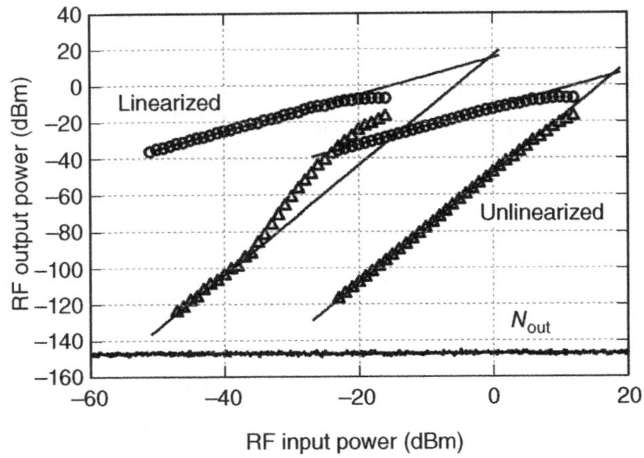

Figure 6.33. Measured fundamental and distortion (symbols) for an IMDD link with and without a predistortion linearizer. The solid lines represent the fits that were used to determine the OIP3s. The output noise power spectral density of −148 dBm/Hz is also shown (Urick et al. 2006).

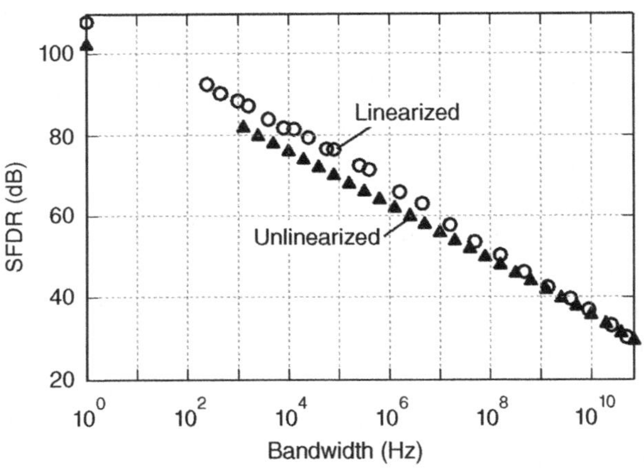

Figure 6.34. Spurious-free dynamic range as a function of electrical bandwidth for the data in Figure 6.33 (Urick et al. 2006).

further stresses the importance of understanding the SFDR metric as presented in Chapter 2.

6.6 PROPAGATION EFFECTS

The majority of the analysis in Chapter 5 on propagation effects was in terms of an intensity-modulated lightwave signal and is therefore directly applicable to IMDD with an external MZM. Important single-channel effects include chromatic dispersion and SBS. The impact of chromatic dispersion on an IMDD link with an MZM is detailed in this section as a compliment to the more general treatment in Chapter 5. The majority of the optical power resides in the carrier at small modulation depth, so SBS will limit the performance an IMDD link as described in Chapter 5. Limited optical power directly affects the maximum I_{dc}, which can be mapped to RF performance using the equations in this chapter. Multichannel impairments for IMDD are as described in Chapter 5. The crosstalk equations for SRS, Equations (5.47) and (5.48), and XPM, Equations (5.52) and (5.53), are directly applicable to an IMDD link employing a quadrature-biased MZM. Given the details in Chapter 5, the remainder of this section concentrates only on specific equations describing chromatic dispersion effects on the MZM-based IMDD format.

The effects of chromatic dispersion on an IMDD link can be quantified by including a propagation constant in the electric field, as described in Section 5.4. The analysis will be exemplified in this section using the chirp-free field[1] given by Equation (6.15), which can be rewritten as

$$E_1 = \gamma' \sum_{m=-\infty}^{\infty} J_{2m+1}\left(\frac{\phi_{\mathrm{rf}}}{2}\right)\cos\left[\omega_m t - \beta(\omega_m)L\right] - \gamma' \sum_{k=-\infty}^{\infty} J_{2k}\left(\frac{\phi_{\mathrm{rf}}}{2}\right)\sin\left[\omega_k t - \beta(\omega_k)L\right]. \tag{6.103}$$

The constant $\gamma' = \gamma\sqrt{g_o/2}$, where γ is given by Equation (6.10), and g_o is the net optical gain for the link and the factor of $1/\sqrt{2}$ results from setting the MZM at quadrature. The propagation length is L. The odd frequency components are $\omega_m = \omega + (2m+1)\Omega$ and the even ones are $\omega_k = \omega + 2k\Omega$. Finally, the mode-propagation constant in Equation (6.103) is defined as (Agrawal 2013)

$$\beta(\omega_n) = \beta_0 + \beta_1(\omega_n - \omega) + \frac{\beta_2}{2}(\omega_n - \omega)^2 + \cdots. \tag{6.104}$$

Equation (6.104) is a Taylor series expansion [Equation (2.10)] about the carrier frequency ω. The β_2 parameter in Equation (6.104) quantifies

[1]A different expression would result if Equation (6.11) were used because of the residual chirp in that field. The reader is referred to works by Devaux et al. (1993), Schiess and Carlden (1994) and Smith et al. (1997) for analysis of chirped MZMs and chromatic dispersion, or Koyama and Iga (1998) for a generalization of chirp in external modulators.

the group velocity dispersion and can be written as (Agrawal 2013)

$$\beta_2 = \frac{-\lambda^2 D}{2\pi c}, \tag{6.105}$$

where λ is the optical wavelength, D is dispersion parameter for the fiber, and c is the speed of light. The parameter D is a characteristic quantity of a fiber and is usually expressed in units of [ps/(nm · km)], describing the time shift per unit wavelength and length (see also Section 5.4).

With the aforementioned definitions, the goal is to calculate the small-signal RF gain for an IMDD link including the effects of chromatic dispersion. To this end, the terms for the optical carrier ($k = 0$), the upper fundamental sideband ($m = 0$), and the lower fundamental sideband ($m = -1$) can be extracted from Equation (6.103) as

$$\begin{aligned} E_1 \approx & -\gamma' \sin(\omega t + \beta_0 L) + \frac{\gamma' \phi_{\mathrm{rf}}}{4} \cos\left(\omega t + \Omega t - \beta_0 L + \beta_1 \Omega L - \frac{\beta_2 \Omega^2 L}{2}\right) \\ & - \frac{\gamma' \phi_{\mathrm{rf}}}{4} \cos\left(\omega t - \Omega t - \beta_0 L - \beta_1 \Omega L - \frac{\beta_2 \Omega^2 L}{2}\right). \end{aligned} \tag{6.106}$$

The small-signal Bessel approximation has been applied to Equation (6.106) such that $J_0(\phi_{\mathrm{rf}}/2) \to 1$ and $J_1(\phi_{\mathrm{rf}}/2) \to \phi_{\mathrm{rf}}/4$. The RF photocurrent can be calculated from Equation (6.106) via the following definition:

$$I(t) = \Re A \left(\frac{\varepsilon}{\mu}\right)^{1/2} \langle E^2(t) \rangle, \tag{6.107}$$

where $\langle\cdot\rangle$ indicates a time average on optical scales. Note that Equation (6.107) applies to a real field such as Equation (6.106); the average optical power is calculated differently than in Equation (6.6), which applies to a complex field. Performing the calculation prescribed by Equation (6.107) on Equation (6.106) yields

$$\begin{aligned} I(t) = & I_{\mathrm{dc}} + I_{\mathrm{dc}} \phi_{\mathrm{rf}} \cos(\beta_1 \Omega L) \cos\left(\frac{\beta_2 \Omega^2 L}{2}\right) \sin(\Omega t) \\ & + I_{\mathrm{dc}} \phi_{\mathrm{rf}} \sin(\beta_1 \Omega L) \cos\left(\frac{\beta_2 \Omega^2 L}{2}\right) \cos(\Omega t), \end{aligned} \tag{6.108}$$

where only first-order terms in ϕ_{rf} were kept (ϕ_{rf}^2 terms dropped) and $I_{\mathrm{dc}} = \Re g_o l_{\mathrm{mzm}} P_o / 2$. Calculating the average output power at the fundamental frequency, $f = \Omega/(2\pi)$, from Equation (6.108) and then taking the ratio of that to the average RF input power give the small-signal RF gain as

$$g = \frac{I_{dc}^2}{V_\pi^2}\pi^2 R_i R_o |H_{pd}|^2 \cos^2\left(\frac{\pi\lambda^2 f^2 DL}{c}\right). \tag{6.109}$$

Equation (6.109) reduces to the RF gain given by Equation (6.26) when $D \times L = 0$ and is of the same well-known form (Schmuck 1995) as Equation (5.16). Equation (6.109) quantifies the fading due to chromatic dispersion experienced in an analog link employing CW intensity modulation of a CW laser source, similarly to Equation (5.16) but including the specific constants associated with an MZM-based link. The first 3-dB point in such a link can be derived from Equation (6.109) as

$$f_{3dB} = \left(\frac{c}{4\lambda^2 DL}\right)^{1/2}. \tag{6.110}$$

Simple unit conversions allow for Equation (6.110) to be written in a convenient form:

$$f_{3dB}[\text{GHz}] = 2.74 \times 10^5 (\lambda^2[\text{nm}] \cdot D[\text{ps/nm/km}] \cdot L[\text{km}])^{-1/2}, \tag{6.111}$$

where the indicated units are those typically employed for each parameter.

The measured response for an IMDD link is plotted in Figure 6.35 and compared to Equation (6.109). The data are normalized to the value of $I_{dc}^2 \pi^2 R_i R_o |H_{pd}|^2 / V_\pi^2$ at 1 GHz. The parameters for the experiment are $D = 16.5$ ps/nm/km, $\lambda = 1551$ nm and $L = 50$ km. As can be seen in Figure 6.35, the theory matches experiment very well; the divergence between theory and measurement as the frequency increases is due to the frequency response of the MZM and photodiode. Equation (6.111) can be employed with these same parameters to determine the fiber-limited 3-dB bandwidth of 6.15 GHz. It would also be possible to operate at the peaks in the response shown in Figure 6.35. However, as shown there and predicted by the functional form of Equation (6.109), the bandwidth centered on any peak decreases as the order of the peak increases. Therefore, such methods may only be applicable in narrowband links.

In addition to fading of the fundamental signal component, chromatic dispersion can induce even-order distortion in MZM-based links. As given by Equation (6.40c), there is no even-order distortion when an MZM is biased precisely at quadrature. However, frequency-dependent

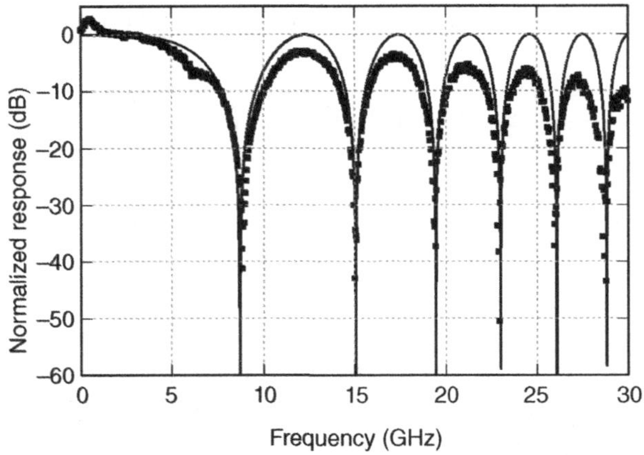

Figure 6.35. Measured (symbols) and calculated (line) normalized response for a 50-km IMDD link with $D = 16.5$ ps/nm/km and $\lambda = 1551$ nm. The roll off in the measured response is due to electro-optic components and not fiber chromatic dispersion (Urick et al. 2011).

phase rotations result in even-order terms in the photocurrent corresponding to the field in Equation (6.103). The second-harmonic content is included by applying the small-signal Bessel approximation to Equation (6.103) and keeping terms up to second order in ϕ_{rf}. The algebra and trigonometry involved in calculating the second-harmonic power as a function of D are more involved than the derivation of Equation (6.109) [see also Equation (13) in McKinney et al. (2011)]; a resulting expression is

$$P_{2\text{f}} = \frac{\phi_{\text{rf}}^4 I_{\text{dc}}^2 R_{\text{o}} |H_{\text{pd}}|^2}{128} \left[\cos\left(\frac{4\pi\lambda^2 f^2 DL}{c} \right) - 1 \right]^2. \tag{6.112}$$

Shown in Figure 6.36 are measured data for a 50-km IMDD link compared to Equation (6.112). The theory and data agree quite well where the chromatic-dispersion-induced second harmonic dominates all other sources of second-harmonic distortion, noting some roll off due to electro-optic components at the higher frequencies. As described previously, MZM bias error will result in even-order distortion; for the data in Figure 6.36, the second harmonic due to MZM bias inaccuracy is near −90 dBm, which is consistent with the measured data between 8 and 9 GHz. When the dispersion-induced second harmonic is nulled, the MZM bias inaccuracy limits performance as shown near 8.5 GHz. However, the MZM-generated second harmonic and dispersion-induced harmonic are out of phase; therefore, they

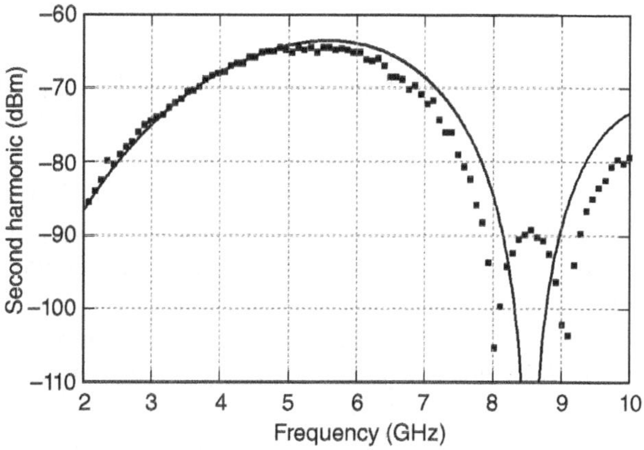

Figure 6.36. Measured (symbols) and calculated (line) second harmonic for a 50-km IMDD link with $D = 17$ ps/nm/km and $\lambda = 1550$ nm (Urick et al. 2011).

will extinguish when their amplitudes match, which is the case just above 8 GHz and just below 9 GHz. Taking the data in Figure 6.36 as an example, an IMDD link that intrinsically exhibits a −90-dBm second harmonic will experience upward of a 25-dB increase in second harmonic distortion after propagation through 50 km of fiber due to chromatic dispersion. The peak second-harmonic level near 5.5 GHz in this example corresponds to a minimal penalty in fundamental response as given by Equation (6.109) for the same parameters. In fact, the 3-dB bandwidth given by Equation (6.111) for these parameters is 6.1 GHz, which can be misleading if the intended operating bandwidth is multioctave. Therefore, high fidelity multioctave long-haul analog link designs must account for dispersion-induced even-order distortion growth in addition to fundamental fading and phase-to-intensity noise conversion. A metric that captures the effect of both fundamental fading and second-order distortion growth is the link OIP2 as a function of the dispersion parameter, which can be obtained by using Equations (6.109) and (6.112) with Equation (2.18):

$$\mathrm{OIP2} = \frac{1}{8} I_{\mathrm{dc}}^2 R_{\mathrm{o}} |H_{\mathrm{pd}}|^2 \sin^{-4}\left(\frac{\pi \lambda^2 f^2 D L}{c}\right). \tag{6.113}$$

In this case, it was assumed that V_π and $|H_{\mathrm{pd}}|$ are equal at the fundamental and the second-order distortion frequencies, a condition that will not always hold. Note also that a factor of four was applied to Equation (6.112) in deriving Equation (6.113) to put the OIP2 in terms of IMD, not harmonic distortion (see Section 2.3.2).

REFERENCES

Ackerman, E. I., "Broad-band linearization of a Mach-Zehnder electrooptic modulator," *IEEE Transactions on Microwave Theory and Techniques*, 47(12), 2271–2279 (1999).

Agrawal, G., *Nonlinear Fiber Optics, 5th edition, Academic Press* (2013).

Betts, G. E., "Linearized modulator for suboctave-bandpass optical analog links," *IEEE Transactions on Microwave Theory and Techniques*, 42(12), 2642–2649 (1994).

Betts, G. E., "$LiNbO_3$ external modulators and their use in high performance analog links," in *RF Photonic Technology in Optical Fiber Links*, W. S. C. Chang (editor), Cambridge (2002).

Bucholtz, F., Urick, V. J., Godinez, M., and Williams, K. J., "Graphical approach for evaluating performance limitations in externally modulated analog photonic links," *IEEE Transactions on Microwave Theory and Techniques*, 56(1), 242–247 (2008).

Chan, E. W. H., Zhang, W., and Minasian, R. A., "Photonic RF phase shifter based on optical carrier and RF modulation sidebands amplitude and phase control," *Journal of Lightwave Technology*, 30(23), 3672–3678 (2012).

Cox, C. H. III and Ackerman, E. I., "Recent advances in high-frequency (>10 GHz) microwave photonic links," in *Optical Fiber Telecommunications VIB*, I. P. Kaminow, T. Li, and A. E. Willner (editors), Academic (2013).

Devaux, F., Sorel, Y., and Kerdiles, J. F., "Simple measurement of fiber dispersion and of chirp parameter of intensity modulated light emitter," *Journal of Lightwave Technology*, 11(12), 1937–1940 (1993).

Hastings, A., Urick, V. J., Sunderman, C., Diehl, J. F., McKinney, J. D., Tulchinsky, D., Devgan, P. S., and Williams, K. J., "Suppression of even-order photodiode nonlinearities in multi-octave photonic links," *Journal of Lightwave Technology*, 26(15), 2557–2562 (2008).

Hutchinson, M. N., Singley, J. M., Urick, V. J., Harmon, S. R., McKinney, J. D., and Frigo, N. J., "Mitigation of photodiode induced even-order distortion in photonic links with predistortion modulation," *Journal of Lightwave Technology*, 32(20), 3885–2892 (2014).

Johnson, L. M. and Roussell, H. V., "Reduction of intermodulation distortion in interferometric optical modulators," *Optics Letters*, 13(10), 928–930 (1988).

Katz, A., "Linearization: reducing distortion in power amplifiers," *IEEE Microwave Magazine*, 2(4), 37–49 (2001).

Kawanishi, T. and Izutsu, M., "Linear single-sideband modulation for high-SNR wavelength conversion," *IEEE Photonics Technology Letters*, 16(6), 1534–1536 (2004).

Kawanishi, T., Sakamoto, T., and Izutsu, M., "High-speed control of lightwave amplitude, phase, and frequency by use of electrooptic effect," *IEEE Journal of Selected Topics in Quantum Electronics*, 13(1), 79–91 (2007).

Korotky, S. K., and de Ridder, R. M., "Dual parallel modulation schemes for low-distortion analog optical transmission," *IEEE Journal on Selected Areas in Communications*, 8(7), 1377–1381 (1990).

Koyama, F. and Iga, K., "Frequency chirping in external modulators," *Journal of Lightwave Technology*, 6(1), 87–93 (1998).

MacDonald, J. A., Kubak, M. V., and Katz, A., "Wideband dynamic range improvement of microwave photonic links," in *IEEE Avionic Fiber-Optics and Photonics Conference*, 67–68 (2005).

McKinney, J. D., Godinez, M., Urick, V. J., Thaniyavarn, S., Charczenko, W., and Williams, K. J., "Sub-10 dB noise figure in a multiple-GHz analog optical link," *IEEE Photonics Technology Letters*, 19(7), 465–467 (2007).

McKinney, J. D., Urick, V. J., and Briguglio, J., "Optical comb sources for high dynamic-range single-span long-haul analog optical links," *IEEE Transactions on Microwave Theory and Techniques*, 59(12), 3249–3257 (2011).

Moslehi, B., Goodman, J. W., Tur, M., and Shaw, H. J., "Fiber-optic lattice signal processing," *Proceedings of the IEEE*, 72(7), 909–930 (1984).

Nazarathy, M., Berger, J., Ley, A. J., Levi, I. M., and Kagan, Y., "Progress in externally modulated AM CATV transmission systems," *Journal of Lightwave Technology*, 11(1), 82–105 (1993).

O'Connor, S. R., Clark, T. R., Novak, D., "Wideband adaptive feedforward photonic link," *Journal of Lightwave Technology*, 26(15), 2810–2816 (2008).

Priest, R. G., "Analysis of fiber interferometer utilizing 3 × 3 fiber coupler," *IEEE Transactions on Microwave Theory and Techniques*, 30(10), 1589–1591 (1982).

Rogge, M. S., Urick, V. J., and Bucholtz, F., "Analysis of an optical channelization technique for microwave applications," NRL Memorandum Report, NRL/MR/5652-07-9061 (2007).

Schiess, M. and Carlden, H., "Evaluation of the chirp parameter of a Mach-Zehnder intensity modulator," *Electronics Letters*, 30(18), 1524–1525 (1994).

Schmuck, H., "Comparison of optical millimeter-wave system concepts with regard to chromatic dispersion," *Electronics Letters*, 31(21), 1848–1849 (1995).

Skeie, H. and Johnson, R. V., "Linearization of electro-optic modulators by a cascade coupling of phase modulating electrodes," *Proceedings of SPIE*, 1583, 153–164 (1991).

Smith, G. H., Novak, D., and Ahmed, Z., "Overcoming chromatic-dispersion effects in fiber-wireless systems incorporating external modulators," *IEEE Transactions on Microwave Theory and Techniques*, 45(8), 1410–1415 (1997).

Urick, V. J., "Long-haul analog photonics principles with applications," Doctoral Dissertation, UMI No. 3255815 (2007).

Urick, V. J., Rogge, M. S., Knapp, P. F., Swingen, L., and Bucholtz, F., "Wideband predistortion linearization for externally-modulated long-haul analog fiber optic-links," *IEEE Transactions on Microwave Theory and Techniques*, 54(4), 1458–1463 (2006).

Urick, V. J., Hastings, A. S., McKinney, J. D., Sunderman, C., Diehl, J. F., Devgan, P. S., Colladay, K., and Williams, K. J., "Photodiode linearity requirements for radio-frequency photonics and demonstration of increased performance using photodiode arrays," in *IEEE Topical Meeting on Microwave Photonics Digest*, 86–89 (2008).

Urick, V. J., Bucholtz, F., McKinney, J. D., Devgan, P. S., Campillo, A. L., Dexter, J. L., and Williams, K. J., "Long-haul analog photonics," *Journal of Lightwave Technology*, 29(8), 1182–1205 (2011).

Urick, V. J., Diehl, J. F., Draa, M. N., McKinney, J. D., and Williams, K. J., "Wideband analog photonic links: some performance limits and considerations for multi-octave implementations," *Proceedings of SPIE*, 8259, 825904-1–14 (2012).

Urick, V. J., Hutchinson. M. N., Singley, J. M., McKinney, J. D., and Williams, K. J., "Suppression of even-order photodiode distortions via predistortion linearization with a bias-shifted Mach-Zehnder modulator," *Optics Express*, 21(12), 14368–14376 (2013).

Weihs, G., Reck, M., Weinfurter, H., and Zeilinger, A., "All-fiber three-path Mach-Zehnder interferometer," *Optics Letters*, 21(4), 302–304 (1996).

Williams, K. J. and Esman, R. D., "Optically amplified downconverting link with shot-noise-limited performance," *IEEE Photonics Technology Letters*, 8(1), 148–150 (1996).

Williams, K. J., Nichols, L. T., and Esman, R. D., "Photodetector nonlinearity limitations on a high-dynamic range 3 GHz fiber optic link," *Journal of Lightwave Technology*, 16(2), 192–199 (1998).

Winzer, P. J., Gnauck, A. H., Chandrasekhar, S., Draving, S., Evangelista, J., and Zhu, B., "Generation and 1,200-km transmission of 448-Gb/s ETDM 56-Gbaud PDM 16-QAM using a single I/Q modulator," in *36th European Conf. on Optical Communication (ECOC)*, 1–3 (2010).

Zhu, G., Liu, W., and Fetterman, H. R., "A broadband linearized coherent analog fiber-optic link employing dual parallel Mach-Zehnder modulators," *IEEE Photonics Technology Letters*, 21(21), 1627–1629 (2009).

CHAPTER 7

EXTERNAL PHASE MODULATION WITH INTERFEROMETRIC DETECTION

7.1 INTRODUCTION

Intensity modulation as described in the previous chapter is directly compatible with a simple photodiode for demodulation. However, optical phase modulation offers a few important advantages over intensity modulation for radio-frequency (RF) signal transport. An electro-optic phase modulator does not require a bias control like a Mach–Zehnder intensity modulator. This front-end simplicity can be particularly attractive when the transmit side of a link must be located in a harsh environment. A phase modulator does not exhibit the intrinsic 3-dB optical loss characteristic of a single-output quadrature-biased Mach–Zehnder modulator (MZM) and therefore makes more efficient use of the input optical power. As discussed in Chapter 5, constant-intensity phase modulation is less susceptible to fiber nonlinearities than intensity modulation. These advantages come at the cost of more-complicated demodulation schemes for phase modulation.

Photodetectors do not respond to changes in optical phase, so optical phase modulation must be converted to intensity modulation for detection. Coherent techniques for phase demodulation include utilization

Fundamentals of Microwave Photonics, First Edition.
Vincent J. Urick Jr, Jason D. McKinney, and Keith J. Williams.

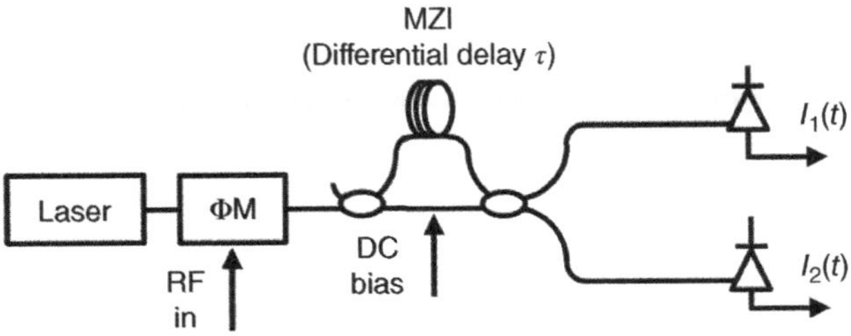

Figure 7.1. Architecture for phase modulation (ΦM) with a Mach–Zehnder interferometer (MZI).

of optical local oscillators (Kalman et al. 1994). A Mach–Zehnder interferometer (MZI) has been demonstrated for analog applications (LaGasse and Thaniyavarn 1997) and is widely employed for decoding of digital phase-shift-keyed signals (Gnauck and Winzer 2005). Various other filter-based mechanisms have also been proposed and demonstrated for optical phase demodulation (Wyrwas and Wu 2009). Commercial needs have resulted in the availability of high quality MZIs for digital phase demodulation, making the MZI a practical candidate for analog phase demodulation. Analog optical phase modulation with interferometric demodulation employing an MZI is the concentration of this chapter, with further review of other phase demodulation techniques provided in the final section.

An analog optical phase-modulated link using an asymmetric MZI for phase-to-intensity conversion is shown in Figure 7.1. A laser is modulated with an electro-optic phase modulator, typically fabricated using $LiNbO_3$, and passed through an MZI with a differential time delay τ. It is assumed that the phase modulator does not exhibit any parasitic intensity modulation (Tavlykaev et al. 2012). A DC bias is applied to the MZI in order to control the time delay with precision on the order of optical scales. It is assumed that the MZI employs ideal, lossless 50/50 couplers as described by Equation (6.2) with $a = 1$ and $K = 1/2$. Differential polarization rotations and losses in the MZI are neglected but could be handled using the formalism introduced by Frigo and Bucholtz (2009). The MZI has two outputs, each of which feeds a photodiode. The two photodiodes are usually configured as a balanced receiver, which can be achieved with direct current subtraction (Figure 6.5) or with an RF hybrid (Figure 6.6). Therefore, this type of link allows for balanced detection with single-fiber transmission span, whereas a MZM-based balanced link requires two phase-matched fibers between transmit and receive ends. As noted previously, balanced detection can result in signal-to-noise ratio (SNR) advantages and can mitigate photodiode even-order distortions (Hastings et al. 2008). The analysis in this section assumes an ideal MZI and linear photodiodes with the goal of calculating the two output photocurrents.

7.2 SIGNAL TRANSFER AND GAIN

This analysis is similar to that conducted for the intensity-modulated link in Chapter 6 and begins with the single-tone response. The single-tone input to the phase modulator is

$$V_{\text{in}}(t) = V_{\text{rf}} \sin(\Omega t), \tag{7.1}$$

where Ω is the angular frequency of the signal. Nonoscillatory (DC) terms are neglected in Equation (7.1), as they give rise to arbitrary constant phase offsets. The optical field at the phase modulator output is then

$$E_0(t) = \gamma e^{i\omega t + \phi(t)}, \tag{7.2}$$

where γ is given by Equation (6.10) with the loss factor for the MZM (l_{mzm}) replaced by the loss factor for the phase modulator ($l_{\phi\text{m}}$) and ω is the optical carrier frequency. The phase shift in Equation (7.2) is simply

$$\phi(t) = \phi \sin(\Omega t) = \frac{\pi V_{\text{rf}}}{V_\pi} \sin(\Omega t), \tag{7.3}$$

where V_π is the voltage required to produce a phase shift of π in the phase modulator. Direct detection of the constant-intensity field given by Equation (7.2) yields a DC photocurrent $I = \Re l_{\phi\text{m}} P_{\text{o}}$, which contains no RF oscillations. However, there are modulation sidebands in the optical domain for the field Equation (7.2). Expansion of Equation (7.2) results in

$$\text{Re}(E_0) = \gamma \sum_{k=-\infty}^{\infty} J_k(\phi) \cos(\omega t + k\Omega t), \tag{7.4}$$

which has RF sidebands at the fundamental frequency and all integer harmonics on both sides of the optical carrier. Qualitatively, the photocurrent created by sidebands on one side of the optical carrier is out of phase with that from the opposite side, and therefore, they cancel each other out (see also Figure 4.9). Recovery of the RF information that has been encoded on the optical phase requires some sort of filtering mechanism. The MZI provides such a mechanism to convert the phase information to intensity information and can be represented by the transfer function

$$\begin{bmatrix} E_1(t) \\ E_2(t) \end{bmatrix} = \frac{\sqrt{l_{\text{mzi}} g_{\text{o}}}}{2} \begin{bmatrix} 1 & i \\ i & 1 \end{bmatrix} \begin{bmatrix} \Gamma(\tau) & 0 \\ 0 & 1 \end{bmatrix} \begin{bmatrix} 1 & i \\ i & 1 \end{bmatrix} \begin{bmatrix} E_0(t) \\ 0 \end{bmatrix}, \quad (7.5)$$

where l_{mzi} is the excess optical loss factor for the MZI and g_{o} is the net optical link gain. The term $\Gamma(\tau)$ is a time-delay operator that acts on a time-dependent field as $\Gamma(\tau)E(t) = E(t - \tau)$. Equation (7.2) can be inserted into Equation (7.5) to calculate the two fields at the MZI output. The two associated photocurrents can be written as

$$I_{1,2}(t) = I_{\text{dc,q}} \mp \cos(\omega\tau)\cos(\Delta\phi) \pm \sin(\omega\tau)\sin(\Delta\phi), \quad (7.6)$$

where $I_{\text{dc,q}} = \Re l_{\text{mzi}} g_{\text{o}} l_{\phi\text{m}} P_{\text{o}}/2$ is the DC photocurrent when the MZI is at quadrature, MZI quadrature is defined as $\omega\tau = (2n + 1)\pi/2$ with n an integer, and $\Delta\phi(t) = \phi(t) - \phi(t - \tau)$. Equation (7.6) is general for any phase-shift waveform $\phi(t)$. Inserting Equation (7.3) for $\phi(t)$ yields

$$\begin{aligned} I_{1,2}(t) = {} & I_{\text{dc,q}} \mp I_{\text{dc,q}} \cos(\omega\tau) J_0(x) \\ & \mp 2I_{\text{dc,q}} \sin(\omega\tau) \sum_{j=0}^{\infty} (-1)^j J_{2j+1}(x) \cos[(2j+1)(\Omega t - \Omega\tau/2)] \\ & \pm 2I_{\text{dc,q}} \cos(\omega\tau) \sum_{k=1}^{\infty} (-1)^k J_{2k}(x) \cos[2k(\Omega t - \Omega\tau/2)], \end{aligned} \quad (7.7)$$

where the argument of the Bessel functions is $x = 2\phi \sin(\pi f \tau)$. The term $\Omega\tau/2$ in the argument of the cosines is inconsequential because it is simply a constant phase offset. However, such terms are important in the case where multiple MZIs are employed, as for linearization (McKinney et al. 2009) or instantaneous frequency measurement (Xu et al. 2011), because they give rise to important differential phase shifts. Equation (7.7) is the desired result, the total RF photocurrents for the two photodiodes. The first line is the DC photocurrent where the two outputs have opposite dependence on MZI bias. The second line in Equation (7.7) is the odd-order RF terms and the third line describes the even-order RF terms. Note that Equation (7.7) is very similar to Equation (6.18), a point that will be revisited in the following section in the context of a two-tone stimulus.

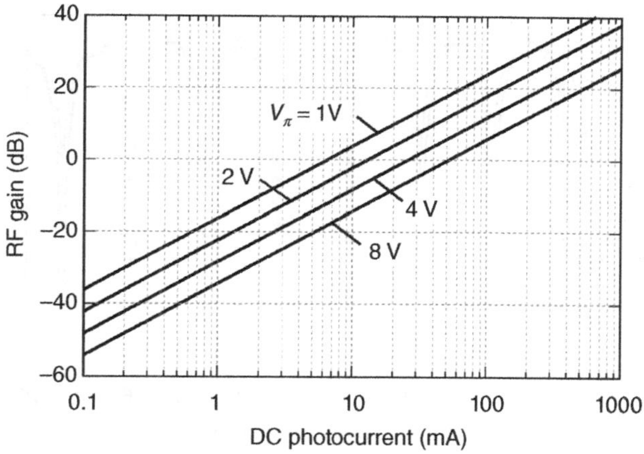

Figure 7.2. Calculated small-signal gain for a phase-modulated link as given by Equation (7.8) for $\pi f\tau = \omega\tau = \pi/2$, $R_i = R_o = 50\,\Omega$, and $H_{pd} = 1/2$.

The current for the fundamental is obtained from the second line in Equation (7.7) with $j = 0$. The small-signal power gain for the fundamental can then be calculated as

$$g = \left[\frac{2\pi I_{dc,q} \sin(\pi f\tau)\sin(\omega\tau)}{V_\pi}\right]^2 R_i R_o |H_{pd}|^2, \tag{7.8}$$

where R_i and R_o are the input and output resistances, respectively, and $H_{pd}(f)$ is the photodiode frequency response. Equation (7.8) is plotted in Figure 7.2 as a function of I_{dc} with four values of V_π for $R_i = R_o = 50\,\Omega$ and $H_{pd} = 1/2$. In Figure 7.2, $\pi f\tau$ and $\omega\tau$ are both some value $(2n+1)\pi/2$ with n an integer, which correspond to the peak RF response and a quadrature MZI bias, respectively. At peak response, a phase-modulated link as described in this section exhibits 6 dB more RF gain than an intensity modulation direct-detection (IMDD) link with the same total average photocurrent and V_π. The frequency response for an analog optical phase-modulated link is shown in Figure 7.3 (Urick et al. 2013). A semiconductor laser was used as the source followed by a commercial 40-GHz phase modulator. The modulator $V_\pi = 2.9$ V at 1.3 GHz and V_π as a function of frequency from 1 to 40 GHz was supplied by the manufacturer. The MZI had a differential time delay of $\tau = 375$ ps and exhibited an optical extinction ratio of about 25 dB (measured without modulation of the laser). One output of the MZI was connected to an optical power meter to monitor bias and the other to a wideband photodiode. The photodiode

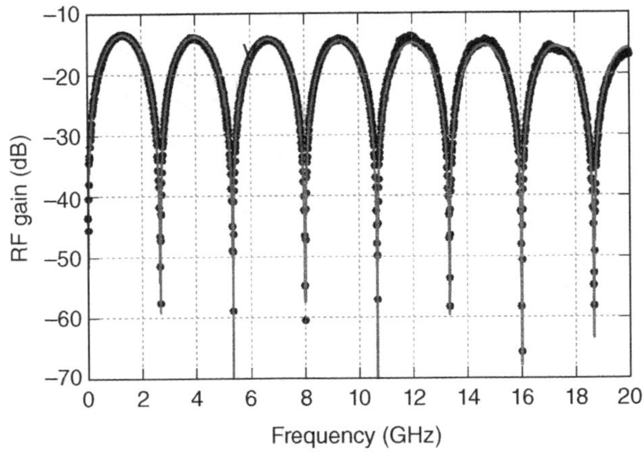

Figure 7.3. Measured (symbols) and calculated (line) small-signal gain for a phase-modulated link (Urick et al. 2013).

package included an impedance-matching circuit with a flat response to beyond 18 GHz where $H_{pd} = 1/2$. The small-signal gain for the link was measured with a network analyzer as shown in Figure 7.3. A single-tone stimulus with an average power of −15 dBm was used for these data. The MZI was biased at quadrature with an average photocurrent of 4.0 mA. The input and output of the link were matched to the source and load, respectively, with $R_i = R_o = 50\,\Omega$. Equation (7.8) is plotted for these parameters in Figure 7.3. Excellent agreement between theory and experiment is observed. The roll off in frequency is dominated by the modulator frequency response except near 20 GHz where the photodiode response drops slightly below the $H_{pd} = 1/2$ used in the calculation. The data between 14 and 16 GHz demonstrate an RF extinction of about 50 dB, which is expected given the optical extinction of about 25 dB.

The response of a phase-modulated link can be tailored to a desired frequency by adjusting the differential delay of the MZI. Shown in Figure 7.4 are normalized measured data for $\tau = 100$ ps and $\tau = 375$ ps, demonstrating the available bandwidth for those delays (Urick et al. 2007). The frequencies at which the response is a maximum are $f_k = (k + 1/2)/\tau$, where k is an integer. Neglecting roll off in components such as the phase modulator or photodiode, the 3-dB bandwidth centered symmetrically around any f_k is $f_{3dB} = 1/(2\tau)$.

Just as in the case of a balanced IMDD link described in Chapter 6, the signal-transfer equations in this present chapter can apply to a balanced- or single-photodiode architecture when the MZI is biased at

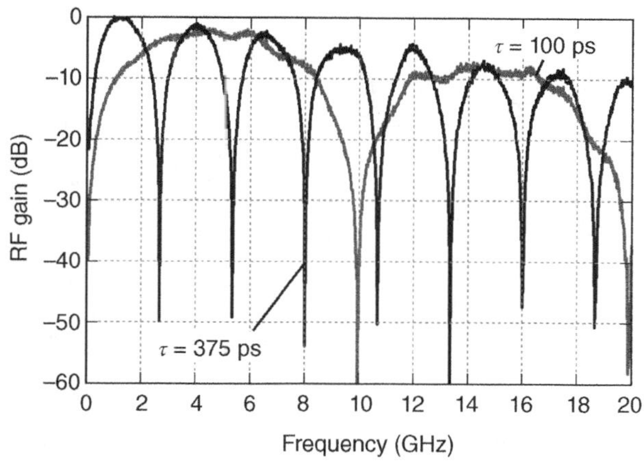

Figure 7.4. Measured small-signal gain for phase-modulated links employing interferometers with $\tau = 375\,\text{ps}$ and $\tau = 100\,\text{ps}$ (Urick 2007a).

quadrature. That is, the signal-transfer equations at MZI quadrature are valid if $I_{\text{dc,q}}$ is the DC photocurrent for the single photodiode case or if $I_{\text{dc,q}}$ is the sum of the two DC photocurrents for balanced photodiodes. As given by Equation (7.8), a balanced link should therefore exhibit 6 dB more gain with the same DC current per photodiode as compared to a single-photodiode link. This point is illustrated in Figure 7.5 where measured data for a single- and balanced-photodiode link employing a MZI with $\tau = 100\,\text{ps}$ are shown (Bucholtz et al. 2006). For these data,

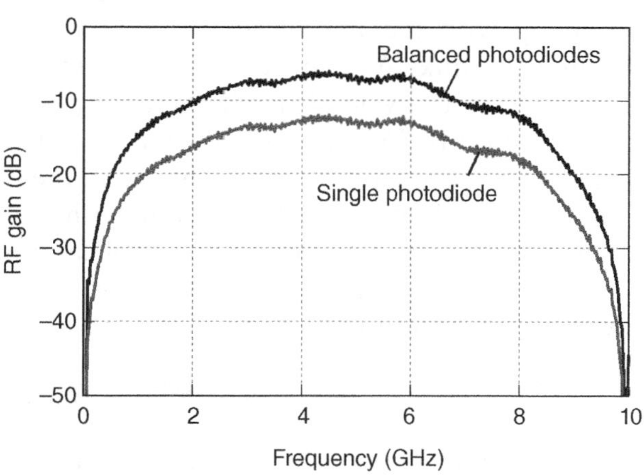

Figure 7.5. Measured small-signal gain for phase-modulated links employing a single photodiode (lower) and balanced photodiodes (upper).

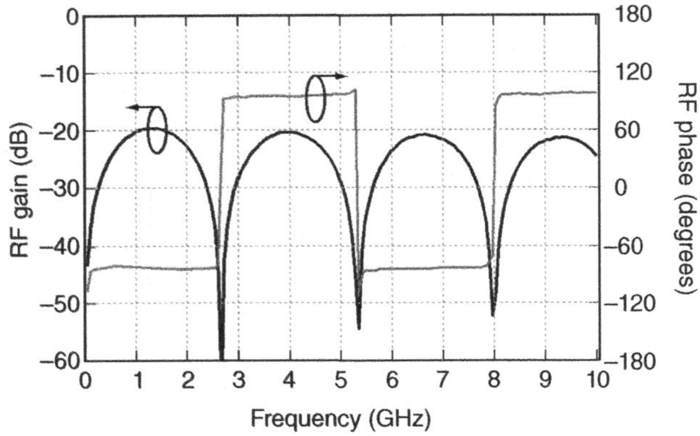

Figure 7.6. Measured small-signal amplitude and phase at the output of a phase-modulated link using a MZI with $\tau = 375\,\text{ps}$ (Urick et al. 2011a).

$R_i = R_o = 50\,\Omega$, $V_\pi = 6.9\,\text{V}$, and $H_{pd} = 1$, all at 5 GHz. The link was biased at quadrature with an average photocurrent of 5 mA per photodiode. As predicted by theory, the balanced link exhibits 6 dB more RF gain than the single photodiode link. Conversely, the gain given by Equation (7.8) matches the measured results if $I_{dc,q} = 5\,\text{mA}$ and $I_{dc,q} = 10\,\text{mA}$ for the single- and balanced-photodiode links, respectively.

Another interesting characteristic of the small-signal, single-tone stimulus for a phase-modulated link employing an MZI is the phase response as a function of frequency. Shown in Figure 7.6 are four cycles of a link employing an MZI with $\tau = 375\,\text{ps}$ (Urick et al. 2011a). The measured data show the measured magnitude squared of the S_{21} parameter along with the measured phase relative to the true time delay through the link. As can be seen in Figure 7.6, the phase flips by 180° from cycle to cycle. This characteristic is predicted by the small-signal response of Equation (7.7) where the sign of $\sin(\pi f \tau)$ changes every period, causing a 180° phase change in the current. This nature of an analog optical phase-modulated link with interferometric detection has potential use in signal processing applications.

A large-signal analysis of Equation (7.7) allows for the calculation of the output power at 1-dB compression, given by

$$P_{\text{out,1dB}} = 2I_{dc}^2 J_1^2(x_{1dB}) R_o |H_{pd}|^2 \approx 0.3587 \cdot I_{dc}^2 R_o |H_{pd}|^2. \qquad (7.9)$$

The right-hand side of Equation (7.9) follows from the solution of Equation (6.33) for $x_{1dB} = 0.9504$, the argument of the Bessel function to compress the signal by 1 dB. Therefore, the output power at 1-dB compression for a phase-modulated link with interferometric detection is equivalent to that for an IMDD link employing an MZM at the same total average photocurrent [Equation (6.32)]. However, the voltage that causes the phase-modulated architecture to compress by 1 dB is affected by the MZI filter function as

$$V_{\mathrm{rf,1dB}} = \frac{0.4752 \cdot V_{\pi}}{\pi \sin(\pi f \tau)}. \tag{7.10}$$

It can be determined by comparing Equations (6.35) and (7.10) that a phase modulator employing an MZI for demodulation at peak response will compress by 1 dB at an input power 6 dB lower than an MZM IMDD link with the same V_{π}.

A two-tone analysis of an analog optical phase-modulated link can be conducted in a manner similar to that employed for IMDD in Chapter 6. The phase shift waveform due to a two-tone drive is

$$\phi(t) = \phi_1 \sin(\Omega_1 t) + \phi_2 \sin(\Omega_2 t), \tag{7.11}$$

where DC components have again been neglected. Equation (7.11) can be inserted into Equation (7.2) to calculate the field at the output of a phase modulator for a two-tone drive. A resulting expression can be written as

$$\mathrm{Re}(E_0) = \gamma \sum_{k=-\infty}^{\infty} \sum_{n=-\infty}^{\infty} J_k(\phi_1) J_n(\phi_2) \cos[(\omega + k\Omega_1 + n\Omega_2)t]. \tag{7.12}$$

As for the field given by Equation (7.4), direct detection of Equation (7.12) does not produce any RF photocurrent. However, there are modulation sidebands in the optical domain (see Figure 4.10). Measured data in support of this claim are shown in Figure 7.7 (Rogge et al. 2007). Shown is a spectrum measured at the output of a phase modulator driven by two equal-amplitude sinusoidal tones at 9 and 12 GHz. The spectrum is centered near 193.4 THz (1550 nm). Dual-sideband modulation at these frequencies is clearly seen in Figure 7.7. Some of the distortion terms can also be resolved. The largest term shown is second-order intermodulation distortion (IMD) at 21 GHz, which is between the two second harmonics at 18 and 24 GHz. Third-order IMD at 30 GHz can also be seen. Other terms at lower offset frequencies are masked by the larger terms due to limited optical spectrum analyzer resolution.

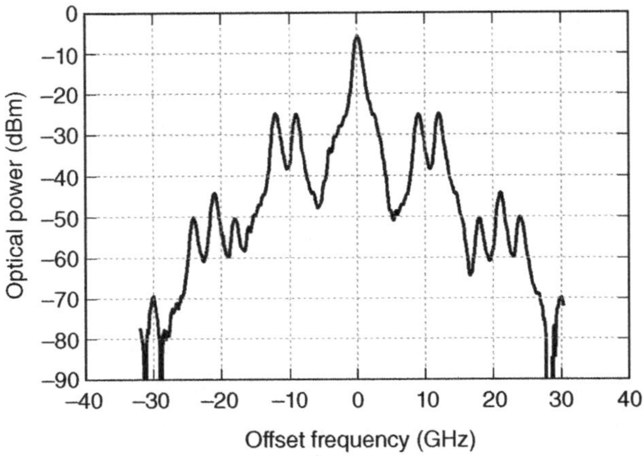

Figure 7.7. Measured optical spectrum at the output of a phase modulator driven with two tones at 9 and 12 GHz (Rogge 2007).

Applying the transfer function given by Equation (7.5) allows for the two photocurrents at the output of the link to be calculated. These photocurrents can be separated into DC, odd-order RF, and even-order RF components as

$$I_{1,2}^{\mathrm{dc}} = I_{\mathrm{dc,q}} \mp I_{\mathrm{dc,q}} \cos(\omega\tau) J_0(x_1) J_0(x_2) \tag{7.13a}$$

$$\begin{aligned}
I_{1,2}^{\mathrm{odd}} = &\pm 2I_{\mathrm{dc,q}} \sin(\omega\tau)\times \\
&\left\{ J_0(x_2) \sum_{k=0}^{\infty} (-1)^k J_{2k+1}(x_1) \cos[(2k+1)\Omega_1 t] \right. \\
&+ J_0(x_1) \sum_{n=0}^{\infty} (-1)^n J_{2n+1}(x_2) \cos[(2n+1)\Omega_2 t] \\
&+ \sum_{k=1}^{\infty} \sum_{n=0}^{\infty} (-1)^{k+n} J_{2k}(x_1) J_{2n+1}(x_2) \cos[2k\Omega_1 t - (2n+1)\Omega_2 t] \\
&+ \sum_{k=0}^{\infty} \sum_{n=1}^{\infty} (-1)^{k+n} J_{2k+1}(x_1) J_{2n}(x_2) \cos[2n\Omega_2 t - (2k+1)\Omega_1 t] \\
&+ \sum_{k=1}^{\infty} \sum_{n=0}^{\infty} (-1)^{k+n} J_{2k}(x_1) J_{2n+1}(x_2) \cos[2k\Omega_1 t + (2n+1)\Omega_2 t]
\end{aligned}$$

$$+\sum_{k=0}^{\infty}\sum_{n=1}^{\infty}(-1)^{k+n}J_{2k+1}(x_1)J_{2n}(x_2)\cos[2n\Omega_2 t+(2k+1)\Omega_1 t]\Bigg\} \tag{7.13b}$$

$$I_{1,2}^{\text{even}} = \mp 2I_{\text{dc,q}}\cos(\omega\tau)\times$$

$$\Bigg\{J_0(x_2)\sum_{k=1}^{\infty}(-1)^k J_{2k}(x_1)\cos[2k\Omega_1 t]$$

$$J_0(x_1)\sum_{n=1}^{\infty}(-1)^n J_{2n}(x_2)\cos[2n\Omega_2 t]$$

$$-\sum_{k=0}^{\infty}\sum_{n=0}^{\infty}(-1)^{k+n}J_{2k+1}(x_1)J_{2n+1}(x_2)\cos[(2k+1)\Omega_1 t-(2n+1)\Omega_2 t]$$

$$-\sum_{k=0}^{\infty}\sum_{n=0}^{\infty}(-1)^{k+n}J_{2k+1}(x_1)J_{2n+1}(x_2)\cos[(2k+1)\Omega_1 t+(2n+1)\Omega_2 t]$$

$$+\sum_{k=1}^{\infty}\sum_{n=1}^{\infty}(-1)^{k+n}J_{2k}(x_1)J_{2n}(x_2)\cos[2n\Omega_1 t-2k\Omega_2 t]$$

$$+\sum_{k=1}^{\infty}\sum_{n=1}^{\infty}(-1)^{k+n}J_{2k}(x_1)J_{2n}(x_2)\cos[2n\Omega_1 t+2k\Omega_2 t]\Bigg\} \tag{7.13c}$$

respectively, where $x_1 = 2\phi_1\sin(\pi f_1\tau)$ and $x_2 = 2\phi_2\sin(\pi f_2\tau)$. Equation (7.13) is very similar to Equation (6.39), the photocurrents for a dual-output MZM IMDD link. The differences between the two are in the Bessel function arguments, and the MZI bias in Equation (7.13) is replaced with the MZM bias in Equation (6.39). Other similarities arise if the limit $x_1 \to x_2$ is considered. That is, the limit where the two tones are equal amplitude and their frequency spacing goes to zero. This limit probes the nonlinearity at a specific frequency point. Inserting $x_1 = x_2$ into Equation (7.13) and calculating the output intercept point for the largest small-signal odd-order term result in

$$\text{OIP3} = 4I_{\text{dc,q}}^2 R_{\text{o}}|H_{\text{pd}}|^2\sin^2(\omega\tau). \tag{7.14}$$

Equation (7.14) does not exhibit a periodic response as the RF gain. At quadrature, Equation (7.14) is equivalent to Equation (6.42). The largest

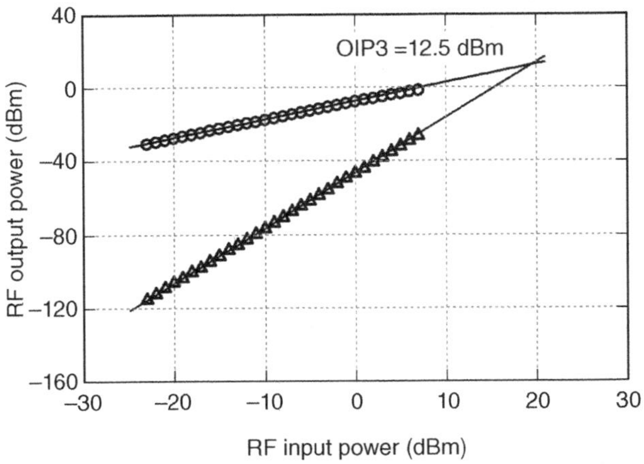

Figure 7.8. Measured OIP3 for a phase-modulated link (Urick 2007a).

small-signal even-order distortion is second order as quantified by

$$\mathrm{OIP2} = \frac{2\sin^4(\omega\tau) I_{\mathrm{dc,q}}^2 R_{\mathrm{o}} |H_{\mathrm{pd}}|^2}{\cos^2(\omega\tau)}. \tag{7.15}$$

Equation (7.15) is identical to Equation (6.47) if $\omega\tau$ and ϕ_{dc} are exchanged. Ideally, there is no even-order distortion when the MZI is biased at quadrature. Shown in Figure 7.8 is the measured OIP3 for a phase-modulated link with quadrature-biased 100-ps MZI (Urick et al. 2007). Two equal-amplitude tones at 4.99 and 5.01 GHz were used at the input to the link, which resulted in IMD terms at 4.97 and 5.03 GHz. The MZI was biased at quadrature, and a balanced-photodiode configuration was employed with $I_{\mathrm{dc,q}} = 10\,\mathrm{mA}$ (sum total). At 5 GHz, $R_{\mathrm{o}} = 50\,\Omega$ and $H_{\mathrm{pd}} = 1$ for this link. Equation (7.14) gives OIP3 = 13.0 dBm for these parameters. As shown in Figure 7.8, the experimental value of OIP3 = 12.5 dBm agrees nicely.

The condition imposed previously, $x_1 \to x_2$, will not always be the case. Therefore, it is interesting to analyze Equation (7.13) outside this limit. Consider the case where two tones at any arbitrary input frequency are used as the stimulus. The output powers at the two fundamentals are given by the first two summations in Equation (7.13b) with $k = n = 0$:

$$P_{f_1} = 2I_{\mathrm{dc,q}}^2 \sin^2(\omega\tau) J_0^2(x_2) J_1^2(x_1) R_{\mathrm{o}} |H_{\mathrm{pd}}|^2, \tag{7.16}$$

$$P_{f_2} = 2I_{\mathrm{dc,q}}^2 \sin^2(\omega\tau) J_0^2(x_1) J_1^2(x_2) R_{\mathrm{o}} |H_{\mathrm{pd}}|^2, \tag{7.17}$$

where the subscripts indicate the frequency for which the power is written. The highest order single-octave distortions are given by the first two double summations in Equation (7.13b):

$$P_{2f_1-f_2} = 2I_{\mathrm{dc,q}}^2 \sin^2(\omega\tau) J_1^2(x_2) J_2^2(x_1) R_\mathrm{o} |H_\mathrm{pd}|^2. \quad (7.18)$$

$$P_{2f_2-f_1} = 2I_{\mathrm{dc,q}}^2 \sin^2(\omega\tau) J_1^2(x_1) J_2^2(x_2) R_\mathrm{o} |H_\mathrm{pd}|^2. \quad (7.19)$$

Overall, the highest order distortions are second-order IMD given by the first two double summations in Equation (7.13c):

$$P_{f_1 \pm f_2} = 2I_{\mathrm{dc,q}}^2 \cos^2(\omega\tau) J_1^2(x_1) J_1^2(x_2) R_\mathrm{o} |H_\mathrm{pd}|^2. \quad (7.20)$$

The validity of Equations (7.16)–(7.20) was determined using the same link that produced the data for Figure 7.3. Two tones at 4.0 and 4.5 GHz were applied to the phase modulator, which had a $V_\pi = 3.1\,\mathrm{V}$ for those frequencies. The effects of the MZI bias on even- and odd-order terms were investigated with a fixed input power level of −15 dBm for each tone. The photodiode was connected to an electrical spectrum analyzer to measure the photodiode output. Data were recorded as a function of bias for the larger fundamental, and the largest single-octave and multioctave IMD. The results are shown in Figure 7.9 (Urick et al. 2013) and are compared to theory as

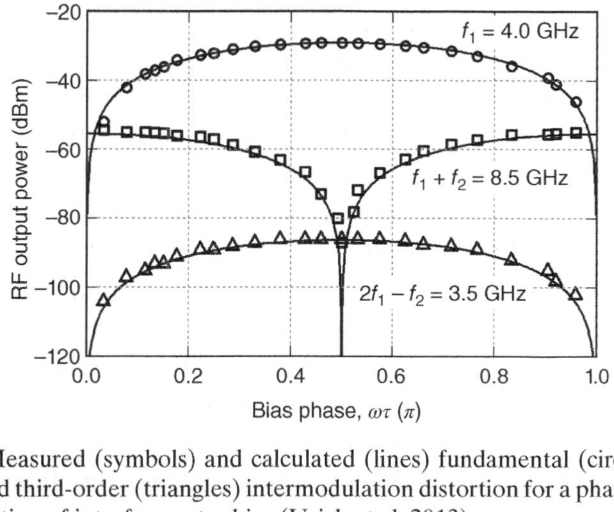

Figure 7.9. Measured (symbols) and calculated (lines) fundamental (circles), second- (squares), and third-order (triangles) intermodulation distortion for a phase-modulated link as a function of interferometer bias (Urick et al. 2013).

given by Equations (7.16), (7.18), and (7.20) demonstrating very good agreement. At quadrature, the odd-order terms are maximized and the second-order term is minimized. The data yield a third-order output intercept point of OIP3 = −0.5 dBm at quadrature. This is comparable with a theoretical OIP3 = −1.0 dBm in the limit that the tone spacing is zero [Equation (7.14)]. A precise expression for the OIP3 with arbitrary tone spacing can be obtained from Equations (7.16) and (7.18). The second order IMD does not vanish at quadrature but is minimized to −87 dBm for a −29 dBm fundamental. This corresponds to a second-order output intercept point of OIP2 = 29 dBm. The apparatus had a measured OIP2 = 67 dBm; the second-order IMD was −125 dBm with −29 dBm fundamentals with the signal sources and spectrum analyzer alone. Therefore, the measurement system is not the cause of the second-order IMD. The response to a sweep of the input power can give more insight into the source of the second-order IMD.

The output components as a function of input power at the two fundamentals are plotted in Figure 7.10 (Urick et al. 2013). The theoretical curves and measured values agree nicely. The calculations for the odd-order terms are given by Equations (7.16)–(7.19). The two fundamentals are offset by $\sin(\pi f_1 \tau)/\sin(\pi f_2 \tau)$ in the small-signal regime. That is, the MZI filter function dictates the response. However, the Bessel functions determine the compression response. Theory predicts zero even-order IMD at a quadrature bias. As noted earlier, the measurement setup was ruled out as the source of the even-order IMD. Bias error in the MZI can be disregarded because the functional form of Equation (7.20) does not match the data. Residual intensity modulation is unlikely for two reasons. Firstly, a previous study measured residual intensity modulation ranging from −101 dBe to −111 dBe for similar commercial phase modulators at 5 dBm input power (Tavlykaev et al. 2012). These levels correspond to a range of −102 dBm to −112 dBm at 5-dBm input for this measurement. Secondly, residual intensity modulation would experience a complementary filter function of the MZI, further reducing the response (Urick et al. 2011a). The theory for the even-order IMD in Figure 7.10 is a product of the two fundamental responses with a constant offset corresponding to a 29-dBm OIP2. This curve is what would be expected if photodiode small-signal distortion was the cause and matches the data well. Thus, the measured even-order IMD is attributed to photodiode distortion.

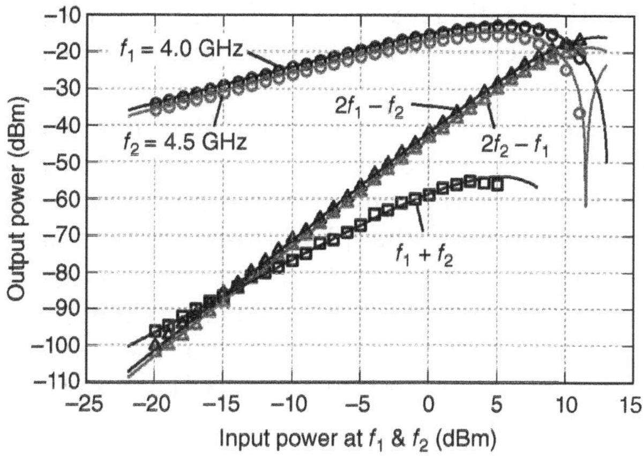

Figure 7.10. Measured (symbols) and calculated (lines) RF power response for a phase-modulated link driven by two tones at 4.0 and 4.5 GHz (Urick et al. 2013).

7.3 NOISE AND PERFORMANCE METRICS

The analysis for noise and the associated performance metrics will assume that the MZI is biased precisely at quadrature. An expression for the output noise power spectral density for an analog optical phase-modulated link employing an MZI for demodulation can be written as

$$N_{\text{out}} = 4\sin^2(\pi f\tau)\frac{I_{\text{dc}}^2}{V_\pi^2}\pi^2 R_{\text{i}}R_{\text{o}}|H_{\text{pd}}|^2 k_{\text{B}}T_{\text{s}} + k_{\text{B}}T_{\text{s}} + 2qI_{\text{dc}}R_{\text{o}}|H_{\text{pd}}|^2 + \text{RIN}\cdot I_{\text{dc}}^2 R_{\text{o}}|H_{\text{pd}}|^2, \quad (7.21)$$

where I_{dc} is DC photocurrent at quadrature. Equation (7.21) is very similar to N_{out} for IMDD as given by Equation (6.49). The only difference between the two equations is the small-signal gain experienced by the input thermal noise. The input thermal noise contribution is contained in the first term of Equation (7.21), which uses the small-signal gain given by Equation (7.8). The second term is due to thermal noise at the link output. The third term in Equation (7.21) quantifies the contribution from shot noise. The final term captures optical noise in excess of the fundamental thermal and shot noise sources as captured by the relative intensity noise (RIN) metric. The DC photocurrent in Equation (7.21) is the sum total DC photocurrent and can be applied to a single- or balanced-photodiode link. However, it will be demonstrated in the following section that the excess RIN may depend on the receiver configuration.

The use of an MZI in a phase-modulated link complicates the treatment of laser noise. Just as the MZI converts phase modulation to intensity modulation, it will convert phase noise into intensity noise. Equation (3.41) quantifies the phase-to-intensity noise conversion through an MZI for a laser whose phase noise is quantified by a Lorentzian lineshape with a full-width half-maximum linewidth $\Delta\nu$. Consider the data in Figure 7.11, which emphasizes the importance of laser selection in a link employing a MZI. Shown are measured RIN spectra for a semiconductor laser with $\Delta\nu = 94\,\text{kHz}$ and an erbium-doped fiber laser with $\Delta\nu < 1\,\text{kHz}$ passed through an MZI with $\tau = 100\,\text{ps}$ (Urick et al. 2007). For both data sets, a balanced photodiode configuration was employed with 5-mA average photocurrent per photodiode. The resulting shot-noise-limited RIN is given by Equation (3.27) with $I_{dc} = 10\,\text{mA}$ as −165 dBc/Hz. Both lasers are approximately shot noise limited, with direct detection at 10 mA average current (Urick et al. 2007). The fiber laser remains shot noise limited when passed through the MZI owing to its narrow linewidth. The spike between 1 and 2 GHz for the fiber laser is due to a competing longitudinal mode. The semiconductor laser exhibits noise well in excess of the shot noise limit when passed through the MZI. As shown in Figure 7.11,

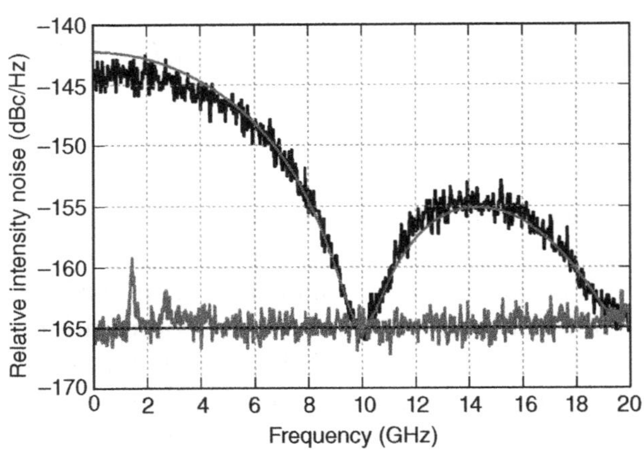

Figure 7.11. Measured relative intensity noise (RIN) spectra for an erbium-doped fiber laser (gray) and a semiconductor laser (black) after passing through an interferometer with $\tau = 100\,\text{ps}$. The calculated (lines) RIN for the semiconductor laser (gray) and shot noise limit (black) are also shown (Urick 2007a).

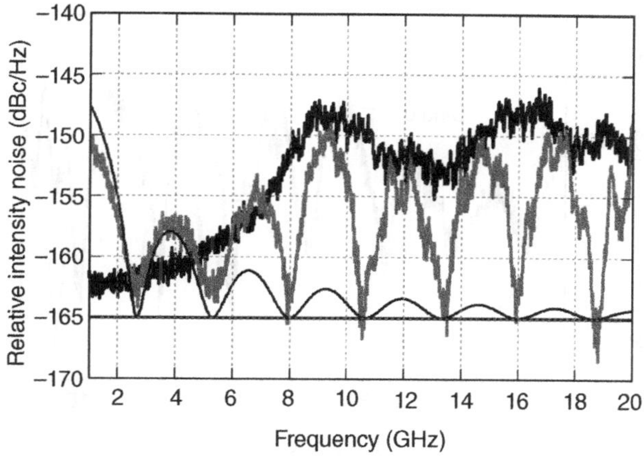

Figure 7.12. Measured data for an external-cavity semiconductor laser showing the direct-detection RIN (black) and the RIN after passing through a MZI with $\tau = 375\,\text{ps}$ with balanced detection (gray). The calculated (lines) RIN due to a laser with $\Delta\nu =$ 3.15 kHz and shot noise limit are also shown (Singley 2011).

this excess noise is described very well by Equation (3.41). Therefore, the laser phase noise must be taken into consideration when employing an interferometric demodulation technique.

Equation (3.41) accounts for laser phase noise described by a Lorentzian linewidth and may not always describe the excess RIN at the output of a MZI. Consider the data in Figure 7.12, which show the direct-detection RIN and the RIN after passing through a 375-ps MZI for an external-cavity semiconductor laser (Singley et al. 2011). Balanced photodiodes operating at 5 mA per photodiode were employed on the MZI output. Also shown in Figure 7.12 is a calculation of Equation (3.41) plus shot noise for the laser's linewidth of 3.15 kHz. The laser linewidth was determined separately via a delayed self-heterodyne experiment. The calculation and measured data agree well until about 5 GHz, after which there is significantly more noise than predicted by Equation (3.41). The additional noise exhibits the filter function of the MZI with the peaks following the shape of the direct-detection RIN curve. Therefore, the noise is attributed to an optical phase noise component associated with the relaxation oscillation. The balanced photodiode curve from Figure 7.12 is plotted again in Figure 7.13 along with a measured spectrum with a single photodiode at 5 mA. The curves agree from 1 to 2 GHz where noise described by the laser linewidth dominates but are drastically different otherwise. The single photodiode is sensitive to both intensity and

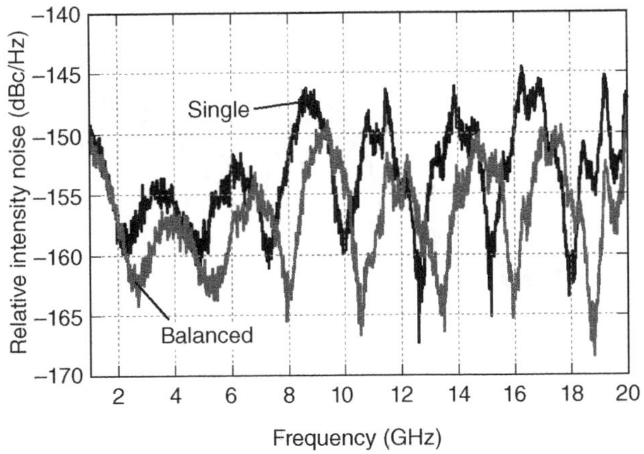

Figure 7.13. Measured RIN for an external-cavity semiconductor after passing through a MZI with $\tau = 375\,\text{ps}$ using a single detector and balanced detection (Singley et al. 2011).

phase noise at the input of the MZI, whereas balanced photodiodes pick up only phase noise at the input of the MZI. The combination of intensity and phase noise with the associated MZI filter functions can be quite different from the phase noise response alone. It is therefore very important to make judicious choices for the receiver configuration when interferometric techniques are employed.

RF performance metrics for an analog optical phase-modulated link with interferometric detection can be determined once the excess RIN and other terms in Equation (7.21) are known. The development is quite similar to that in the previous chapter and uses the definitions established in Chapter 2. The RF noise factor can be written as

$$F = \frac{N_{\text{out}} V_\pi^2}{4\pi^2 I_{\text{dc}}^2 \sin^2(\pi f \tau) R_{\text{i}} R_{\text{o}} |H_{\text{pd}}|^2 k_{\text{B}} T_{\text{s}}}. \tag{7.22}$$

Equation (7.22) is plotted in Figure 7.14, first as a function of I_{dc} for various V_π values with RIN = 0 and then for various RIN levels with a fixed V_π. The curves in Figure 7.14 assume that $\sin^2(\pi f \tau) = 1$ and use the values $R_{\text{i}} = R_{\text{o}} = 50\,\Omega$ and $H_{\text{pd}} = 1/2$. The functional dependence of noise factor with frequency varies with the type of noise at the link output. Inserting Equation (7.21) into Equation (7.22) can give some insight into the trends. For example, the noise factor is independent of frequency if input thermal noise dominates. If shot noise and/or output thermal noise are the main contributing factors, then the noise factor will follow the inverse of the small-signal gain curve.

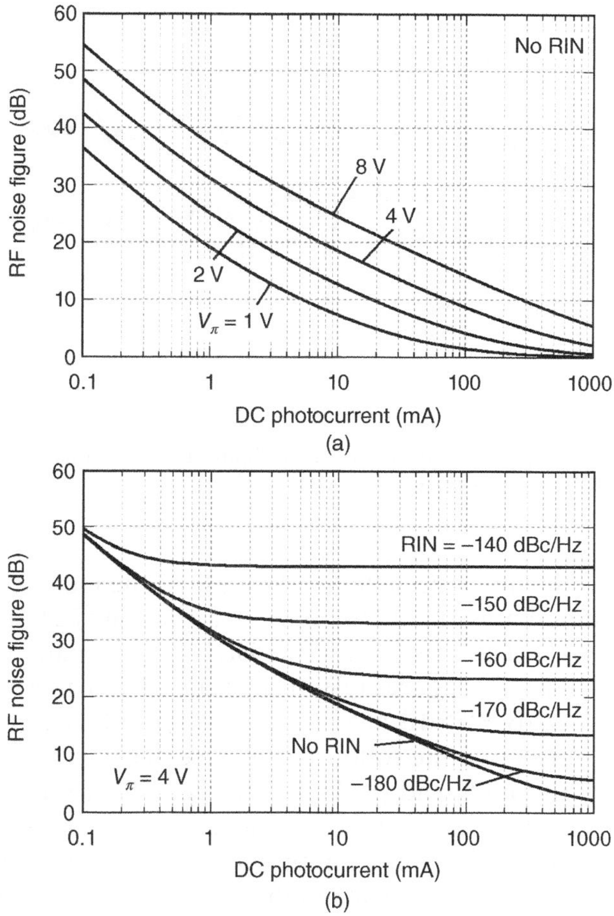

Figure 7.14. Calculated noise figure as given by Equation (7.22) with $f\tau = 1/2$, $R_i = R_o = 50\,\Omega$, and $H_{pd} = 1/2$. Shown are (a) the effect of V_π with RIN = 0 and (b) the effect of RIN with $V_\pi = 4\,V$.

A RIN-limited noise factor can be more complicated as shown in Figure 7.15. The data in Figure 7.15 include the measured gain and measured noise figure as a function of frequency for a phase-modulated link (Urick et al. 2011b). For these data, a narrow-linewidth fiber laser was used as the source with a 2-W polarization-maintaining erbium-doped fiber amplifier (EDFA) before the phase modulator, which had a $V_\pi = 1.72\,V$ at 12 GHz. A quadrature-biased MZI with $\tau = 375\,ps$ was placed before a pair of custom balanced photodiodes with $|H_{pd}|^2$ amounting to −10 dB at 12 GHz. The total average detected photocurrent was 18 mA (9 mA per diode). At 12 GHz, the measured $G = -1$ dB and NF = 17.5 dB. Equation (7.8) yields a calculated

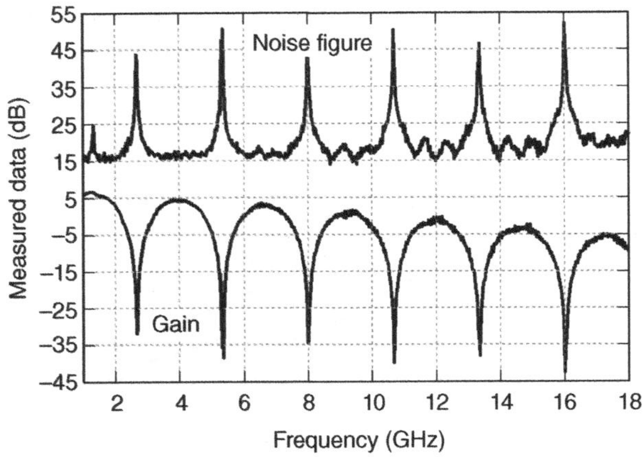

Figure 7.15. Measured gain (bottom) and noise figure (top) for a phase-modulated link using a MZI with $\tau = 375\,\text{ps}$ (Urick et al. 2011b).

$G = 0.3\,\text{dB}$ for the link parameters, very close to the measured value. A shot-noise-limited noise figure as given by Equation (7.22) is calculated as 9 dB at 12 GHz. The noise due to the EDFA is the main contribution to the measured noise figure, resulting in an EDFA noise penalty of $\text{NP} = 17.5\,\text{dB} - 9\,\text{dB} = 8.5\,\text{dB}$ at 12 GHz. The noise figure as a function of frequency does not quite follow the inverse of the gain as a function of frequency. The nulls and peaks in the noise figure and gain plots, respectively, occur in the same locations. However, as can be seen in Figure 7.15, the periodic minima in the noise figure curve are flattened compared to the periodic maxima in the gain curve. This is because the EDFA-generated noise exhibits a spectrum with peaks that mirror the gain response. The nulls in the EDFA noise spectrum are not quite as sharp as those for the gain spectrum due to limited extinction of the noise. Finally, the spike in noise figure near 1 GHz is due to laser noise originating from a competing mode in the laser cavity.

Equations (7.9) and (7.14) can be used with Equation (7.21) to specify the 1-dB compression dynamic range (CDR) and SFDR_3 for a phase-modulated link employing an MZI. Expressions for these two metrics are identical to those for IMDD given by Equations (6.53) and (6.56), respectively, with the only caveats being that N_{out} is given by Equation (7.21), the MZI is biased at quadrature, and the two-tone test is conducted in the limit that the tone spacing is very small. Theoretical plots of the single-tone $\text{CDR}_{1\text{dB}}$ and two-tone SFDR_3 for a phase-modulated link are shown in Figures 7.16 and 7.17, respectively, exemplifying the dependence on I_{dc}, V_{π}, and excess

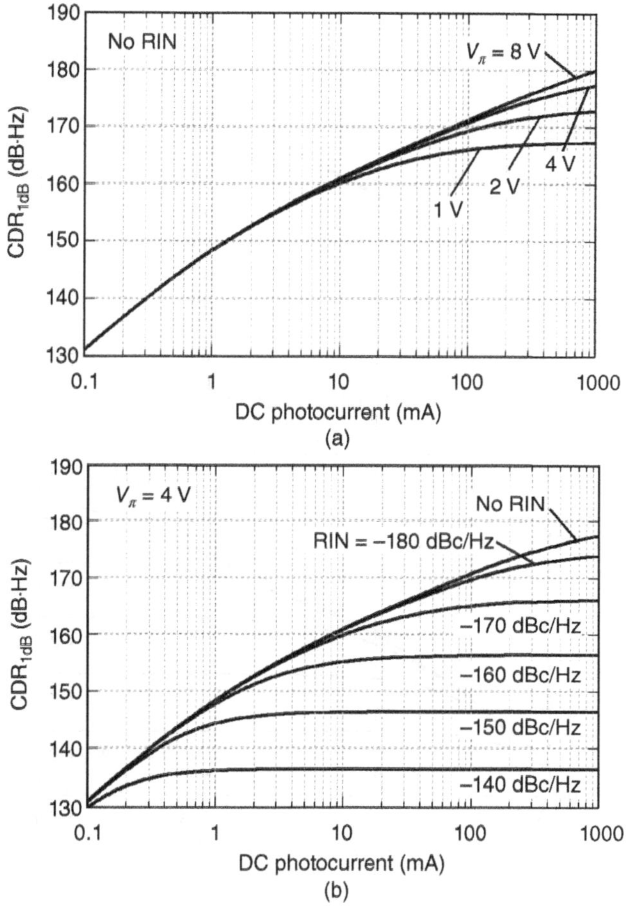

Figure 7.16. Calculated compression dynamic range for $f\tau = 1/2$, $R_i = R_o = 50\,\Omega$, $B = 1$ Hz, and $H_{pd} = 1/2$. Shown are (a) the effect of V_π with RIN = 0 and (b) the effect of RIN with $V_\pi = 4$ V.

RIN. The calculations in this case assume that the MZI is biased at quadrature, the RF output is at peak [$\sin(\pi f\tau) = 1$], $R_i = R_o = 50\,\Omega$, $H_{pd} = 1/2$, and $B = 1$ Hz. As compared to curves corresponding to IMDD, Figures 6.16 and 6.17, the CDR_{1dB} and $SFDR_3$ for the two formats are equivalent when output thermal noise, shot noise, or excess RIN determines the performance. However, as has been seen in this and the previous chapter, the excess RIN can be drastically different for the two links. Owing to increased gain in the phase-modulated link, the maximum CDR_{1dB} and $SFDR_3$ are smaller for a phase-modulated link at the peak response for the same V_π. The maximum obtainable

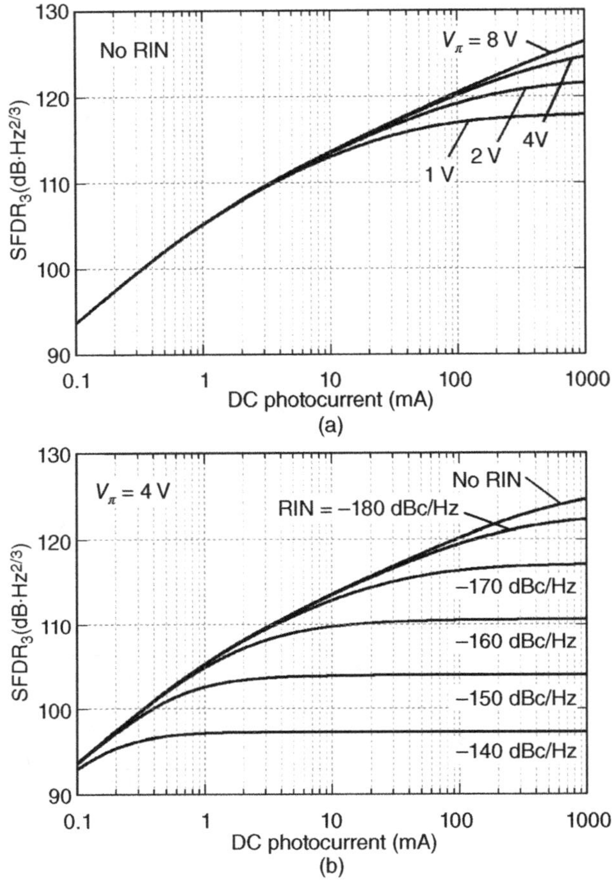

Figure 7.17. Calculated spurious-free dynamic range for $f\tau = 1/2$, $R_i = R_o = 50\,\Omega$, $B = 1$ Hz, and $H_{pd} = 1/2$. Shown are (a) the effect of V_π with RIN = 0 and (b) the effect of RIN with $V_\pi = 4$ V.

dynamic range is that for the input thermal noise limit, given by the first term in Equation (7.21). Inserting this term into Equations (6.52) and (6.55) gives the maximum dynamic ranges for the phase-modulated link as

$$\mathrm{CDR}_{1\mathrm{dB,max}} = \frac{(0.1129)V_\pi^2}{\sin^2(\pi f\tau)\pi^2 R_i k_B T_s B}. \tag{7.23}$$

$$\mathrm{SFDR}_{\max} = \left(\frac{V_\pi^2}{\sin^2(\pi f\tau)\,\pi^2 R_i k_B T_s B}\right)^{2/3}. \tag{7.24}$$

Other metrics for a phase-modulated link with interferometric detection can be derived following the procedures in Chapter 6, including SNR and single-sideband phase and amplitude noise. Similarly, Equation (7.21) can be separated into its constituent components to derive the shot noise and RIN limits as was done for IMDD in Sections 6.3.2 and 6.3.3, respectively. These exercises are fairly straightforward and left for the interested reader.

7.4 LINEARIZATION TECHNIQUES

As noted in Section 7.3, there are numerous similarities in the signal transfer functions for IMDD with an external MZM and phase modulation with interferometric demodulation. Therefore, it should not come as a surprise that linearization techniques similar to those described in Chapter 6 are applicable to phase modulation. Three particular techniques are described in this section, largely qualitatively, for linearization of phase-modulated links. Firstly, a technique to mitigate photodiode-generated even-order distortion is discussed. Two techniques to linearize the intrinsic link response are then described. The first of these uses two wavelengths at perpendicular polarizations, whereas the other uses a single laser but multiple interferometers with varied differential delays. Linearization techniques for phase modulation employing multiple MZIs can be analogous to methods employing multiple MZMs. However, multiple MZIs on the receiver-side of a link can be easier to implement because they do not need to include the modulation mechanism, only differential delays and a relatively slow bias control. Arrays of MZMs are thus restricted to materials that afford both optical and RF waveguides, whereas multiple MZIs are open to a wider range of materials, as they require an optical waveguide only. This point makes integration of MZI arrays for phase modulation linearization more feasible than integration of MZM arrays for microwave photonics applications.

The technique to linearize photodiode even-order distortion in an IMDD link as described in the latter half of Section 6.4 can have a drawback in long point-to-point links. Namely, a feedback loop from the front- to the back-end of the link may be required to maintain the cancellation condition. While this would not be a concern in delay lines, where the MZM and photodiode are co-located, a long feedback loop would not be practical in point-to-point links. A phase-modulated link with an MZI has the distinct advantage of having the MZI and

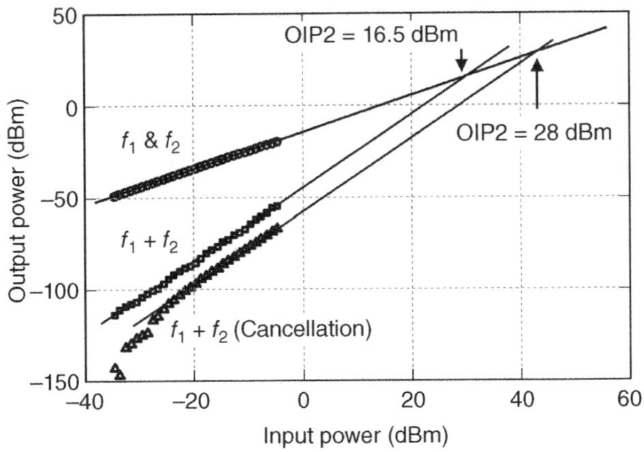

Figure 7.18. Measured data demonstrating suppression of photodiode second-order distortion in a phase-modulated link via predistortion with a MZI (Hutchinson et al. 2014).

photodiodes in the same location. As will be shown briefly in the following section, this architecture can be used to implement the photodiode linearization in much the same manner as an MZM-based link. The analysis follows the same procedure as in Section 6.4, using Equation (6.101) for the photodiode response but replacing Equation (6.40) with Equation (7.13) for the ideal link response (Hutchinson et al. 2014). The main result is the power for the IMD2 given by

$$P_2 = \frac{2\pi^4 V^4 I_{\mathrm{dc,q}}^2 \sin^2(\pi f_1 \tau)\sin^2(\pi f_2 \tau) R}{V_\pi^4}[\cos(\omega\tau) + 2a_2 I_{\mathrm{dc,q}} \sin^2(\omega\tau)]^2, \tag{7.25}$$

which also contains the cancellation condition in terms of the MZI bias. A data set demonstrating the cancellation is shown in Figure 7.18. Plotted are data showing an 11.5-dB improvement in OIP2 at the cancellation condition compared to quadrature, with suppression of IMD2 upwards of 30 dB at low input powers (Hutchinson et al. 2014).

Electro-optic modulators constructed from $LiNbO_3$ are polarization sensitive, where the ratio of V_π for orthogonal polarizations is approximately three. Johnson and Roussell (1988) used this property in an MZM to linearize the intrinsic link transfer function. A similar technique was demonstrated by Haas et al. (2008) using the architecture in Figure 7.19 to linearize a phase-modulated link. The concept is fairly straightforward. Two wavelengths at orthogonal polarizations (TE and TM) are launched into a phase modulator. The amount of optical power

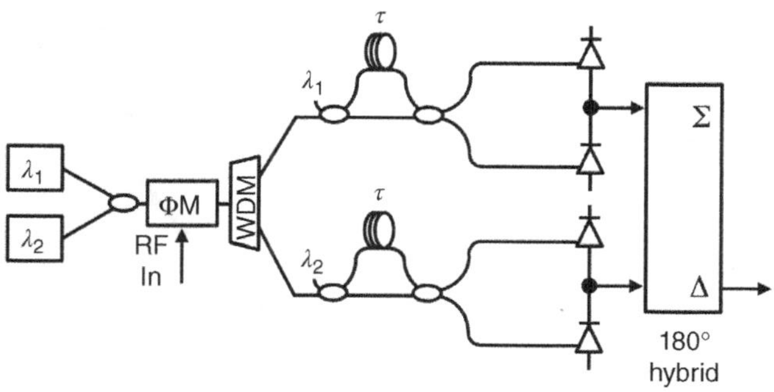

Figure 7.19. Architecture for linearization of a phase-modulated link employing two lasers at different wavelengths.

in each polarization is adjusted to match the amplitude of their respective third-order terms. A wavelength-division multiplexer (WDM) is then employed to separate the signals into two channels, which are then subtracted to eliminate the third-order nonlinearity. Assuming that the TM mode is more efficiently modulated, the linearization condition can be derived as (Haas et al. 2008)

$$P_{\mathrm{TM}} = \gamma^3 P_{\mathrm{TE}} = \left(\frac{V_{\pi,\mathrm{TM}}}{V_{\pi,\mathrm{TE}}}\right)^3 P_{\mathrm{TE}}, \tag{7.26}$$

where $V_{\pi,\mathrm{TM}}$ and $V_{\pi,\mathrm{TE}}$ are the V_π for each polarization and P_{TM} and P_{TE} are the average received optical powers for each polarization. The cost of the third-order linearization is a reduction of fundamental output power in the amount of $\gamma^2(\gamma^2 - 1)^2$, or −10.6 dB for a typical situation where $V_{\pi,\mathrm{TM}} \approx V_{\pi,\mathrm{TE}}/3$. A data set demonstrating this technique is shown in Figure 7.20 (Haas et al. 2008). As shown, the nonlinearized link exhibits 12 dB more gain than the linearized link. Both links exhibit the same noise floor of −155 dBm/Hz and, thus, the linearized link has a higher noise figure. However, the distortion for the linearized link is a fifth-ordered function of the input power with an OIP5 = −11.5 dBm as compared to the nonlinearized OIP3 = −4 dBm. The resulting dynamic range for the linearized link is larger in 1 Hz, SFDR = 115 dB · $\mathrm{Hz}^{4/5}$ as compared to the nonlinearized SFDR = 101 dB · $\mathrm{Hz}^{2/3}$, both at 5-GHz center frequency.

A phase-modulated link can be linearized by employing two MZIs with unmatched differential delays. This technique was first demonstrated by McKinney et al. (2009) using an architecture such as that shown in Figure 7.21. A single laser and phase modulator are employed

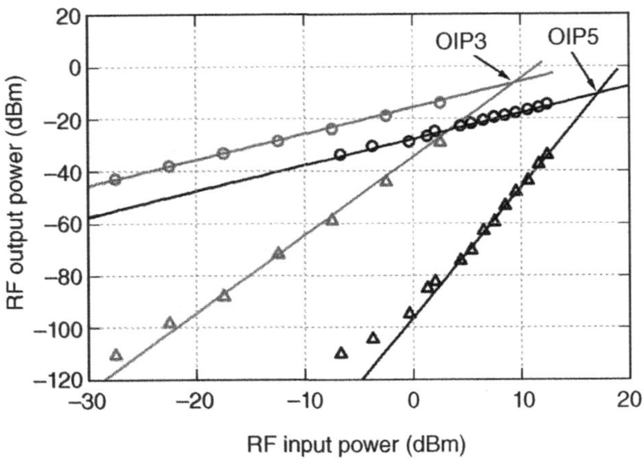

Figure 7.20. Measured data for the architecture in Figure 7.19, showing suppression of third-order intermodulation distortion. The unlinearized data are shown in gray and the linearized data are shown in black (Haas et al. 2008).

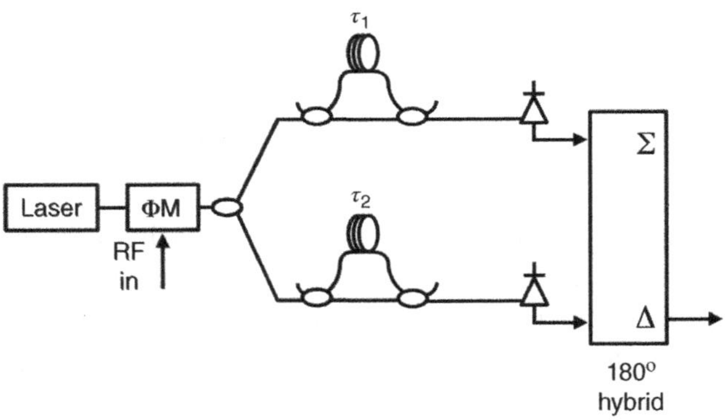

Figure 7.21. Architecture for linearization of a phase-modulated link by using different frequency responses of two unmatched MZIs.

on the front end of the link, whereas the back end is split into two channels having two MZIs. The outputs of the unmatched MZIs are subtracted with an RF hybrid. The different frequency responses afforded by the two MZIs allow for suppression of third-order distortion according to the linearization condition (McKinney et al. 2009)

$$\left|\frac{\sin(2f\tau_2)}{\sin(2f\tau_1)}\right|^3 = \frac{I_{\mathrm{dc},1}}{I_{\mathrm{dc},2}}, \tag{7.27}$$

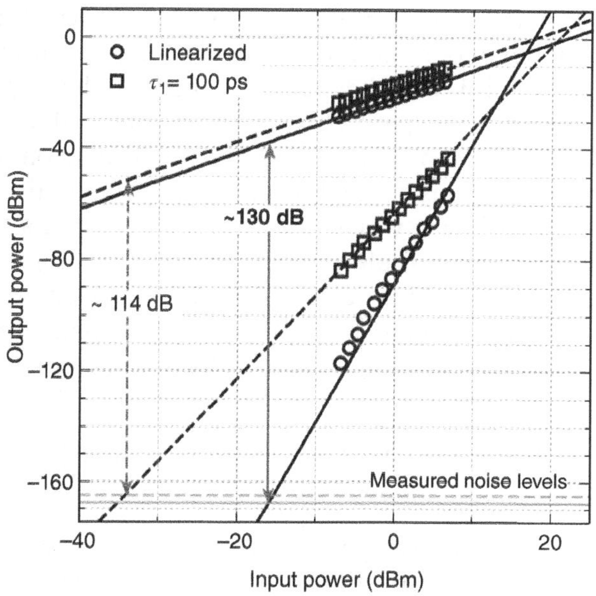

Figure 7.22. Measured data for the architecture in Figure 7.21 with $\tau_1 = 100\,\text{ps}$, $\tau_2 = 375\,\text{ps}$, and $f = 1\,\text{GHz}$ (circles) compared against a single-MZI link with $\tau = 100\,\text{ps}$ (squares) (McKinney et al. 2009).

where the subscripts correspond to the two different channels, and it is assumed that the frequency response of each channel is determined by the MZI delay only. Data demonstrating this technique are shown in Figure 7.22 (McKinney et al. 2009). For these experiments, $\tau_1 = 100\,\text{ps}$, $\tau_2 = 375\,\text{ps}$, and $f = 1$ GHz. As shown in Figure 7.22, the linearized link is limited by fifth-order distortion resulting in $\text{SFDR} = 130\,\text{dB} \cdot \text{Hz}^{4/5}$ as compared to the nonlinearized $\text{SFDR} = 114\,\text{dB} \cdot \text{Hz}^{2/3}$. This increase in SFDR is over a relatively narrow bandwidth and comes at the cost of decreased gain and increased noise figure.

7.5 PROPAGATION EFFECTS

Some of the propagation effects on a phase-modulated link are basically as presented in Chapter 5. For example, SBS impairments on a small-signal phase-modulated link are essentially identical to the IMDD case, because the majority of the power resides in the optical carrier. However, optical phase modulation is a constant-intensity format and is therefore less susceptible to many fiber nonlinearities than an intensity-modulated link. Nonetheless, chromatic dispersion can convert phase modulation to intensity modulation, which in turn can induce intensity-dependent nonlinearities. This can occur in links even if the net dispersion is low and depends strongly on the dispersion map through the transmission span. This section begins with an analysis of phase modulation in the presence of chromatic dispersion, building on the treatments in Sections 5.4 and 6.6. A largely qualitative discussion of crosstalk data follows, which illustrates the robustness of phase modulation to some fiber nonlinearities.

It is instructive to consider direct detection of a phase-modulated signal after passing through a dispersive fiber span. To that end, Equation (7.4) can be rewritten including the propagation constant as

$$\mathrm{Re}(E_0) = \gamma \sum_{k=-\infty}^{\infty} J_k(\phi)\cos[\omega_k t - \beta(\omega_k)L], \tag{7.28}$$

where $\omega_k = \omega + k\Omega$, β is given by Equation (6.104), and L is propagation distance. Taking steps similar to those in moving from Equation (6.103) to (6.109), the small-signal fundamental gain associated with direct detection of Equation (7.28) is

$$g = \frac{I_{\mathrm{dc}}^2}{V_\pi^2}\pi^2 R_{\mathrm{i}} R_{\mathrm{o}} |H_{\mathrm{pd}}|^2 \sin^2\left(\frac{\pi\lambda^2 f^2 DL}{c}\right). \tag{7.29}$$

The form of Equation (7.29) is as predicted by Equation (5.17). In the case of $D=0$ (or $D\times L=0$), there is no fundamental output power as is expected for pure phase modulation. However, when $f=\sqrt{(2n+1)c/(2\lambda^2 DL)}$ where n is an integer, Equation (7.29) is equivalent to the IMDD link gain [Equation (6.26)]. Such phase modulation to intensity modulation conversion is well documented (Chraplyvy et al. 1986, Chi and Yao 2008). Thus, chromatic dispersion can be viewed as a mechanism to "demodulate" phase modulation. Shown in Figure 7.23 is a plot of Equation (7.29) against measured data to further illustrate this point.

As described in Chapter 5, the peaks in the response shown in Figure 7.23 correspond to frequencies where phase modulation has been converted into intensity modulation via chromatic dispersion. Likewise, the nulls are where the link output is pure phase modulation. Further examination of the phasor diagrams in Chapter 4 lends additional insight into this response, including what amounts to a quadrature-type relationship between intensity modulation and phase modulation. This relationship was described generally in Chapter 5

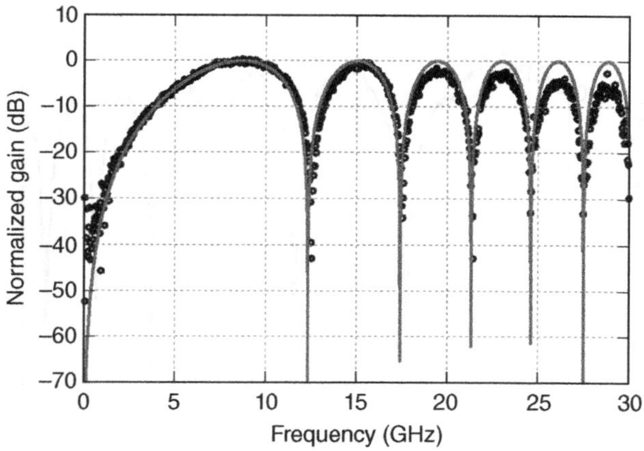

Figure 7.23. Measured (symbols) and calculated [line, Equation (7.29) normalized responses for a phase-modulation, direct-detection link with $L = 50$ km, $\lambda = 1551$ nm, and $D = 16.5$ ps/nm/km (Urick 2007b).

[Equations (5.16) and (5.17)] and is explicitly shown for external modulators by comparing Equations (6.109) and (7.29). To reiterate the point, consider the measured data plotted in Figure 7.24, demonstrating interleaved RF gain curves for external phase- and intensity-modulated links through the same fiber span. This relationship provides insight into the differences between phase and intensity modulation and also into the effects on chromatic dispersion on a lightwave signal. Furthermore, the complimentary nature of a dispersive link to phase and intensity modulation has been exploited to overcome chromatic dispersion and for signal processing capabilities. It was first proposed by Urick and Bucholtz (2006) and later by Cui et al. (2011, 2012) to launch the same analog information encoded on the amplitude and phase of a laser in order to mitigate signal fading induced by chromatic dispersion [a similar technique was also shown in Zhang et al. (2012)]. The resulting link response can be equalized to some extent but does include some ripple due to phase terms not captured in the amplitude data shown in Figure 7.24. In addition, the complimentary nature shown in Figure 7.24 can be used to enhance RF instantaneous frequency measurement capabilities (Urick et al. 2011a) as described in Section 10.3.2.

The MZI used in the link shown in Figure 7.1 converts phase modulation to intensity modulation as described by Equation (7.8). Furthermore, it is easily shown that the receiver is sensitive to only phase modulation at the MZI input when it is biased at quadrature

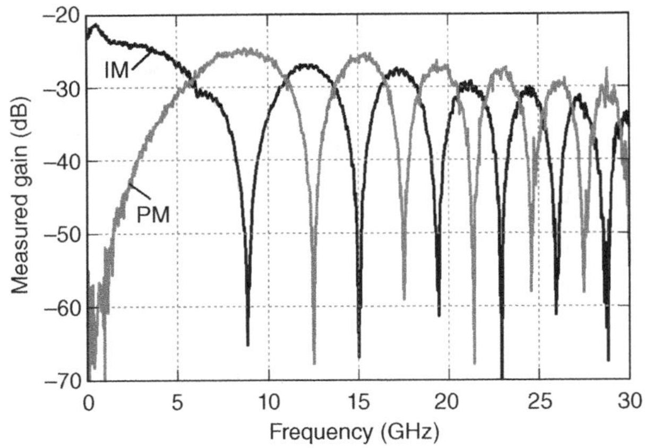

Figure 7.24. Comparison of measured RF gain for phase modulation (gray) and intensity modulation (black) with direct detection through a 50-km span of standard single-mode fiber. A quadrature-type relationship is shown with the amplitude mismatch at higher frequencies due to differences in the modulator responses (Urick 2007b).

and balanced detection is employed (see Section 10.3.2 for a further discussion of the response for an asymmetric MZI). As described previously, chromatic dispersion can be viewed as a mechanism that converts phase modulation into intensity modulation after passing through a fiber span. When an MZI employing balanced detection is placed at the output of such a link, the response will be a combination of the filter function due to the MZI and that for chromatic dispersion. The result can be derived using Equations (7.5) and (7.28) (Urick 2007) or by inspection using Equation (7.8) with the analysis in Section 5.4. In either case, the result is:

$$g = \left[\frac{2\pi I_{\mathrm{dc,q}} \sin(\pi f \tau)}{V_\pi}\right]^2 R_{\mathrm{i}} R_{\mathrm{o}} |H_{\mathrm{pd}}|^2 \cos^2\left(\frac{\pi\lambda^2 f^2 DL}{c}\right). \tag{7.30}$$

Again, Equation (7.30) is valid only for quadrature bias and balanced detection. Equation (7.30) is plotted with respect to measured data in Figure 7.25 using the same fiber span that produced the data in Figure 7.23. Consider the response near 15 GHz, where a deep null is shown in Figure 7.25. Now, the 100-ps MZI has a peak response to phase modulation at 15 GHz. However, the data in Figure 7.23 shows a peak near 15 GHz, indicating that chromatic dispersion has converted the phase modulation to intensity modulation at that frequency. The null in Figure 7.25 is thus explained because the balanced MZI rejects

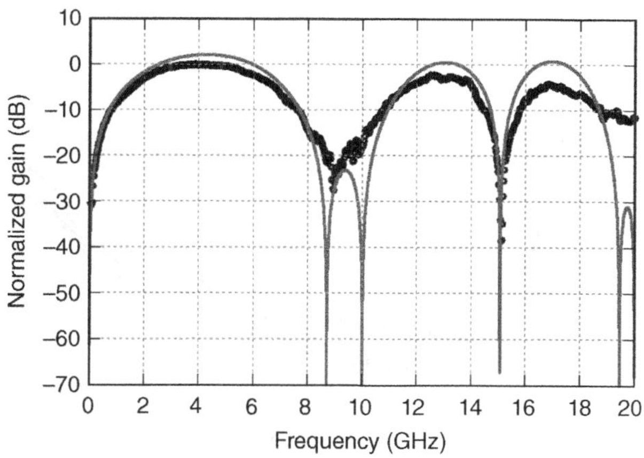

Figure 7.25. Measured (symbols) and calculated [line, Equation (7.30)] normalized responses for a quadrature-biased phase-modulation, interferometric-detection link with $\tau = 100$ ps, $L = 50$ km, $\lambda = 1551$ nm, and $D = 16.5$ ps/nm/km (Urick 2007b).

intensity modulation at its input. Other similar observations can be made by comparing these two figures. In fact, chromatic dispersion compensation receivers have been implemented using a balanced MZI and direct detection simultaneously to recover the phase and intensity components of the modulation, respectively, in dispersive analog links (Urick and Bucholtz 2005).

Phase modulation is less susceptible than intensity modulation to many fiber nonlinearities, because a phase-modulated field is of constant intensity. This advantage is a major selling point for multiple-channel phase modulation links in both digital (Gnauck and Winzer 2005) and analog (Bucholtz et al. 2004) applications. However, through the process described previously, chromatic dispersion can convert phase modulation to intensity modulation. When this occurs, interchannel crosstalk can be induced through various mechanisms such as cross-phase modulation (XPM). As described in Section 5.7, intensity fluctuations will be converted to phase fluctuations through the XPM process. Such phase fluctuations may then be converted back to intensity fluctuations via chromatic dispersion. The complimentary nature of phase and intensity modulation in the presence of dispersion (Figure 7.24) then dictates the states of the signal and crosstalk as they propagate the fiber. Thus, the dispersion map, not only the net dispersion, is an important parameter for multichannel links. Generally speaking though, phase-modulated links exhibit inherently lower crosstalk than intensity-modulated links in terms of XPM because (i)

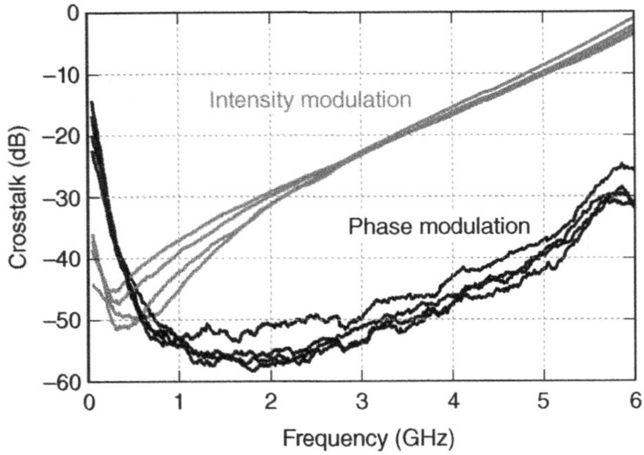

Figure 7.26. Comparison of measured crosstalk for intensity (gray) and phase modulation (black) through a 33.5-km link with ±12.5- and ±25-GHz channel spacing. The phase modulation architecture employed an MZI having a peak response at 5 GHz (Rogge et al. 2005).

XPM will not be induced until dispersion first converts phase modulation to intensity modulation and (ii) XPM-induced crosstalk and the signal will be in a quadrature-type relationship. Data are plotted in Figure 7.26 to emphasize this point. Shown is measured crosstalk for an IMDD link compared to a phase-modulated link with a 100-ps MZI, both propagated through the same 33.5-km fiber span (Rogge et al. 2005). At the peak response of the MZI (5 GHz), the phase-modulated link demonstrates 25 dB lower crosstalk.

7.6 OTHER TECHNIQUES FOR OPTICAL PHASE DEMODULATION

This chapter has concentrated on optical phase modulation that is converted to intensity modulation via an MZI for detection with a photodiode. Such interferometric approaches were among the first techniques demonstrated for decoding information contained in the phase of optical fields. Early examples include the use of birefringent crystals to achieve a differential delay between two polarization states each carrying the same phase-modulated signal (Harris 1964), a balanced free-space MZI (Kaminow 1964), and a fiber MZI (Sorin et al. 1992) as described in this chapter. As mentioned previously, the use of MZIs for microwave photonics applications can leverage component technology developed for demodulation of digital phase-encoded signals. For example, a single interferometer is used for differential-phase-shift-keyed (DPSK) formats (Gnauck and Winzer 2005), and two parallel interferometers are required for differential quadrature phase-shift keying (DQPSK) demodulation (Winzer and Essiambre 2006). However, there are numerous other mechanisms that can be used for optical phase demodulation besides a path-imbalanced interferometer, many of which stem from a resurgence in coherent fiber optics communications over the last decade. Such coherent systems, where precise knowledge of the optical phase is required (Seeds 1995), were heavily researched in the early days of optical fiber communications (Yamamoto and Kimura 1981) but were set aside for some time due to the advent of the EDFA. In one way or another, the various coherent techniques involve converting phase-encoded information into intensity modulation such that a photodetector can pass the information to the electrical domain. In this section, the most prevalent phase demodulation techniques besides an MZI are reviewed.

As described in this chapter, an MZI serves as a filter to convert phase modulation into intensity modulation. Many other types of filter functions can be employed to this end, and a generalized approach to describing the resulting signal as a function of an arbitrary filter function has been presented by Gasulla and Capmany (2011). Shown in Figure 7.27 is an architecture used by Gasulla and Capmany (2011) to describe a phase-modulated link where two complimentary optical filters are placed before a balanced detector pair (a single filter could just as easily be used in this formalism). The RF performance metrics can be derived for such a link, where the resulting link gain contains a filter function itself that describes the phase-to-intensity conversion caused by the optical filter (Gasulla and Capmany 2011, Wyrwas 2012). Equation (7.8) serves as an example where the filter function in the RF domain is $4\sin^2(\pi f\tau)\sin^2(\omega\tau)$. In this case, the MZI can be viewed as a single-tap finite impulse response filter (FIR). Any filter that converts

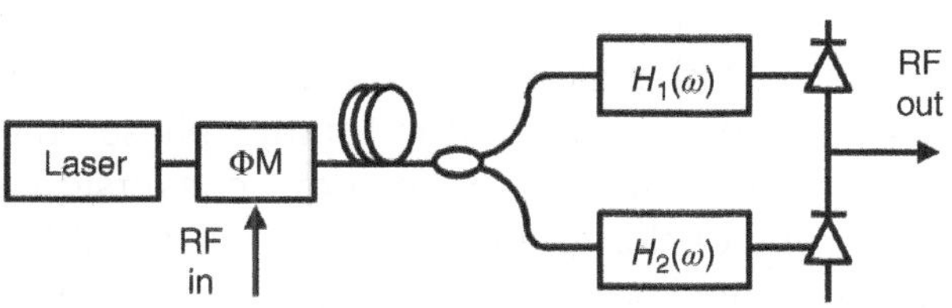

Figure 7.27. A generalized phase-modulated link using two complimentary optical filter functions, $H_1(\omega)$ and $H_2(\omega)$, and balanced photodiodes for demodulation.

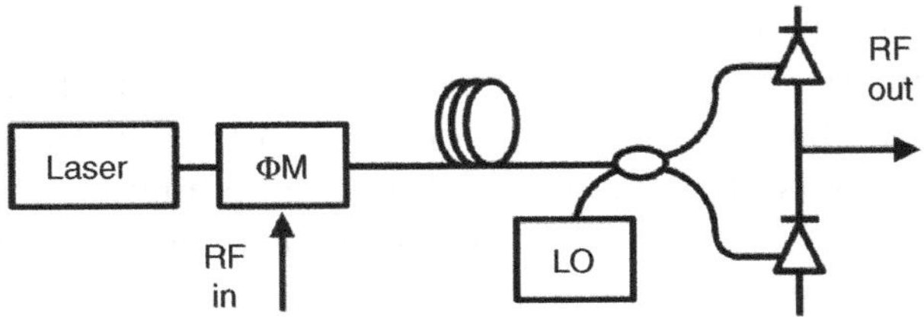

Figure 7.28. A phase-modulated link with an optical local oscillator.

phase to intensity fluctuations can be employed for phase demodulation. Higher order FIR filters can provide a more linear response than a single MZI, such as a serial cascade of MZIs (Wyrwas et al. 2012). Infinite impulse response (IIR) filters have also been demonstrated such as a ring-assisted MZI (Wyrwas et al. 2012). Equation (7.29) describes the effect of chromatic dispersion on a phase-modulated signal, which can be viewed as a demodulation mechanism with an RF filter function of $\sin^2(\pi\lambda^2 f^2 DL/c)$. This has been used explicitly by Gasulla and Capmany (2012) for demodulation of phase-encoded radio-over-fiber links. Optical filtering can also be combined with RF downconversion. Examples are provided by Pagan et al. (2011) and Pagan and Murphy (2013), where fiber Bragg gratings are used to suppress certain frequency components in a phase-modulated spectrum that is then downconverted with a second phase modulator and electrical local oscillator (LO).

The use of optical LOs for phase demodulation was heavily researched for cable television applications from the mid-1980s and into the 1990s (Kalman et al. 1994). Heterodyne detection with a separate laser as shown in Figure 7.28 is the most basic approach. A problem with this technique is that the weighted convolution of the signal laser lineshape and the LO laser will be imposed on the intermediate frequency (IF) signal much similarly to the laser characterization setup shown in Figure 3.17(a). More recently, the use of optical phase locked loops (OPLLs) has been utilized where the LO laser tracks the signal laser's phase (Ramaswamy et al. 2008). The architecture for such a link is shown in Figure 7.29, depicting a feedback loop using a second phase modulator for the phase tracking. Note that the LO can be a separate laser for downconversion to an IF or an unmodulated copy of the original signal laser. In addition, OPPLs have also been combined with other phase demodulation techniques such as MZIs (Zhang et al. 2008). A major limitation of these types of links is that the bandwidth is limited to about 1/(10× feedback

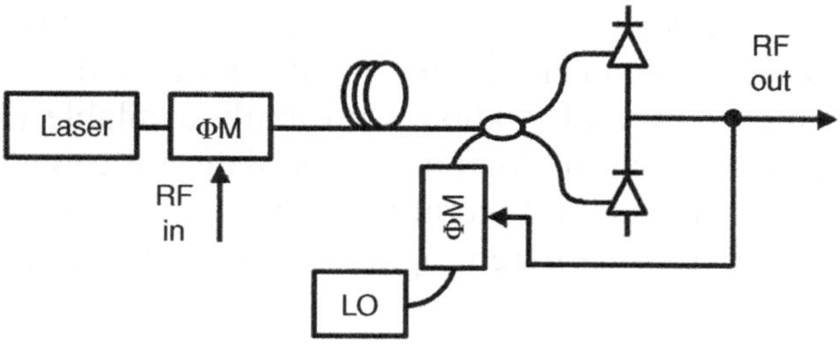

Figure 7.29. A phase-modulated link with an optical phase locked loop.

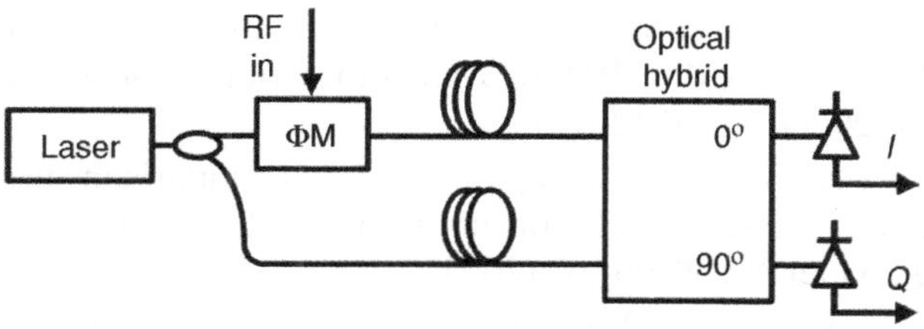

Figure 7.30. A coherent in-phase/quadrature (I/Q) phase-modulated optical link.

loop delay) (Johansson et al. 2010). Thus, very short loops are required for wideband microwave photonics applications, which makes a strong case for integrated OPLLs (Ramaswamy et al. 2008, Krishnamachari et al. 2011).

Resurgence in coherent digital telecommunications research during the first decade of the 21st century (Ip et al. 2008) resulted in advanced in-phase/quadrature (I/Q) optical modulation schemes. High capacity digital formats encode different information on the amplitude and phase of each of two polarization states (Bosco et al. 2011), which can be demodulated with two optical 90° hybrids and four balanced photodiode pairs (Savory 2010). Such receivers are used in other digital links such as "continuous" phase modulation (Detwiler et al. 2011) and radio-over-fiber links (Caballero et al. 2011). Microwave photonics applications using optical 90° hybrids have been studied by Clark et al. (2010) with links such as shown in Figure 7.30, the operation of which is as follows. A laser is split into two paths, one of which is phase modulated by an RF signal. The two paths are then recombined via an optical 90° hybrid from which in-phase and quadrature components are recovered. (Alternatively, a separate optical LO could be used in place of the unmodulated signal.) The RF signal can then be linearly reconstructed by digitizing and processing the in-phase and quadrature photocurrents (Clark et al. 2010). The receiver technology for such techniques is receiving much attention, with integrated silicon optical hybrids and high power photodiodes presently available (Joshi et al. 2013). These and similar analog phase modulation receivers will most likely evolve into the next generation of microwave photonic links, leveraging the industrial coherent optical receiver infrastructure.

REFERENCES

Bosco, G., Curri, V., Carena, A., Poggiolini, P., and Forghieri , "On the performance of Nyquist-WDM terabit superchannels based on PM-BPSK, PM-QPSK, PM-8QAM or PM-16QAM subcarriers," *Journal of Lightwave Technology*, 29(1), 53–61 (2011).

Bucholtz, F., Urick, V. J., and Campillo, A. L., "Comparison of crosstalk for amplitude and phase modulation in an analog fiber optic link," in *IEEE MWP Dig.*, paper MC-9, 66–69 (2004).

Bucholtz, F., Urick, V. J., and Williams, K. J., "Performance of analog photonic links employing phase modulation," in *OSA COTA Conference*, Whistler, BC, Canada, paper CFA6 (2006).

Caballero, A., Zibar, D., and Monroy, I. T., "Performance evaluation of digital coherent receivers for phase-modulated radio-over-fiber links," *Journal of Lightwave Technology*, 29(21), 3282–3292 (2011).

Chi, H. and Yao, J., "Power distribution of phase-modulated microwave signals in a dispersive fiber-optic link," *IEEE Photonics Technology Letters*, 20(4), 315–317 (2008).

Chraplyvy, A. R., Tkach, R. W., Buhl, L. L., and Alferness, R. C., "Phase modulation to amplitude modulation conversion of CW laser light in optical fibres," *Electronics Letters*, 22(8), 409–411 (1986).

Clark, T. R., O'Connor, S. R., and Dennis, M. L., "A phase-modulation I/Q-demodulation microwave-to-digital photonic link," *IEEE Transactions on Microwave Theory and Techniques*, 58(11), 3039–3058 (2010).

Cui, Y., Xu, K., Dai, J., Dai, Y., Wu, J., and Lin, J., "Compensation of the chromatic dispersion in long-reach optical links incorporating parallel electro-optic phase and intensity modulators," in *IEEE MWP Dig.*, 270–273 (2011).

Cui, Y., Xu, K., Dai, J., Sun, X., Dai, Y., Ji, Y., and Lin, J., "Overcoming chromatic-dispersion-induced power fading in ROF links employing parallel modulators," *IEEE Photonics Technology Letters*, 24(14), 1173–1175 (2012).

Detwiler, T. F., Searcy, S. M., Ralph, S. E., and Basch, B., "Continuous phase modulation for fiber-optic links," *Journal of Lightwave Technology*, 29(24), 3659–3671 (2011).

Frigo, N. J. and Bucholtz, F., "Geometrical representation of optical propagation phase," *Journal of Lightwave Technology*, 27(15), 3283–3293 (2009).

Gasulla, I. and Capmany, J., "Analytical model and figures of merit for filtered microwave photonic links," *Optics Express*, 19(20), 19758–19774 (2011).

Gasulla, I. and Capmany, J., "Phase-modulated radio over fiber multimode links," *Optics Express*, 20(11), 11710–11717 (2012).

Gnauck, A. H. and Winzer, P. J., "Optical phase-shift-keyed transmission," *Journal of Lightwave Technology*, 23(1), 115–130 (2005).

Haas, B. M., Urick, V. J., McKinney, J. D., and Murphy, T. E., "Dual-wavelength linearization of optically phase-modulated analog microwave signals," *Journal of Lightwave Technology*, 26(15), 2748–2753 (2008).

Harris, S. E., "Demodulation of phase-modulated light using birefringent crystals," *Proceedings of the IEEE*, 52(7), 823–831 (1964).

Hastings, A. S., Urick, V. J., Sunderman, C., Diehl, J. F., McKinney, J. D., Tulchinksy, D. A., Devgan, P. S., and Williams, K. J., "Suppression of even-order photodiode nonlinearities in multioctave photonic links," *Journal of Lightwave Technology*, 26(15), 2557–2562 (2008).

Hutchinson, M. N., Singley, J. M., Urick, V. J., Harmon, S. R., McKinney, J. D., and Frigo, N. J., "Mitigation of photodiode induced even-order distortion in photonic links with predistortion modulation," *Journal of Lightwave Technology*, 32(20), 3885–2892 (2014).

Ip, P., Lau, A. P. T., Barros, D. J. F., and Kahn, J. M., "Coherent detection in optical fiber systems," *Optics Express*, 16(2), 753–791 (2008).

Johansson, L. A., Krishnamachari, U., Ramaswamy, A., Klamkin, J., Ristic, S. Chou, H. F., Coldren, L. A., Rodwell, M., and Bowers, J. E., "Linear coherent optical receivers," in *Optical Fiber Communications Dig.*, paper OThF5 (2010).

Johnson, L. M. and Roussell, H. V., "Reduction of intermodulation distortion in interferometric optical modulators," *Optics Letters*, 3(10), 928–930 (1988).

Joshi, A., Datta, S., Rue, J., Rajagopalan, S., and Lemke, S., "High power-handling linear integrated coherent photoreceivers for RF photonics," in *Proceedings of SPIE*, 8720, 872011-17 (2013).

Kalman, R. F., Fan, J. C., and Kazovsky, L. G., "Dynamic range of coherent analog fiber-optic links," *Journal of Lightwave Technology*, 12(7), 1263–1277 (1994).

Kaminow, I. P., "Balanced optical discriminator," *Applied Optics*, 3(4), 507–510 (1964).

Krishnamachari, U., Ristic, S., Ramaswamy, A., Johansson, L. A., Chen, C. H., Klamkin, J., Bhardwaj, A., Rodwell, M. J., Bowers, J. E., and Coldren, L. A., "Integrated coherent receiver for linear optical phase demodulation," *Microwave and Optical Technology Letters*, 53(10), 2343–2345 (2011).

LaGasse, M. J. and Thaniyavarn, S., "Bias-free high-dynamic-range phase-modulated fiber-optic link," *IEEE Photonics Technology Letters*, 9(5), 681–683 (1997).

McKinney, J. D., Colladay, K., and Williams, K. J., "Linearization of phase-modulated analog optical links employing interferometric demodulation," *Journal of Lightwave Technology*, 27(9), 1212–1220 (2009).

Pagan, V. R. and Murphy, T. E., "Phase-modulated radio-over-fiber systems," in *OFC Technical Dig.*, paper OW1D.5 (2013).

Pagan, V. R., Haas, B. M., and Murphy, T. E., "Linearized electrooptic microwave downconversion using phase modulation and optical filtering," *Optics Express*, 19(2), 883–895 (2011).

Ramaswamy, A., Johansson, L. A., Klamkin, J., Chou, H. F., Sheldon, C., Rodwell, M. J., Coldren, L. A., and Bowers, J. E., "Integrated coherent receivers for high-linearity microwave photonic links," *Journal of Lightwave Technology*, 26(1), 209–216 (2008).

Rogge, M. S., Urick, V. J., Bucholtz, F., Williams, K. J., and Knapp, P., "Comparison of crosstalk in amplitude- and phase-modulated hyperfine WDM fiber optic links," in *CLEO Technical Dig.*, paper CMH2 (2005).

Rogge, M. S., Urick, V. J., and Bucholtz, F., "Analysis of an optical channelization technique for microwave applications," NRL Memorandum Report, NRL/MR/5652-07-9061 (2007).

Savory, S. J., "Digital coherent optical receivers: algorithms and subsystems," *IEEE Journal of Selected Topics in Quantum Electronics*, 16(5), 1164–1179 (2010).

Seeds, A. J., "Coherent techniques in analogue signal transmission," in *Analog Optical Fibre Communications*, B. Wilson, Z. Ghassemlooy, and I. Darwazeh (editors), The Institute of Electrical Engineers, 33–50 (1995).

Singley, J. M., Diehl, J. F., and Urick, V. J., "Laser noise considerations for phase modulated links," *IEEE AVFOP Conf. Dig.*, paper ThD5 (2011).

Sorin, W. V., Chang, K. W., Conrad, G. A., and Hernday, P. R., "Frequency domain analysis of an optical FM discriminator," *Journal of Lightwave Technology*, 10(6), 787–793 (1992).

Tavlykaev, R., Wiberg, A. O. J., Gopalakrishnan, G. K., and Radic, S., "Nonlinear distortions in EO phase modulators," *IEEE Photonics Technology Letters*, 24(19), 1778–1780 (2012).

Urick, V. J., "Long-haul analog photonics principles with applications," Doctoral Dissertation, UMI No. 3255815 (2007).

Urick, V. J. and Bucholtz, F., "Compensation of arbitrary chromatic dispersion in Analog links using a modulation-diversity receiver," *IEEE Photonics Technology Letters*, 17(4), 893–895 (2005).

Urick, V. J. and Bucholtz, F., "Modulation diversity for chromatic dispersion compensation in analog photonic links," NRL Memorandum Report, NRL/MR/5650-06-8977 (2006).

Urick, V. J., Bucholtz, F., Devgan, P. S., McKinney, J. D., and Williams, K. J., "Phase modulation with interferometric detection as an alternative to intensity modulation with direct detection for analog-photonic links," *IEEE Transactions on Microwave Theory and Techniques*, 55(9), 1978–1985 (2007).

Urick, V. J., McKinney, J. D., Diehl, J. F., and Singley, J. M., "Simultaneous optical phase and intensity modulation for analog signal processing," in *IEEE AVFOP Conf. Dig.*, paper TuD4 (2011a).

Urick, V. J., McKinney, J. D., and Williams, K. J., "Fiber-optic links with all-photonic RF gain and low RF noise figure," in *IEEE Intl. Microwave Symp.*, paper WE2C-1 (2011b).

Urick, V. J., McKinney, J. D., Diehl, J. F., and Williams, K. J., "Equations for Two-Tone Analog Optical Phase Modulation With an Asymmetric Interferometer," *IEEE Photonics Technology Letters*, 25(15), 1527–1530 (2013).

Winzer, P. J. and Essiambre, R. J., "Advanced modulation formats for high-capacity optical transport networks," *Journal of Lightwave Technology*, 24(12), 4711–4728 (2006).

Wyrwas, J. M., "Linear, Low noise microwave photonic systems using phase and frequency modulation," Doctoral Dissertation, University of California at Berkeley, Technical Report No. UCB/EECS-2012-89 (2012).

Wyrwas, J. M. and Wu, M. C., "Dynamic range of frequency modulated direct-detection analog fiber optic links," *Journal of Lightwave Technology*, 27(24), 5552–5562 (2009).

Wyrwas, J. M., Peach, R., Meredith, S., Middleton, C., Rasras, M. S., Tu, K. Y., Earnshaw, M. P., Pardo, F., Cappuzzo, M. A., Chen, E. Y., Gomez, L. T., Klemens, F., Keller, R., Bolle, C., Zhang, L. Buhl, L., Wu, M. C., Chen, Y. K., and DeSalvo, R., "Linear phase-and-frequency-modulated photonic links using optical discriminators," *Optics Express*, 20(24), 26292–26298 (2012).

Xu, K., Dai, J., Duan, R., Dai, Y., Li, Y., Wu, J., and Lin, J., "Instantaneous microwave frequency measurement based on phase-modulated links with interferometric detection," *IEEE Photonics Technology Letters*, 23(18), 1328–1330 (2011).

Yamamoto, Y. and Kimura, T., "Coherent optical fiber transmission systems," *IEEE Journal of Quantum Electronics*, 17(6), 919–935 (1981).

Zhang, J., Hone, N. A., and Darcie, T. E., "Phase-modulated microwave-photonic link with optical-phase-locked-loop enhanced interferometric phase detection," *Journal of Lightwave Technology*, 26(15), 2549–2556 (2008).

Zhang, H., Pan, S., Huang, M., and Chen, X., "Polarization-modulated analog photonic link with compensation of the dispersion-induced power fading," *Optics Letters*, 37(5), 866–868 (2012).

CHAPTER 8

OTHER ANALOG OPTICAL MODULATION METHODS

External intensity modulation with a Mach–Zehnder modulator (MZM) was covered in detail in Chapter 6 followed by a thorough, but relatively less extensive, treatment of external phase modulation in Chapter 7. Both techniques are rooted in external phase modulation, with the MZM converting integrated phase modulation into intensity modulation via optical interference within the modulator structure. Phase modulated links use coherent techniques to convert the phase encoding to an intensity modulation. As described in Chapter 4, pure optical intensity modulation and phase modulation exhibit a quadrature-type relationship. However, other properties of light may be modulated in order to encode radio-frequency (RF) information onto a lightwave carrier, such as polarization or frequency.

There are numerous physical mechanisms that can be exploited in materials to achieve optical modulation including electrooptic, electroabsorption, acoustooptic, or magnetooptic effects (Kaminow and Turner 1966, Kaminow 1975, Chen 1970, Alferness 1982). Many materials exhibit such properties, and there are numerous architectures to leverage the effects for analog applications (Dagli 1999, Chen and Murphy 2012). For example, external intensity modulation can be achieved with optical branches having variable reflection coefficients

Fundamentals of Microwave Photonics, First Edition.
Vincent J. Urick Jr, Jason D. McKinney, and Keith J. Williams.

at material boundaries (Naitoh et al. 1978, Solgaard et al. 1992). Directional coupler modulators (Schmidt and Alferness 1979, Alferness 1981, Thaniyavarn 1986) can produce intensity modulation by varying the coupling ratio between optical waveguides. Electroabsorption modulators (EAMs) (Chang 2002, Liu et al. 2003) can provide direct external intensity modulation by varying the absorption coefficient of a material, typically a semiconductor. Such EAMs have potential for photonic integration but have issues with their optical dynamic range, multioctave RF distortion, and optical insertion loss.

The electrooptic crystal $LiNbO_3$ is the most mature technology for commercial external modulators (Wooten et al. 2000). The III–V semiconductors have been extensively studied as optical modulators (Bennett et al. 1990, Walker 1991), including GaAs (Wang and Lin 1988, Walker et al. 2013, Shin and Dagli 2013) and InP (Betty 2012). Silicon optical modulators (see Section 10.6) are also an active research topic. More potential for optical modulation lies in the form of electrooptic polymers (Lee et al. 2002, van Eck 2002, Hung and Fetterman 2005, Melikyan et al. 2014). Regardless of the technique or material platform, the RF metrics of concern remain the same as briefly exampled in this chapter for a variety of methods.

This chapter provides an introductory treatment of five analog optical modulation techniques. Direct modulation of laser intensity and frequency is covered in Section 8.1. The next three sections address other methods using an MZM. An MZM can be low biased to implement a suppressed carrier modulation; this and a brief survey of other suppressed-carrier techniques are provided in Section 8.2. Section 8.3 focuses on single-sideband (SSB) modulation as can be achieved with a dual-electrode MZM. Sampled analog links, which typically employ one or more MZMs, are covered in Section 8.4. Finally, analog optical polarization modulation is the topic of Section 8.5. This chapter does not concentrate on the materials that are used to implement a particular modulation. The interested reader is referred to the cited literature in the terse survey mentioned previously for such information and/or for treatments of methods not addressed in this chapter.

8.1 DIRECT LASER MODULATION

The most straightforward concept for encoding RF information onto light is via direct modulation of a laser's intensity or frequency by modulating its pump. The former presents a simple analog link comprising a laser and photodetector, while the latter requires some sort of discriminator to convert the frequency modulation into intensity modulation. The bandwidth for direct modulation is tied to the lifetime of the laser gain medium, which precludes solid-state lasers for high frequency analog photonics in favor of faster semiconductor lasers. Direct modulation of semiconductor lasers practically coincides with the advent of such lasers [see Paoli and Ripper (1970) and the references therein] and remains in use today. Typical laser diodes used in direct modulation include the distributed-feedback (DFB) laser diode, the Fabry–Perot laser diode, and the vertical cavity surface emitting laser (VCSEL) (Cox 2004). Direct modulation of a semiconductor transistor laser (Walter et al. 2004) has also been considered for microwave photonics applications (Iezekiel 2014). This section will review basic properties of direct modulation analog photonic links, covering intensity and frequency modulation in Sections 8.1.1 and 8.1.2, respectively.

8.1.1 Direct Intensity Modulation

A basic diagram of a direct intensity modulation analog photonic link using a laser diode is shown in Figure 8.1. A bias current and modulation signal are applied to the laser through the inductive and capacitive ports of a bias tee, respectively. A matching circuit with resistance R_{in} can be employed between the signal source and the laser, as the laser series resistance, R_l, is typically much less than 50 Ω. Although impedance matching does not maximize the modulation depth of the laser current, it is often employed to minimize reflections. The bias current is well above the laser threshold current, and the applied RF voltage results in modulation of the laser output power. The slope of the curve describing the laser output power as a function of applied current is termed the slope efficiency:

$$s_l = \frac{dP_{\text{o}}}{dI_{\text{in}}} = \frac{\eta_{\text{d}}\hbar\omega}{q}, \tag{8.1}$$

where η_{d} is the differential quantum efficiency, $\hbar$ is Planck's constant divided by 2π, ω is the angular laser frequency, and q is the electronic charge constant. The modulated light is then propagated to a photodiode through a fiber optic link with net optical gain factor g_{o} (or loss when $g_{\text{o}} < 1$) and demodulated with a photodetector. The photodetector efficiency is quantified by the previously used responsivity, $\Re = \eta q/(\hbar\omega)$, where η is the photodetector quantum efficiency.

The analysis in this case is straightforward and uses basic concepts. A more thorough treatment of analog intensity modulation including

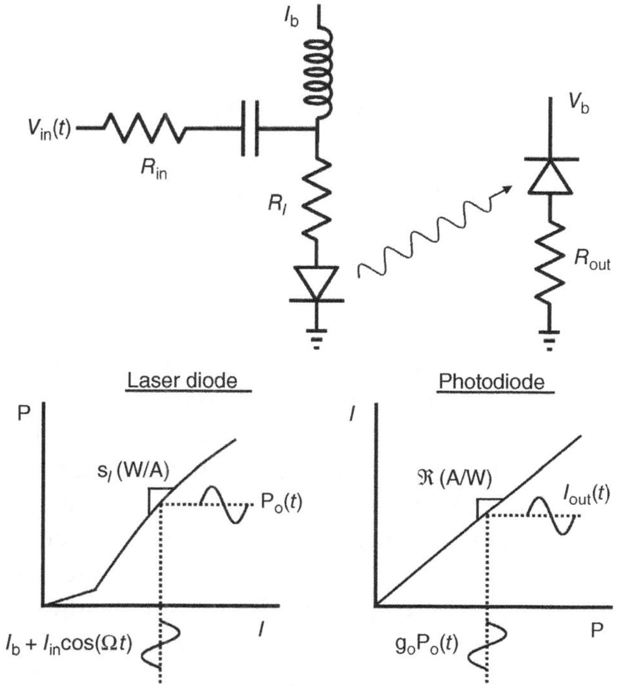

Figure 8.1. Simple circuit and basic parameters for a direct intensity modulation with direct detection analog photonic link.

detailed circuit analysis is provided by Cox et al. (1990) and Cox (2004). In this case, it is assumed that the RF driving signal source is matched to the laser circuit shown in Figure 8.1, which is in turn matched to the output on the photodiode side of the link. Thus, $R_{in} + R_l = R_{out} = R$. The laser diode is assumed to be an ideal current sink above threshold. It is further assumed for this analysis that the laser is single mode, does not distort the modulation (s_l is independent of current), and has a flat response to RF modulation (frequency dependence will be handled later). Given these conditions, the optical power out of laser can be written as

$$P_o = P_o(I_b) + s_l \frac{V_{in}(t)}{R}, \tag{8.2}$$

where I_b is the bias current. Using a sinusoidal input signal, $V_{in} = V_{rf}\sin(\Omega t)$, and neglecting propagation effects, the output photocurrent is then

$$I_{out} = \Re g_o \left[P_o\left(I_b\right) + s_l \frac{V_{rf}}{R} \sin(\Omega t) \right], \tag{8.3}$$

The RF output power from Equation (8.3) is

$$P_{\mathrm{rf,out}} = \frac{V_{\mathrm{rf}}^2}{2R}(s_l g_{\mathrm{o}} \Re)^2. \tag{8.4}$$

Finally, the RF gain for the link is obtained as

$$g \equiv \frac{P_{\mathrm{rf,out}}}{P_{\mathrm{rf,in}}} = (s_l g_{\mathrm{o}} \Re)^2, \tag{8.5}$$

where $P_{\mathrm{rf,in}} = V_{\mathrm{rf}}^2/(2R)$ is the RF input power. With $g_{\mathrm{o}} = 1$ in Equation (8.5), the intrinsic gain for a matched direct modulation link results, which is often employed for link design (Cox 2004). This intrinsic gain is sometimes described as being independent of the average received photocurrent (I_{dc}). It is more precise to state that Equation (8.5) is independent of the laser output power, as long as g_{o} and $\Re$ are not affected by changes in laser power. In terms of received photocurrent, $g \propto I_{\mathrm{dc}}^2$, where $I_{\mathrm{dc}} = \Re g_{\mathrm{o}} P_{\mathrm{o}}(I_{\mathrm{b}})$ can be adjusted via g_{o}.

Shown in Figure 8.2 are two measured RF gain curves for analog optical links employing directly modulated semiconductor laser diodes. The data in this case are for an 18-GHz Fabry–Perot laser and a 25-GHz DFB laser. For both curves, the lasers were pumped at about 100 mA, and the output is attenuated such that the average received photocurrent was 2 mA. As can be seen in Figure 8.2, both lasers exhibit a relatively flat response until about 15 GHz and then roll off to

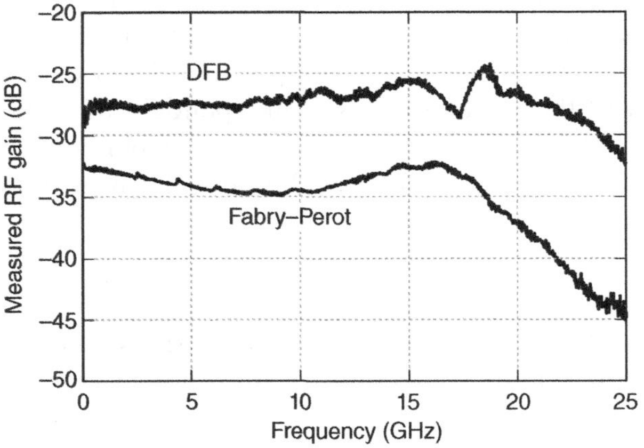

Figure 8.2. Measured response for two directly modulated semiconductor laser diodes, a distributed feedback (DFB) laser and a Fabry–Perot laser.

their respective bandwidths. As described in the following section, this type of behavior for semiconductor lasers is well understood.

A thorough analysis of direct modulation of semiconductor laser diodes entails studying the laser rate equations describing photon and carrier densities. A popular work on the topic is provided by Coldren et al. (2012), some results of which are borrowed in this chapter to describe the frequency response. The frequencies at which a diode laser can be modulated are limited by the photon and carrier lifetimes, τ_p and τ_c, respectively. Coldren et al. (2012) derived a fairly simple expression for the normalized RF gain for a laser diode:

$$g(f) \propto \left[\left(\frac{f^2}{f_r^2} - 1\right)^2 + \frac{f^2}{f_r^2}\left(2\pi f_r \tau_p + \frac{1}{2\pi f_r \tau_c}\right)^2\right]^{-1}, \tag{8.6}$$

where f_r is the resonance frequency. The resonance frequency was introduced in Equations (1.6) and (3.32); in this case, it is written in an expanded form as (Coldren et al. 2012)

$$f_r = \frac{1}{2\pi}\sqrt{\frac{AP_o}{\tau_p \alpha_m h\nu V_l}}, \tag{8.7}$$

where A is the differential optical gain coefficient, P_o is the unmodulated optical output power at frequency ν, h is Planck's constant, V_l is the volume of the laser cavity, and α_m is a loss term associated with the cavity mirrors. This last term can be written as $\alpha_m = \ln(1/r_1 r_2)/L$, where r_1 and r_2 are the amplitude reflection coefficients for the mirrors on either side of the cavity of length L. The dependence of f_r on P_o is shown in Figure 8.3 using the same parameters as Coldren et al. (2012): $\tau_p = 2$ ps, $\tau_c = 3$ ns, $A = 5 \times 10^{-16}\,\text{cm}^2$, $\alpha_m = 60\,\text{cm}^{-1}$, $h\nu = 1.5$ eV (corresponding to 827-nm light), and $V_l = 5 \times 0.25 \times 200\,\mu\text{m}^3$. Note that the bandwidth of the laser increases with optical output power, which is usually limited in practice by the amount of heat that can be dissipated. Typically, the operation of direct modulation links is restricted to frequencies below f_r; however, some work has been performed on modulating above f_r, albeit with limited fractional bandwidth, by using coupled cavities (Lau 2011). Another important feature of the direct modulation curve given by Equation (8.6) is the roll off in the limit $f \gg f_r$, which is $1/f^4$ or −40 dB/decade.

There are numerous other tradeoffs for an analog link employing direct modulation. As described by Figure 8.3, the bandwidth is

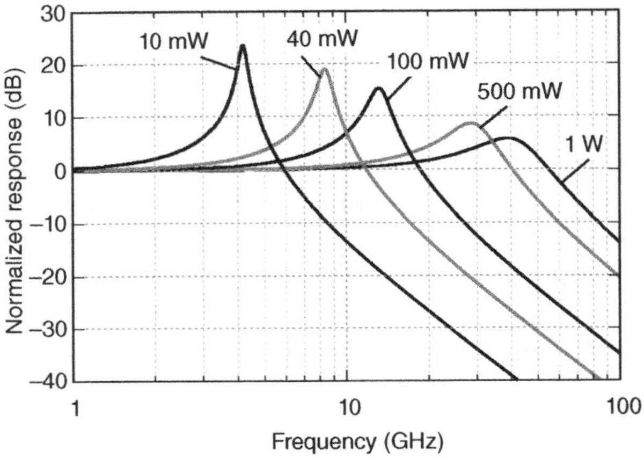

Figure 8.3. Calculation of Equation (8.6) for various values of P_o.

optimized at a high bias current. However, RF gain and noise figure are better at lower biases, where the slope efficiency is largest and the output noise is minimized. The optimum bias point in terms of spurious-free dynamic range is typically somewhere in between (Cox 2004), and the trade space there is not as well defined analytically. As opposed to an MZM-based intensity-modulated direct detection (IMDD) link, a major downside of direct modulation is that both even- and odd-order distortions are present in the output. These nonlinearities tend to be functions of frequency and bias current, which further complicate matters. These properties of current-modulated laser diodes are well known (Darcie et al. 1985) and leave direct modulation very undesirable for high fidelity multioctave applications. Some system-level techniques have been applied to this problem, such as balanced detection of two laser diodes driven in a push–pull configuration (Marpaung et al. 2009) and feedforward linearization (Ismail et al. 2007). However, this balancing method has limitations in and of itself including the need for 180° phase shift on the modulation side and two phase-matched fiber runs. Likewise, feedforward techniques can be hardware intensive.

Another major distinction between MZM-based and direct IMDD links is that the latter exhibits strong chirp, or residual frequency modulation, whereas a properly designed MZM has minimal chirp. Chirp is not an issue in an ideal direct detection scenario because the frequency fluctuations will not be sensed by a photodetector. However, the chirp from a current-modulated laser diode can cause significant second- and third-order distortion when propagation effects are present, such a chromatic dispersion and nonlinear refractive index (Phillips et al. 1991). However, the chirp of a current-modulated semiconductor laser can be employed to encode RF information onto the laser frequency.

8.1.2 Direct Frequency Modulation

A "frequency-modulated" photonic link typically refers to direct modulation of a laser's frequency by modulating the pump for the medium (Epworth 1985). This is in contrast to external phase modulation as described in Chapter 7, although the term "indirect frequency modulation" has been employed to describe an electronic integrator coupled to an external phase modulator (Kalman et al. 1994). Direct detection of a frequency-modulated lightwave will not recover the modulating signal. Rather, some sort of discrimination technique must be employed such as those describe in Section 7.6 for optical phase demodulation. For example, a finite impulse response filter can be used to convert frequency modulation into intensity modulation (Fiksman et al. 1993). Wyrwas and Wu (2009) provides a good review of optical frequency demodulation techniques in the context of microwave photonics. In this case, the frequency response of laser diode frequency modulation is compared to intensity modulation without any further discussion on demodulation of the former.

The frequency modulation response of a laser diode has been detailed by Coldren et al. (2012). In this section, equations using pieces from a few sources and the parameters in Yabre and Le Bihan (1995) are used to illustrate some of the differences between direct frequency and direct intensity modulation. An expression for the normalized intensity modulation can be written as

$$|\mathrm{AM}|^2 = (1 + \Gamma A \tau_\mathrm{p} N_\mathrm{tr})^2 \left[\left(\frac{f^2}{f_r^2} - 1 \right)^2 + \left(\frac{2\pi f \varepsilon}{A} \right)^2 \right]^{-1}, \tag{8.8}$$

where Γ is the optical confinement factor, N_tr is the transparency carrier density, and ε is the gain compression damping coefficient. Equation (8.8) is similar to that given by Yabre and Le Bihan (1995) but using the frequency dependence given by Darcie et al. (1985). The frequency dependence of Darcie et al. (1985) exhibits the same high frequency response as Equation (8.6) but contains fewer terms than more complicated treatments (Wang et al. 1993). The magnitude-squared frequency deviation can be written in terms of Equation (8.8) as (Yabre and Le Bihan 1995)

$$|\mathrm{FM}|^2 = \frac{m_i^2 |\mathrm{AM}|^2 \alpha^2 f^2}{4}, \tag{8.9}$$

where m_i is the current modulation depth and α is the linewidth enhancement factor.

Equations (8.8) and (8.9) are plotted in Figure 8.4. The calculations in this case use the same parameters as Yabre and Le Bihan (1995): $(1 + \Gamma A \tau_p N_{tr}) = 1.4235$, $f_r = 5.46$ GHz, $\varepsilon / A = 0.00645$ ns, $m_i = 0.2$, and $\alpha = 6$. For these values, the frequency modulation depth is higher than that of intensity modulation at all frequencies above 2 GHz. However, Equation (8.9) accounts for carrier effects only; current modulation of a laser diode modulates the temperature, which in turn causes additional frequency modulation. Such thermal effects dominate the frequency modulation depth in the low frequency range (Kobayashi et al. 1982), typically up to a few hundred megahertz or even approaching 1 GHz. Therefore, the frequency modulation in a current-modulated semiconductor laser will exceed the intensity modulation at almost all frequencies. Another important difference between the two curves in Figure 8.4 is the slope in the limit $f \gg f_r$, −20 dB/decade for frequency modulation as compared to −40 dB/decade for intensity modulation. Although not shown explicitly in Figure 8.4, it is often the case that the intensity modulation peak occurs slightly below the relaxation oscillation frequency, whereas the frequency modulation peak is more closely aligned with f_r (Coldren et al. 2012).

Direct intensity and frequency modulation present the simplest transmitters for lightwave systems. However, a more complicated receiver for the latter typically will result in direct intensity modulation being the more attractive of the two. Direct frequency modulation can take advantage of receive-end linearization techniques such as cascaded optical filters. Neither of these direct modulation techniques is particularly enticing for high fidelity, multioctave applications because of the intrinsic even- and odd-order distortion. The relative nonexistent even-order distortion in an MZM gives external modulation a marked advantage for such applications. A constant-intensity pure frequency modulation has advantages in long-reach applications, as described in Chapter 7 in the context of external phase modulation. However, the strong coupling between intensity and frequency variations in direct modulation techniques negates this advantage for direct frequency modulation. Studies

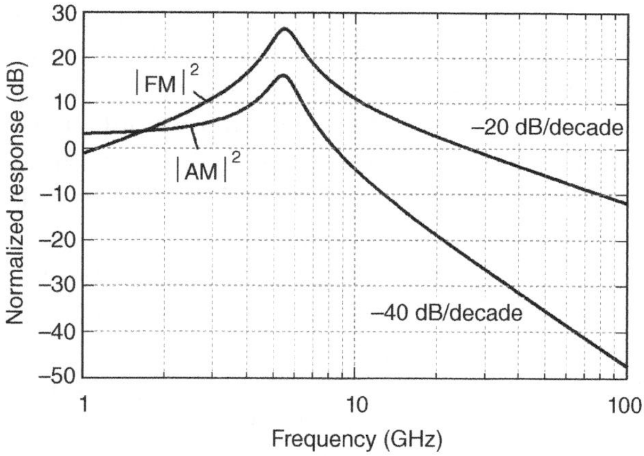

Figure 8.4. Calculation of Equations (8.8) and (8.9), showing a comparison of normalized intensity modulation and the magnitude-squared frequency deviation for a current-modulated laser diode.

comparing direct and external modulation techniques have been completed for analog applications (Stephens and Joseph 1987, Cox et al. 1990), which cite other tradeoffs than those listed in this chapter.

8.2 SUPPRESSED CARRIER MODULATION WITH A LOW BIASED MZM

Most analog optical links described to this point are typically operated at low modulation depths, with the majority of the power residing in the optical carrier. However, suppressed-carrier methods offer some advantages for analog links. Higher RF power conversion efficiency can be achieved with suppressed carriers at high modulation depth. Suppressed-carrier techniques are very attractive for long-reach applications, where distribution of the optical power among numerous frequency components separated beyond the stimulated Brillouin scattering (SBS) bandwidth can significantly enhance the effective SBS threshold. The concentration of this section is a carrier-suppression technique using a low biased MZM. Sampled analog links, another suppressed-carrier method, are covered in Section 8.4. Numerous other suppressed carrier techniques have been used for optical modulation, as briefly mentioned in the following section. A conceptually simple technique, which can be difficult in practice, is to employ an optical filter to notch the carrier after modulation has occurred. This has been demonstrated with various optical filters including a fiber Mach–Zehnder interferometer (LaGasse et al. 1994), a Fabry–Perot filter (Esman and Williams 1995), and a Sagnac interferometer (Frankel and Esman 1998). Other MZM-based approaches have also been implemented. For example, a low biased MZM with a coherent optical local oscillator has been demonstrated by Middleton and DeSalvo (2010). Class-AB methods using two complementary biased MZMs have been studied by Darcie and Driessen (2006). Other interesting results have been obtained in suppressed-carrier frequency-/phase-modulated architectures (Darcie et al. 2007, Driessen et al. 2008) and polarization-encoded formats (Bull et al. 2006).

The concept of a suppressed-carrier link employing a low biased MZM is relatively straightforward. Consider the MZM transmission curves shown in Figure 8.5, which might arise from the individual outputs in a link as shown in Figure 6.6. When biased at quadrature, both outputs have the same average output power. However, the DC bias can be shifted to pass the majority of the average power out of one port, leaving the other port "low biased." In this case, the modulation depth on the low biased port is much larger than that for the high biased port. As will be shown in the following section, the low biased output can provide some advantages from an RF perspective. The analog metrics for a low biased MZM architecture are well known (Farwell et al. 1993, Ackerman et al. 1993). The analysis in this section is brief and assumes an ideal MZM with a transfer matrix given by Equation (6.13). A parameter that can be important in low biased links is the MZM extinction, assumed in this case to be infinite. Nonideal

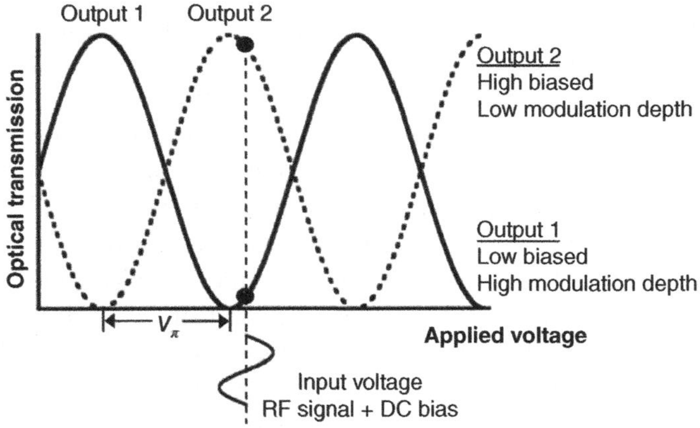

Figure 8.5. Transfer characteristic for a dual-output MZM with notes on different bias points.

extinction can be handled with a multiplicative term on the diagonal of the phase matrix in Equation (6.13) (Kim and Gnauck 2002, Urick et al. 2011a). Given these assumptions, the small-signal RF gain as a function of bias can be derived from Equation (6.18) as

$$g = \frac{I_{\mathrm{dc,q}}^2}{V_\pi^2}\pi^2 R_\mathrm{i} R_\mathrm{o} |H_\mathrm{pd}|^2 \sin^2\phi_\mathrm{dc}, \tag{8.10}$$

where $I_{\mathrm{dc,q}}$ is the DC photocurrent when the MZM is at quadrature. The RF noise factor is obtained using Equation (6.60) with Equation (8.10):

$$F = 1 + \frac{V_\pi^2}{R_\mathrm{i}\pi^2 \sin^2\phi_\mathrm{dc}}\left[\frac{1}{I_{\mathrm{dc,q}}^2 R_\mathrm{o}\left|H_\mathrm{pd}\right|^2} + \frac{2q(1-\cos\phi_\mathrm{dc})}{I_{\mathrm{dc,q}} k_\mathrm{B} T_\mathrm{s}} + \frac{\mathrm{RIN}\left(1-\cos\phi_\mathrm{dc}\right)^2}{k_\mathrm{B} T_\mathrm{s}}\right]. \tag{8.11}$$

The treatment in this case concerns primarily Equations (8.10) and (8.11). It is worth noting that some of the best-reported RF noise figures for microwave photonic links have employed low biased links where Equation (8.11) is minimized (Roussell et al. 2007, Karim and Davenport 2007, Urick et al. 2011b). The two major tradeoffs are that a low biased link makes inefficient use of the input optical power and introduces significant even-order distortion. The effects of the latter were analyzed for small deviations from quadrature in Chapter 6, specifically Equations (6.46)–(6.48) and (6.59)–(6.61). Those equations can be readily applied in this case as well. Although some hardware-intensive structures have been demonstrated for even-order linearization in low biased links (Devgan et al. 2009), it is generally accepted that low biasing an MZM is useful only for single-octave applications.

Shown in Figure 8.6 are calculations of Equations (8.10) and (8.11) as a function of MZM bias along with a plot of I_dc as given by the first line in Equation (6.18). The following values were employed for the plots: $I_{\mathrm{dc,q}} = 100$ mA, $V_\pi = 4$ V, $H_\mathrm{pd} = 1/2$, $R_\mathrm{i} = R_\mathrm{o} = 50\ \Omega$, and RIN = 0 ($-\infty$dBc/Hz). At quadrature ($0.5\pi$ in units for Figure 8.6), the values for G and NF match that in Figures 6.7 and 6.15(a), respectively, for a fundamentally limited link (i.e., no excess RIN). The RF gain is maximized at quadrature, which is the most efficient bias point in terms

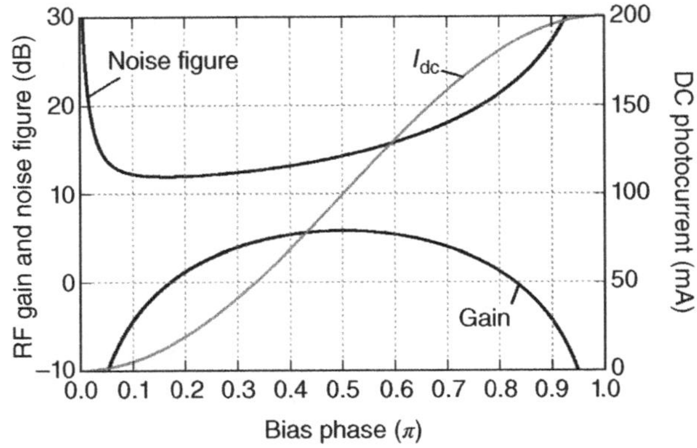

Figure 8.6. Plots of RF gain and RF noise figure (left axis) and average photocurrent (right axis) as a function of MZM bias phase. For these calculations, $I_{dc,q} = 100$ mA, $V_\pi = 4$ V, $H_{pd} = 1/2$, $R_i = R_o = 50\ \Omega$, and RIN = 0.

of RF power transfer, and $G \to -\infty$ dB ($g = 0$) for both peak and null bias. Furthermore, G is symmetric about quadrature. There is significant asymmetry in the NF curve, however. The minimum value for NF occurs below quadrature, at a level of 12 dB for $\phi_{dc} = 0.165\pi$, as compared to NF = 14 dB at quadrature. The RF gain at $\phi_{dc} = 0.165\pi$ is significantly reduced to a level of –0.2 dB as compared to 6 dB at quadrature. The percentage of optical power wasted is quite high for $\phi_{dc} = 0.165\pi$, at about 93.5% compared to 50% at quadrature. Finally, a substantial penalty is paid in multioctave dynamic range by low biasing. For $\phi_{dc} = 0.165\pi$, the SFDR in 1 Hz is limited by second-order distortion to $\text{SFDR}_2 = 75$ dB [Equation (6.59)] as compared to $\text{SFDR}_3 = 120.5$ dB in 1 Hz at quadrature [Equation (6.56)].

The noise figure improvement of a low biased link can outweigh the tradeoffs listed previously. Shown in Figure 8.7 is the calculated minimum RF noise figure as a function of the DC photocurrent when the MZM is biased at quadrature. As noted previously, the received photocurrent for the minimum NF point is much less than at quadrature; $I_{dc,q}$ is used as a parameter so that Figure 8.7 can be compared directly with Figure 6.15(a), which shows the NF at quadrature for the same parameters. Figure 8.7 was plotted using Equation (8.11) with $H_{pd} = 1/2$, $R_i = R_o = 50\ \Omega$, and RIN = 0.

There are subtleties associated with the effects of noise terms beyond shot and thermal noise in Equation (8.11) as captured by RIN. Firstly, consider the case where the source of excess RIN lies before the MZM, such as would arise from laser noise or an optical preamplifier. In this

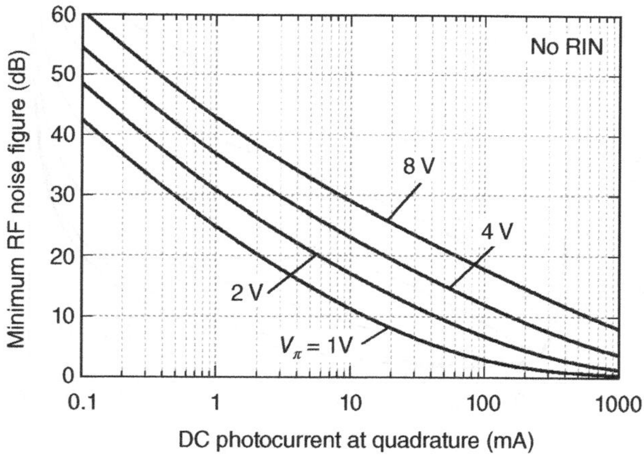

Figure 8.7. Plots of the lowest RF noise figure possible in a bias-shifted MZM link exhibiting only thermal and shot noise as a function of the average photocurrent at quadrature. Equation (8.11) was employed with $H_{pd} = 1/2$, $R_i = R_o = 50\ \Omega$, and RIN = 0.

case, the level of the RIN in Equation (8.11) is not a function itself of ϕ_{dc}. Rather, the amount of noise translated to the link output due to the RIN varies with ϕ_{dc} as captured by the last term in Equation (8.11). Effects of this type of pre-MZM RIN are examined in Figure 8.8. Shown in Figure 8.8(a) are calculations of Equation (8.11) for various RIN levels with $I_{dc,q} = 100$ mA, $V_\pi = 4$ V, $H_{pd} = 1/2$, and $R_i = R_o = 50\ \Omega$. At quadrature, the penalties imposed by RIN match those in Figure 6.15(b) for $I_{dc} = 100$ mA. However, the minimum noise figure is less affected by RIN, although some practical concerns do arise as the RIN increases. As the RIN increases, the location of the minima migrates to lower bias points. In this case, the effects of limited MZM extinction can become a concern. The minima also become sharper as RIN increases, requiring a more precise bias voltage to maintain the noise figure level. The impact of pre-MZM RIN on noise figure is shown in Figure 8.8(b) as a function of average photocurrent at quadrature, further demonstrating differences as compared to Figure 6.15(b). Note that the change relative to a fundamentally limited link for RIN < −170 dBc/Hz is negligible on the scale shown in Figure 8.8(b).

The analysis of a low biased link is further complicated by optical parameters that are a function of MZM optical output power, as these effects must be included into the aforementioned equations. Such is the case when saturable optical amplifiers are placed after the MZM, as has been studied for an erbium-doped fiber amplifier (EDFA) (Urick

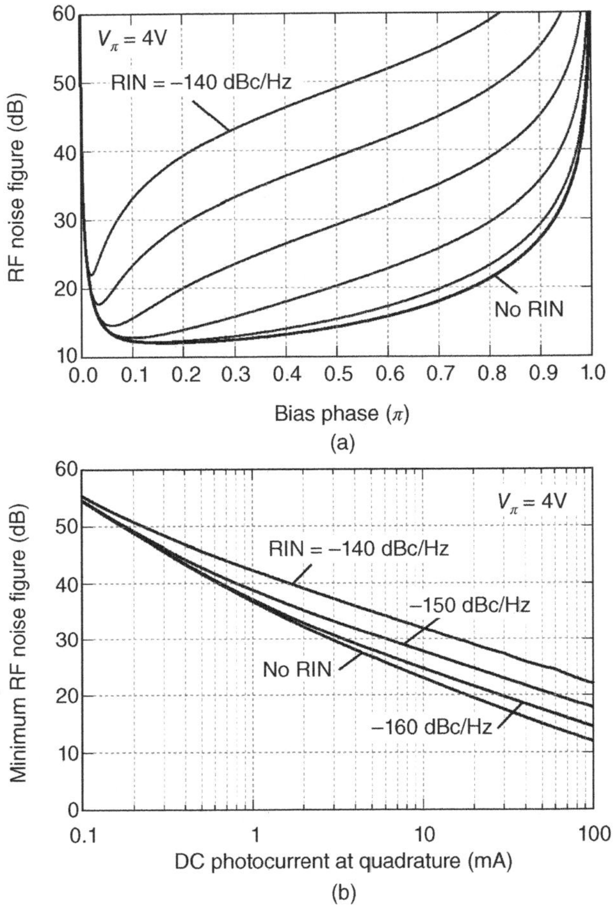

Figure 8.8. (a) Calculated RF noise figure as a function of bias with $I_{dc,q} = 100$ mA and various RIN levels. From top down: RIN = −140 dBc/Hz, −150 dBc/Hz, −160 dBc/Hz, −170 dBc/Hz, −180 dBc/Hz. The bottom curve is plotted without excess RIN. (b) Minimum RF noise figure as a function of quadrature photocurrent for various RIN levels. For both plots, $V_\pi = 4$ V, $H_{pd} = 1/2$, and $R_i = R_o = 50\ \Omega$.

et al. 2009). Equation (8.10) can be modified with a multiplicative term that describes the optical gain as a function of input power resulting in $g \propto g_o^2(\phi_{dc})$. The RIN in Equation (8.11) would also be a function of ϕ_{dc} in this case. Data exemplifying such a situation are shown in Figures 8.9 and 8.10 (Urick et al. 2009). The measured and calculated RF gain for an MZM-based link with an EDFA immediately after the MZM is shown in Figure 8.9. The calculation in this case employs an empirical equation for $g_o(\phi_{dc})$ and also accounts for limited MZM

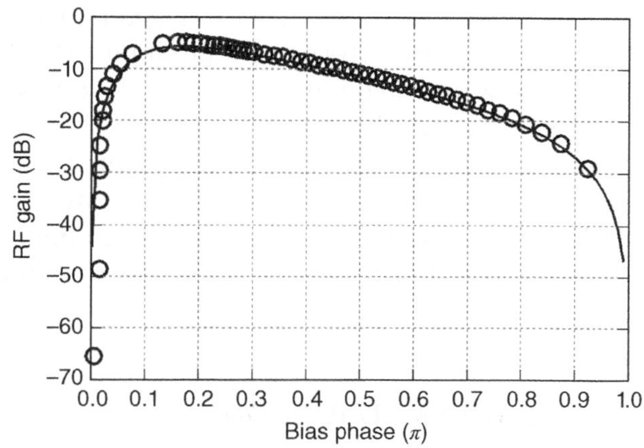

Figure 8.9. Measured (symbols) and calculated gain as a function of MZM bias in a link employing a post-MZM EDFA (Urick et al. 2009).

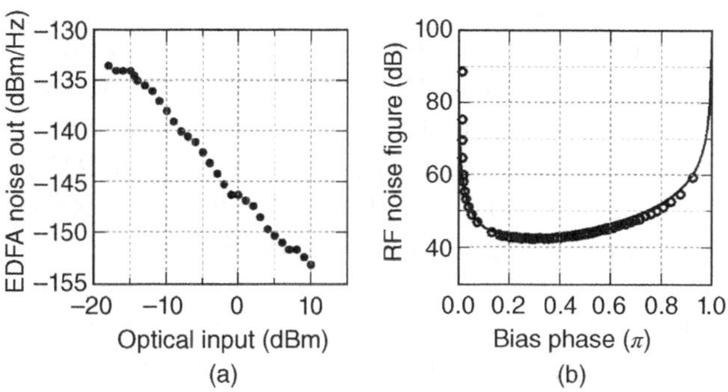

Figure 8.10. (a) Measured output noise power spectral density as a function of injected optical power into EDFA used in a low biased link (b) measured RF noise figure at the output of said link (symbols) shown against theory (line) using the data in (a) (Urick et al. 2009).

extinction (Urick et al. 2009). The gain is asymmetric about quadrature, with the peak gain occurring at a low bias point just above 0.1π, well away from quadrature. As described by Urick et al. (2009), this is because the EDFA provides more optical gain at low biases owing to its compression characteristic as a function of optical input power. That is, the MZM output power at quadrature pushes the EDFA into compression, whereas low bias points access more optical gain from the amplifier. Shown in Figure 8.10(a) is the measured noise power spectral density at the output of the same EDFA as a function of optical input power. As described in Section 3.5.1, the output noise decreases as the seed to the EDFA is increased, owing to less spontaneous emission. However, as shown in Figure 8.10(b), the minimum RF noise figure occurs at a relatively low optical input power, near 0.3π bias. The theory plotted in Figure 8.10(b) uses the data from Figure 8.10(a), which can be used to calculate RIN(ϕ_{dc}) for the EDFA from the equations in Section 3.5.1. Given the tradeoffs listed at the beginning of this section, low-biased-MZM links can supply solutions for some applications, with perhaps more work warranted on the effects of other nonlinear optical elements after the MZM.

8.3 SINGLE-SIDEBAND MODULATION

As described in previous chapters, chromatic dispersion can cause signal fading in dual-sideband intensity-modulated links. This phenomenon can be viewed as a cancellation of the photocurrents due to the upper and lower sideband. Thus, a SSB modulation scheme is relatively impervious to such chromatic-dispersion-induced fading. A fairly straightforward method to implement a SSB modulation scheme was introduced by Smith et al. (1997) using an architecture such as shown in Figure 8.11. A dual-electrode MZM is required, where the input RF signal is split equally and applied to each arm of the MZM with one RF signal shifted by 90°. A DC voltage is applied to one arm to bias the MZM at quadrature. When these conditions are precisely upheld, pure SSB modulation is achieved. However, the suppression of the unwanted sideband is limited in practice by the amplitude matching of the two modulation paths and achieving an ideal frequency-independent 90° phase shift. The latter is particularly challenging over very wide RF bandwidths, which is usually implemented with a 90° RF hybrid. Although the configuration shown in Figure 8.11 is the most common SSB modulation technique,

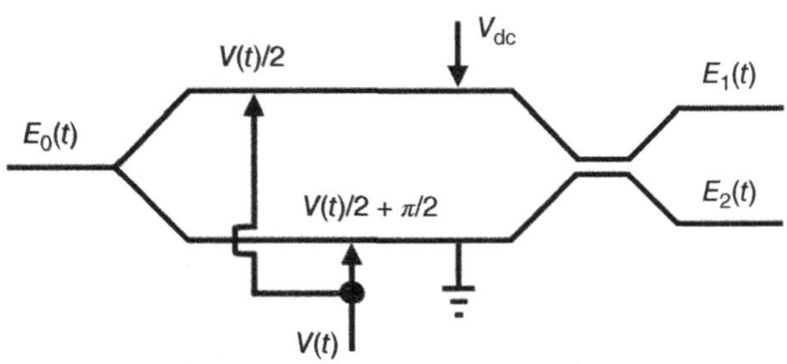

Figure 8.11. Configuration of a Mach–Zehnder modulator to achieve single-sideband modulation.

other schemes have been implemented. For example, linearized SSB has been demonstrated with a coherent technique (Lim et al. 2007) and with a DPMZM such as shown in Figure 6.32 (Kawanishi and Izutsu 2004).

Analysis of the architecture in Figure 8.11 can be conducted using scalar field transmission matrices as introduced in Chapter 6. It is assumed in this case that the amplitudes are precisely matched, the 90° phase shift is achieved, and the MZM is exactly at quadrature. The signal $V(t) = V_{rf} \sin \Omega t$ will be used, and the bias is set at $V_{dc} = V_\pi/2$. Let the fields at the output of such a structure be

$$\begin{bmatrix} E_1 \\ E_2 \end{bmatrix} = \frac{\sqrt{l_{mzm}}}{2} \begin{bmatrix} 1 & i \\ i & 1 \end{bmatrix} \begin{bmatrix} \exp(i\pi/2 + i\phi_{rf} \sin \Omega t) & 0 \\ 0 & \exp(i\phi_{rf} \cos \Omega t) \end{bmatrix} \begin{bmatrix} 1 & i \\ i & 1 \end{bmatrix} \begin{bmatrix} E_0 \\ 0 \end{bmatrix}, \tag{8.12}$$

where l_{mzm} is the MZM optical loss and $\phi_{rf} = \pi V_{rf}/V_\pi$. The input field E_0 is as described by Equation (6.5). Equation (8.12) can be reduced to

$$\begin{bmatrix} E_1 \\ E_2 \end{bmatrix} = \frac{\gamma}{2} e^{i\omega t} \begin{bmatrix} \exp(i\pi/2 + i\phi_{rf} \sin \Omega t) - \exp(i\phi_{rf} \cos \Omega t) \\ i \exp(i\pi/2 + i\phi_{rf} \sin \Omega t) + i \exp(i\phi_{rf} \cos \Omega t) \end{bmatrix}, \tag{8.13}$$

where γ is defined by Equation (6.10). The Jacobi–Anger expansions (Appendix IV) can be used to write the fields in Equation (8.13) as infinite sums of Bessel functions. Taking only the terms that correspond to the optical carrier and the first-order modulation sidebands results in

$$\mathrm{Re}(E_1) = -\frac{\gamma J_0(\phi_{rf})}{2}[\sin(\omega t) + \cos(\omega t)] + \gamma J_1(\phi_{rf}) \sin(\omega t - \Omega t). \tag{8.14}$$

$$\mathrm{Re}(E_2) = -\frac{\gamma J_0(\phi_{rf})}{2}[\sin(\omega t) + \cos(\omega t)] - \gamma J_1(\phi_{rf}) \cos(\omega t + \Omega t). \tag{8.15}$$

As can be seen from Equations (8.14) and (8.15), one output of the MZM will contain the lower fundamental sideband with the upper fundamental sideband on the other output.

Shown in Figure 8.12 are measured S_{21} data for a SSB link demonstrating its tolerance to chromatic dispersion in a link using 50 km of SMF-28 ($D \approx 16.5$ ps/nm/km). For these data, a dual-electrode MZM was driven by the outputs of a 90° RF hybrid and biased at quadrature. For comparison, the response of an IMDD link using an external MZM is also shown. Both responses are normalized to their low frequency response and not corrected for component roll off. As can be seen in Figure 8.12, the SSB modulation does not experience any of the signal fading that is characteristic of the IMDD link.

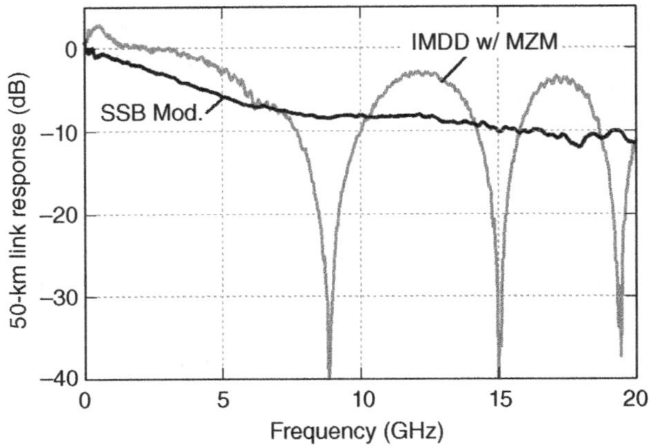

Figure 8.12. Measured responses through 50 km of standard single mode fiber for a single-sideband modulator (black) and a standard MZM (gray).

8.4 SAMPLED ANALOG OPTICAL LINKS

As discussed in Chapter 6, the most prevalent link architecture is the IMDD analog link employing an external MZM. In the conventional link format, the input RF signal is impressed onto a continuous-wave (CW) optical carrier, transmitted over a given length of optical fiber, and recovered from the modulated optical intensity via a photodetector (see Figures 6.4–6.6). Unique capabilities may be achieved when the CW laser in these architectures is replaced with a pulsed optical source such as a mode-locked laser (active or passive) (Carruthers and Duling 1996, Tamura et al. 1992), an optical comb generator (Wu et al. 2010), or a CW laser and low biased MZM driven with an RF pulse (McKinney and Williams 2009, McKinney and Urick 2010). In such an architecture, the pulsed optical carrier is analogous to an electronic analog-to-digital converter (ADC), in that the incoming RF signal is discretely sampled at the temporal location of each optical pulse with the pulse energy representing the instantaneous amplitude of the input RF signal. Subsequent quantization may be accomplished via a photonic sample-and-hold (Juodawlkis et al. 2001) or electronic digitization of the temporally demultiplexed pulse stream (Ghelfi et al. 2014) to achieve photonic analog-to-digital conversion (Valley 2007). The concentration in this case is the situation where the modulated optical pulse train is directly detected with a photodiode, that is, a sampled analog optical link architecture (McKinney and Williams 2009). Such an optical link architecture enables several capabilities including wideband subsampling or downconversion (McKinney and

Urick 2010, Pile and Taylor 2012) as well as a means to mitigate SBS (see Section 5.5) in long-haul analog links. Similar architectures have also been employed for "compressive sampling" (Nichols and Bucholtz 2011, Clark et al. 2012). Basic sampled link performance and these capabilities will be discussed further in the following section, with the exception of compressive sampling.

The primary concern in this section is the small-signal transmission performance of a sampled link (i.e., link gain). Note, however, that the RF performance of a sampled link (dynamic range, noise figure) in terms of average photocurrent and modulator half-wave voltage is well described by conventional IMDD link theory, as discussed in Chapter 6. In this case, it is assumed that the output of the pulsed optical source is modulated using an MZM as in a conventional externally modulated IMDD link. The first step in describing a sampled link is to consider the modulated photocurrent at the output of the photodiode. For a quadrature-biased modulator, the output photocurrents may be written as (McKinney and Williams 2009)

$$i(t) = p(t)[1 \pm v_{\text{in}}(t) * h_{\text{mzm}}(t)] * h_{\text{pd}}(t). \tag{8.16}$$

where $p(t)$ is the time-domain intensity profile of the sampling optical pulse train, $v_{\text{in}}(t)$ is the input RF voltage applied to the MZM, $h_{\text{pd}}(t)$ is the impulse response of the photodiode, and $*$ denotes convolution. The impulse response of the MZM is defined as

$$h_{\text{mzm}}(t) = \frac{1}{2\pi}\int_{-\infty}^{\infty} \frac{\pi}{V_\pi(\Omega)} \exp(i\Omega t)d\Omega, \tag{8.17}$$

where $V_\pi(\Omega)$ is the frequency-dependent half-wave voltage of the MZM. Note that small-signal conditions have been assumed: $|v_{\text{in}}(t) * h_{\text{mzm}}(t)| \ll 1$ for a generalized input voltage $v_{\text{in}}(t)$. For a single-tone sinusoidal input voltage, one would achieve a similar expression by evaluating Equation (6.20) for $\phi_{\text{rf}} = \pi V_{\text{rf}}/V_\pi \ll 1$. The (double-sided) RF power spectrum is determined by taking the magnitude squared of the Fourier transform of the modulated photocurrent in Equation (8.16). The resulting RF power spectrum is given by

$$P_{\text{rf}}(\Omega) = \Re^2 \frac{\pi^2}{2} \left| P(\Omega) * \frac{V_{\text{in}}(\Omega)}{V_\pi(\Omega)} \right|^2 |H_{\text{pd}}(\Omega)|^2 R_{\text{o}}. \tag{8.18}$$

In this expression, $\Re$ is the DC responsivity of the photodiode, $P(\Omega)$ is the Fourier transform of the intensity envelope of the sampling pulse train, $H_{pd}(\Omega)$ is the frequency response of the photodiode normalized to the DC responsivity, and R_o is the load resistance seen by the photodiode. For a perfectly periodic train of identical pulses, the sampling pulse train may be expressed as

$$p(t) = \widetilde{p}(t) * \sum_{n=-\infty}^{\infty} \delta(t - nT), \tag{8.19}$$

where the pulse repetition period is given by T, and the shape of a single pulse in the train is given by $\widetilde{p}(t)$. In the frequency domain, the Fourier transform of the time-domain intensity comprises a periodic comb of frequencies with spacing $\Omega_{rep} = 2\pi/T$ weighted by the Fourier transform of a single pulse in the train $[\widetilde{P}(\Omega)]$

$$P(\Omega) = \widetilde{P}(\Omega) \times \sum_{n=-\infty}^{\infty} \delta(\Omega - n\Omega_{rep}). \tag{8.20}$$

In the case where the link input is frequency limited to satisfy the sampling theorem (i.e., an antialiasing filter is used), the frequency-dependent RF gain is readily found from the RF power spectrum in Equation (8.18) by dividing by the input RF power $|V_{in}(\Omega)|^2/(2R_i)$, where R_i is the input resistance of the MZM. The RF gain of the sampled link is then given by

$$g(\Omega) = \left[\frac{I_{dc}}{V_\pi(\Omega)}\right]^2 \pi^2 R_i R_o |H_{pd}(\Omega)|^2. \tag{8.21}$$

In this case, $I_{dc} = \Re P_{avg}$ where P_{avg} is the average optical power of the pulsed source. Comparing Equation (8.21) with Equation (6.26), it can be seen that the sampled link gain in this case is equivalent to that for a conventional IMDD link. Figure 8.13 shows the measured gain at a frequency of 190 MHz for a 1-GHz sampled link as a function of photocurrent compared to the theoretical value calculated from Equation (8.21) for a half-wave voltage of $V_\pi \approx 2$ V. As is evident, there is excellent agreement between the measured data and the theoretical values—the roll off near $I_{avg} = 20$ mA occurs when the sampling pulse energy is sufficient to bleach the absorbing region in the photodiode (Miller et al. 1981, Juodawlkis et al. 2002). As noted previously, in this mode of operation, the other RF performance metrics (SFDR, CDR, and NF) may also be calculated using conventional IMDD link theory (Chapter 6). The interested reader may consult McKinney and Williams (2009) for

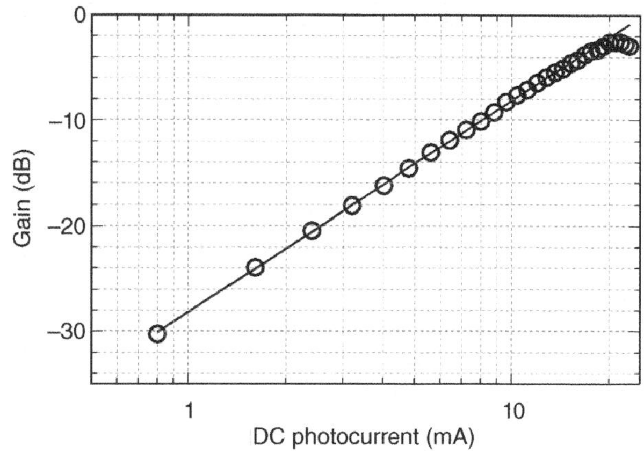

Figure 8.13. RF gain as a function of average photocurrent for a 1-GHz sampled link with an RF input signal at 190 MHz (McKinney and Williams 2009).

the analysis and measurement of these metrics in a sampled link architecture.

8.4.1 RF Downconversion Via Sampled Analog Optical Links

When the RF input to the sampled link is no longer frequency limited to satisfy the sampling theorem (McKinney and Urick 2010, Pile and Taylor 2012), aliasing occurs and the link operates in a downconverting mode. This mode of operation is frequently used in optical ADC architectures (Juodawlkis et al. 2003, Kim et al. 2008, Ghelfi et al. 2014). In this case, the optical sidebands arising from one element of the optical comb that represents the sampling pulse train in the frequency domain beat with an offset element of the comb at the photodiode. This gives rise to an aliased version of the input RF signal. A wideband demonstration of this mode of operation was provided by McKinney and Urick (2010). In this work, a 1-GHz optical pulse train was generated by driving a low biased MZM with a step recovery diode (SRD) to produce nominally 30-ps full-width half-maximum (FWHM) duration pulses as shown in Figure 8.14. As shown, the pulse shape is calculated on the basis of the drive voltage from the SRD and the optical transmission function of the MZM.

Downconverting functionality of the 1-GHz sampled link is shown in Figure 8.15. The input frequency to the link was varied in the 0–10-GHz range, and the output RF signal was measured within the fundamental Nyquist band ($0 < f < 500$ MHz). To illustrate the aliasing operation, several measured spectra are shown in Figure 8.14 as the

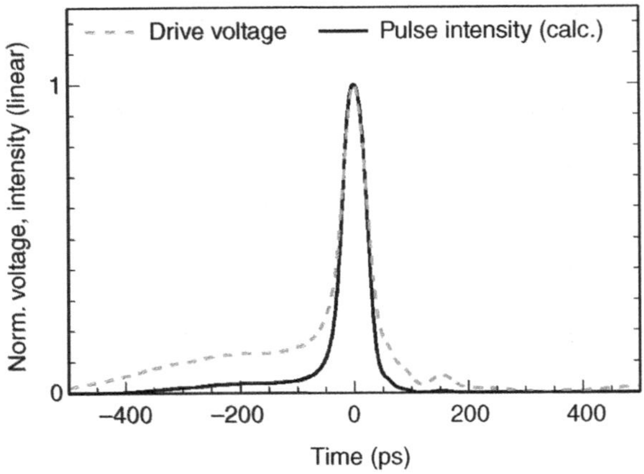

Figure 8.14. Calculated sampling pulse (black) generated via a low biased MZM driven with a step-recovery diode in a 1-GHz sampled link. The measured drive voltage is shown by the dashed gray curve (McKinney and Urick 2010).

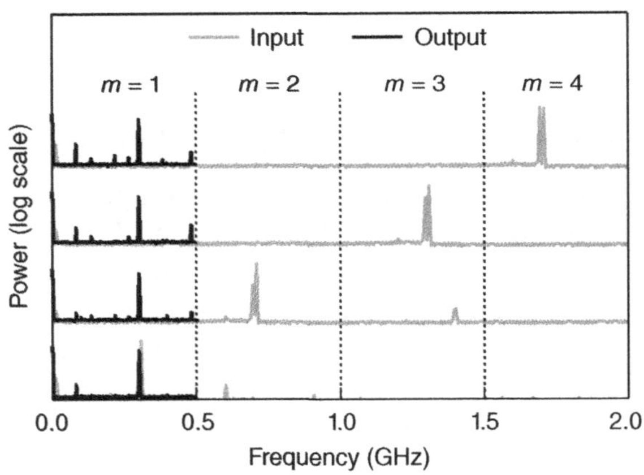

Figure 8.15. Measured spectra demonstrating downconversion (aliasing) in a 1-GHz sampled analog link. Input spectra are shown in gray, and the output spectra are shown in black (McKinney and Urick 2010).

input frequency is tuned across the first four Nyquist bands (frequency bands of 500 MHz—one half the sample rate—indexed by the integer m). In this case, the input signal frequencies (f = 300, 700, 1300, 1700 MHz) are chosen such that all alias to a frequency of 300 MHz within the fundamental Nyquist band (m = 1). In contrast to ADC architectures where the signal input frequencies are band limited to a particular Nyquist zone such that the input signal frequency may be determined from the aliased signal, the wideband sampled link clearly introduces ambiguity. Resolving this ambiguity to enable the use of sampled analog optical links in wideband receiver architectures is an area of continuing research.

Several items are of key importance when a sampled analog link operates in the downconverting mode. First and foremost, if the link is to achieve large bandwidth (in the sense that high frequency signals may be efficiently downconverted), the optical pulses should be sufficiently short. This may be understood by considering that the energy of a particular sampling pulse transmitted through the modulator (representing the sample of the RF input signal at a particular time) is given by the product of the pulse intensity profile and the modulator transmission function, integrated over the optical pulse duration. For a sinusoidal signal, as the input frequency increases, this integration occurs over a larger portion of the input signal period that leads to a decrease in modulation depth (the modulation depth approaches zero in the limit where the pulse duration exceeds the period). This effect leads to a decrease in the RF gain as the input signal frequency increases. Mathematically, this effect manifests itself as the magnitude squared of the Fourier transform of the time-domain intensity profile [Equation (8.20)] multiplying the RF gain as given by Equation (8.21). To illustrate the consequences of long sampling pulse durations, the deviation of the link gain from that predicted by Equation (8.21) for the link described by McKinney and Urick (2010) is shown in Figure 8.16. As the frequency of the input signal is increased, there is substantial roll off in the measured RF gain—at a frequency of 10 GHz,

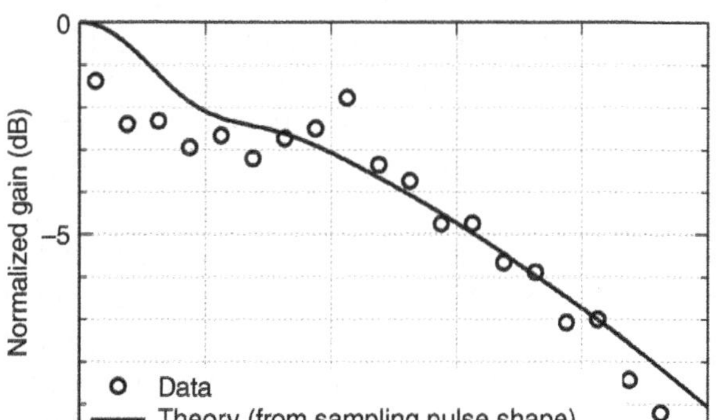

Figure 8.16. Measured gain variation in a 1-GHz subsampled link (circles) as compared to calculation determined by the magnitude spectrum of the sampling pulse intensity (curve) (McKinney and Urick 2010).

the gain has decreased by an order of magnitude from that predicted by Equation (8.21). For comparison, the magnitude squared of the Fourier transform of the sampling pulse intensity (calculated from the time-domain intensity profile shown in Figure 8.14) is also shown in Figure 8.16. As is evident, the gain variation is well described by the bandwidth of the optical sampling pulse. The sampling pulses should, therefore, be as short as possible to maintain gain uniformity. From a component point of view, it is very interesting to note that, when the RF output is taken from the fundamental Nyquist band, the photodiode need not possess a particularly large bandwidth. Because the downconversion operation is the result of an optical-domain mixing process, the photodiode frequency response needs only to be uniform over the fundamental Nyquist bandwidth. This means that downconversion of frequencies approaching 100 GHz might be successfully achieved using photodiodes with bandwidths of less than 1 GHz. However, the optical modulator must show a desirable response across the input frequency band of interest. In terms of Equation (8.21), the frequency dependence of H_{pd} is important only at the intermediate-frequency (IF) output, whereas that for V_π corresponds the RF input signal. Commercial modulators are, however, approaching 100 GHz bandwidths that may bring very wideband downconverting sampled links to fruition (see Section 10.5).

8.4.2 Mitigation of Stimulated Brillouin Scattering with Sampled Links

There are certain applications, long-haul antenna remoting for example, where the required fiber span may well exceed several tens of kilometers. In such applications, the fiber loss dictates that optical amplification must be used in order to overcome the intrinsic optical loss of the transmission fiber and to maintain an adequate signal level. However, placing amplification stages throughout the span is not tractable in some applications. For such long links, the optical launch power is also limited by SBS (Section 5.5). These limits necessitate optical amplification at the terminus of the link that can lead to unacceptably high link noise figures arising from the deleterious effects of spontaneous emission from the optical amplifier (Section 3.5.1). Sampled links have been shown to improve the noise figure and dynamic range in such links by allowing an increase in optical launch power, which is not afforded in conventional CW links. This technique exploits the fact that SBS is a relatively narrowband effect. As described in Section 5.5, the input power within bandwidth of the SBS gain will excite the SBS process above the threshold power given by Equation (5.37). For spans exceeding 10 km, this severely limits the optical launch power. It should be noted, however, that the Brillouin waves excited by two optical frequencies offset by more than the Brillouin bandwidth are independent. Thus, by distributing the desired optical power over an array of optical carriers with a frequency spacing exceeding the Brillouin bandwidth, the overall launch power may be increased. In a sampled analog link, the pulsed optical carrier may be described in the frequency domain as a comb of optical carriers spaced by the repetition rate of the pulse train. When the repetition rate is much higher than the SBS bandwidth, the available launch power then scales linearly with the number of lines in the optical comb. For a link employing a back-end EDFA, this means that more power may be injected into the amplifier, thereby reducing the required gain to achieve a particular photocurrent. Because the amount of spontaneous emission decreases with the amplifier gain, this means the noise penalty accrued also decreases as more power is injected into the amplifier. The end result is that, for a fixed average current, the link noise figure will decrease as more optical power is launched into the link.

To illustrate the noise figure and dynamic range enhancements that may be achieved in a sampled long-haul analog link architecture, McKinney et al. (2011) used a 3-GHz repetition rate optical comb source as the carrier in a 50-km optical link. The comb source was based on cascaded intensity and phase modulation, allowing the comb bandwidth to be varied using the amplitude of the applied phase modulation. Several example optical combs are shown by the gray curves in Figure 8.17. In this case, the number of features N is defined as

$$N = 1 + \frac{\Delta f_{\mathrm{rms}}}{f_{\mathrm{rep}}}, \tag{8.22}$$

where f_{rms} is the root-mean-square (RMS) bandwidth of the comb envelope, and f_{rep} is the combline spacing. Each comb was modulated with an RF signal in the 50–750-MHz range and launched into a 50-km optical link. The transmitted and reflected optical powers from the link, as well as the link gain and noise figure, were then measured as a function of the number of comb lines. To illustrate the increase in optical power reaching the link output as the number of comblines is increased, the measured transmitted and reflected optical powers are shown in Figure 8.18 for the $N = 1$ (CW laser) and $N = 14$ combs. The reflected powers clearly show an increase in SBS threshold for the $N = 14$ comb. The threshold is observed to increase from $P_{\mathrm{sbs}} \approx 6.3$ dBm for the case of CW excitation to approximately $P_{\mathrm{sbs}} \approx 18.4$ dBm for the $N = 14$ case.

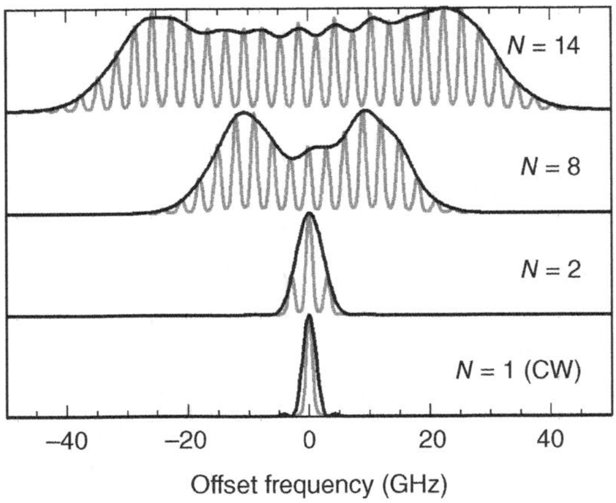

Figure 8.17. Representative optical combs showing RMS features ranging from 1 to 14. The measured optical spectra are shown in gray, and the comb envelopes are shown in black (McKinney et al. 2011).

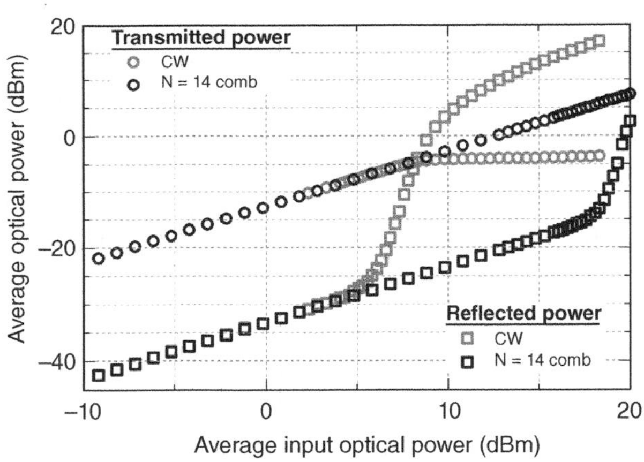

Figure 8.18. Comparison of the transmitted (circles) and reflected (squares) optical power for a 50-km fiber link as the number of comb elements is increased from $N = 1$ (gray) to $N = 14$ (black) (McKinney et al. 2011).

This threshold increase of $\Delta P_{\text{sbs}} = 12.1$ dB agrees quite well with the $10\log(N) = 11.5$ dB increase predicted on the basis of the number of features in the optical comb. This translates into increased total transmitted power for the $N = 14$ element comb, as is readily seen from the transmitted power curves.

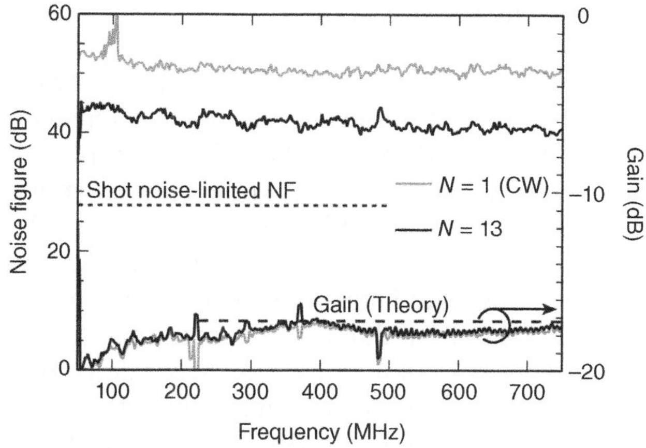

Figure 8.19. Comparison of the gain and noise figure for a 50-km link employing $N = 1$ (gray) and $N = 13$ (black) element optical combs for a fixed average photocurrent of 10 mA (McKinney et al. 2011).

Figure 8.19 illustrates the RF impact of increased launch power achieved via broad optical combs by comparing the gain and noise figure achieved using $N = 1$ and $N = 13$ element optical combs. In both cases, the average photocurrent was held fixed at 10 mA by controlling the gain of a post-link EDFA. As expected, the measured RF gain remains fixed in the $G \approx -17$-dB range as determined by the average photocurrent and MZM half-wave voltage ($V_\pi \approx 6$ V). The noise figure, however, shows a substantial decrease as the number of comb elements is increased. The noise figure for the $N = 13$ element comb is NF $\approx$ 41 dB as compared to NF $\approx$ 50 dB for the $N = 1$ (CW) case. The presence of the EDFA does impart a 13-dB penalty over the shot-noise-limited noise figure of NF $\approx$ 28 dB [Equation (6.73)]; the link output power before the EDFA is, however, only approximately 3× lower than the power needed to achieve an average photocurrent of 10 mA directly, without the need for the post-link EDFA. Therefore, it is feasible to reach shot-noise-limited performance with a 3× increase in the number of comb lines, well within reach, as the current state-of-the-art is 38 lines (10 GHz spacing) within 1-dB power variation (Wu et al. 2010). The link SFDR for the $N = 13$ element comb case is SFDR = 105.5 dB (1-Hz bandwidth), which is only roughly 8 dB below the shot-noise-limited value [Equation (6.77)] and 6 dB better than the $N = 1$ case (McKinney et al. 2011).

8.5 POLARIZATION MODULATION

The polarization of light offers another physical property to encode RF information. Early polarization-encoded optical architectures include suggestions by Niblack and Wolf (1966). Fiber-coupled polarization modulators are commercially available today. For example, Bull et al. (2004) have demonstrated a mode converter in an AlGaAs ridge waveguide that can be used as a polarization modulator. A representative polarization-modulated link is shown in Figure 8.20. A laser impinges on a polarization modulator similar to that described by Bull et al. (2004), which is essentially a TE-to-TM mode converter. In this case, it is assumed that the arc of modulation is along the equator of the Poincaré sphere (see Figure 5.19). The polarization modulated signal then traverses a length of non-polarization-maintaining fiber. The output of the fiber span is passed through a polarization controller that is achieved with a rotating half-wave plate (HWP) placed between two independent rotating quarter-wave plates (Lefevre 1980). The polarization controller serves to align the polarization before passing to a second rotating HWP, which in turn serves as a "bias" mechanism for the modulation between TE and TM modes. The polarization beam splitter (PBS) then converts the polarization modulation into intensity modulation that can be recovered with a photodiode(s).

The analysis for a polarization modulated link obviously cannot be carried out with scalar fields. In this case, a transfer function using Jones matrices (see Section 5.9) is employed:

$$\begin{bmatrix} E_1 \\ E_2 \end{bmatrix} = \frac{\sqrt{l_o}}{2} \begin{bmatrix} \cos 2\varphi & \sin 2\varphi \\ \sin 2\varphi & -\cos 2\varphi \end{bmatrix} \begin{bmatrix} 1 & -i \\ -i & 1 \end{bmatrix} \begin{bmatrix} e^{i\delta(t)} & 0 \\ 0 & e^{-i\delta(t)} \end{bmatrix} \begin{bmatrix} 1 \\ 1 \end{bmatrix} E_0, \quad (8.23)$$

where E_0 is given by Equation (6.5), E_1 and E_2 are the outputs of the PBS, l_o is the net optical loss, φ is the orientation of the HWP with respect to the x-axis, and δ is the time-dependent retardance imposed by the modulator. This last term is of the form $\delta(t) = \pi/2 + \pi V(t)/V_\pi$, where $V(t)$ is the modulating signal. The 2×2 matrices in Equation (8.23) describe, from right to left, the modulator (Campillo and Bucholtz 2006), the polarization controller with both QWPs at 0° with respect to the x-axis and the HWP at 22.5°, and the second rotatable HWP (Kliger et al. 1990, Damask 2005). As has been demonstrated by Hutchinson et al. (2014), the transfer matrix given by Equation (8.23) leads to photocurrents in form of Equations (6.18) and (7.7) for external intensity and phase modulation, respectively. In

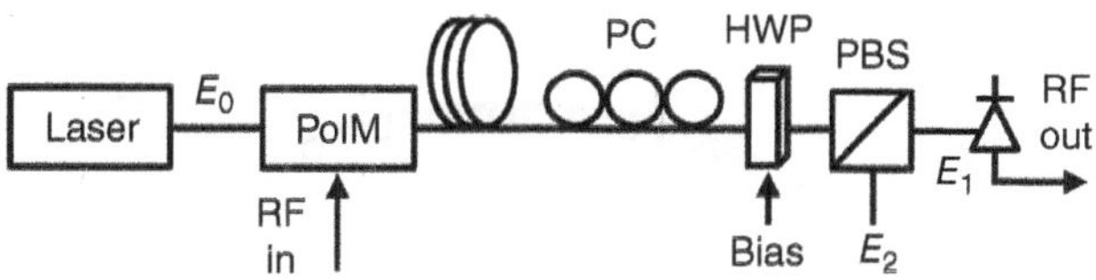

Figure 8.20. Block diagram for a polarization-modulated fiber optic link. Shown are a TE-to-TM mode converter polarization modulator (PolM), a polarization controller (PC), a rotatable half-wave plate (HWP), and a polarization beam splitter (PBS).

fact, the precise form of the photocurrents for the two outputs of the PBS is given by Equation (6.18) with ϕ_{dc} replaced by -2φ. Thus, the RF metrics for the polarization modulated link in Figure 8.20 are very much similar to those presented in Chapter 6, neglecting propagation effects. Included in this similarity is the ability to employ balanced detection of the two complimentary outputs of the PBS shown in Figure 8.20.

As an example, consider an expression for the small signal RF gain as a function of the HWP orientation:

$$g = \frac{I_{dc,q}^2}{V_\pi^2}\pi^2 R_i R_o |H_{pd}|^2 \sin^2(2\varphi), \tag{8.24}$$

which mirrors Equation (8.10). Equation (8.24) is plotted against measured data for a 35-GHz polarization-modulated link in Figure 8.21. For those data, $I_{dc,q} = 2.5$ mA, $V_\pi = 3.7$ V, $R_i = R_o = 50\ \Omega$, and $H_{pd} = 1/2$. As can be seen in Figure 8.21, the measured data and calculation agree nicely, and the curve exhibits the same shape as in Figure 8.6.

Despite the similarities demonstrated previously, many of the propagation impairments for a polarization-modulated link are quite different from that for other modulation formats. The effects described in Section 5.9 are most important; PDL and PMD can significantly distort a polarization-modulated link. The effects of DRS and SBS remain basically the same as described in Chapter 5. Chromatic dispersion effects can be similar but, as described by Campillo and Bucholtz (2006), depend strongly on the axis of modulation and the receiver configuration. Shown in Figure 8.22 are measured data demonstrating the effects of chromatic dispersion on the link configuration in Figure 8.20. Plotted there are the back-to-back link response (showing the response of the modulator and photodiode) against that through 50 km of standard single-mode fiber. In this case, the original modulation was originally oriented along the equator of the Poincaré sphere. The

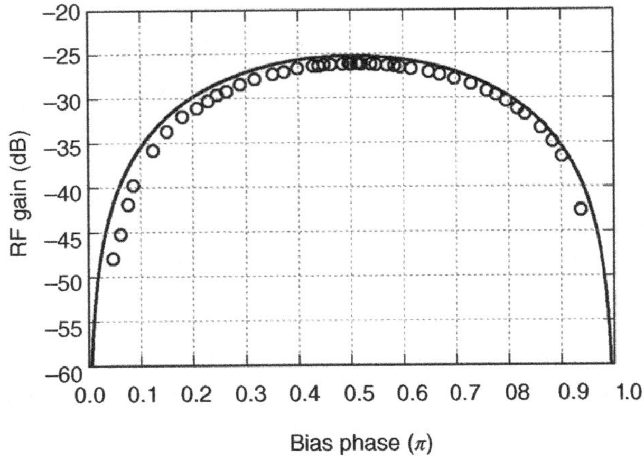

Figure 8.21. Measured (symbols) and calculated (line) RF gain as a function of orientation of the HWP for the architecture shown in Figure 8.20 (Hutchinson et al. 2014).

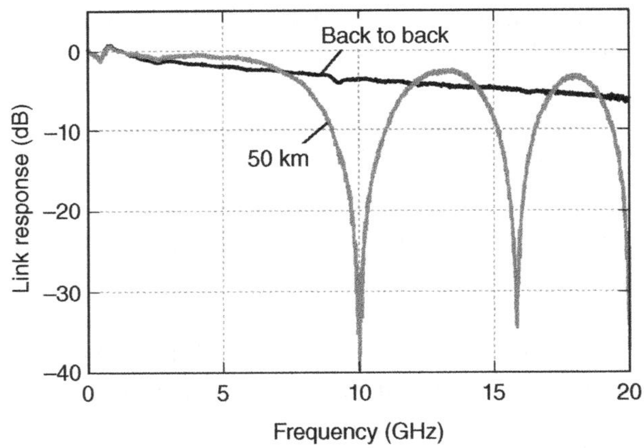

Figure 8.22. Measured link response for a polarization modulated link with (gray) and without (black) a 50-km fiber span in place (Campillo 2006).

periodic response shown in Figure 8.22 is similar, but not identical, to that for a 50-km IMDD link depicted in Figure 8.12. The same fiber span was used in both cases. On comparison, it is evident that the null locations in the response are significantly different. As can be seen, the polarization-modulated link is slightly more tolerant to chromatic dispersion, with the first-order null occurring at a higher frequency. A phasor representation such as that presented in Chapter 4 can aid in a qualitative understanding of the effects of dispersion on polarization modulation, if a fourth dimension is added to account for polarization.

A detailed mathematical analysis of the effects can be carried out in Stokes space but is too lengthy to include in this chapter. However, the materials presented in this chapter and in Chapter 5 provide the basic tools to understand thorough treatments such as those presented by Campillo and Bucholtz (2006).

Polarization modulation is ideally a constant-intensity format and therefore should enjoy some advantages over intensity modulation in terms of multichannel crosstalk. Indeed, this is typically the case. Propagation effects in multichannel polarization-encoded systems have been studied extensively for polarization multiplexed systems (Nelson et al. 2001, Xie 2011), digital polarization-shift-keyed links (Benedetto and Poggiolini 1992, Carena et al. 1998, Lepley et al. 2000, Siddiqui et al. 2002), and analog applications (Campillo 2006). The impairments due to SRS are of the same order of magnitude as described in Chapter 5 (Campillo 2006). However, nonlinearities stimulated by intensity fluctuations such as XPM are not as detrimental in polarization modulated links. However, XPolM can directly impact a polarization-modulated link, which can act in concert with PDL and PMD. As described theoretically and experimentally by Campillo (2006), these effects dominate at microwave frequencies. Measured crosstalk data for analog polarization- and intensity-modulated links are compared in Figure 8.23. Both links traversed the same fiber span, 25 km of Corning Metrocor with $D = -8$ ps/nm/km and $A_{\mathrm{eff}} = 60\ \mu\mathrm{m}^2$. In both cases, the modulated channel was at 1549.3 nm and the crosstalk was measured on an unmodulated 1550.9-nm channel. The launch power at both

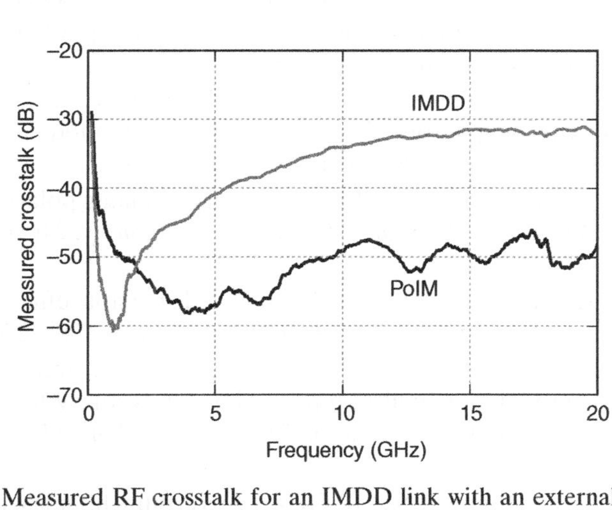

Figure 8.23. Measured RF crosstalk for an IMDD link with an external MZM (gray) and a polarization modulated link (black) (Campillo 2006).

wavelengths was 4 mW. As can be seen in Figure 8.23, the crosstalk for the polarization-modulated link is significantly lower at higher frequencies. At low frequency, the crosstalk is comparable as dictated by SRS. The low crosstalk in the IMDD link near 1 GHz is attributed to cancellation of two sources of crosstalk, most likely SRS and XPM. The interested reader is referred to Campillo (2006) for a more quantitative analysis.

REFERENCES

Ackerman, E., Wanuga, S., Kasemset, D., Daryoush, A. S., and Samant, N. R., "Maximum dynamic range operation of a microwave external modulation fiber-optic link," *IEEE Transactions on Microwave Theory and Techniques*, 41(8), 1299–1306 (1993).

Alferness, R. C., "Guided-wave devices for optical communication," *IEEE Journal of Quantum Electronics*, 17(6), 946–959 (1981).

Alferness, R. C., "Waveguide electrooptic modulators," *IEEE Transactions on Microwave Theory and Techniques*, 30(8), 1121–1137 (1982).

Benedetto, S. and Poggiolini, P., "Theory of polarization shift keying modulation," *IEEE Transactions on Communications*, 40(4), 708–721 (1992).

Bennett, B. R., Soref, R. A., and del Alamo, J. A., "Carrier-induced change in refractive index of InP, GaAs, and InGaAsP," *IEEE Journal of Quantum Electronics*, 26(1), 113–122 (1990).

Betty, I., "Indium phosphide-based electro-optic modulators," in *Broadband Optical Modulators*, A. Chen and E. Murphy (editors), CRC Press (2012).

Bull, J. D., Jaeger, N. A. F., Kato, H., Fairburn, M., Reid, A., and Ghanipour, P., "40 GHz electro-optic polarization modulator for fiber optic communication systems," *Proceedings of SPIE*, 5577, 133–143 (2004).

Bull, J. D., Darcie, T. E., Zhang, J., Kato, H., and Jaeger, N. A. F., "Broadband class-AB microwave-photonic link using polarization modulation," *IEEE Photonics Technology Letters*, 18(9), 1073–1075 (2006).

Campillo, A. L., "Interchannel nonlinear crosstalk in analog polarization modulated WDM systems," *Journal of Lightwave Technology*, 24(3), 1186–1193 (2006).

Campillo, A. L. and Bucholtz, F., "Chromatic dispersion effects in analog polarization-modulated links," *Applied Optics*, 45(12), 2742–2748 (2006).

Carena, A., Curri, V., Gaudino, R., Greco, N., Poggiolini, P., and Benedetto, S., "Polarization modulation in ultra-long haul transmission systems: a promising alternative to intensity modulation," in *ECOC Technical Digest*, 429–430 (1998).

Carruthers, T. F. and Duling, I. N., "10-GHz, 1.3 ps erbium fiber laser employing soliton pulse shortening," *Optics Letters*, 21(23), 1927–1929 (1996).

Chang, W. S. C., "Multiple quantum well electroabsorption modulators for RF photonic links," in *RF Photonic Technology in Optical Fiber Links*, W. S. C. Chang (editor), Cambridge, 165–202 (2002).

Chen, F. S., "Modulators for optical communications," *Proceedings of the IEEE*, 58(10), 1440–1457 (1970).

Chen, A. and Murphy, E. (editors), *Broadband Optical Modulators*, CRC Press (2012).

Clark, T. R. Jr., Callahan, P. T., and Dennis, M. L., "Wideband photonic compressive sampling analog-to-digital converter: architecture, device requirements and applications," *Proceedings of SPIE*, 8259 (2012).

Coldren, L. A., Corzine, S. W., and Mashanovitch, M. L., *Diode Lasers and Photonic Integrated Circuits*, Wiley (2012).

Cox, C. H. III, *Analog Optical Links*, Cambridge (2004).

Cox, C. H. III, Betts, G. E., and Johnson, L. M., "An analytical and experimental comparison of direct and external modulation in analog fiber-optic links," *IEEE Transactions on Microwave Theory and Techniques*, 38(5), 501–509 (1990).

Dagli, N., "Wide-bandwidth lasers and modulators for RF photonics," *IEEE Transactions on Microwave Theory and Techniques*, 47(7), 1151–1171 (1999).

Damask, J. N., *Polarization Optics in Telecommunications*, Springer (2005).

Darcie, T. E. and Driessen, P. F., "Class-AB techniques for high-dynamic range microwave-photonic links," *IEEE Photonics Technology Letters*, 18(8), 929–931 (2006).

Darcie, T. E., Tucker, R. S., and Sullivan, G. J., "Intermodulation and harmonic distortion in InGaAsP lasers," *Electronics Letters*, 21(16), 665–666 (1985).

Darcie, T. E., Zhang, J., Driessen, P. F., and Eun, J. J., "Class-B microwave-photonic link using optical frequency modulation and linear frequency discriminators," *Journal of Lightwave Technology*, 25(1), 157–164 (2007).

Devgan, P. S., Diehl, J. F., Urick, V. J., Sunderman, C. E., and Williams, K. J., "Even-order harmonic cancellation for off-quadrature biased Mach-Zehnder modulator with improved RF metrics using dual wavelength inputs and dual outputs," *Optics Express*, 17(11), 9028–9039 (2009).

Driessen, P. F., Darcie, T. E., and Zhang, J., "Analysis of a class-B microwave-photonic link using optical frequency modulation," *Journal of Lightwave Technology*, 26(15), 2740–2747 (2008).

van Eck, T., "Polymer modulators for RF photonics," in *RF Photonic Technology in Optical Fiber Links*, W. S. C. Chang (editor), Cambridge, 203–229 (2002).

Epworth, R. E., "Optical frequency modulation system," U.S. Patent No. 4,561,119 (1985).

Esman, R. D. and Williams, K. J., "Wideband efficiency improvement of fiber optic systems by carrier subtraction," *IEEE Photonics Technology Letters*, 7(2), 218–220 (1995).

Farwell, M. L., Chang, W. S. C., and Huber, D. R., "Increased linear dynamic range by low biasing the Mach-Zehnder modulator," *IEEE Photonics Technology Letters*, 5(7), 779–782 (1993).

Fiksman, G., Gross, R., Fan, J., and Kazovsky, L., "Performance optimization of directly modulated FM-SCM systems with optical discriminator," *IEEE Photonics Technology Letters*, 5(7), 845–848 (1993).

Frankel, M. Y. and Esman, R. D., "Optical single-sideband suppressed-carrier modulator for wide-band signal processing," *Journal of Lightwave Technology*, 16(5), 859–863 (1998).

Ghelfi, P., Laghezza, F., Scotti, F., Serafino, G., Capria, A., Pinna, S., Onori, D., Porzi, C., Scaffardi, M., Malacarne, A., Vercesi, V., Lazzeri, E., Berizzi, F., and Bogoni, A., "A fully photonics-based coherent radar system," *Nature*, 507(3), 341–345 (2014).

Hung, Y. C. and Fetterman, H. R., "Polymer-based directional coupler modulator with high linearity," *IEEE Photonics Technology Letters*, 17(12), 2565–2567 (2005).

Hutchinson, M. N., Singley, J. M., Urick, V. J., Harmon, S. R., McKinney, J. D., and Frigo, N. J., "Mitigation of photodiode induced even-order distortion in photonic links with predistortion modulation," *Journal of Lightwave Technology*, 32(20), 3885–3892 (2014).

Iezekiel, S., "Microwave-photonic links based on transistor-lasers: small-signal gain analysis," *IEEE Photonics Technology Letters*, 26(2), 183–186 (2014).

Ismail, T., Liu, C. P., Mitchell, J. E., and Seeds, A. J., "High-dynamic-range wireless-over-fiber link using feedforward linearization," *Journal of Lightwave Technology*, 25(11), 3274–3282 (2007).

Juodawlkis, P. W., Twichell, J. C., Betts, G. E., Hargreaves, J. J., Younger, R. D., Wasserman, J. L., O'Donnell, F. J., Ray, K. G., and Williamson, R. C., "Optically sampled analog-to-digital converters," *IEEE Transactions on Microwave Theory and Techniques*, 49(10), 1840–1853 (2001).

Juodawlkis, P. W., Hargreaves, J. J., and Twichell, J. C., "Impact of photodetector nonlinearities on photonic analog-to-digital converters," in *CLEO Technical Digest*, 11–12 (2002).

Juodawlkis, P. W., Hargreaves, J. J., Younger, R. D., Titi, G. W., and Twichell, J. C., "Optical down-sampling of wide-band microwave signals," *Journal of Lightwave Technology*, 21(12), 3116–3124 (2003).

Kalman, R. F., Fan, J. C., and Kazovsky, L. G., "Dynamic range of coherent analog fiber-optic links," *Journal of Lightwave Technology*, 12(7), 1263–1277 (1994).

Kaminow, I. P., "Optical waveguide modulators," *IEEE Transactions on Microwave Theory an Techniques*, 23(1), 57–70 (1975).

Kaminow, I. P. and Turner, E. H., "Electrooptic light modulators," *Applied Optics*, 5(10), 1612–1628 (1966).

Karim, A. and Davenport, J., "Low noise figure microwave photonic link," in *IEEE Microwave Symposium Digest*, 1519–1522 (2007).

Kawanishi, T. and Izutsu, M., "Linear single-sideband modulation for high-SNR wavelength conversion," *IEEE Photonics Technology Letters*, 16(6), 1534–1536 (2004).

Kim, H. and Gnauck, A. H., "Chirp characteristics of dual-drive Mach-Zehnder modulator with a finite DC extinction ratio," *IEEE Photonics Technology Letters*, 14(3), 298–300 (2002).

Kim, J., Park, M. J., Perrott, M. H., and Krtner, F. X., "Photonic subsampling analog-to-digital conversion of microwave signals at 40 GHz with higher than 7-ENOB resolution," *Optics Express*, 16(21), 16509–16515 (2008).

Kliger, D. S., Lewis, J. W., and Randall, C. E., *Polarized Light in Optics and Spectroscopy*, Academic (1990).

Kobayashi, S., Yamamoto, Y., Ito, M., and Kimura, T., "Direct frequency modulation in AlGaAs semiconductor lasers," *IEEE Journal of Quantum Electronics*, 18(4), 582–595 (1982).

LaGasse, M. J., Charczenko, W., Hamilton, M. C., and Thaniyavarn, S., "Optical carrier filtering for high dynamic range fibre optic links," *Electronics Letters*, 30(25), 2157–2158 (1994).

Lau, K. Y., *Ultra-High Frequency Linear Fiber Optics Systems*, Springer (2011).

Lee, M., Katz, H. E., Erben, C., Gill, D. M., Gopalan, P., Heber, J. D., and McGee, D. J., "Broadband modulation of light by using an electro-optic polymer," *Science*, 298(11), 1401–1403 (2002).

Lefevre, H. C., "Single-mode fibre fractional wave devices and polarisation controllers," *Electronics Letters*, 16(20), 778–780 (1980).

Lepley, J. J., Ellison, J. G., Edirisinghe, S. G., Siddiqui, A. S., and Walker, S. D., "Excess penalty impairments of polarization shift keying transmission format in presence of polarization mode dispersion," *Electronics Letters*, 36(8), 736–737 (2000).

Lim, C., Nirmalathas, A., Lee, K. L., Novak, D., and Waterhouse, R., "Intermodulation distortion improvement for fiber-radio applications incorporating OSSB + C modulation in an optical integrated-access environment," *Journal of Lightwave Technology*, 25(6), 1602–1612 (2007).

Liu, B., Shim, J., Chiu, Y. J., Keating, A., Piprek, J., and Bowers, J. E., "Analog characterization of low-voltage MQW traveling-wave electroabsorption modulators," *Journal of Lightwave Technology*, 21(12), 3011–3019 (2003).

Marpaung, D., Roeloffzen, C., and van Etten, W., "Enhanced dynamic range in a directly modulated analog photonic link," *IEEE Photonics Technology Letters*, 21(24), 1810–1812 (2009).

McKinney, J. D. and Urick, V. J., "Radio-frequency down-conversion via sampled analog links," NRL Memorandum Report, NRL/MR/5650-10-9275 (2010).

McKinney, J. D. and Williams, K. J., "Sampled analog links," *IEEE Transactions on Microwave Theory and Techniques*, 57(8), 2093–2099 (2009).

McKinney, J. D., Urick, V. J., and Briguglio, J., "Optical comb sources for high dynamic-range single-span long-haul analog optical links," *IEEE Transactions on Microwave Theory and Techniques*, 59(12), 3249–3257 (2011).

Melikyan, A., Alloatti, L., Muslija, A., Hillerkuss, D., Schindler, P. C., Li, J., Palmer, R., Korn, D., Muehlbrandt, S., van Thourhout, D., Chen, B., Dinu, R., Sommer, M., Koos, C., Kohl, M., Freude, W., and Leuthold, J., "High-speed plasmonic phase modulators," *Nature Photonics*, 8(3), 229–233 (2014).

Middleton, C. and DeSalvo, R., "Improved microwave photonic link performance through optical carrier suppression and balanced coherent heterodyne detection," *Proceedings of SPIE*, 7700, 770008 (2010).

Miller, A., Miller, D. A. B., and Smith, S. D., "Dynamic non-linear processes in semiconductors," *Advances in Physics*, 30(6), 697–800 (1981).

Naitoh, H., Muto, K., and Nakayama, T., "Mirror-type optical branch and switch," *Applied Optics*, 17(1), 101–104 (1978).

Nelson, L. E., Nielsen, T. N., and Kogelnik, H., "Observation of PMD-induced coherent crosstalk in polarization-multiplexed transmission," *IEEE Photonics Technology Letters*, 13(7), 738–740 (2001).

Niblack, W. K. and Wolf, E. H., "Polarization modulation and demodulation," U.S. Patent No. 3,284,632 (1966).

Nichols, J. M. and Bucholtz, F., "Beating Nyquist with light: a compressively sampled photonic link," *Optics Express*, 19(8), 7339–7348 (2011).

Paoli, T. L. and Ripper, J. E., "Direct modulation of semiconductor lasers," *Proceedings of the IEEE*, 58(10), 1457–1465 (1970).

Phillips, M. R., Darcie, T. E., Marcuse, D., Bodeep, G. E., and Frigo, N. J., "Nonlinear distortion generated by dispersive transmission of chirped intensity-modulated signals," *IEEE Photonics Technology Letters*, 3(5), 481–483 (1991).

Pile, B. C. and Taylor, G. W., "Performance of subsampled analog optical links," *Journal of Lightwave Technology*, 30(9), 1299–1305 (2012).

Roussell, H. V., Regan, M. D., Prince, J. L., Cox, C. H., Chen, J. X., Burns, W. K., Betts, G. E., Ackeman, E. I., and Campbell , "Gain, noise figure and bandwidth-limited dynamic range of a low-biased external modulation link," in *IEEE International Topical Mtg. on Microwave Photonics Digest*, 84–87 (2007).

Schmidt, R. V. and Alferness, R. C., "Directional coupler switches, modulators, and filters using alternating $\Delta\beta$ techniques," *IEEE Transactions on Circuits and Systems*, 26(12), 1099–1108 (1979).

Shin, J. H. and Dagli, N., "Ultralow drive voltage substrate removed GaAs/AlGaAs electro-optic modulators at 1550 nm," *IEEE Journal of Selected Topics in Quantum Electronics*, 19(6), 3400408 (2013).

Siddiqui, A. S., Edirisinghe, S. G., Lepley, J. J., Ellison, J. G., and Walker, S. D., "Dispersion-tolerant transmission using a duobinary polarization-shift keying transmission scheme," *IEEE Photonics Technology Letters*, 14(2), 158–160 (2002).

Smith, G. H., Novak, D., and Ahmed, Z., "Technique for optical SSB generation to overcome dispersion penalties in fibre-radio systems," *Electronics Letters*, 33(1), 74–75 (1997).

Solgaard, O., Ho, F., Thackara, J. I., and Bloom, D. M., "High frequency attenuated total internal reflection light modulator," *Applied Physics Letters*, 21(23), 2500–2502 (1992).

Stephens, W. E. and Joseph, T. R., "System characterization of direct modulated and externally modulated RF fiber-optic links," *Journal of Lightwave Technology*, 5(3), 380–387 (1987).

Tamura, K., Haus, H. A., and Ippen, E. P., "Self-starting additive pulse modelocking erbium fiber ring laser," *Electronics Letters*, 28(24), 2226–2228 (1992).

Thaniyavarn, S., "Modified 1x2 directional coupler waveguide modulator," *Electronics Letters*, 22(18), 941–942 (1986).

Urick, V. J., Godinez, M. E., Devgan, P. S., McKinney, J. D., and Bucholtz, F., "Analysis of an analog fiber-optic link employing a low-biased Mach-Zehnder modulator followed by an erbium-doped fiber amplifier," *Journal of Lightwave Technology*, 27(12), 2013–2019 (2009).

Urick, V. J., Bucholtz, F., McKinney, J. D., Devgan, P. S., Campillo, A. L., Dexter, J. L., and Williams, K. J., "Long-haul analog photonics," *Journal of Lightwave Technology*, 29(8), 1182–1205 (2011a).

Urick, V. J., McKinney, J. D., Diehl, J. F., and Williams, K. J., "Fiber-optic links with all-photonic RF gain and low RF noise figure," in *IEEE Microwave Symposium Digest*, (2011b).

Valley, G. C., "Photonic analog-to-digital converters," *Optics Express*, 15(5), 1955–1982 (2007).

Walker, R. G., "High-speed III-V semiconductor intensity modulators," *IEEE Journal of Quantum Electronics*, 27(3), 654–667 (1991).

Walker, R. G., Cameron, N. I., Zhou, Y., and Clements, S. J., "Optimized gallium arsenide modulators for advanced modulation formats," *IEEE Journal of Selected Topics in Quantum Electronics*, 19(6), 3400912 (2013).

Walter, G., Holonyak, N., Feng, M., and Chan, R., "Laser operation of a heterojunction bipolar light-emitting transistor," *Applied Physics Letters*, 85(20), 4768–4770 (2004).

Wang, S. Y. and Lin, S. H., "High speed III-V electrooptic waveguide modulators at $\lambda = 1.3$ mm," *Journal of Lightwave Technology*, 6(6), 758–771 (1988).

Wang, J., Haldar, M. K., and Mendis, F. V. C., "Formula for two-carrier third-order intermodulation distortion in semiconductor laser diodes," *Electronics Letters*, 29(15), 1341–1343 (1993).

Wooten, E. L., Kissa, K. M., Yi-Yan, A., Murphy, E. J., Lafaw, D. A., Hallemeier, P. F., Maack, D., Attanasio, D. V., Fritz, D. J., McBrien, G. J., and Bossi, D. E., "A review of lithium niobate modulators for fiber-optic communications systems," *IEEE Journal of Selected Topics in Quantum Electronics*, 6(1), 69–82 (2000).

Wu, R., Supradeepa, V. R., Long, C. M., Leaird, D. E., and Weiner, A. M., "Generation of very flat optical frequency combs from continuous-wave lasers using cascaded intensity and phase modulators driven by tailored radio frequency waveforms," *Optics Letters*, 35(19), 3234–3236 (2010).

Wyrwas, J. M and Wu, M. C., "Dynamic range of frequency modulated direct-detection analog fiber optic links," *Journal of Lightwave Technology*, 27(24), 5552–5562 (2009).

Xie, C., "Impact of nonlinear and polarization effects in coherent systems," *Optics Express*, 19(26), B915–B930 (2011).

Yabre, G. and Le Bihan, J., "Intensity modulation technique using a directly frequency-modulated semiconductor laser and an interferometer," *Journal of Lightwave Technology*, 13(10), 2093–2098 (1995).

CHAPTER 9

HIGH CURRENT PHOTODETECTORS

As discussed in Chapter 2, the signal fidelity of analog photonic links is characterized by noise figure and dynamic range. Expressions for intensity-modulated direct-detection (IMDD) links were derived in Chapter 6 including those for link gain [Equation (6.27)], OIP3 [Equation (6.43)], noise figure [Equations (6.51) and (6.73)], CDR [Equations (6.54) and (6.75)], and third-order SFDR [Equations (6.57) and (6.77)]. These IMDD link metrics all improve with output photocurrent, so long as the photodetector itself does not contribute to overwhelming distortion. Similarly, metrics from links using several of the other modulation techniques covered in Chapters 7 and 8 also improve with increasing output photocurrent. The majority of the work in previous chapters assumed that the transfer function of the photodiode, H_{pd}, was constant and independent of photodetector average current. As will be described in this chapter, this is not always the case. While Chapter 4 concentrated on the linearity of photodetectors, this chapter deals with obtaining high radio-frequency (RF) output current from a photodetector. Often, but not necessarily, a high current photodiode can achieve large output intercept points. Thus, there are many cases where a high current photodiode can also achieve high linearity. This being noted, the present chapter provides a detailed

Fundamentals of Microwave Photonics, First Edition.
Vincent J. Urick Jr, Jason D. McKinney, and Keith J. Williams.

discussion of the internal photodiode limitations associated with generating high photocurrents. A discussion of amplitude compression in Section 9.1 will be followed by the limitations imposed by series resistance in Section 9.2, as well as discussions related to the thermal and space-charge limits in Sections 9.3 and 9.4, respectively. Finally, Section 9.5 will discuss the large-signal characteristics of photodiodes in terms of their RF power-conversion efficiency. This chapter concludes with a brief summary of photodetector performance in Section 9.6.

9.1 PHOTODETECTOR COMPRESSION

Many types of input optical fields can be considered as they relate to photodiode compression, including pulsed or continuous-wave (CW). However, with the widespread use of conventional IMDD links (i.e., those employing a CW laser), photodiode performance under IMDD modulation conditions is an important operational regime. The sinusoidal transfer function of a Mach–Zehnder modulator (MZM)-based IMDD link was treated previously (see Figure 6.2). With the MZM biased at quadrature, an input sinusoid modulates the output intensity of a laser that is transmitted to a photodetector. The characteristics of the optical waveform are a large DC component combined with small-signal (AC) modulation components at the fundamental stimulus frequency and harmonics. Since optical power is converted directly to current in a photodetector, the photocurrent will likewise contain a large average current with a smaller AC component. Thus, the response of a photodetector should be characterized under high DC current levels with small modulation depth in order to evaluate its utility in an MZM-based IMDD link.

Response compression in a photodiode was observed by Williams et al. (1994). For these measurements, the modulation depth from a pair of phase-locked lasers was adjusted by controlling the amplitudes of each laser. Similar compression data were reported by Williams et al. (1994) and are plotted in Figure 9.1 for average currents of 800 and 1000 μA relative to the response of the photodiode at an average photocurrent of 100 μA. As can be seen in the 800-μA curve, the response quickly compresses by 1–3 dB for frequencies up to 1 GHz and then remains compressed at relatively the same level for frequencies up to 24 GHz. At the higher compression level (1000 μA), the response is seen to recover slightly (~0.7 dB) from 2 to 24 GHz. Similar frequency response characteristics were observed in the pulsed domain by Kuhl et al. (1992). The response compression observed was

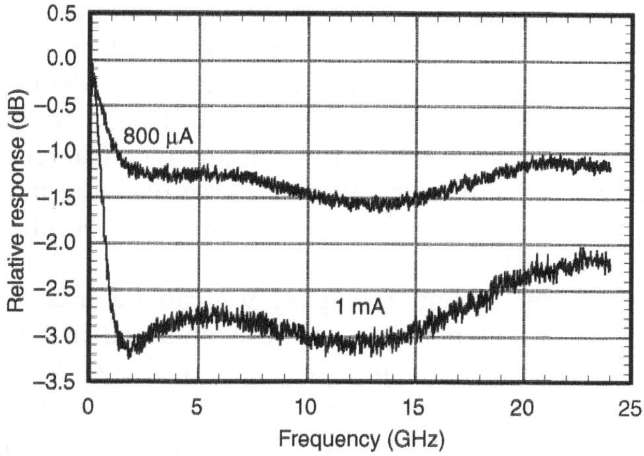

Figure 9.1. Small-signal relative frequency response of a −5-V biased photodiode. Average photocurrents of 800 and 1000 μA plotted relative to the photodiode response at 100 μA (Williams 1994).

attributed to space-charge buildup within the depletion region and will be discussed in more detail in Section 9.4. An important feature to note in this case is that the response compression appeared to be primarily caused by the average (not the peak) photocurrent.

As described in Chapter 4, the measurement system used by Williams et al. (1994) was later refined in Davis et al. (1996) to provide more accurate and independent control between the DC and AC components of the photocurrent. Figure 9.2 outlines a basic setup for photodetector compression characterization. A CW and an RF modulated laser at different wavelengths are combined and passed to a photodiode. The wavelengths are close enough to provide good overlap in field strength within the photodiode, but not too close as

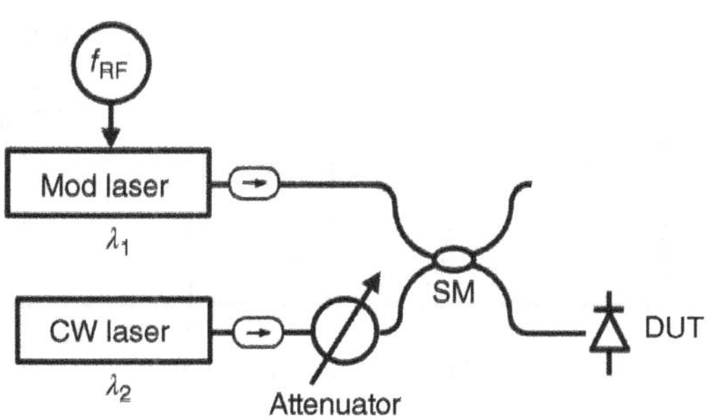

Figure 9.2. Photodiode response compression measurement apparatus. A small-signal modulated laser is combined with an adjustable power CW laser for independent control of photodiode DC current. SM: single-mode.

to cause other mixing products that could contribute to measurement uncertainties. Wide wavelength separations (>100 nm) can lead to significant differences in charge densities within the depletion region as demonstrated by Williams et al. (1998), which can lead to erroneous results. The RF-modulated laser can be of any type (direct or external modulation), but the link gain must be stable over the measurement period to ensure that small responsivity changes in the photodetector on the order of 0.1 dB can be accurately observed.

The measurement system in Figure 9.2 can be used to perform compression measurements such as those obtained in Figure 9.1 by applying a small-signal-modulated laser to a device under test (DUT) and measuring the S_{21} amplitude response with the CW laser off. This measurement can then be repeated at higher values of average current by adjusting the CW laser power incident on the DUT and plotting the response relative to the low power result. A family of curves as in Figure 9.1 can then be generated, which contain data as a function of both frequency and average current at a fixed photodiode bias voltage. An alternative measurement (Williams and Esman 1996) has become popular, as it leads to a direct measure of the current-handling capability of any given photodetector. For this measurement, the frequency is fixed and the photodetector response is measured until the response compresses by −1.0 dB for a given applied bias voltage. Subsequent data points are taken by an adjustment of the applied bias voltage. Families of curves for various DUT parameters can be plotted as shown in Figure 9.3 (Williams and Esman 1999). The data show a

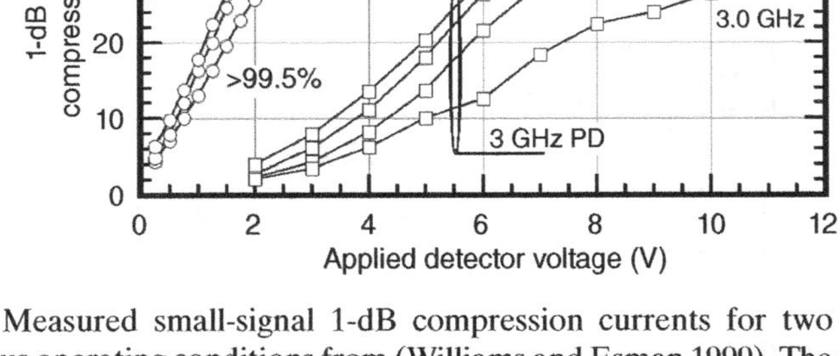

Figure 9.3. Measured small-signal 1-dB compression currents for two photodiodes under various operating conditions from (Williams and Esman 1999). The curves on the right show the frequency dependence of a 3-GHz photodiode. The curves on the left demonstrate the effect of fill factor (%) for a 25-GHz photodiode operated at 10 GHz.

typical photodetector characteristic of having higher small-signal compression currents at higher bias voltages. The frequency dependencies in Figure 9.3 are also typical, where lower frequencies offer higher compression currents for the same applied bias voltage. Also plotted is a series of curves for a 25-GHz photodetector (Williams and Esman 1999) where the compression-current dependence on optical intensity is shown. In this case, the current handling improves as the light is distributed more evenly across the absorbing area of the photodetector. Curves such as those in Figure 9.3 will again be used in Section 9.5 to quantify the efficiency performance of power photodiodes.

9.2 EFFECTS DUE TO FINITE SERIES RESISTANCE

Space-charge effects (see Section 9.4) were among the first mechanisms identified to explain photodetector nonlinearities and response compression (Williams et al. 1994). At that time, most measurements on high current photodetectors yielded maximum 1-dB small-signal compression currents below 60 mA [see for example (Williams and Esman 1996, Williams et al. 1998, Williams and Esman 1999) and (Shimizu et al. 2000)] with the exception of the 150-mA result obtained by Davis et al. (1996) in a 295-MHz bandwidth photodetector. In this photodiode structure, first suggested by Pearsall et al. (1981) to reduce dark currents, the InGaAs absorbing layers were p-doped and only nonabsorbing intrinsic InP material was used for the depletion layer. In addition to lower dark currents, the use of InP in the depletion layer allowed for increased operating bias voltages and power dissipation levels due to the higher thermal conductivity of InP. In this style of photodiode, minority electrons generated within the absorbing undepleted InGaAs must travel into the depletion layer either by minority-carrier diffusion or by minority-carrier drift velocity components. The latter is associated with a small electric field supplied by the majority-carrier drift current in this region.

The work of Li et al. (2003a) extended the concept of using undepleted absorbing layers as a way to improve quantum efficiency while keeping the depletion layer thin to aid in the power dissipation properties of the photodetector epilayer design as had been suggested by Williams and Esman (1999). Small-signal compression current measurements on these devices resulted in up to 110 mA at 1 GHz and 57 mA at 10 GHz (7–9-V bias voltages) with p-side illumination through a ring contact. Initial simulations on this structure suggested

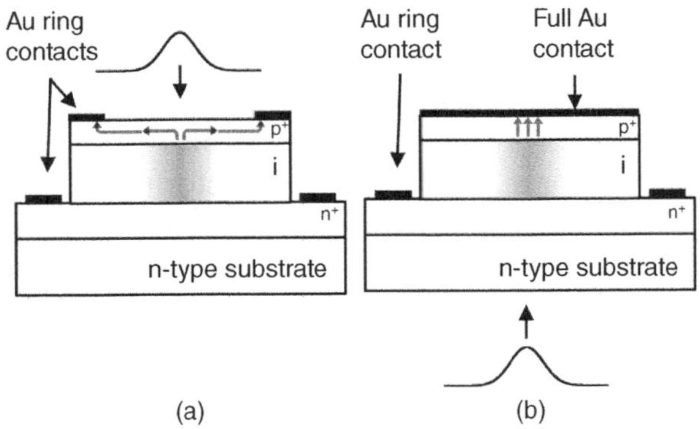

Figure 9.4. Cross-sections of mesa-style photodiodes with a ring-type p-contact (a) and a full metal p-contact (b). Both structures have ring-type n-contacts. Absorption is depicted by the gray areas within the depletion regions due to Gaussian beam illumination.

drastically higher space-charge-limited compression current performance, but measured data did not support this hypothesis. Ultimately, it was discovered that the lateral p-resistance (reverse biased) within the p-doped layer lowered the voltage across the depletion region and was responsible for the deviation between modeled and measured results. This was validated by further processing the same wafer, but with full metal p-contacts to reduce series resistance and illuminating the device through the n-side of the structure. Measurement data for these devices demonstrated 199-mA operation at only a 2-V bias voltage (Li et al. 2003b). A systematic study between front- and back-side illumination in this epilayer design was completed the following year, where the 60-mA barrier at microwave frequencies was significantly exceeded, yielding over 100 mA small-signal compression current at 10 GHz under only 3-V bias conditions (Li et al. 2004a).

To study the effects of excess reverse series resistance due to lateral p-doped resistance in ring contact photodiodes, the following qualitative analysis is provided. Figure 9.4 shows the cross-section of two p–i–n photodiodes. One is p-side illuminated through a ring contact, and the other is n-side illuminated through the substrate using a full metal p-contact. The equivalent circuit for both photodiode structures can be represented by the circuit in Figure 9.5. In both structures of Figure 9.4, absorption occurs along the center axis of the device, and for simplicity, it is assumed that absorption occurs only within the i-layer (the results are similar if p- or n-layer absorption is considered). Carriers that are absorbed within the intrinsic region drift to their respective

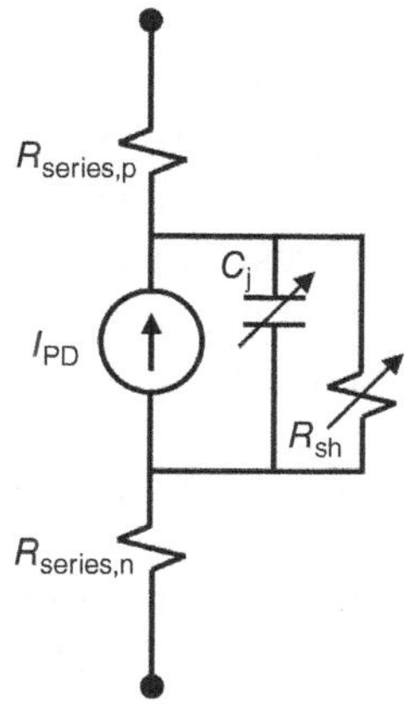

Figure 9.5. Equivalent circuit diagram for the photodiodes in Figure 9.4 including contributions to the series resistance from both the lateral n- and p-doped contact layers.

undepleted doped layers where they become majority carriers. Once there, however, they must still travel toward the metal contacts. For both structures, the electrons travel from the center toward the ring-style n-metal contacts. Since the n-contact geometry is identical for these two devices, the series resistance due to electrons, $R_{series,n}$, is the same. The difference lies in resistance attributed to the holes, $R_{series,p}$. For the ring-style p-contact, the holes must travel a path length that is dramatically further than that for the device with the full metal p-contact. This is due to the p-doped contact layers being very thin (few hundred nanometers). Geometrically, it can be shown that resistivity improves by a factor that is at least the ratio of the device physical diameter divided by the incident optical illumination radius. This improvement can be as much as 5× to 10×.

Unfortunately, it is very difficult to measure the reverse series resistance in a photodiode, making it hard to ascertain precisely the improvement between the two configurations in Figure 9.4, although it is possible to calculate the reverse resistance for a given structure. Such calculations require sufficient information for the hole and electron mobilities of the various layers, their doping profiles, and details on the specific geometry. A forward resistance measurement can help to identify if a structure has the potential for high current operation, as forward I-V measurements place a lower bound on the reverse series resistance. Consider Figure 9.4 for a qualitative explanation. The contribution to forward series resistance is determined by the material and the conductivity between the p- and n-contacts as the carriers travel the lowest resistance path between the two contacts. The reverse resistance is higher because the origination of the carriers due to absorption is always further from the direct path between the contacts for both photodetector designs in Figure 9.4. Nevertheless, measurements of the forward resistance can be very helpful to screen possible devices for high current operation.

Let the current through an ideal p–n junction diode be expressed as

$$I_D \propto \exp[(qV - E_g)/k_B T], \tag{9.1}$$

where E_g is the bandgap of the p–n junction, q is the electronic charge, k_B is Boltzman's constant, T is the absolute temperature, and V is the voltage across the depletion region. The total voltage drop across the diode as represented by Figure 9.5 can be expressed as

$$V_D = I_D R_s + \frac{E_g}{q} + \frac{kT}{q}\ln(I_D), \tag{9.2}$$

where R_s represents the total forward series resistance including both the n- and p-doped layers, and the proportionality constant in Equation (9.1) has been set to unity in the proper units. Equation (9.2) is plotted in Figure 9.6 for a p–n junction at room temperature with $E_g = 0.8\,\text{eV}$ and $R_s = 5\ \Omega$. Since p–n junctions have a nonlinear I–V relationship, only the differential resistance, dV/dI, can be measured (also shown in Figure 9.6). Notice how the differential resistance decreases as the forward voltage increases and ultimately approaches

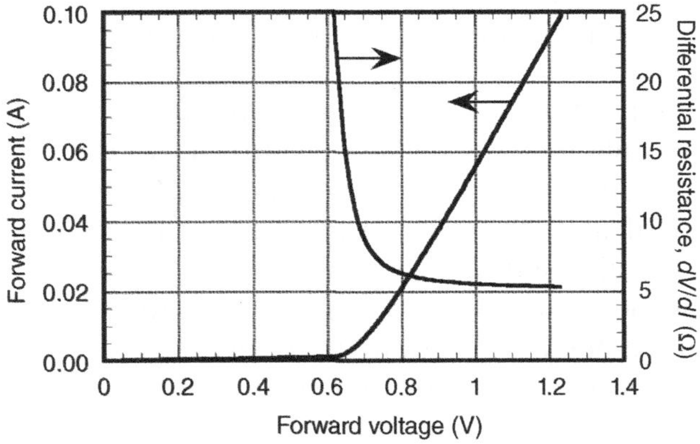

Figure 9.6. Plot of I-V characteristics of an ideal p–n junction with $E_g = 0.8\,\text{eV}$ and a 5 Ω series resistance according to Equation (9.2). Also plotted is the differential resistance of the I–V curve.

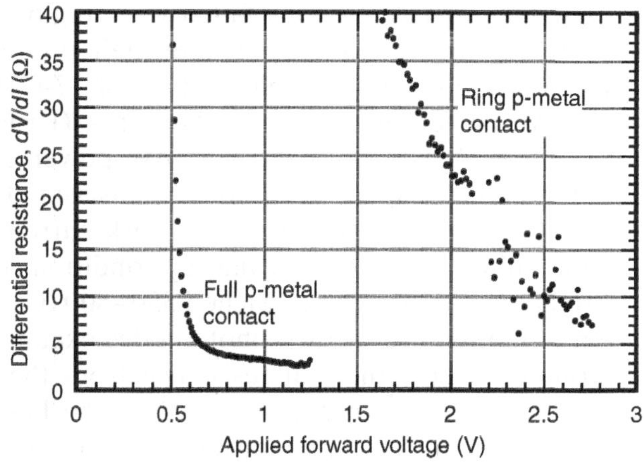

Figure 9.7. Measured forward differential resistance for mesa photodiodes fabricated from the same epitaxial material but with different p-contact metallizations (see also Figure 9.4).

the series resistance, R_s, at high applied bias voltages. Differential forward resistance measurements reported by Tulchinsky et al. (2004) and plotted in Figure 9.7 clearly demonstrate the advantages of full p-metal contacts as opposed to ring p-metal contacts (see Figure 9.4). In Figure 9.7, the forward series differential resistance is below 3 Ω at applied voltages as low as 1.0 V for the full p-contact metallizations, whereas values above 10 Ω are observed even as applied bias voltages exceeded 2.5 V for the ring p-metal contact designs.

Other reasons for minimizing the series resistance in high current photodiodes are added thermal loading and the voltage drop that occurs for the depletion region. For a series resistance of 10 Ω, a 100-mA photodetector would experience a 1-V drop in depletion region bias voltage. This voltage drop would also result in an additional 100 mW of thermal load within the photodetector. Many photodiodes fail under thermal loads of between 200 and 400 mW as will be highlighted in the next section. In such cases, consuming 25–50% of the maximum available thermal capacity of the device due to the series resistance is undesirable.

9.3 THERMAL LIMITATIONS

Many power electronic devices are limited ultimately by thermal loading or the maximum temperature within the device, and high current photodiodes are no exception. Paslaski et al. (1996) published data showing that several photodiodes failed over regions of maximum photocurrent and applied voltages where their product (power dissipated) was a constant. For example, failure of 17-GHz devices occurred at 260-mW power dissipation. Other data (Paslaski et al. 1994) suggested failures between 25 and 400 mW power dissipation, depending on photodetector style. Theoretically, Paslaski et al. (1996) determined that a thermally induced runaway dark current could be responsible for the failure and that the runaway condition could occur if internal material peak temperatures reached 550–800 °C.

A basic thermal treatment of p–i–n photodiodes examined contributions to the photodetector thermal resistance from the depletion region and substrate (Williams and Esman 1999). The basic analysis technique was to determine the temperature rise from a solution to Laplace's equation. For example, consider the photodiode geometry and equivalent heat circuit shown in Figure 9.8, which shows a mesa-style photodiode between two heat sinks, one removing heat from the p-side of the structure and one removing heat through the n-type substrate/heat sink. An equivalent thermal resistance model is shown in Figure 9.8(b). For the special case of one-dimensional heat flow between surfaces of constant temperature with a uniform

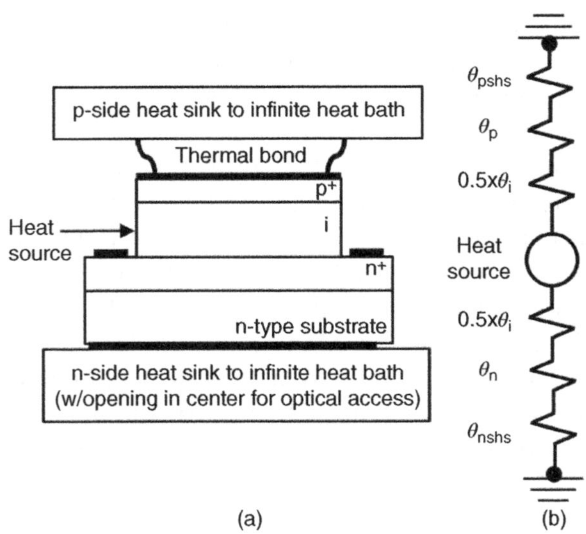

Figure 9.8. Mesa photodiode structure sandwiched between p-side and n-side heat sinks to remove heat from both surfaces (a). The heat circuit represents the various contributions to the thermal resistance from each layer or interface, assuming the only heat source is from current that is uniformly distributed in the depletion region (no series resistance) (b).

heat source on only one surface, the solution for the temperature between any two surfaces can be described with the use of the thermal resistance, θ, as

$$\theta = \frac{\Delta T}{P} = \frac{\Delta x}{kA}, \tag{9.3}$$

where ΔT is the temperature difference between the surfaces, P is the total power of the heat source, Δx is the distance between the surfaces, A is the surface area, and k is the thermal conductivity of the material. Equation (9.3) can be very useful in describing the thermal resistance of the absorber region of a photodiode that is coupled to a heat sink on one side (p- or n-side but not both). One-dimensional heat flow is a good approximation as the absorber is usually very thin relative to its diameter, resulting in very little lateral heat flow. It is important to first discuss the temperature rise of the depletion or intrinsic layer itself (θ_{i}), as it is common to most photodetector structures, including lumped waveguide, traveling-wave designs, and surface illuminated p–i–n photodiodes. To treat the entire structure fully, thermal resistance contributions from the other layers as shown in Figure 9.8 must also be accounted for. This includes thermal resistance terms from both the p-doped (θ_{p}) and n-doped (θ_{n}) layers, as well as contributions from the p-side (θ_{pshs}) and n-side (θ_{nshs}) heat sink paths, when applicable.

In a single dimension, it can be shown that the thermal resistance of a region with a uniform distribution of heat generation throughout its volume is equivalent to having all the heat generated at a plane in the absorber center. Thus, the thermal resistance for a depleted intrinsic InGaAs absorber with thickness d_1 and heat source radius r is given by Equation (9.3) as

$$\theta_{\mathrm{i}} = \frac{d_1}{2k\pi r^2}, \tag{9.4}$$

where it is assumed that the heat is extracted from only one surface. The thermal conductivity of many semiconductors decreases with increasing temperature, making Equation (9.4) valid only for small temperature changes where k can be assumed to be a constant. For many semiconductor materials applicable to long-wavelength photodiodes (InP, InGaAs, and InGaAsP), the temperature dependence of k can be approximated as (Adachi 1993)

$$k(T) = \frac{k(T_0)T_0^c}{T^c}, \tag{9.5}$$

where T is the temperature, c is a constant in the range of $1.2 \leq c \leq 1.55$, and T_0 is a temperature where k is equal to $k(T_0)$. Equation (9.5) is good approximation for k in the range of T_0 from 150 to 600 K, with $c = 1.375$ and 1.55 for InGaAs and InP, respectively. To calculate the internal temperature rise, the temperature dependence of k must be taken into account to solve Laplace's equation. Since the only temperature-dependent term in Equation (9.4) is k, Joyce (1975) has shown that a Kirchoff transformation can be used to yield the nonlinear solution immediately from the linear solution that assumes constant k. Using the approach in Williams and Esman (1999) with a Kirchoff transformation, an expression for the temperature rise can be written as

$$T = \left[\frac{T_0 + (1-c)\,\theta_0 P}{T_0^c}\right]^{\frac{1}{1-c}}, \tag{9.6}$$

where T_0 is the heat sink temperature and θ_0 is the thermal resistance as calculated by Equation (9.4) at T_0.

Equation (9.6) is plotted in Figure 9.9 for heat source diameters of 10, 15, and 25 μm and photodiode depletion layer thicknesses of 1.0 and 1.5 μm. As can be seen from the calculation, the temperature increase can be many hundreds of degrees for power dissipation levels as low as 100–400 mW in the smaller geometries. This is consistent with the

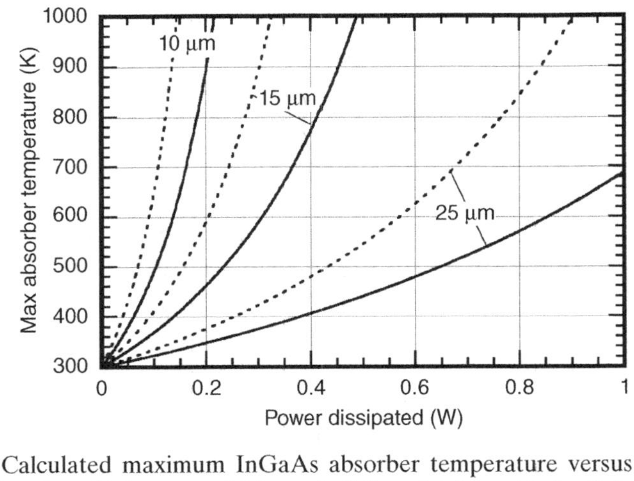

Figure 9.9. Calculated maximum InGaAs absorber temperature versus power dissipated for absorber thicknesses of 1.0 μm (solid) and 1.5 μm (dashed) for heat source diameters of 10, 15, and 25 μm The substrate side of the absorber surface is assumed to be at 300 K (Williams and Esman 1999).

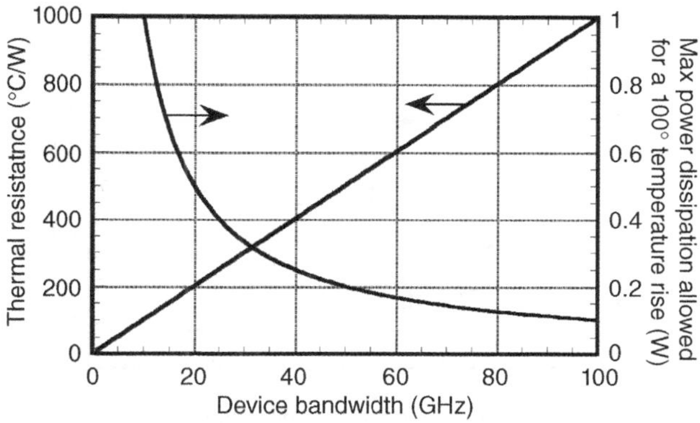

Figure 9.10. Calculated thermal resistance for capacitively limited bandwidth photodiodes using InGaAs depletion regions. Also plotted is the maximum power dissipation allowed within the depletion region for a 100° peak temperature rise.

failure data and runaway thermal analysis of Paslaski et al. (1996) for mesa devices with similar bandwidth and power dissipation levels.

As detailed by Equation (9.4), the thermal resistance is proportional to thickness and inversely proportional to heat source area. For optimal devices, the power should be spread out over the entire physical surface area to limit both the thermal resistance and space charge densities. For Gaussian beam illumination, the heat source diameter is often made slightly smaller than the physical device diameter to capture most of the light to maintain quantum efficiency (Williams and Esman 1999). Regardless of the illumination condition, the thermal resistance as described by Equation (9.4) will be inversely proportional to the device capacitance. For capacitively limited bandwidth devices, this places a lower limit on the thermal resistance of approximately 10 °C/(W-GHz) for devices made with InGaAs-only depletion regions, which in turn determines the maximum power that a certain bandwidth device can attain. Figure 9.10 is a plot of the thermal resistance as a function of bandwidth with the corresponding maximum allowable power dissipation in the device assuming a maximum temperature rise of 100°. Allowing the peak temperature to rise within any semiconductor device generally leads to reduced reliability. However, limited reliability data for power photodiodes at high temperatures exists (see Paslaski et al. 1996), necessitating such studies before widespread deployment of power photodiodes can be considered.

The limitations imposed by Equation (9.4) for capacitively limited bandwidth devices does allow for some optimization. The calculations in Figure 9.10 assume that heat is only being removed from a single

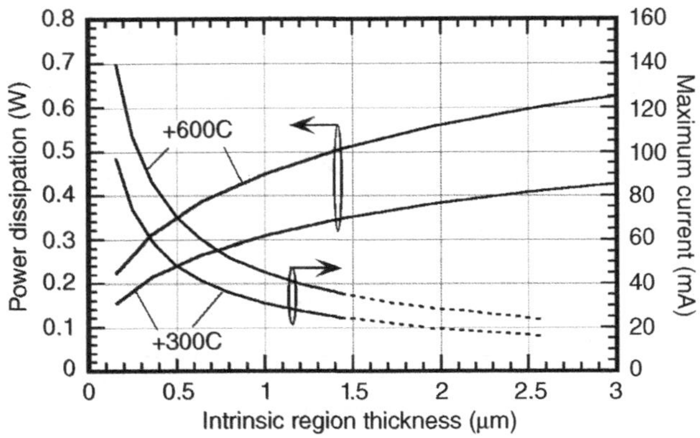

Figure 9.11. Calculated power dissipation that results in a peak absorber temperature increase of 300° and 600° for a circular RC-limited 20 GHz bandwidth photodiode (Williams and Esman 1999). An InGaAs absorber on an InP substrate is assumed. Also plotted is the maximum photocurrent for a photodiode on the basis of the calculated power dissipated and the applied voltage necessary to achieve an average electric field strength of 100 kV/cm. The dashed lines indicated extrapolated results.

side of the depletion region. If heat can be removed from both the p- and n-side surfaces as depicted in Figure 9.8, then the 10 °C/(W-GHz) limit for InGaAs depletion regions can be reduced by 50%. While the thermal resistance due to fixed capacitance becomes dependent only on k, there is still freedom to make the device thinner and smaller in diameter or thicker and larger in diameter so long as the ratio of area/thickness is a constant. Thermal calculations at constant electric field strength are plotted in Figure 9.11. The calculations show that while maximum current does suffer slightly from confining the carriers to a smaller mode volume by making the devices thinner and smaller in diameter, the power dissipation increases dramatically. This allows the electric field to be increased to overcome the reduction in maximum current. Williams and Esman (1999) concluded that high current photodiodes would benefit from the thermal perspective by using thinner depletion regions. Thin depletion regions will also be shown to mitigate the limitations to current handling due to space-charge as outlined in the next section. Another possible solution to lower the thermal resistance for a given bandwidth surface-illuminated device is to use a depletion layer material with a better thermal conductivity, such as InP, Ge, or GaSb. These materials have 10–20 times the thermal conductivity of InGaAs but have other issues related to achieving high current handling that have largely been unexplored.

9.4 SPACE-CHARGE EFFECTS

The responsivity degradations due to depletion region space-charge in photodiodes were reported by Williams et al. (1994). These simulations showed that under high density illumination conditions, the photo-generated space charge caused the depletion region electric field to collapse near the n-side of the depletion region. This caused the carriers to accumulate in the low field region, thus slowing them sufficiently to where they no longer contribute to the high frequency response. Responsivity reductions such as shown in Figure 9.1 are a consequence of this effect. Space charge has also been considered to affect the resistance and reactance of a photodiode significantly (Hollis 2011). A number of solutions have been proposed for alleviating the effects of photo-generated space-charge. Such ideas can be broadly categorized into two areas: reducing the optical intensity in the depletion layer and engineering the depletion layer geometry, material, and doping profile. Because the space-charge density in the depletion region is responsible for the electric field collapse, reducing the power density of the light is an obvious solution. In a surface-illuminated photodiode, this is accomplished by spreading the light out as uniform as possible over the photosensitive area. Analysis and data for this approach are covered by Williams and Esman (1999). An alternative is to use a traveling-wave photodetector structure (TWPD) and distribute the absorption over the length of a waveguide—an idea that has been studied by several groups [see Hietala et al. (1995), Lin et al. (1996), and Jasmin et al. (1997)].

Engineering the geometry, material, and/or doping profile of photodetector structures has been studied extensively for mitigation of space-charge effects. Such techniques include dual-depletion-region photodiodes (Effenberger and Joshi 1996), uni-traveling-carrier photodiodes (Ishibashi et al. 1997), and partially-depleted-absorber (PDA) photodiodes (Li et al. 2002). In a dual-depletion-region (DDR) photodiode, a depletion region having two separate materials is employed. This photodiode structure was first proposed by Effenberger and Joshi (1996) as a way to optimize the electron and hole transit times to increase photodetector bandwidth. The structure was revisited by Williams (2002) in terms of mitigating the effects of space-charge, a summary of which is provided in the following section. Another concept is to use a nonabsorbing depletion-region material such as InP in combination with a p-doped absorber. Pearsall et al. (1981) was the first to suggest this structure to lower dark currents, and Davis et al. (1996)

used this same approach to improve current handling. Ishibashi et al. (1997) later labeled the device a uni-traveling-carrier (UTC) photodiode because only higher velocity electrons serve as carriers. Analysis by Williams (2002) showed that this structure exhibits issues when compared to other two-carrier designs if the carrier velocities are saturated. Variations in the UTC design, referred to as modified UTC (MUTC) (Jun et al. 2006), have achieved high performance levels but are structurally very much similar to DDR photodiodes. Kato (1999) has suggested that because the depletion region can be very thin in a UTC photodiode, the electrons that are injected from the p-type absorber can traverse the depletion region ballistically before scattering reduces (saturates) their velocity. Such structures are now termed near-ballistic UTC (NBUTC) photodetectors and are promising for high frequency applications. Finally, the concept of reducing the intrinsic region thickness was proposed by Williams and Esman (1999) and implemented by Li et al. (2003a, 2004a). In these works, a thin intrinsic InGaAs layer and a doped p-type InGaAs absorber were employed to aid in improving the quantum efficiency. This structure was termed a PDA photodiode, as only a portion of the absorber is depleted. The performance achieved in these various structures will be discussed in Section 9.6. The remainder of this section will examine calculations concerning differences between various depletion region configurations.

The basic structure to be studied is shown in Figure 9.12, a p–i–n photodiode comprising an intrinsic region that contains both InGaAs and InP. By varying the thickness of either material, the full spectrum of PDA (100% InGaAs, 0% InP), UTC (0% InGaAs, 100% InP), and DDR (variable proportions) photodetector designs can be analyzed. Illumination from either the p- or n-side can be considered. Effects due to space charge are contained in the solutions to the carrier continuity equations (see Equations 4.6 and 4.7) and Poisson's equation (also Equation 4.5):

$$\nabla \cdot E = \frac{q}{\varepsilon}(p - n + N_{\mathrm{d}} - N_{\mathrm{a}}). \tag{9.7}$$

It is important in this case to review the basic space-charge properties of a p–i–n photodiode as explained in Chapter 4, Figure 4.21.

The carrier density and electric field data in Figure 4.21 are estimates based on simplifying assumptions as presented by Kato (1999). As explained in Chapter 4, the hole and electron densities are approximately linear functions over the depletion region. The electric field (Equation 9.7) can be directly integrated with p(x) and n(x) given by

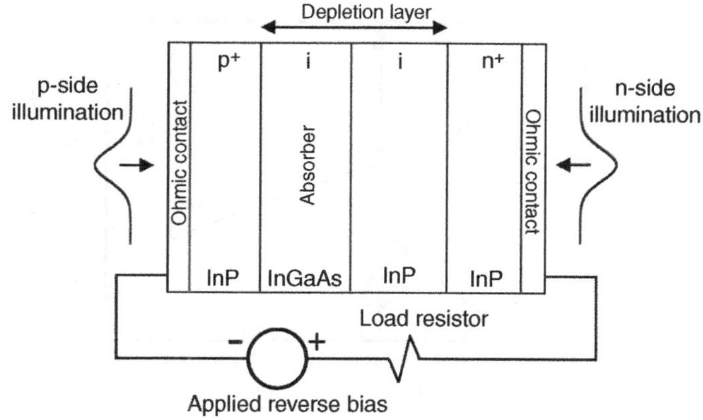

Figure 9.12. Simplified structure of a p–i–n photodiode containing two distinct materials in the depletion layer.

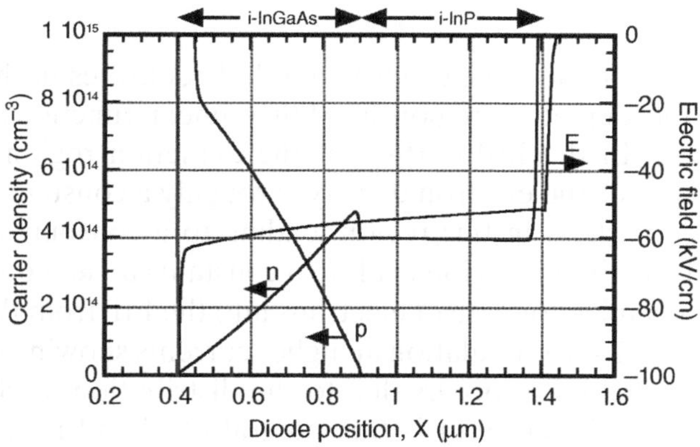

Figure 9.13. Calculated electric field and carrier densities at 1 mA photocurrent for an n-side illuminated dual-depletion region p–i–n photodiode. The percentage of i-InGaAs in the depletion region was 50%, the applied bias was −5 V, the total depletion layer thickness was 1 μm, the physical diameter was 25 μm, and the spot size was 15 μm (uniform intensity) (Williams 2002).

Figure 4.21(a) and the additional condition that the total integral of the electric field be equal to the applied photodiode terminal voltage. The electric field will thus take a parabolic shape given linear functions for the carrier densities. If no series resistance is assumed, then the electric field can be calculated for different current densities as shown in Figure 4.21(b).

Full simulations are required to model device performance more accurately. Figures 9.13 and 9.14 depict examples of calculations at low and high currents for a DDR photodiode containing 50% InGaAs

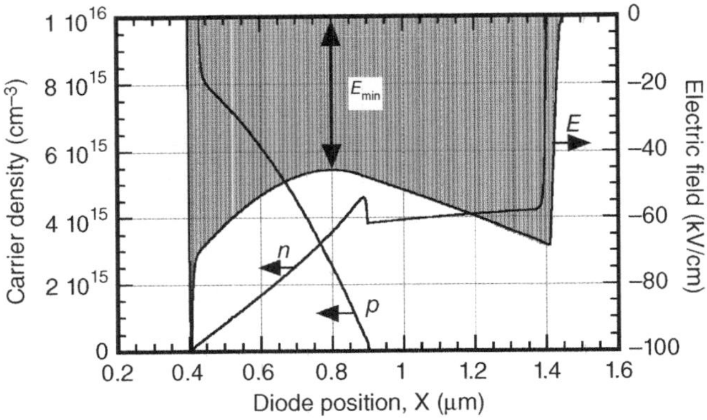

Figure 9.14. Calculated electric field and carrier densities at 10 mA photocurrent for the same photodiode in Figure 9.13. The shaded portion represents the area under the electric field curve, which is equal to the applied voltage (Williams 2002).

and 50% InP. Note how the electron and hole densities in the InGaAs portion of the depletion region are fairly linear functions as shown in Figure 4.21. In the InP portion of the depletion region where no absorption occurs, the electron density is roughly a constant due to the constant electric field in this region leading to a constant velocity. A constant velocity in this region leads to a constant carrier concentration due to a steady injection rate of electrons into the InP from the InGaAs region. Figure 9.14 is a simulation at higher currents showing the change in the electric field that occurs due to the distribution of charge. The shaded region in Figure 9.14 is the integral of E and is equal to the applied voltage when the series resistance is small. To determine the capacity of any particular structure to tolerate high currents, simulations are performed at higher and higher photocurrents until the electric field minimum decreases below a few kilovolts per centimeter. At this level, the carriers will begin to build up and lead to response compression. Additional insight can be gained by observing the electric field as the current increases and the impact of the ratio of electron-to-hole velocities. Recall from Figure 4.20 that the electrons maintain their high velocity even at low fields, while hole velocities tend to decrease more monotonically toward zero from their saturated value to zero field.

Figure 9.15 plots the maximum current before the electric field minimum decreases to below a few kilovolts per centimeter in the depletion region for devices with varying InGaAs/InP ratios at both p- and n-side illumination conditions. Also plotted in Figure 9.15 is the ratio of the

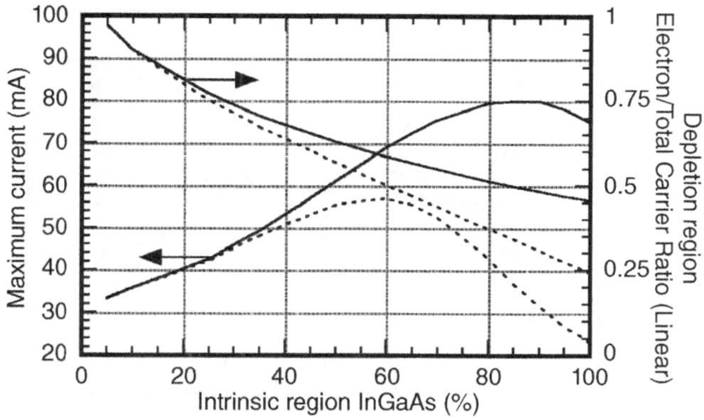

Figure 9.15. Simulated maximum photocurrent for p-side (solid) and n-side (dashed) illuminated dual-depletion region photodiodes. The applied bias was −5 V, the total depletion layer thickness was 1 μm, the physical diameter was 25 μm, and the spot size was 15 μm (uniform). Also shown is the ratio of electron density to total carrier density in the depletion region as a function of i-InGaAs percentage for p-side (solid) and n-side (dashed) illuminated dual-depletion region photodiodes operating at 20 mA (Williams 2002).

number of electrons to the total number of carriers in the depletion region. Note how the peak in maximum current for both illumination conditions occurs when this ratio is 50%. This is the ratio where there are an equal number of holes and electrons in the depletion region. Therefore, Williams (2002) concluded that optimum high current performance is achieved when a space-charge balance is achieved within the depletion region.

The results of Figure 9.15 outline the benefits of having the integral of the term $(p - n)$ in Equation (9.7) be equal to zero over the depletion region. For these calculations, the term $(N_a - N_d)$ in Equation (9.7) was assumed to be zero everywhere in the depletion region due to the intrinsic material assumption. Williams (2002) suggested doping the depletion region to increase current handling performance by charge compensation. The idea is basically to make the terms $(p - N_a)$ and $(n - N_d)$ equal to zero at the designed operating current. Under these conditions, the electric field at the operating current would be a constant, similar to the dark electric field for an undoped depletion region material ($N_a = N_d = 0$). Figure 9.16 depicts electric field calculations at various current levels using this charge compensation scheme. The same photodiode layer structure as used in Figure 9.14 was employed, but with different intrinsic region doping values, where $(p - N_a) = (n - N_d) = 0$ at 40 mA. Note how the electric field at 40 mA is

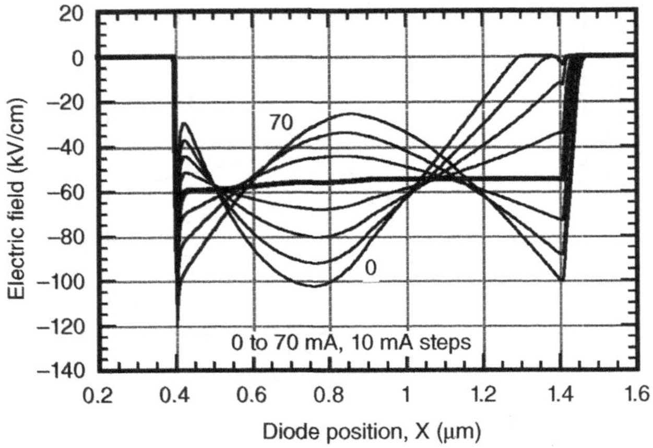

Figure 9.16. Calculated electric field for various photocurrents for an n-side illuminated dual-depletion region p–i–n photodiode. The layer structure is the same as in Figure 9.14 but with charge compensation employed, where $(p-N_a) = (n-N_d) = 0$ at 40 mA.

relatively flat, the design conditions for this charge compensation simulation. From an alternative point of view, charge compensation allows the electric field under dark conditions to be enhanced in locations where the increasing flow of current would cause the photo-generated space charge to decrease the field. Figure 9.16 shows exactly this as the dark electric field is increased in the depletion region just to the p-side of the midpoint of the depletion region. This is the same location where the field begins to reduce at high currents (see Figure 9.14). Charge compensation should be useful to approximately double the available current level performance due to space-charge limits, given that the doping profiles can be accurately controlled.

9.5 PHOTODETECTOR POWER CONVERSION EFFICIENCY

Efficiency is a metric that cannot be ignored, particularly in today's world where new sources of energy and energy conservation are considered in nearly every field. Indeed, "green" techniques are a topic of present research for industrial optical telecommunications (Tucker 2011, Kilper and Tucker 2013) and wireless applications (Xiong et al. 2011, Ngo 2013). Efficiency is critical in high power analog applications such as wireless RF transmission, radar, and electromagnetic warfare because of the amount of heat that must be dissipated. This is especially true in dense antenna arrays where the amount of physical area is severely restricted. In such arrays, it is most difficult to place electronic power amplifiers behind every antenna element; rather, high power amplifier outputs might be distributed to numerous antenna elements incurring losses in the division and propagation. High power photodetectors may present a solution in such applications where a photodetector can be placed directly behind each antenna element, serving as a booster or perhaps even replacing electronic amplifiers. Systems using low power photodiodes for antenna array calibration have already been demonstrated (Dorsey et al. 2010). There are numerous advantages of high power photodetectors for array applications including the potential for integration directly with the antenna, the possibility of minimizing metallic components in the array backplane, and high power conversion efficiency (PCE). In such cases, the PCE of the photodetector circuit is the main concern because the inefficiencies associated with the generation of the driving optical power can be remoted via fiber to a location where it is easier to dissipate heat. This section provides an introduction to the PCE of RF photodiodes for antenna array applications.

A detailed analysis of photodiode PCE was first conducted by Williams and Esman (1999) and later refined by Gliese et al. (2010). The treatment in this case follows the development in the latter as it is applicable to a general driving waveform as opposed to the sinusoidal drive considered by Williams and Esman (1999). The equivalent circuit shown in Figure 9.17 is used for the analysis, which is similar to that shown in Figure 1.17 but excluding the shunt and matching resistances. The photodetector itself is represented by an ideal current source with a parallel capacitance and a series resistance, $R_s = 1/\alpha$, where α is the slope of curve plotting photodetector compression current as a function of bias voltage (Williams and Esman 1999). An external reverse bias (V_{bias}) is applied through an inductive port. The output of the photodetector is capacitively coupled to a load with resistance R_{load}. The DC photocurrent (I_{dc}) flows only through the series resistance, whereas the RF portion of the photocurrent (I_{rf}) flows through the series and load resistances. The PCE of such a circuit is defined as

$$\eta_\Omega \equiv \frac{P_\Omega}{P_{bias} + P_{opt}} = \frac{P_\Omega}{-I_{dc}V_{bias} + I_{dc}/\Re}, \qquad (9.8)$$

where P_Ω is the average RF power delivered to the external load, P_{bias} is the power resulting from the reverse bias voltage, P_{opt} is the input optical power, and $\Re$ is the photodetector responsivity. The negative

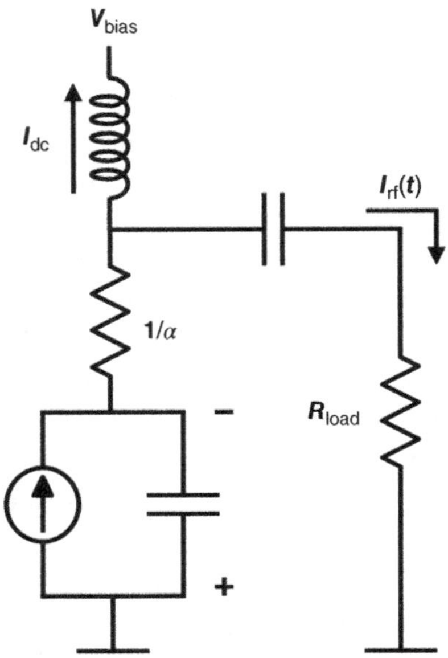

Figure 9.17. Equivalent circuit of a photodetector for analysis of power conversion efficiency. The photodetector is represented as a current source with a parallel capacitance and a series resistance ($1/\alpha$). Also shown are the externally applied reverse bias (V_{bias}), the average photocurrent (I_{dc}), and the RF photocurrent (I_{rf}) delivered to a load resistance (R_{load}).

sign in the term $-I_{dc}V_{bias}$ is a result of the sign of V_{bias}, which is negative for a reverse bias on the photodiode following the convention in Gliese et al. (2010). The inclusion of P_{opt} in Equation (9.8) is important when comparing photodetector performance to other competing technologies, as opposed to a "wall-plug efficiency" that excludes the optical input (Sakai et al. 2010). In order to ensure linear operation, the bias voltage must be greater than the sum of the photodetector threshold voltage (V_{th}), the DC voltage, and the peak RF voltage. Mathematically, this condition can be represented as

$$V_{bias} \leq -V_{th} - \frac{I_{dc}}{\alpha} - \max[I_{rf}(t)] \times \left(R_{load} + \frac{1}{\alpha}\right). \tag{9.9}$$

Now, taking the equality in Equation (9.9) as the bias voltage corresponding to the maximum PCE, Equation (9.8) can be rewritten as

$$\eta_{\Omega} = \frac{P_{\Omega}}{\frac{I_{dc}^2}{\alpha} + \max[I_{rf}(t)] \times I_{dc}\left(R_{load} + \frac{1}{\alpha}\right) + I_{dc}V_{th} + \frac{I_{dc}}{\Re}}. \tag{9.10}$$

Equation (9.10) is general for any driving waveform I_{rf}. The parameters α, V_{th}, and $\Re$ are directly related to the photodetector itself. Although they contain the photodiode responsivity, the I_{dc} and I_{rf} terms can be viewed as being determined by the driving waveform.

It is instructive to consider the case where the photodiode is driven by a pure sinusoid, which can be likened to Class A operation of transistor amplifiers. In this case, $I_{rf}(t) = mI_{dc} \sin \Omega t$, $P_{\Omega} = m^2 I_{dc}^2 R_{load}/2$, and $\max[I_{rf}(t)] = mI_{dc}$, where m is the modulation depth. Such a driving waveform can be achieved in practice with small-signal (linear) modulation using an MZM or heterodyning of two lasers separated by angular frequency Ω. Inserting the aforementioned values into Equation (9.10) and rearranging terms yield

$$\eta_{\Omega} = \left[\frac{2}{m} + \frac{2(m+1)}{\alpha m^2 R_{load}} + \frac{1}{m} \sqrt{\frac{2}{P_{load} R_{load}}} \left(\frac{1}{\Re} + V_{th} \right) \right]^{-1}. \quad (9.11)$$

When comparing results in the literature, it is important to note that Equation (9.11) is identical to that derived by Gliese et al. (2010) [his Equation (23)], which corrects erroneous terms inserted by Tulchinsky et al. (2008) in his Equation (5). Equation (9.11) is plotted in Figure 9.18 using four sets of parameters. For all plots, 100% modulation depth is assumed ($m = 1$), $R_{load} = 50\,\Omega$, and a slightly negative threshold is considered, $V_{th} = -0.4\,V$. The top curve in Figure 9.18 shows the PCE for a

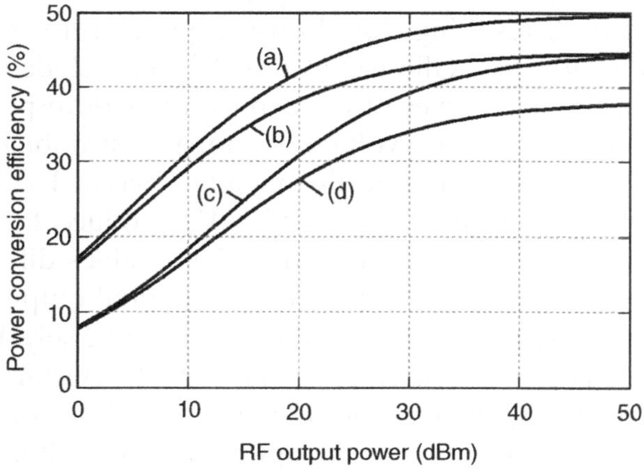

Figure 9.18. Calculations of Equation (9.11) for various conditions. For all curves, $m = 1$, $R_{load} = 50\,\Omega$, and $V_{th} = -0.4\,V$. The remaining parameters are (a) $\alpha \to \infty$ and $\Re = 1\,A/W$, (b) $\alpha = 0.35\,A/V$ and $\Re = 1\,A/W$, (c) $\alpha = 0.35\,A/V$ and $\Re = 0.5\,A/W$, (d) $\alpha = 0.13\,A/V$ and $\Re = 0.5\,A/W$.

nearly ideal photodiode with $\alpha \rightarrow \infty$ and $\Re = 1$ A/W. As can be seen, the PCE asymptotically approaches 50%, the theoretical limit for sine-wave modulation. Comparatively, the second curve down demonstrates the effect of increasing the series resistance by setting $\alpha = 0.35$ A/V and keeping all other parameters fixed. Note that the top two curves are similar at low P_Ω but diverge to different asymptotic limits at high P_Ω. Comparing the second and third curves shows that $\Re$ affects the PCE at low P_Ω but does not change the asymptotic limit. All of these trends can be easily determined by inspection of Equation (9.11). Finally, the PCE for a realistic device having $\alpha = 0.13$ A/V and $\Re = 0.5$ A/W as described by Tulchinsky et al. (2008) is plotted in the fourth curve.

The PCE for an ideal photodetector can be increased beyond 50% by driving it with waveforms other than a sinusoid. For example, $\eta_\Omega \rightarrow 81\%$ can be achieved by driving with a square wave (Gliese et al. 2010). This is intuitive in that the ratio of the power for the fundamental sinusoidal component of a square wave to the average current is much greater than that for a pure sinusoid of the same power. The increased efficiency comes at the cost of considerable distortion, that is, significant power at higher frequency components. Gliese et al. (2010) has demonstrated that an MZM can be used to shape the driving waveform, a technique that results in a maximum $\eta_\Omega = 67\%$ for an ideal photodiode.

The analysis of photodetector PCE when driven by an MZM is readily conducted using Equations (6.18) and (9.10). Equation (6.18) gives the waveform-based parameters I_{dc} and I_{rf} as functions of MZM DC bias voltage and RF drive voltage, which can be inserted into Equation (9.10). A resulting closed-form expression for η_Ω is readily obtained but lengthy [see Gliese et al. (2010)] and excluded in this case for brevity. Partial derivatives of the expression with respect to MZM DC bias voltage and RF drive voltage are not as easily obtained. Gliese et al. (2010) determined numerically that a DC bias of $V_{dc} = 0.494 \cdot V_\pi$ and an RF drive voltage of $V_{rf} = 0.572 \cdot V_\pi$ produce the maximum PCE for an ideal photodiode. Note that these values differ from the condition that produces the maximum fundamental output power for an ideal MZM-based link, $V_{rf} = 0.586 \cdot V_\pi$ and precisely $V_{dc} = V_\pi/2$, as given by Equation (6.28). Gliese et al. (2010) determined that the V_{dc} that maximizes η_Ω changes for nonideal photodetector parameters but that $V_{rf} = 0.586 \cdot V_\pi$ remains the optimum drive voltage. Shown in Figure 9.19 are measured data and calculation for a 68-µm photodiode having $V_{th} = -0.36$ V, $\alpha = 0.33$ A/V, and $\Re = 0.5$ A/W (Gliese et al. 2010). The additional parameters for Figure 9.19 are $V_{bias} = -1.95$ V, $R_{load} = 50\,\Omega$, $\Omega/(2\pi) = 300$ MHz, $V_{dc} = 0.47 \cdot V_\pi$, and $V_{rf} = 0.57 \cdot V_\pi$.

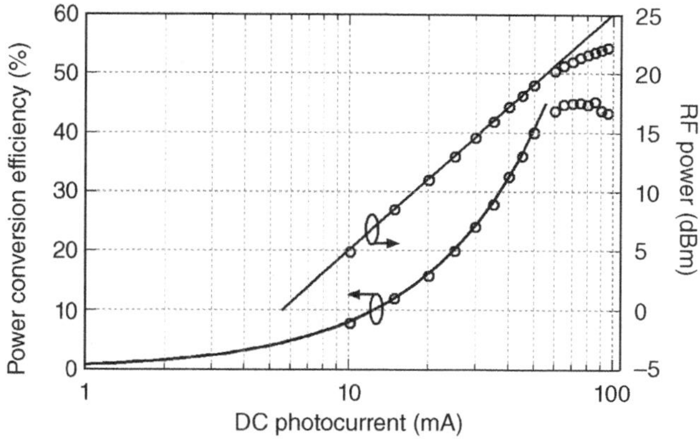

Figure 9.19. Measured (symbols) and calculated (lines) power conversion efficiency and RF output power for a 68-μm partially-depleted-absorber (PDA) photodiode.

The average photocurrent was adjusted by increasing the incident optical power while maintaining the aforementioned values. The theory and measurement agree nicely until the photodiode output power begins to compress. After the onset of compression, the PCE diverges from the calculation, peaks, and then decreases with increasing I_{dc}. The peaking of photodiode PCE in compression is due to signal clipping of the photodiode, a well-known phenomenon in transistor amplifiers under Class AB operation.

The clipping of a waveform by operating a photodiode in compression is shown explicitly in Figure 9.20. Plotted is the measured load voltage for an 80-μm diameter photodiode at $I_{dc} = 103\,\text{mA}$ operated at 0.5-dB compression at the fundamental due to photodiode-induced clipping (Gliese et al. 2010). The photodiode exhibited similar parameters to the 68-μm diameter photodiode, $V_{th} = -0.36\,\text{V}$, $\alpha = 0.33\,\text{A/V}$, and $\Re = 0.5$ A/W. The bias, load, and driving waveform parameters in this case were $V_{bias} = -2.87\,\text{V}$, $R_{load} = 50\,\Omega$, $\Omega/(2\pi) = 300\,\text{MHz}$, $V_{dc} = 0.47 \cdot V_\pi$, and $V_{rf} = 0.57 \cdot V_\pi$. For comparison, a calculated "uncompressed" MZM-shaped waveform is also shown in Figure 9.20. The calculated voltage was obtained by inserting the aforementioned parameters into Equation (6.18). For the curve in Figure 9.20, the expansion was carried out to the tenth harmonic (i.e., up to $m=4$ for odd terms and $k=5$ for even terms). Significant clipping of the measured waveform as compared to the calculated MZM-shaped waveform can be observed, but this clipping is advantageous for the PCE. The measured $\eta_\Omega = 53.5\%$ for the data in Figure 9.20, a 7% increase above that for the calculated waveform (Gliese et al. 2010).

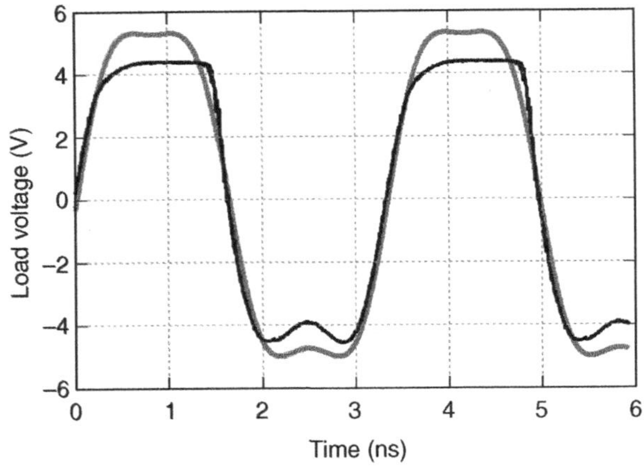

Figure 9.20. Measured output waveform (black) for an 80-μm partially-depleted -absorber (PDA) photodiode at $I_{dc} = 103\,\text{mA}$ compared to the calculated input waveform (gray). The measured waveform was taken at 0.5-dB photodiode compression, whereas the theory represents the MZM-shaped input waveform.

The attained PCEs for the two photodiodes discussed previously are shown in Figure 9.21. The data point for the 80-μm diameter photodiode corresponds to the waveform in Figure 9.20, whereas the curve for the 68-μm diameter photodiode uses the data from Figure 9.19. Both photodiodes achieved maximum PCE near 0.5 dB photodiode compression: $\eta_\Omega = 45.5\%$ at $I_{dc} = 62\,\text{mA}$ and $P_\Omega = 21.1\,\text{dBm}$ for the 68-μm photodiode, and $\eta_\Omega = 53.5\%$ at $I_{dc} = 103\,\text{mA}$ and $P_\Omega = 24.4\,\text{dBm}$ for the 80-μm photodiode (Gliese et al. 2010). Increasing the photodiode compression beyond 1 dB had no benefit in either case. Nonetheless, the results in Figure 9.21 demonstrate that high power, high efficiency photodiodes hold promise for dense antenna array applications, possibly even competing with traditional broad-bandwidth RF amplification technology.

9.6 STATE OF THE ART FOR POWER PHOTODETECTORS

The state of the art for high power photodetectors is continuously changing as it relates to microwave photonics applications. This section provides an overview of recent results but will probably be dated by the time this work is in press. The interested reader can undertake a literature survey starting with the references listed in this chapter. The highest photocurrents reported for singular devices in the microwave range are presently about 200 mA, as opposed to the 60-mA barrier from about a decade ago noted in Section 9.2. Output powers of

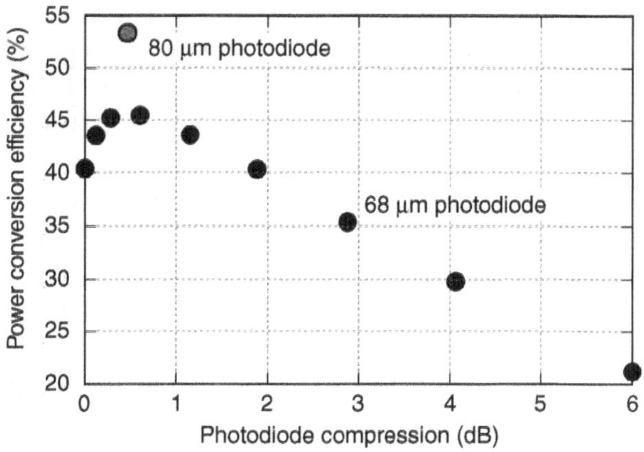

Figure 9.21. Measured power conversion efficiency for an 80-μm diameter photodiode (gray) and a 68-μm diameter photodiode (black).

about 1 W are achievable at microwave frequencies, dropping to about 100 mW in the millimeter wave. At frequencies above 1 GHz, reported OIP3s approach 60 dBm (Pan et al. 2009, Li et al. 2011) and OIP2s reach nearly 70 dBm (Hastings et al. 2009). For microwave photonics applications, the majority of high power photodiodes use InGaAs on InP substrates, but other materials have also been considered such as Ge on Si, leveraging the increased thermal conductivity in Si (Ramaswamy et al. 2010).

As listed in Section 9.4, there are numerous photodetector designs that have been employed for high power applications. Presently, the most prevalent are the PDA p–i–n and various forms of UTC implementations. For example, a backside-illuminated PDA achieved 29-dBm output power at 5 GHz (Itakura et al. 2009). In a UTC device, the absorption region is totally p-doped and only electrons serve as active carriers. Ito et al. (2004) provided a survey of results for p–i–n and UTC photodetectors, demonstrating that the latter provides significantly more output power at millimeter-wave frequencies. For example, a single UTC photodiode sourced 17 dBm at 40 GHz (Li et al. 2004b). MUTC designs use partially p-doped absorption regions and are essentially a DDR design. Comparisons between a MUTC and an UTC photodiode were made by Chtioui et al. (2012), but the difference in depletion layer thickness for the two structures makes it difficult to ascertain a generalization between the MUTC and UTC performance. Various doping densities have been studied in MUTC photodetectors (Li et al. 2010), which have produced output powers as high as 0.75 W at 15 GHz (Li et al. 2011). The highest frequency power photodetectors have employed additional p-type charge in the collection layer to increase the electron drift velocity. Termed NBUTC photodetectors, such devices have demonstrated 10-dBm output power at 110 GHz (Shi et al. 2010). Other high frequency designs include photodiodes with other variants on the DDR design, which have been demonstrated at W-Band (Rymanov et al. 2014).

In addition to improving performance at the device level, many architectural techniques have been used to achieve high RF output power from a microwave-photonic receiver. For singular photodetectors, beam shaping has been used to distribute the incident optical power more uniformly as opposed to direct fiber coupling (Joshi et al. 2008). Multiple photodetectors have also been combined into a single receiver architecture. Such photodetector arrays include parallel sets with direct current summation (Urick et al. 2008) and RF power combining (Itakura et al. 2010) and serial arrays with traveling-wave current combination (Kuo et al. 2011). The last has achieved output powers as high as 18 dBm at 95 GHz (Kuo et al. 2011). High power integrated balanced photodetectors have also been demonstrated (Houtsma et al. 2011, Zhou et al. 2013). For example, two balanced InGaAs photodiodes on InP have achieved an output power of 1.5 W at 8 GHz, operating at a total photocurrent of 320 mA (Zhou et al. 2013). These types of results demonstrate that a combination of device-level improvements and architectural innovations can be used to fulfill the continually increasing need for high power and highly linear photodetectors.

REFERENCES

Adachi, S., "Thermal conductivity of InGaAs," in *Properties of Lattice-Matched and Strained Indium Gallium Arsenide*, P. Bhattacharya, (editor), IEE Inspec, 35–40, (1993).

Chtioui, M., Lelarge, F., Enard, A., Pommereau, F., Carpentier, D., Marceaux, A., van Dijk, F., and Achouche, M., "High responsivity and high power UTC and MUTC GaInAs-InP photodiodes," *IEEE Photonics Technology Letters*, 24(4), 318–320 (2012).

Davis, G. A., Weiss, R. E., LaRue, R. A., Williams, K. J., and Esman, R. D., "A 920-1650-nm high-current photodetector," *IEEE Photonics Technology Letters*, 8(10), 1373–1375 (1996).

Dorsey, W. M., McDermitt, C. S., Bucholtz, F., and Parent, M. G., "Design and performance of frequency selective surface with integrated photodiodes for photonic calibration of phased array antennas," *IEEE Transactions on Antennas and Propagation*, 58(8), 2588–2593 (2010).

Effenberger, F. J. and Joshi, A. M., "Ultrafast, dual-depletion region, InGaAs/InP p-i-n detector," *Journal of Lightwave Technology*, 14(8), 1859–1864 (1996).

Gliese, U., Colladay, K., Hastings, A. S., Tulchinsky, D. A., Urick, V. J., and Williams, K. J., "RF power conversion efficiency of photodiodes driven by Mach-Zehnder modulators," *IEEE Transactions on Microwave Theory and Techniques*, 58(11), 3359–3371 (2010).

Hastings, A. S., Tulchinsky, D. A. and Williams, K. J., "Photodetector nonlinearities due to voltage-dependent responsivity," *IEEE Photonics Technology Letters*, 21(21), 1642–1644 (2009).

Hietala, V. M., Vawter, G. A., Brennan, T. M., and Hammons, B. E., "Traveling-wave photodetectors for high-power, large-bandwidth applications," *IEEE Transactions on Microwave Theory and Techniques*, 43(9), 2291–2298 (1995).

Hollis, M. A., "Space-charge impedance in photodiodes," *IEEE Photonics Technology Letters*, 23(18), 1307–1309 (2011).

Houtsma, V., Hu, T., Weimann, N. G., Kopf, R., Tate, A., Frackoviak, J., Reyes, R., Chen, Y. K., and Zhang, L., "A 1 W linear high-power InP balanced unit-traveling carrier photodetector," in ECOC Technical Digest (2011).

Ishibashi, T., Shimizu, N., Kodama, S., Ito, H., Nagatsuma, T., and Furuta, T., "Uni-traveling-carrier photodiodes," in OSA TOPS on Ultrafast Electronics and Optoelectronics, 13, 83–87 (1997).

Itakura, S., Sakai, K., Nagatsuka, T., Akiyama, T., Hirano, Y., Ishimura, E., Nakaji, M., and Aoyagi, T., "High-current backside-illuminated InGaAs/InP p-i-n photodiode," in IEEE Intl. Mtg. on Microwave Photonics Digest, 1–4 (2009).

Itakura, S., Sakai, K., Nagatsuka, T., Ishimura, E., Nakaji, M., Otsuka, H., Mori, K., and Hirano, Y., "High-current backside-illuminated photodiode array module for optical analog links," *Journal of Lightwave Technology*, 28(6), 965–971 (2010).

Ito, H., Kodama, S., Muramoto, Y., Furuta, T., Nagatsuma, T., and Ishibashi, T., "High-speed and high-output InP-InGaAs unitraveling carrier photodiodes," *IEEE Journal of Selected Topics in Quantum Electronics*, 10(4), 709–727 (2004).

Jasmin, S., Vodjdani, N., Renaud, J. C., and Enard, A., "Diluted- and distributed-absorption microwave waveguide photodiodes for high efficiency and high power," *IEEE Transactions on Microwave Theory and Techniques*, 45(8), 1337–1341 (1997).

Joshi, A., Datta, S., Becker, D., "GRIN lens coupled top-illuminated highly linear InGaAs photodiodes," *IEEE Photonics Technology Letters*, 20(17), 1500–1502 (2008).

Joyce, W.B., "Thermal resistance of heat sinks with temperature-dependent conductivity," *Solid-State Electronics*, 18, 321–322 (1975).

Jun, D. H., Jang, J. H., Adesida, I., and Song, J. I., "Improved efficiency-bandwidth product of uni-traveling carrier photodiode structures using an undoped photo-absorption layer," *Japanese Journal of Applied Physics*, 45(4B), 3475–3478 (2006).

Kato, K., "Ultrawide-band/high-frequency photodetectors," *IEEE Transactions on Microwave Theory and Techniques*, 47(7), 1265–1281 (1999).

Kilper, D. C. and Tucker, R. S., "Energy-efficient telecommunications," in *Optical Fiber Telecommunications VIB*, Academic (2013).

Kuhl, D., Hieronymi, F., Bottcher, E. H., Wolf, T., Bimberg, D., Kuhl, J., and Klingenstein, M., "Influence of space charges on the impulse response of InGaAs metal–semiconductor-metal photodetectors," *Journal of Lightwave Technology*, 10(6), 753–759 (1992).

Kuo, F. M., Chou, M. Z., and Shi, J. W., "Linear-cascade near-ballistic unitraveling-carrier photodiodes with and extremely high saturation current-bandwidth product," *Journal of Lightwave Technology*, 29(4), 432–438 (2011).

Li, X., Li, N., Zheng, X., Demiguel, S., Campbell, J. C., Tulchinsky, D., and Williams, K. J., "High-speed high-saturation-current partially depleted absorber photodetectors," in Lasers and Electro-Optics Society Annual Meeting, Post Deadline Paper (2002).

Li, X., Li, Ning, Zheng, X., Demiguel, S., Campbell, J. C., Tulchinsky, D. A., Williams, K. J., "High-saturation-current InP-InGaAs photodiode with partially depleted absorber," *IEEE Photonics Technology Letters*, 15(9), 1276–1278 (2003a).

Li, X., Demiguel, S., Li, N., Campbell, J. C., Tulchinsky, D. L., and Williams, K. J., "Backside illuminated high saturation current partially depleted absorber photodetectors," *Electronics Letters*, 39(20), 1466–1467 (2003b).

Li, X., Li, N., Demiguel, S., Campbell, J. C., Tulchinsky, D., and Williams, K. J., "A comparison of front- and backside-illuminated high-saturation power partially depleted absorber photodetectors," *IEEE Journal of Quantum Electronics*, 40(9), 1321–1325 (2004a).

Li, N., Li, X., Demiguel, S., Zheng, X., Campbell, J. C., Tulchinsky, D. A., Williams, K. J., Isshiki, T. D., Kinsey, G. S., and Sudharsansan, R., "High-saturation-current charge-compensated InGaAs-InP uni-traveling -carrier photodiode," *IEEE Photonics Technology Letters*, 16(3), 864–866 (2004b).

Li, Z., Pan, H., Chen, H., Beling, A., and Campbell, J. C., "High-saturation-current modified uni-traveling-carrier photodiode with cliff layer," *IEEE Journal of Quantum Electronics*, 46(5), 626–632 (2010).

Li, Z., Fu, Y., Piels, M., Pan, H., Beling, A., Bowers, J. E., and Campbell, J. C., "High-power high-linearity flip-chip bonded modified uni-traveling carrier photodiode," *Optics Express*, 19(26), B385–B390 (2011).

Lin, L. Y., Wu, M. C., Itoh, T., Vang, T. A., Muller, R. E., Sivco, D. L., and Cho, A. Y., "Velocity-matched distributed photodetectors with high-saturation power and large bandwidth," *IEEE Photonics Technology Letters*, 8(10), 1376–1378 (1996).

Ngo, H. Q., "Energy and spectral efficiency of very large multiuser MIMO systems," *IEEE Transactions on Communications*, 61(4), 1436–1449 (2013).

Pan, H., Li, Zhi, Beling, A. and Campbell, J. C., "Measurement and modeling of high-linearity modified unitraveling carrier photodiode with highly-doped absorber," *Optics Express*, 17(22), 20221–20226 (2009).

Paslaski, J., Chen, P.C., Chen, J.S., and Bar-Chaim, N., "High-power microwave photodiode for high-dynamic-range analog transmission," in Optical Fiber Communications Technical Digest, paper ThG5, 208–209 (1994).

Paslaski, J., Chen, P.C., Chen, J.S., Gee, C.M., and Bar-Chaim, N., "High-power microwave photodiode for improving performance of RF fiber optic links," in Proceedings of SPIE, Photonics and Radio Frequency, 2844, 110–119 (1996).

Pearsall, T. P., Piskorski, M., Brochet, A., and Chevrier, J., "A $Ga_{0.47}In_{0.53}As$/InP heterophotodiode with reduced dark current," *IEEE Journal of Quantum Electronics*, 17(2), 255–259 (1981).

Ramaswamy, A., Piels, M., Nunoya, N., Yin, T., and Bowers, J. E., "High power silicon-germanium photodiodes for microwave photonic applications," *IEEE Transactions on Microwave Theory and Techniques*, 58(11), 3336–3343 (2010).

Rymanov, V., Stohr, A., Dulme, S., and Tekin, T., "Triple transit region photodiodes (TTR-PDs) providing high millimeter wave output power," *Optics Express*, 22(7), 7550–7558 (2014).

Sakai, K., Ishimura, E., Nakaji, M., Itakura, S., Hirano, Y., and Aoyagi, T., "High-current back-illuminated partially depleted-absorber p-i-n photodiode with depleted nonabsorbing region," *IEEE Transactions on Microwave Theory and Techniques*, 58(11), 3154–3160 (2010).

Shi, J. W., Kuo, F. M., Wu, C. J., Chang, C. L., Liu, C. Y., Chen, C. Y., and Chyi, J. I., "Extremely high saturation current-bandwidth product performance of a near-ballistic uni-traveling-carrier photodiode with a flip-chip bonding structure," *Journal of Quantum Electronics*, 46(1), 80–86 (2010).

Shimizu, N., Miyamoto, Y., Hirano, A., Sato, K., and Ishibashi, T., "RF saturation mechanism of InP/InGaAs uni-traveling-carrier photodiode," *Electronics Letters*, 36(8), 750–751 (2000).

Tucker, R. S., "Green optical communications," *IEEE Journal of Selected Topics in Quantum Electronics*, 17(2), 245–274 (2011).

Tulchinsky, D. A., Li, X., Li, N., Demiguel, S., Campbell, J. C. and Williams, K. J., "High-saturation current wide-bandwidth photodetectors," *IEEE Journal of Selected Topics in Quantum Electronics*, 10(4), 702–708 (2004).

Tulchinsky, D. A., Boos, J. B., Park, D., Goetz, P. G., Rabinovich, W. S., and Williams, K. J., "High-current photodetectors as efficient, linear, and high-power RF output stages," *Journal of Lightwave Technology*, 26(4), 408–416 (2008).

Urick, V. J., Hastings, A. S., McKinney, J. D., Sunderman, C., Diehl, J. F., Devgan, P. S., Colladay, K., and Williams, K. J., "Photodiode linearity requirements for radio-frequency photonics and demonstration of increased performance using photodiode arrays," in IEEE Topical Meeting on Microwave Photonics Digest, 86–89 (2008).

Williams, K. J., Esman, R. D., and Dagenais, M., "Effects of high space-charge fields on the response of microwave photodetectors," *IEEE Photonics Technology Letters*, 6(5), 639–641 (1994).

Williams, K. J., "*Microwave Nonlinearities in Photodiodes*," Doctoral Dissertation, UMI (1994).

Williams, K.J. and Esman, R.D., "Optically amplified downconverting link with shot-noise-limited performance," *IEEE Photonics Technology Letters*, 8(1), 148–150 (1996).

Williams, K. J., Esman, R. D., Wilson, R. B., and Kulick, J. D., "Differences in p-side and n-side illuminated p-i-n photodiode nonlinearities," *IEEE Photonics Technology Letters*, 10(1), 132–134 (1998).

Williams, K. J. and Esman, R. D., "Design considerations for high-current photodetectors," *Journal of Lightwave Technology*, 17(8), 1443–1454 (1999).

Williams, K.J., "Comparisons between dual-depletion-region and uni-traveling -carrier p-i-n photodetectors," *IEE Proceedings-Optoelectronics*, 149(4), 131–137 (2002).

Xiong, C., Li, G. Y., Zhang, S., Chen, Y., and Xu, S., "Energy- and spectral-efficiency tradeoff in downlink OFDMA networks," *IEEE Transactions on Wireless Communications*, 10(11), 3874–3886 (2011).

Zhou, Q., Cross, A. S., Fu, Y., Beling, A., Foley, B. M., Hopkins, P. E., and Campbell, J. C., "Balanced InP/InGaAs photodiodes with 1.5-W output power," *IEEE Photonics Journal*, 5(3), 6800307 (2013).

CHAPTER 10

APPLICATIONS AND TRENDS

An introduction to some of the applications of microwave photonics was offered in Chapter 1. Being such an applied field, any comprehensive work would be lacking without providing more detail than given there. The present chapter is intended to expose the reader to some of the application opportunities of the technologies described in the preceding chapters. In many instances, the descriptions are terse and rely heavily on the references. An entire book could certainly be dedicated to the wide range of applications of microwave photonics. Indeed, there is a large volume of work that covers implementations of analog fiber optic technology. Zmuda and Toughlian (1994) provided a unique work on radio-frequency (RF) photonics as it relates to radar, a primarily military application. The basics of lightwave systems carrying video for legacy cable television have been given by Phillips and Darcie (1997). A book edited by Wilson et al. (1995) reviews analog fiber optics for industrial telecommunications applications with a concentration on modulation techniques. A more recent commercial application of microwave photonics to radio over fiber (RoF), a term typically associated with wireless networks, has been detailed by Novak et al. (2013). Some less-obvious applications have also been studied, such as microwave photonics for biomedical applications (Daryoush 2009) and

Fundamentals of Microwave Photonics, First Edition.
Vincent J. Urick Jr, Jason D. McKinney, and Keith J. Williams.

photonic methods to transmit and receive RF signals simultaneously (Cox and Ackerman 2009, 2013). There are numerous challenges in transitioning microwave-photonic technology for many applications (Capmany et al. 2013a) but the technology has widespread promise nonetheless. The intent in this case is not to duplicate the existing literature but rather provide descriptions of some less-reviewed and/or emerging application opportunities for microwave photonics.

10.1 POINT-TO-POINT LINKS

Perhaps, the most obvious application of an analog fiber optic link is for RF signal transport and routing from one location to another. As described in Chapter 1, there are numerous advantages of using fiber optics as opposed to copper for such links. Wide bandwidth and frequency-independent low propagation loss are major advantages of photonics, particularly at high frequencies and/or long lengths. Optical fiber does not emit nor does it pick up RF radiation, making it desirable for low noise or covert tactical links. Such electromagnetic interference (EMI) mitigation can sometimes be achieved in copper cables with the tradeoff of expensive and unwieldy shielding. Photonics can also protect one side of a link from large bursts of energy on the distal end, such as from a lightning strike or a high power electronic attack. There are many applications where an analog-to-digital converter (ADC) is used to process analog signals. An ADC can be very power-consuming, particularly when processing large bandwidths. Analog fiber optic links are often used to relocate an ADC further from an antenna where heat can be removed more easily, especially in dense array applications. An analog fiber link can pass the entire bandwidth of an antenna to a remote location, where advanced signal processing and storage can be implemented. The EMI from an ADC can also impair high sensitivity receivers when placed near an antenna feed, making analog remoting more desirable. Furthermore, the reliability of analog optical links can surpass their digital counterparts; in radio-astronomy applications, it has been shown that the mean time between failures (MTBF) of an analog fiber optic link can far exceed that of a digital link (Montebugnoli et al. 2005). The RF performance of an analog fiber optic point-to-point link depends heavily on the propagation length. A recent survey (Urick et al. 2013) demonstrates that noise figures below 10 dB and SFDRs well above 120 dB in 1 Hz are achievable, albeit in relatively short links. However, as detailed in the following examples, the performance of microwave photonic point-to-point links can vary as widely as their specific utilities.

The architectures for fiber optic links that were employed for short-distance remoting of radar signals are shown in Figure 10.1 (Roman et al. 1998). These links, the first of their kind published, demonstrated the ability to remote the AN/SPQ-9B Advanced Development Model (ADM) X-Band radar. As shown in Figure 10.1, two fiber optic links were connected through an RF circulator to the radar antenna, one path for transmit and the other for receive. At the time of this work, the AN/SPQ-9B ADM radar required particularly stringent phase noise performance, which the fiber optic links were

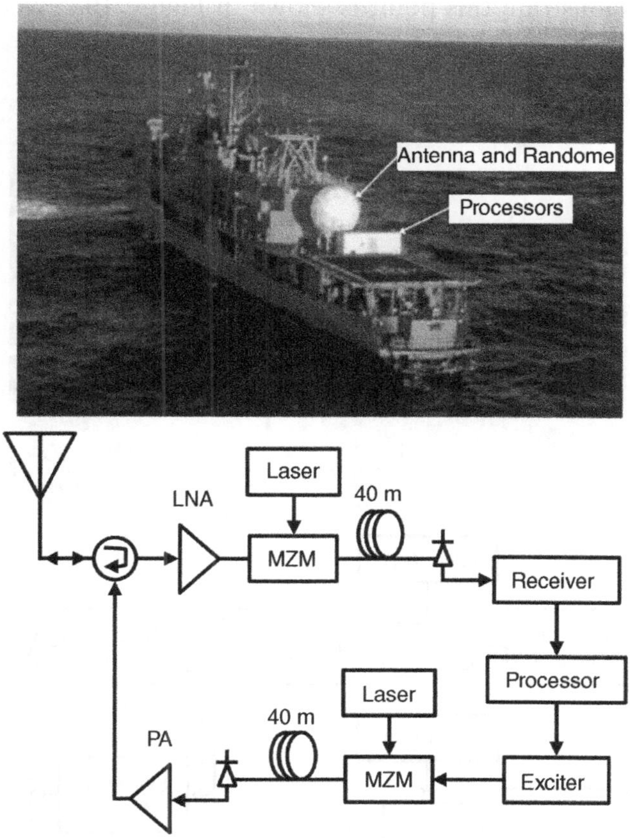

Figure 10.1. Architecture for two fiber optic links employed to remote the transmitted and received signals from an AN/SPQ-9B Advanced Development Model X-Band radar. The photograph depicts locations of the radar antenna and processing hardware aboard the Self Defense Test Ship out of Port Hueneme, CA, USA (Roman et al. 1998).

able to achieve. Direct and external intensity modulation links were tested during the design phase for this application. External intensity modulation with an MZM (see Chapter 6) proved superior in terms of RF noise figure and phase noise. The primary reasons for using fiber optics in this implementation were size and weight requirements at the antenna. Although not shown explicitly in Figure 10.1, the system employs up and down conversion at the exciter and receiver blocks, respectively. These relatively short fiber links allowed for the hardware associated with such frequency translation functions to be located in the processing station, where the size and weight restrictions were not as severe.

Short-run fiber optic links can be employed to provide wideband RF functionality, as opposed to a passive link as described previously. An example to illustrate this point is shown in Figure 10.2, where a fiber

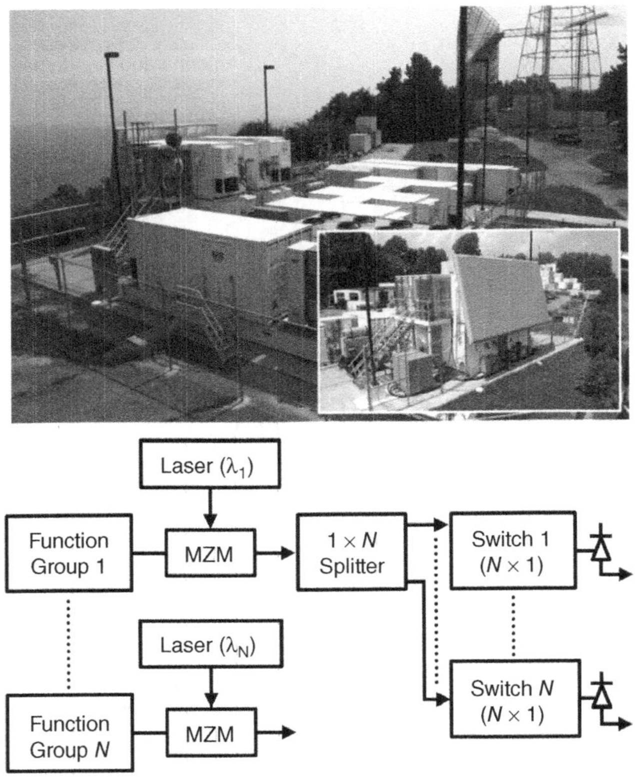

Figure 10.2. Diagram for an $N \times N$ photonically enabled crossbar switch as used in the Advanced Multifunction Radio Frequency Concept (AMFRC) project. The photograph shows the AMRFC test bed at the Chesapeake Bay Detachment Test Facility, MD, USA.

optic distribution network used for the Advanced Multifunction Radio Frequency Concept (AMRFC) Program is depicted (Tavik et al. 2005). The distribution network in Figure 10.2 functions as a crossbar switch, where any one of the N inputs can be routed to any one of the N outputs. In this implementation, each RF input is modulated onto a different optical wavelength, split into N outputs via a fiber optic coupler, and fed into an $N \times 1$ fiber optic switch. The AMFRC system manages numerous signal chains including communications, radar, and electronic warfare. As implemented for AMFRC, the inputs from four 6 to 18-GHz function groups can be routed independently to four transmit subarray quadrants or a single input can be distributed to the entire array. The fiber optic structure allows for active signal distribution at baseband RF instead of electronic routing in the intermediate frequency (IF) path, where the implementation would be more complicated and hardware intensive. Although a relatively small-scale example compared to the massive signal routing capability of modern telecommunication networks (Alferness 2012), the configuration shown in Figure 10.2 demonstrates the utility of fiber optics for providing functionality that is not practical in the electronic domain.

As described in Chapter 1, the advantages of fiber optics over copper coaxial transmission lines typically increase with the modulation frequency. However, fiber optics can provide solutions even at relatively low modulation rates. An example of such a link is shown in Figure 10.3, which is employed to remote HF signals (3–30 MHz) from each antenna in a seven-element array to a processing station 2.9 km away. This task could have been completed with coaxial cable, but system requirements dictated that a mid-span RF amplifier was required to offset loss. Since DC power for such amplifiers was not readily available, a fiber solution was employed. As shown in Figure 10.3, the link design (Urick et al. 2008a) included some unique features to meet stringent RF performance requirements for the array tasking. The input HF signal is passed through a preamplifier before being modulated onto the optical intensity via a dual-output MZM. Each MZM output is split once more for a total of four fibers per link. The receiver adds the photocurrents from the fibers on common MZM outputs and then subtracts the composite photocurrents from each MZM arm. The four fiber spans serve to circumvent SBS (see Chapter 5), allowing four times the optical power handling of a single fiber. Furthermore, the increase in output power eliminates the need for optical amplification in this link design even at distances well above 7 km (Urick et al. 2008a). As detailed in Chapter 6, a 4× increase (6 dB) in output power translates to 12 dB

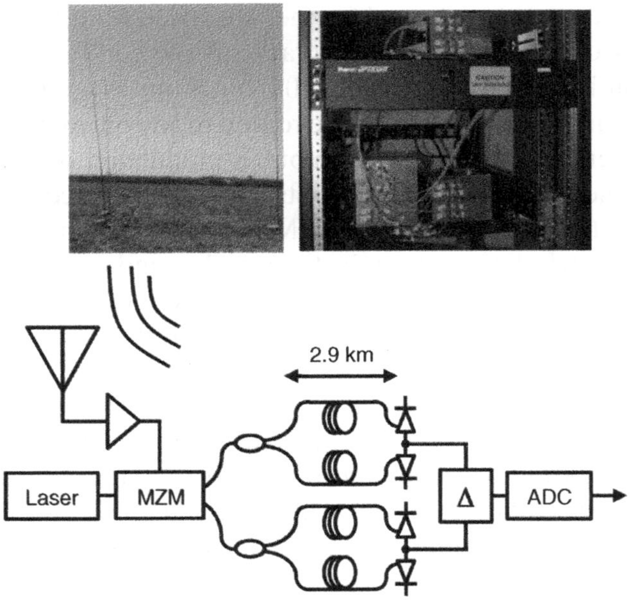

Figure 10.3. Block diagram for a 2.9-km fiber optic link for HF antenna remoting. The photographs show the antenna field and the receiver rack where the link is installed at the Southwest Research Institute, TX, USA (Urick et al. 2008b).

more RF output power. Furthermore, the balanced detection serves to reject common mode noise in the link and allows for cancellation of photodiode even-order distortion. The added cost of cabling 28 fibers (seven antennas × four fibers per antenna) as opposed to seven fibers was an acceptable trade given the performance benefits. Phase matching 28 fibers is more cumbersome than trimming seven fibers but is an achievable endeavor at 30 MHz.

The measured noise figure for the link shown in Figure 10.3 is plotted in Figure 10.4. Data are shown there from preliminary laboratory measurements at the Naval Research Laboratory, Washington, DC, USA and after installation at the Southwest Research Institute, TX, USA. The laboratory and field data agree quite well with the increased noise figure at the lower frequencies attributed to increased environmental noise in the field. The links were third-order limited at 30 MHz where SFDR = 117 dB·Hz$^{2/3}$ without the preamplifier and SFDR = 110 dB·Hz$^{2/3}$ with the amplifier in place (Urick et al. 2008a). This scenario demonstrates the classic tradeoff between sensitivity (noise figure) and dynamic range in a microwave photonic link employing a preamplifier. The installed links were configured with

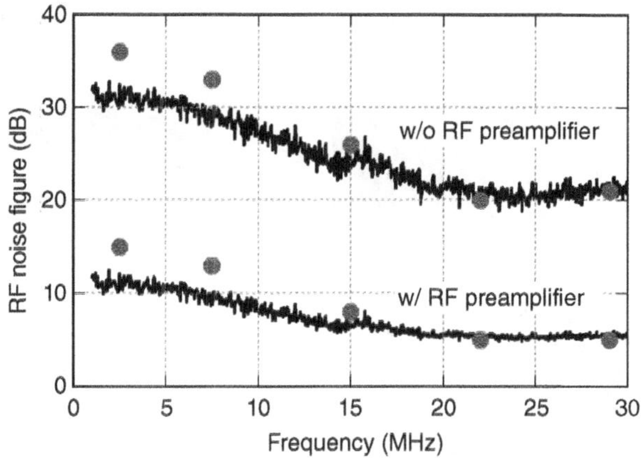

Figure 10.4. Measured RF noise figure for an HF antenna remoting link showing laboratory results (lines) and data after installation (symbols) with and without an RF preamplifier (Urick et al. 2008b).

the RF preamplifiers, trading lower SFDR for increased sensitivity, and operate continuously to connect the seven antennas to a signal processing station.

There are situations where it is advantageous to operate an analog fiber optic link at an IF band as opposed to the full RF bandwidth. Such a link is shown in Figure 10.5, which is employed to remote the 100-m Green Bank Telescope (GBT) from the processing and control systems in the Jansky Laboratory (Jewell 2000). The two sites are 1.8 km apart and both located at the National Radio Astronomy Observatory (NRAO), Green Bank, WV, USA. The RF operational bandwidth of the GBT is about 100 MHz–115 GHz, and eight IF links such as shown in Figure 10.5 are available, each with 7 GHz of bandwidth (Jewell 2000). These IF bandwidths more than suit the tasking for the GBT and are wide enough that a fiber solution outperforms an RF link. An additional advantage of this analog fiber optic link is that the EMI from advanced digital signal processing of the radio signals can be isolated in the RF shielded Jansky Laboratory.

The link examples discussed to this point do not employ optical amplification stages. However, optical amplifiers are typically necessary for transmission distances beyond 10 km or so. A 20-km link employing a single EDFA on the output is shown in Figure 10.6 (Urick et al. 2006). The link employed a DFB semiconductor laser, an MZM with $V_\pi = 2.8\,\text{V}$ at 9 GHz and an 18-GHz p–i–n photodiode with an RF impedance-matching circuit ($H_{\text{pd}} = 1/2$). A specialty fiber span was

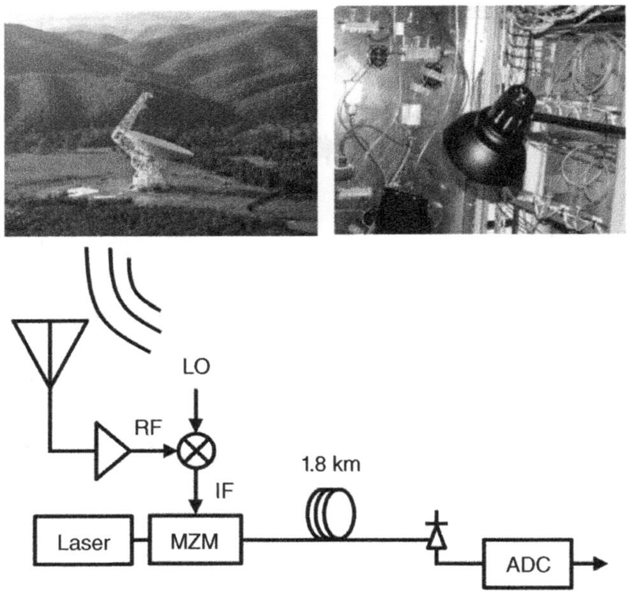

Figure 10.5. Diagram for an intermediate frequency (IF) analog photonic link used to remote the Green Bank Telescope (GBT) from a processing station. The photographs show the GBT with its 100-m dish and some of the fiber components in receiver room. (Photographs obtained from http://www.gb.nrao.edu/gallery/).

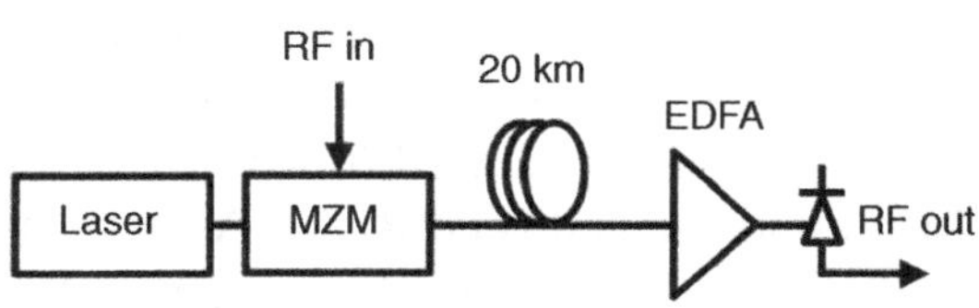

Figure 10.6. Diagram for a 20-km link with a booster amplifier.

employed to mitigate SBS and chromatic dispersion simultaneously. The fiber span consisted of two types of fiber having SBS frequencies separated by 50 MHz and having nearly opposite dispersion (Urick et al. 2006). The measured SBS threshold for the link was in excess of 13 dBm and the launch power was set much lower at 6 dBm (with the MZM at quadrature). The EDFA at the output was set to achieve $I_{dc} = 9\,\text{mA}$, and the link was characterized at 9 GHz (Urick et al. 2006). The measured small-signal RF gain for the link is shown in Figure 10.7. The value of −13 dB at 9 GHz agrees closely with that predicted by Equation (6.27) with the aforementioned parameters. The effect of chromatic dispersion on the link frequency response is negligible.

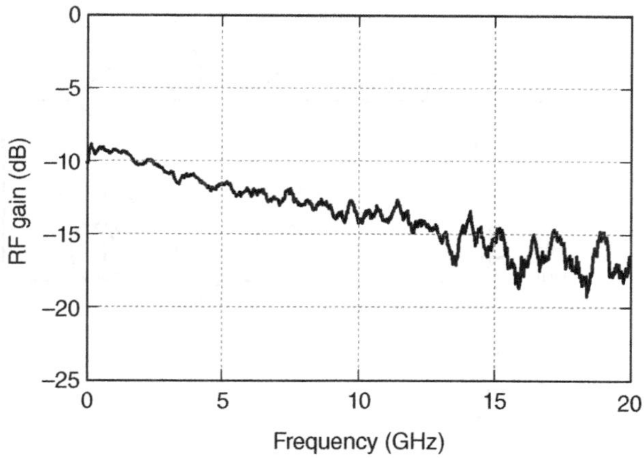

Figure 10.7. Measured small-signal RF gain for the link shown in Figure 10.6 (Urick et al. 2006).

The first null due to dispersion was measured at 46 GHz; this result can be used along with Equations (6.109) and (6.110) to calculate the 3-dB bandwidth (due to dispersion alone) of the fiber as 33 GHz. The measured SFDR for the link was 102 dB in a 1-Hz bandwidth over the 6–12-GHz octave. This link gives an example of a high performance medium-length link employing a passive repeater-less transmission span. However, mid-span gain stages can be used to improve the RF performance of long-distance point-to-point links significantly.

A 110-km link employing mid-span amplification via EDFAs is shown in Figure 10.8. The link uses the IMDD format with an external MZM, and the transmission span consists of four dispersion- and SBS-managed 27.5-km sections (Urick et al. 2013). Although similar fiber techniques were used in the repeater-less link mentioned previously (Figure 10.6), the mid-span amplification in this case enabled a better SFDR of 103 dB in 1 Hz at 18 GHz. The measured RF gain as a function of frequency for the link is compared to the back-to-back case in Figure 10.9. As shown, the fiber does not degrade the response to 30 GHz. In fact, the measured net chromatic dispersion ($D \times L$) for the fiber is only 8 ps/nm at 1550 nm, which yields 62 GHz as the 3-dB bandwidth by way of Equation (6.111).

It is not always practical to employ discrete mid-span optical amplification in long-haul links. A back-end booster amplifier such as shown in Figure 10.6 suffices if the optical signal-to-noise ratio (OSNR) is adequate after propagation through the span. However, distributed Raman amplification is a good alternative in many applications. As described

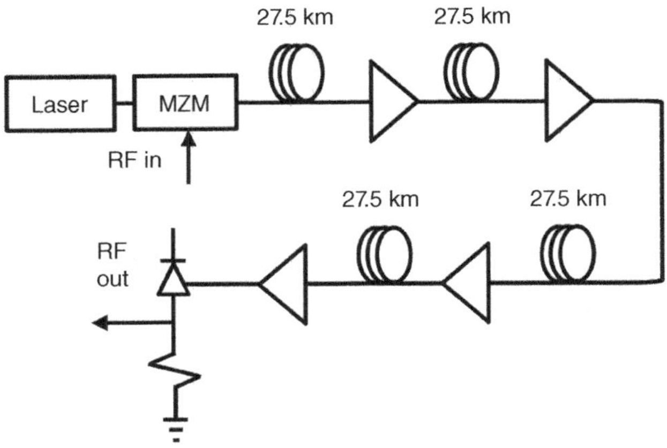

Figure 10.8. Architecture for a 110-km link employing mid-span EDFAs.

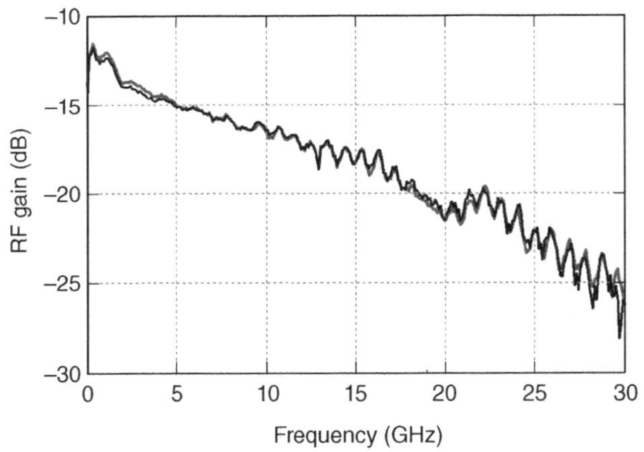

Figure 10.9. Measured small-signal gain for the link in Figure 10.8 (black) compared to the back-to-back case (gray) without the transmission span (Urick et al. 2013).

in Chapter 3, Raman gain can be achieved by using pump light that can be supplied from a remote location. An example of such a link is shown in Figure 10.10, a 105-km phase-modulated implementation. The signal laser was at 1551 nm, which was Raman amplified via a pump laser at 1455 nm supplied from the back end of the link. The phase modulation was demodulated with a quadrature-biased MZI with a path imbalance of 100 ps. The first pass band of the MZI response had a 3-dB bandwidth of ±2.5 GHz centered at 5 GHz (see Chapter 7). The efficacy of the link for high fidelity signal transport was demonstrated using quadrature amplitude modulation (QAM) signals running at

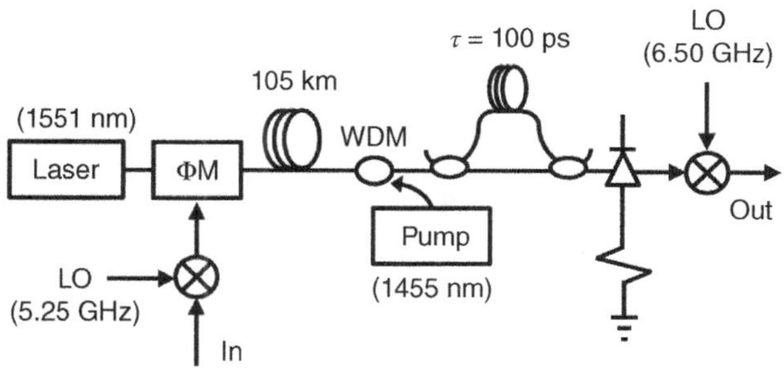

Figure 10.10. Diagram for a 105-km phase-modulated link employing Raman amplification.

TABLE 10.1 Results for 105-km Link Experiments (Urick et al. 2004)

Format	Data Rate (Gb/s)	EVM (%rms)	MER (dB)	BER
16-QAM	4	4.04	25.3	$<10^{-12}$
32-QAM	5	3.92	25.8	3.95×10^{-10}
64-QAM	6	3.77	24.8	4.46×10^{-5}

1 Gsymbol/s. The baseband QAM signal was upconverted to 5.25 GHz to match the MZI filter function. The output of the link was downconverted to 1.25 GHz for processing. The results for 16-, 32-, and 64-QAM are shown in Table 10.1, and a representative constellation is shown in Figure 10.11 (Urick et al. 2006). The measured error vector magnitude (EVM) is used to calculate the modulation error ratio (MER) and bit error rate (BER). The BERs shown in Table 10.1 are "raw" in that no forward error correction (FEC) was employed. Although the 105-km link was originally designed for traditional analog signal transport, its capability of carrying high order digital modulation on RF subcarriers demonstrates the versatility of analog fiber optic systems.

10.2 ANALOG FIBER OPTIC DELAY LINES

An extremely powerful tool afforded by microwave photonics is a fiber optic delay line (FODL). The concept of a FODL is simple: light is modulated by an RF waveform(s) and a length of fiber is used to store the information. The delay or storage time in the fiber is given by Equation (5.3) and amounts to 4.90 μs/km for typical fiber. The low optical loss in fiber over a very wide bandwidth, typically 0.2 dB/km or 0.04 dB/μs, makes a photonic delay line much more feasible than an

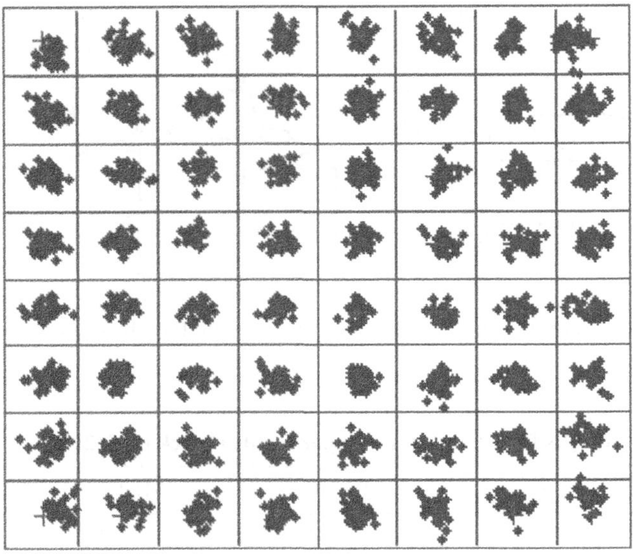

Figure 10.11. Measured constellation diagram for 64-QAM at 1 Gsymbol/s after transmission through a 105-km link (Urick et al. 2004).

RF coaxial cable for any appreciable delay length. Electrical power consumption, instantaneous bandwidth, and low latency are also characteristics in favor of an analog FODL as opposed to digital electronics when the bandwidth is greater than 1 GHz or so. One downside of an analog FODL as compared to digital memory is that the former does not provide continuously accessible memory. However, there are a number of applications where this is not a concern.

As opposed to a point-to-point link, a FODL serves as a "black box" that delays an RF signal for a certain amount of time. The physical size of a FODL is driven largely by the amount of fiber employed, but there are numerous applications where rack-mount units are feasible. However, the size of a FODL can be minimized by using a recirculating structure (Nguyen et al. 2014), similar to that described in Chapter 3 for laser linewidth characterization. Shown in Figures 10.12 and 10.13 are two applications of a FODL in electromagnetic warfare (EW) scenarios. A FODL is used as a buffer for a cued receiver in the architecture in Figure 10.12. In this type of system, an incoming RF spectrum is channelized into some number of channels for coarse, high speed signal detection. The channelization can be conducted in the RF domain or, as described in Section 10.3, in the optical domain. The output of the channelizer cues a high performance ADC that is tuned to the frequency of interest. The cueing process can take a considerable amount of time,

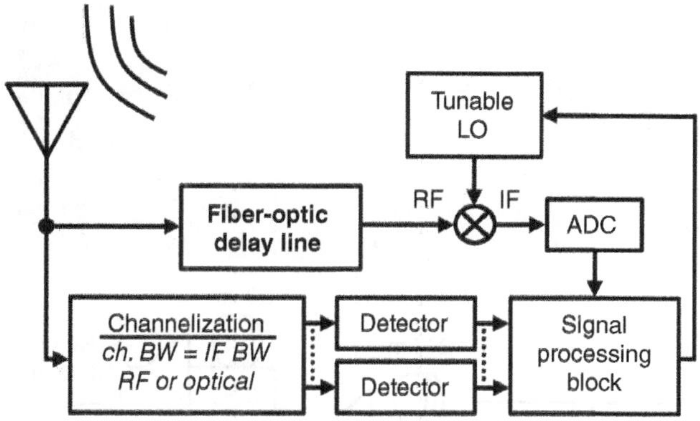

Figure 10.12. A cueing-receiver architecture employing a FODL.

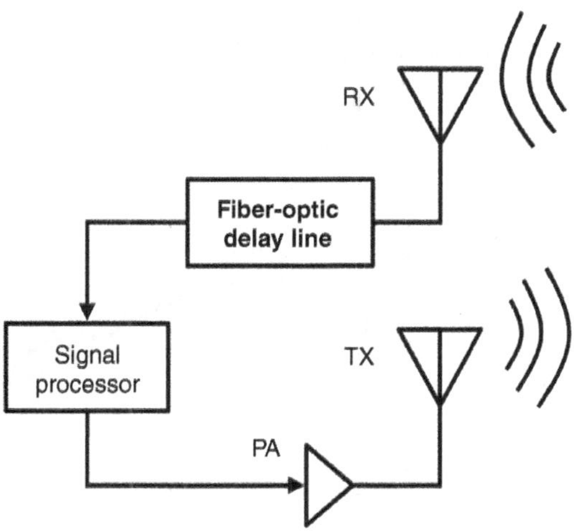

Figure 10.13. An architecture using a FODL to emulate range in radar testing.

thus the utility of the FODL is to preserve a time-delayed copy of the received spectrum. The requirements on the FODL in this implementation can be quite stringent, driven by the sensitivity and dynamic range of the ADC, as well as the time needed for the cueing process. A FODL can also be used to emulate range in radar testing such as depicted in Figure 10.13. In this scenario, a transmitted radar signal impinges on a target at a fixed range and the returned signal is passed through a FODL before processing. A fixed-length FODL can allow for testing situations beyond what is feasible in a chamber or limited test range.

An image of the front panel for a state-of-the-art FODL is shown in Figure 10.14 along with a block diagram for the internal structure (Urick

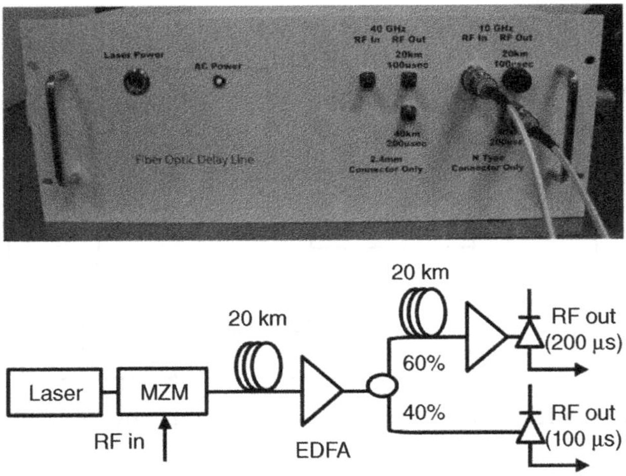

Figure 10.14. Photograph and block diagram for a 40-GHz FODL switchable between 100- and 200-μs delays. The amplifiers are EDFAs (Urick et al. 2012).

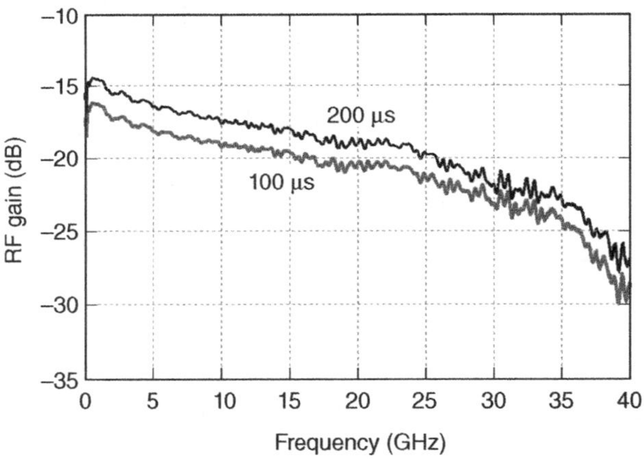

Figure 10.15. Measured small-signal RF gain for the FODL in Figure 10.14 in the 100- and 200-μs configurations (Urick et al. 2012).

et al. 2012). The entire unit is housed in a 21 × 19 × 7″ rack-mount enclosure, which is powered by 120-V AC and consumes about 17 W at room temperature. The modulation format for the FODL is IMDD with an external MZM. As shown in Figure 10.14, the architecture offers two possible delay settings: 100 and 200 μs. The delay lines employ SBS and chromatic-dispersion management (Urick et al. 2012). Both delays are available with over 40 GHz of bandwidth as demonstrated by the measured data in Figure 10.15. As can be seen, the small-signal RF gain rolls off across the band as determined by MZM and photodiode frequency responses (not fiber dispersion). The noise figure rises across the band in a similar manner with values at 40 GHz of NF = 51 dB and NF = 54 dB for 100 and 200 μs, respectively. As described Equation (3.71), this 3-dB difference is expected for a loss-compensated system employing similar EDFAs. This FODL was characterized extensively at X-Band (Urick et al. 2012), where the SFDR ranges from 103 to 107 dB · $Hz^{2/3}$ and the SSB phase noise above 1 kHz offset varies from −135 to −145 dBc/Hz,

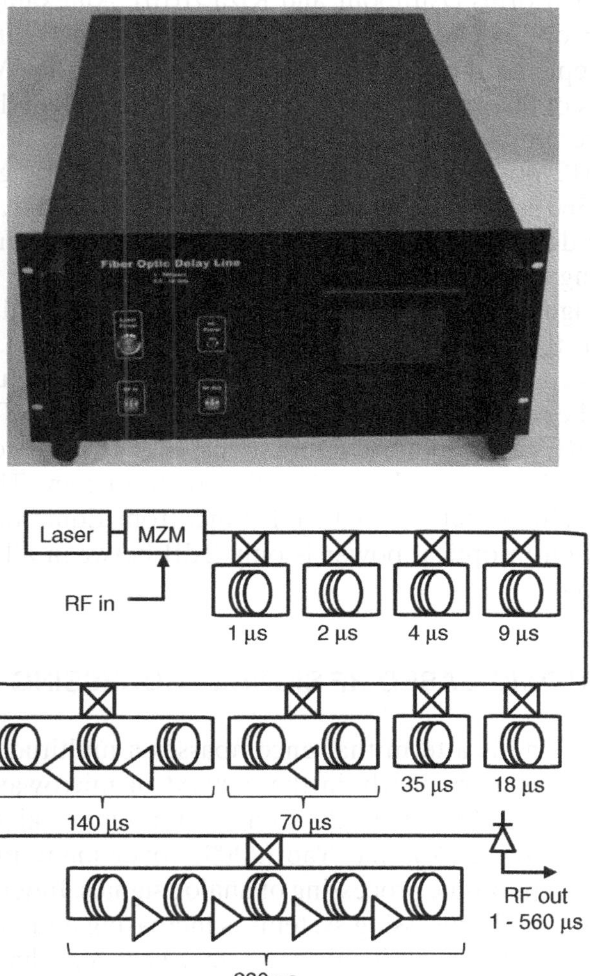

Figure 10.16. Image and block diagram for a BiFODL with delays tunable between 1 and 560 μs with 1-μs steps. The amplifiers shown are EDFAs (Singley et al. 2014).

depending on the configuration. Although the FODL in Figure 10.14 was originally designed for radar testing, its performance is representative of what can be expected from high end fixed-length FODLs for other applications.

As compared to a fixed-length FODL, units with adjustable delay offer more versatility with the tradeoffs of added cost and complexity. A series of delays and switches in a binary arrangement provides a large number of possible paths in a hardware-efficient manner. Such architectures were first implemented in the late 1980s using fiber optic components (Goutzoulis et al. 1989) and more recently in smaller integrated platforms (Fathpour and Riza 2010). An example of such a binary fiber optic delay line (BiFODL) is shown in Figure 10.16. The BiFODL depicted there (Singley et al. 2014) uses the IMDD format and consists of fiber spans, each twice the preceding length. Fiber optic switches are employed to insert or remove each delay. As a result, there are 512 possible paths through the BiFODL ranging between 1.1 μs and 560 μs in 1.1-μs steps. In general, a BiFODL architecture with a base delay of x and N total delay lines will result in 2^N possible paths ranging from x to $x \cdot 2^N$ in steps of x. The EDFAs in the BiFODL shown in Figure 10.16 were specially designed to equalize, as much as practical, the RF performance for each delay setting. That is, the EDFAs were arranged to be in saturation for all configurations with nominally the same optical gain and output noise power. The measured bounds on RF gain and noise figure across all BiFODL configurations are shown in Figure 10.17 as a function of frequency. The maximum variation in either metric is 4 dB inferring, from Equation (3.55), that the fluctuation in optical power is only 2 dB between all 512 possible configurations.

10.3 PHOTONIC-BASED RF SIGNAL PROCESSING

Signal processing is a term that encompasses a multitude of methods and applications, particularly in the context of microwave photonics. Review articles by Minasian (2006) and Minasian et al. (2013), Capmany et al. (2005, 2006), and Yao (2009) cover many principles and applications of photonic processing of analog signals. Indeed, the role of photonics in signal processing systems is increasing and, in some cases, moving toward all-optical signal processing, a concept that is also being considered for industrial telecommunications (Willner et al. 2014). This section provides an overview of four particular functions afforded by microwave photonics: wideband channelization of an RF environment,

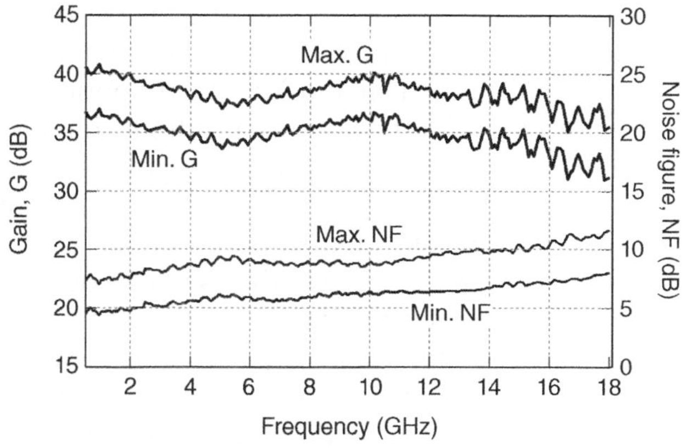

Figure 10.17. Measured maximum and minimum RF gain and noise figure values for the BiFODL in Figure 10.16 with the addition of an RF preamplifier (Singley et al. 2014).

instantaneous measurement of the frequency from a single RF emitter, downconversion, and phased-array beamforming.

10.3.1 Wideband Channelization

Wideband spectral analysis is an application where a photonic implementation exhibits potential advantages over an all-electronic solution. Consider the case where 100 GHz of bandwidth must be channelized into 500-MHz portions in real time. This is a realistic problem for modern electronic support (ES) systems. A possible electronic solution in this case would require 200 fixed LOs, 200 bandpass filters, and 200 ADCs. The cost and power consumption of such a solution present a major scalability problem. Furthermore, the RF gain ripple of the constituent components over such a large bandwidth can inhibit its implementation. However, photonics offers wide bandwidth with manageable gain ripple, scalability, and relatively low power consumption. A possible photonic solution would be a cueing receiver architecture as shown in Figure 10.12, where photonics is used to direct a tunable electronic LO. Such a solution can still provide "real-time" information on ES time scales, typically on the order of a second (Spezio 2002). Numerous photonic techniques have been presented as solutions for wideband RF energy detection. The majority is based on optical filtering (Agarwal et al. 2011) such as can be achieved with a variety of multiplexing technologies used by the telecommunication industry. However, there is a disparity between the filter bandwidths required in telecommunications and those needed for effective RF channelization. Some methods used for microwave-photonic channelization include fiber-based finite impulse response (FIR) filters (Williamson and Esman 2008, Dai and Yao 2010), optical-comb-based FIR filters (Hamidi et al. 2010), fiber-loop infinite impulse response (IIR) filters (Chan 2013), arrays of fiber Bragg gratings (Hunter et al. 2005, Wang and Yao 2013), frequency-selective SBS amplification (Zou et al. 2013), parametric mixing (Bres et al. 2011), injection locking of oscillators (Urick et al. 2009), the use of cryogenically cooled crystals (Merkel et al. 2013) such as Tm:YAG (Mohan et al. 2007), and compressive sampling[1] (Nichols and Bucholtz 2011, McKenna et al. 2013). Particularly interesting implementations of photonics-based FIR RF filters have been suggested using tapered fibers (Villarruel et al. 2006, Mothe and Bin 2011) and multicore fibers (Gasulla and Capmany 2012). Similar techniques can be conceived leveraging the telecommunications-based infrastructures in space-division multiplexing with photonic-crystal fibers (Russell 2006) and multicore fibers (Zhu et al. 2012), or mode-division multiplexing with multimode fibers (Ho and Kahn 2014). Rather than concentrating on a particular technique, some requirements for a filter-based RF channelization are reviewed in this section.

An ideal optical filter for RF channelization would exhibit a flat-top bandpass matching the desired channel bandwidth with very steep skirts and very high out of band rejection. Shown in Figure 10.18 are the responses for two channels of a 12.5-GHz filter intended for RF channelization applications (Rogge et al. 2005). As demonstrated, the ideal filter response is hardly realized, and this is typically the case in narrow-bandwidth optical filters. Nonetheless, the filter in Figure 10.18 has a channel-to-channel bandwidth (BW) of 12.5 GHz and free-spectral range (FSR) of 100 GHz, the latter being the spacing between repetitions of the filter bandpass. Although not shown, the filter has eight channels to fill the FSR continuously. An optical filter of the type shown in Figure 10.18 can be used with a variety of the modulation formats discussed in previous chapters. In most implementations, each filter output would feed a photodiode for detection. Dual-sideband formats can limit the unambiguous frequency range to FSR/2, whereas single-sideband formats may mitigate this issue. Regardless of the modulation format, optical LOs can be placed in

[1] Compressive sampling in this regard (Clark et al. 2012) is a subset of the larger field of photonic analog-to-digital conversion (ADC). While ADC is not covered explicitly in this book, the reader is referred to Juodawlkis et al. (2001), Valley (2007), and Jalali et al. (2013) for descriptions of the associated photonic techniques.

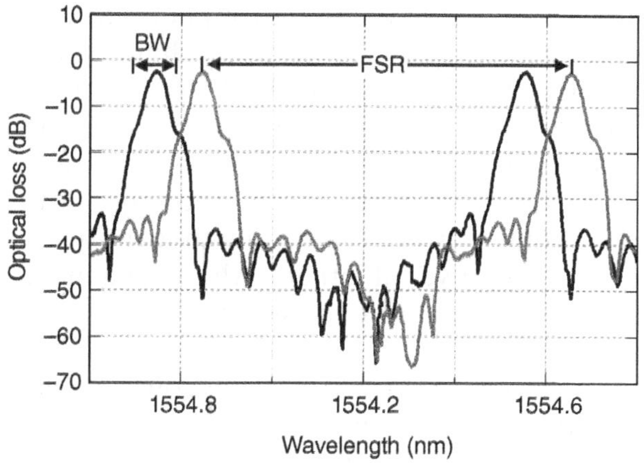

Figure 10.18. Example of an optical filter that might be used for RF channelization with the bandwidth (BW) and free-spectral range (FSR) shown (Rogge et al. 2005).

every channel to increase the received single power as opposed to a modulation sideband only. Alternatively, a modulated carrier can be placed in every filter channel, although this technique is hardware intensive. Regardless of the format, the maximum spurious-free dynamic range of the system will usually be tied to the filter rejection, which is a function of frequency as shown in Figure 10.18. Crossovers in adjacent channels present a problem because two channels will yield the same result, but this issue can be addressed with back end processing. The sensitivity in channelization applications can be a particularly difficult metric to achieve, such as when optics is forced to compete with electronic ES systems having noise figures on the order of a few decibels. These issues become increasingly difficult as the bandwidth of the filter is narrowed. The problem of channelizing 100 GHz into 500 MHz channels with −71 dBm sensitivity (6-dB noise figure with 10-dB margin) is often cited as an application opportunity for photonic techniques. Less stringent requirements include channelizing UHF-through Ku-Bands into 10 MHz with −90 dBm sensitivity, E-Band into 750 MHz with −65 dBm sensitivity, and F-Band into 5 GHz with −55 dBm sensitivity (Ridgway 2013).

10.3.2 Instantaneous Frequency Measurement

A subset of photonic techniques has emerged to provide fast estimation of a received frequency, as opposed to a full-scale channelization as described previously. Often referred to as photonic-assisted instantaneous frequency measurement (IFM), such techniques rely on a pair of complimentary filter functions (Zou et al. 2009a), from which an amplitude comparison function (ACF) can be constructed. The value of the ACF then unambiguously determines the frequency over some bandwidth by mapping amplitude to frequency. The major downside of such IFM techniques is that they are useful only to identify a single emitter whose power dominates over all others. Although a niche application, these techniques are relevant to some systems such as an early-warning EW receiver.

The methods employed to generate a complimentary photonic filter pair for IFM are widely varied. A simple path-imbalanced MZI will yield complimentary RF outputs when preceded by an MZM operated at null bias (Dai et al. 2010). A phase-modulated link with interferometric demodulation such as described in Chapter 7 can be used to construct an ACF, when two MZIs of different path imbalances are used (Xu et al. 2011). In fact, the architecture shown in Figure 7.21, excluding the RF hybrid, could be employed to this end. As an example, the normalized responses of two such links with $\tau_1 = 100$ and $\tau_2 = 375$ ps are shown in Figure 10.19(a) along with the corresponding ACF in 10.19(b). As described by Equation (7.8), the response for each link is given by $|S_{21}|^2 = A_{1,2}\sin^2(\pi f \tau_{1,2})$, where A is the normalization constant. Thus, the ACF $= \sin^2(\pi f \tau_1)/\sin^2(\pi f \tau_2)$, which is plotted along with the measured data in Figure 10.19(b). As described in previous chapters and also by Urick and Bucholtz (2005), phase and intensity modulation exhibit complimentary RF power responses when

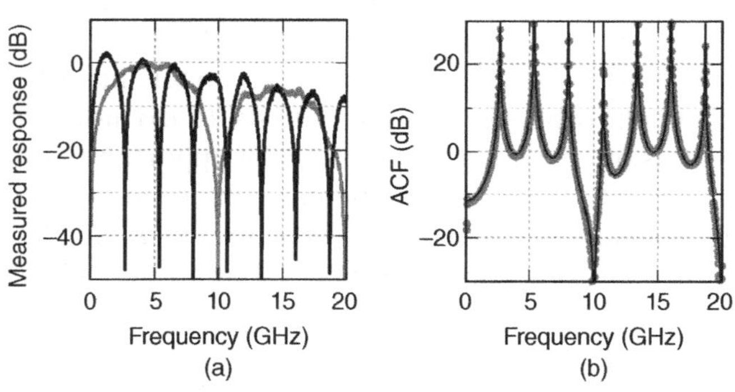

Figure 10.19. (a) Measured normalized responses for phase modulation through interferometers with differential delays of 100 ps (gray) and 375 ps (black) (Urick et al. 2007) (b) The corresponding amplitude comparison function (ACF, gray symbols) plotted with theory (solid black line).

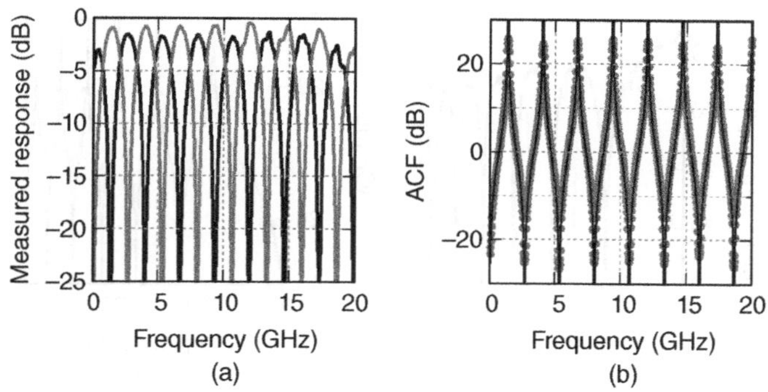

Figure 10.20. (a) Measured normalized response of phase modulation (gray) and intensity modulation (black) through an interferometer with a 375-ps path imbalance. (b) Amplitude comparison function (ACF, gray symbols) compared to theory (solid black line) (Urick et al. 2011).

passed through a dispersive medium. This property has been exploited by Zhou et al. (2009) and Zou et al. (2009b) for RF IFM. Finally, although most photonic-assisted IFM systems have been implemented with bulk fiber optic components, integrated IFM techniques have also been studied (Marpaung 2013, Fandino and Munoz 2013).

The complimentary nature of the RF responses of optical phase and intensity modulation through a dispersive medium and an MZI is particularly attractive for IFM (Urick et al. 2011). The latter is exemplified with the data set in Figure 10.20. Shown in Figure 10.20(a) are the normalized responses for phase and intensity modulation through a quadrature-biased MZI with a differential delay of $\tau = 375\,\text{ps}$. The phase response is as given previously; it can be shown that the response of an intensity-modulated link employing a quadrature-biased MZM through an MZI is $|S_{21}|^2 = A\cos^2(\pi f \tau)$, where A is the normalization parameter (Harmon et al. 2013). The resulting $\text{ACF} = \tan^2(\pi f \tau)$, which has an unambiguous detection bandwidth of $1/(2\tau)$ as plotted in Figure 10.20(b). A similar situation is encountered with a dispersive medium. As given by Equations (5.16) and (5.17), the responses of intensity and phase modulation through a dispersive medium can be written as $|S_{21}|^2 = B\cos^2(\pi D\lambda^2 L f^2/c)$ and $|S_{21}|^2 = C\sin^2(\pi D\lambda^2 L f^2/c)$, respectively, where B and C are normalization parameters. The corresponding normalized $\text{ACF} = \tan^2(\pi D\lambda^2 L f^2/c)$. The detection bandwidth using this response decreases with f because of the f^2 dependence. The largest unambiguous detection band is from DC to $f = [c/(2D\lambda^2 L)]^{1/2}$. These characteristics are exampled in Figure 10.21 for $D = 16.5\,\text{ps/nm/km}$, $\lambda = 1550\,\text{nm}$, and $L = 50\,\text{km}$.

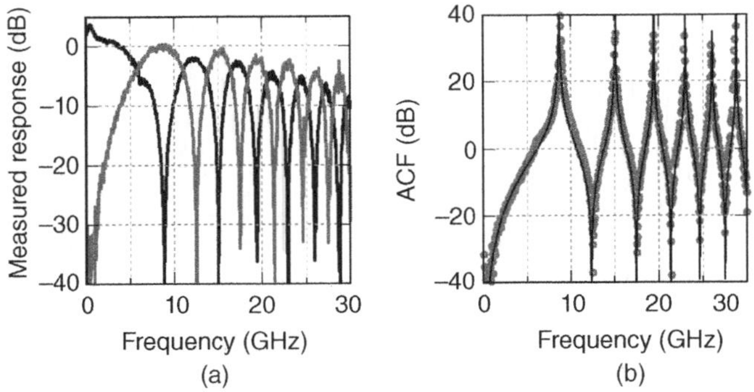

Figure 10.21. (a) Measured normalized response of intensity modulation (black) and phase modulation (gray) through 50 km of fiber with a dispersion parameter of 16.5 ps/nm/km at 1550 nm. (b) The amplitude comparison function (ACF, gray symbols) against theory (black line) (Urick et al. 2011).]

10.3.3 Downconversion

Photonic downconversion techniques address some issues with all-RF downconversion, such as RF/LO isolation, nonlinearity, scalability, and efficiency. Numerous classes of photonic links incorporating downconversion have been demonstrated (Pappert et al. 2002), including those with an LO at the output and a traditional RF mixer or a photodiode used as the mixing element, injection of an electronic LO directly into a laser or into an external modulator, and those using an optical LO. These methods are depicted as block diagrams in Figure 10.22. Shown in Figure 10.22(a) is an electronic mixer, which may also employ RF and IF filters as shown in Figure 1.18. A photodiode can be used to replace an RF mixer, leveraging increased power conversion efficiency. As shown in Figure 10.22(b), direct modulation of a laser's amplitude or frequency by the RF signal and an LO can result in downconversion to an IF after the optical-to-electrical (O/E) process. Note that the O/E blocks in Figure 10.22 may contain some conditioning, such as an MZI, depending on the modulation format employed. As described in previous chapters, external modulation offers numerous advantages over direct modulation, a trend that remains true for photonic downconversion. As opposed to a single modulator, dual serial modulation stages [Figure 10.22(c)] can provide some advantages. This was demonstrated explicitly by Gopalakrishnan et al. (1993) by comparing two serial quadrature-biased MZMs to a single MZM biased for maximum transmission. Optical sources can also be employed as the LO in photonic downconversion. Shown in Figure 10.22(d) is an example of where the frequency difference between two lasers serves as the LO frequency.

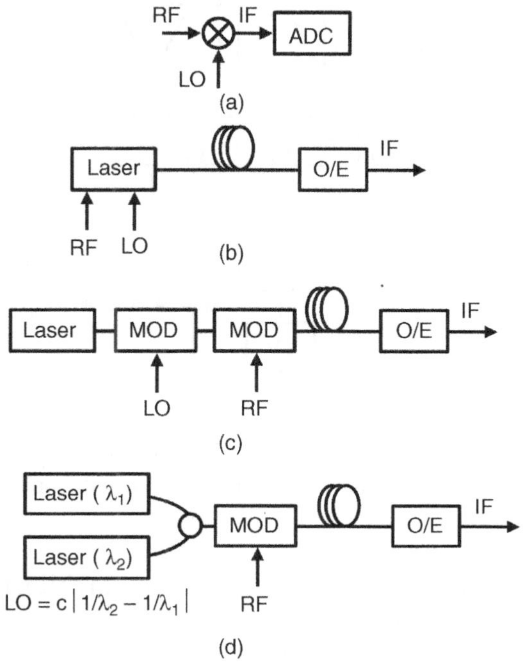

Figure 10.22. Block diagrams for down-converting methods showing (a) a traditional RF mixer, (b) a direct-modulation photonic link, (c) external modulation photonic link, and (d) a dual-laser photonic link.

Of all the photonic downconversion techniques, serial external intensity modulation incorporating an electronic LO has proven particularly useful, whether implemented with dual-output (Williams and Esman 1996) or single-output (Helkey et al. 1997) MZMs. Demonstrations of these architectures achieved an SFDR of 64 dB in 500 MHz (Karim and Devenport 2008), leveraging state-of-the-art modulator technology and linearization techniques. Coherent techniques offer potential for better performance, including serial phase modulation using an optical hybrid for demodulation (Clark et al. 2010). As described in Section 8.4, sampled links can also be employed for downconversion (McKinney and Urick 2010).

10.3.4 Phased-Array Beamforming

Analog phased array beamforming was identified early in the history of microwave photonics as a promising application of the technology (Zmuda and Toughlian 1994, Frigyes and Seeds 1995) with system demonstrations conducted in the 1990s (Ng et al. 1991,

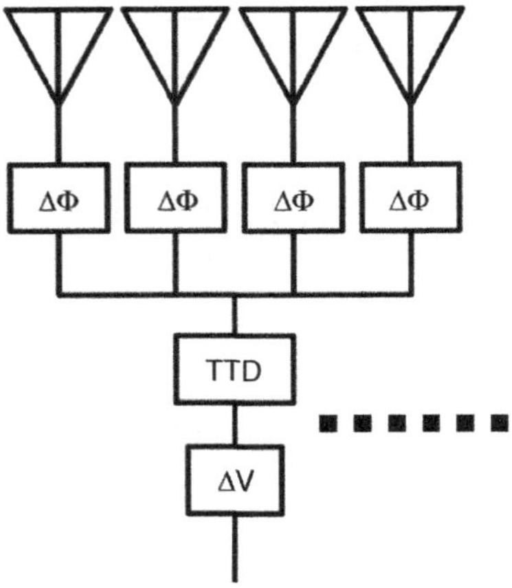

Figure 10.23. An example of a subelement in a phased array employing discrete true-time-delay (TTD) modules, RF phase shifters (ΔΦ), and amplitude control (ΔV).

Frankel et al. 1998). A representative array architecture is shown in Figure 10.23 employing discrete variable true-time delay (TTD) modules and RF phase shifters (Mailloux 1984). The TTD modules impose an RF phase that scales with frequency, whereas the phase shifters are ideally frequency independent. The biggest issue with these components is scalability, as the number of antenna elements in an array can number in the hundreds. Numerous photonic techniques have been proposed for TTD and RF phase shifting; the prevalent ones are reviewed in this chapter.

The low loss in fiber makes it attractive for TTD applications. Stretching lengths of fiber offer a variable TTD but require significant force and excessive amounts of fiber to achieve large tunable delay. Mechanical fiber stretchers are also quite slow and therefore not very practical. Even so, fiber spooled on piezoelectric (PZT) rings have been considered (Herczfeld et al. 1987). Switchable fiber delays offer an alternate solution, although they do not provide continuously tunable delay. Such switched delay lines can be realized in parallel or serial arrangements (Goutzoulis and Davies 1994), with the latter being more scalable. Therefore, a BiFODL as described previously in Section 10.2 has found utility for beamforming (Goutzoulis et al. 1989). Other fiber-based techniques for TTD include using chromatic dispersion in single-mode fiber (Esman et al. 1992, 1993) and modal dispersion in multimode fiber (Yang et al. 1987). Free-space techniques have also been demonstrated for photonic-enabled TTD (Dolfi et al. 1991, Schermer et al. 2011) but are typically not scalable to large arrays.

The significance of optical-based TTD is evident in the varied versions of its implementation, including outside-the-box ideas such as slow light based on SBS (Chin et al. 2010), the use of movable liquid metal mirrors in fiber (Schermer et al. 2013), and nanocomposites (Ricchiuti et al. 2014).

Numerous photonic solutions for RF phase shifting have emerged because wideband, frequency-independent electronic RF phase shifters can be difficult to implement. For example, an array of N fiber optic delay lines with differential path delays spaced by λ_{RF}/N, where λ_{RF} is the center wavelength, can be used to control RF phase if their relative amplitudes can be continuously tuned (Draa et al. 2012). The minimum number of paths for full 360° control in this implementation is $N=3$ (Yilmaz et al. 2003), allowing for control of three phasors in the I–Q plane from which any resultant phasor can be constructed. However, this technique suffers from limited bandwidth because the delay lines are precisely phased for only one frequency; deviation from that frequency quickly spoils the phasor basis. Two nested MZMs and an RF hybrid can also be used for a wideband I–Q phase shifter (Coward et al. 1993). In these and all amplitude-controlled I–Q approaches, phase-to-amplitude conversion must be carefully considered. Aside from the conceptually straightforward I–Q devices, more exotic photonic phase shifters have been proposed such as those based on SBS (Loayssa and Lahoz 2006) and slow light in SOAs (Xue et al. 2010). An elegant solution to photonic RF phase shifting was presented by Chan et al. (2012), employing an integrated dual-parallel MZM (DPMZM). This technique is attractive because it is wideband, can be quickly tuned, exhibits minimal RF amplitude variations, and is scalable to large channel counts. These characteristics are of utmost importance when considering photonic solutions for RF phase control in beamforming applications.

10.4 PHOTONIC METHODS FOR RF SIGNAL GENERATION

Microwave photonics affords numerous unique methods to generate RF signals for applications such as metrology, radar, local oscillators for electronic support systems, and clocks for analog-to-digital conversion. A straightforward approach for photonic signal generation is photodetection of two laser signals separated at the desired RF frequency, a process that can be stabilized via a phase-locked loop (Williams et al. 1989) or injection locking (Yao 2009). As described in Chapter 4, the former is often employed for photodiode characterization

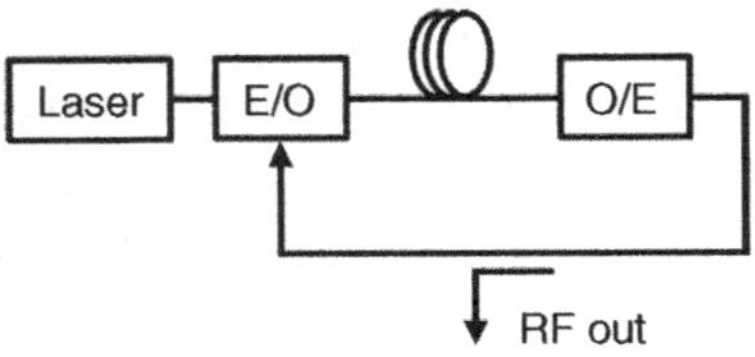

Figure 10.24. Basic block diagram for an optoelectronic oscillator (OEO).

(Draa et al. 2011). Other common methods include mixing of selected sidebands from an externally modulated laser or a mode-locked laser (Yao 2009) and sideband injection locking (Schneider et al. 2013). As described in Section 10.5, these techniques are used to generate millimeter-wave signals for wireless and radio-astronomy applications. In this section, two sophisticated methods for photonics-based RF signal generation are reviewed. An optoelectronic oscillator (OEO) is capable of generating spectrally pure RF signals with great potential for many applications. Photonic-enabled arbitrary waveform generation (AWG) allows for wideband signal generation applicable to communications and radar.

An OEO is a self-seeding ring oscillator that is basically a photonic link fed back on itself. Shown in Figure 10.24 is a functional block diagram for an OEO. A laser serves as the "pump" for an OEO, which further comprises a hybrid optical and electrical ring cavity. As shown in Figure 10.24, the O/E conversion of a photonic link is connected to the electrical-to-optical (E/O) block, thus forming the ring. Self-seeded oscillation can occur if the loop exhibits net RF gain. The low loss in the optical path (typically fiber) allows for very long cavity lengths and therefore a very high quality-factor (Q) for the oscillator. In the microwave photonics community, the OEO is usually credited to seminal work by Yao and Maleki (1994, 1996a, 1996b), but the OEO structure was also proposed by Neyer and Voges (1981, 1982). A primary focus of the OEO has been for low noise, low cost oscillators for the applications noted previously. In addition to being an RF oscillator, the OEO has also been demonstrated as an optical pulsed source (Yao et al. 2000). Other utilities of this novel oscillator include signal processing functionality such as clock recovery and spectral analysis (Devgan 2013).

The measured single-sideband (SSB) phase noise spectrum for an OEO is shown in Figure 10.25 (Okusaga et al. 2013). The data there are relative to a 10-GHz output from the cavity, which was about 6 km in length. The noise level below 10 kHz is quite low after which spurious

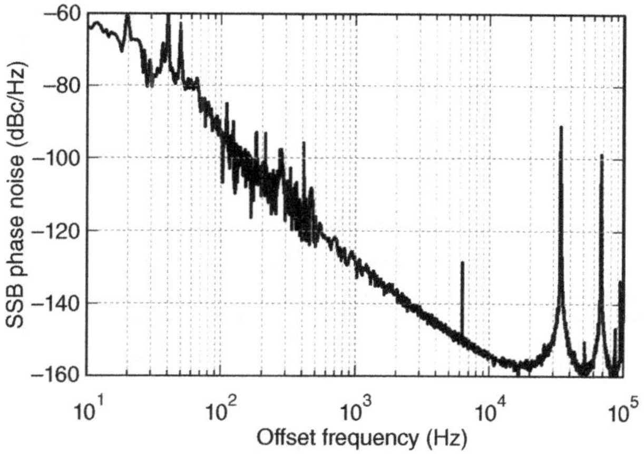

Figure 10.25. Measured single-sideband (SSB) phase noise for the output of a ~6-km OEO oscillating at 10 GHz (Okusaga et al. 2013).

cavity modes can be seen. The first spurious mode is at about 333 kHz, corresponding to the 6-km cavity length. Since its inception, there has been considerable work devoted to understanding the spectrum for an OEO including analytical studies (Yao 2002, Matsko et al. 2009, Okusaga et al. 2011a), computational models (Levy et al. 2009, 2010), and treatments of noise (Eliyahu et al. 2008, Okusaga et al. 2011b). The main tradeoff for a low noise OEO is between the Q of the cavity ($\propto \tau$) and the mode spacing ($\propto 1/\tau$), where τ is the loop delay. Numerous architectures have been employed to balance this tradeoff, all of which use some sort of filtering mechanism to obtain single-frequency operation. For example, an OEO with two loops of different lengths can be used to select a single mode (Yao and Maleki 2000). Injection locking two OEOs to a common mode has also been proven as a viable technique (Zhou and Blasche 2005). Coupling an OEO to a mode-locked laser has also been demonstrated (Yao et al. 2000, Salik et al. 2007). Noise reduction via frequency modulation has produced some of the best noise levels to date (Eliyahu et al. 2008, Okusaga et al. 2013). Small form-factor OEOs have also been implemented by coupling with whispering-gallery mode resonators (Savchenkov et al. 2010). Although other architectures have certainly been considered, this listing covers the most prevalent methods for making low noise OEOs. Examples of the SSB phase noise from such OEOs are shown in Figure 10.26. For all these data, the oscillation frequency is near 10 GHz (X-Band), a frequency often used for radar systems. Spur levels are not included in the selected data points, which are given for every decade

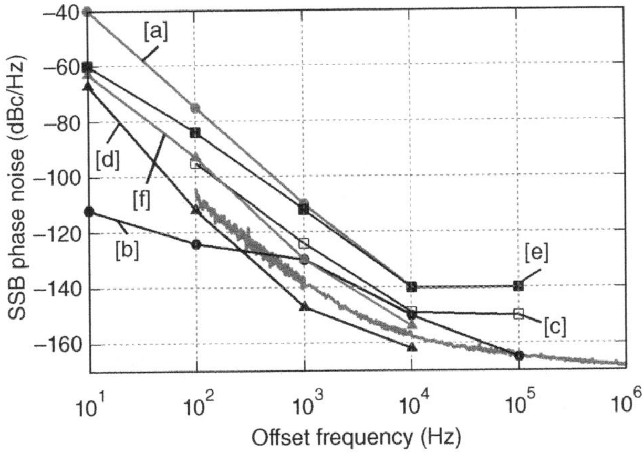

Figure 10.26. Reported results for single sideband (SSB) phase noise of various ~10-GHz OEOs. Shown are results for (a) multiloop OEO (Yao and Maleki 2000), (b) dual injection-locked OEOs (Zhou and Blasche 2005), (c) OEO coupled with a mode-locked laser (Salik et al. 2007), (d) OEO with optical frequency modulation (Eliyahu et al. 2008), (e) whispering-gallery mode based OEO (Savchenkov et al. 2010), and (f) 6-km OEO with optical frequency modulation (Okusaga et al. 2013). For comparison, a trace for a state-of-the-art electronic oscillator (PSI) is also shown.

from the reported results. A curve for a state-of-the-art electronic oscillator is also shown, demonstrating that an OEO is capable of good performance in a potentially more compact package.

Active research areas on the OEO other than those listed previously are wide ranging. Achieving the oscillation condition in an OEO without RF amplification is an important goal, with the intention to eliminate phase noise added by RF amplifiers. To this end, OEOs with all-optical gain have been demonstrated with optical amplifiers (Devgan et al. 2007, 2009) or high power lasers (Nelson et al. 2007, Zhou et al. 2008, Loh et al. 2012). Many of the OEO implementations employ intensity modulation with an external MZM as described in Chapter 6, particularly the early demonstrations. However, numerous other O/E and E/O methods have been demonstrated in an OEO including direct modulation of semiconductor lasers (Sung et al. 2009), phase modulation (Sakamoto et al. 2006, Li and Yao 2012), and polarization modulation (Pan and Yao 2009, Tang et al. 2012). Implementing more of the OEO components in the optical domain has promise for reducing the form factor. For example, intracavity optical filters for mode selection have been considered as an alternative to RF filters (Strekalov et al. 2003). The OEO becomes even more attractive for millimeter-wave applications, with demonstrations having been completed in Ka-Band near 40 GHz (Chang et al. 2002, van Dijk et al. 2008). Integrating the components in an OEO (Yao et al. 2005) is an area receiving much attention, with the promise of providing a miniature, low cost, and low noise oscillator unrivaled by other competing technology.

Photonic techniques can be employed for wideband AWG as detailed by McKinney and Weiner (2013) and the references therein. Such photonic synthesis of RF waveforms is a powerful technique, leveraging the tremendous bandwidth afforded by optical processing (Weiner 2009). A typical photonic AWG structure entails a pulsed laser source such as a mode-locked laser that is input into an optical pulse shaper followed by an O/E block to produce an electrical signal. While the O/E converter does require a high speed photodiode, a high speed E/O is not necessarily needed in a photonic AWG system. The pulsed laser source provides numerous optical frequency components that are essentially filtered by the pulse shaper to produce a tailored RF waveform. There are numerous ways to implement an optical pulse shaper; two predominant methods are the Fourier transform optical pulse shaper (Weiner 2000, 2011) and the discrete space-to-time optical pulse shaper (Leaird and Weiner 2001, McKinney et al. 2003). Two particularly interesting architectures using photonic AWG are summarized in this section: a balanced-detection polarization pulse shaping technique and a photodiode characterization apparatus based on optical combs with pulse shaping.

Consider the apparatus shown in Fig. 10.27 (McKinney 2010). In this case, approximately 100-fs pulses from a passively mode-locked fiber laser (50 MHz repetition rate, ~4.4 THz bandwidth) are dispersed in 5 km of single-mode fiber. After amplification with an EDFA, a

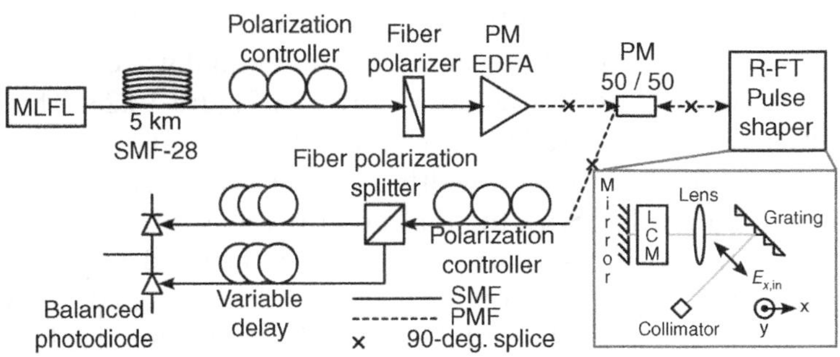

Figure 10.27. Architecture for generating background-free wideband waveforms. SMF: single-mode fiber, PMF: polarization-maintaining fiber, MLFL: mode-locked fiber laser, R-FT: reflective Fourier transform, LCM: liquid crystal modulator.

user-defined structure is impressed on the power spectrum of the short pulse via a reflective Fourier transform pulse shaper (inset, Weiner 2000). In this particular architecture, the optical pulse shaper imposes the user-defined pattern onto the polarization state of the incoming pulse using a liquid-crystal-on-silicon (LCOS) modulator array. In a manner similar to that described in Section 8.5, this polarization modulation is converted to intensity modulation via a fiber polarization splitter, and the modulated intensity is then detected with a high speed photodiode. When the incoming light is polarized at 45° relative to the liquid crystal axis and the bias phase of the modulator is set to $\phi_b = \pi/2$, the intensity transmission function of the pulse shaper can be represented by the Jones matrix:

$$M_p = \frac{1}{2}\begin{bmatrix} 1 - \sin\phi_s & 1 + \sin\phi_s \\ 1 + \sin\phi_s & 1 - \sin\phi_s \end{bmatrix}, \tag{10.1}$$

where ϕ_s is the phase function applied to the LCM array. Defining the desired temporal waveform as $s(t)$, the two orthogonal polarization states give rise to currents of the form

$$I_\pm = \frac{1}{2}w(t) \times [1 \pm s(t)]. \tag{10.2}$$

where $w(t)$ is the overall waveform envelope arising from the underlying shape of the optical power spectrum. If these currents are considered independently, it is seen that the detected signal is positive definite with the user-defined modulation superposed on a background pedestal resulting from the envelope $w(t)$ multiplying the constant term in both expressions. However, balanced detection, $I_+ - I_- = w(t) \times s(t)$, can be used to suppress the common-mode pedestal and obtain a true bipolar oscillatory waveform. In the frequency-domain, this removes undesirable low frequency content present in the individual waveforms.

As an example, consider the data shown in Figure 10.28. The waveform in this case was tailored to be a bandwidth-limited pulse exhibiting an approximately 2-GHz bandwidth at a center frequency of 5 GHz. As shown in Figure 10.28(a), the dashed gray line shows the waveform derived from the positive-polarity photodiode, the solid gray line shows the negative-polarity waveform, and the solid black curve shows the waveform achieved via balanced detection. The pedestal has been suppressed by more than an order of magnitude (~20 mV to < 2 mV near 1 ns) and the amplitude of the oscillatory waveform doubles. The RF power spectra corresponding to the time-domain waveforms are shown

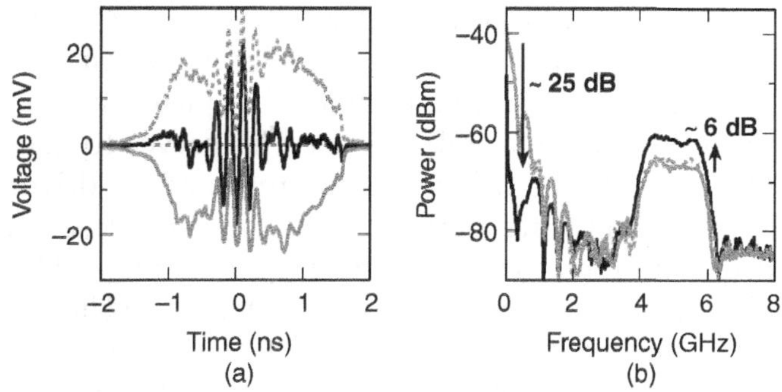

Figure 10.28. Data obtained using the apparatus shown in Figure 10.29 including (a) time-domain waveforms and (b) the associated power spectral densities in the frequency domain (McKinney 2010). The dashed gray is the output from the positive-polarity photodiode, solid gray represents the negative-polarity photodiode, and the black corresponds to balanced detection.

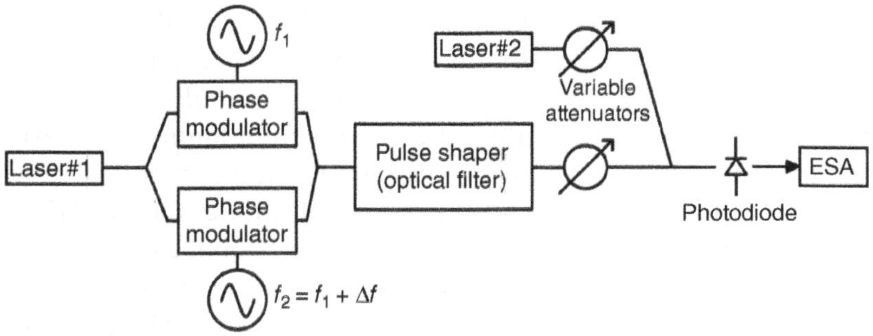

Figure 10.29. Apparatus for measuring photodiode harmonic distortion. ESA: electrical spectrum analyzer.

in Figure 10.28(b). Each single-ended output exhibits the desired spectral structure about 5 GHz but with substantial extraneous frequency content below 1 GHz arising from the waveform pedestal. This undesired frequency content is greatly suppressed in the balanced detection spectrum, where below 400 MHz the waveform exhibits greater than 25 dB reduction in undesired frequency content. Over the bandwidth of the desired signal, the balanced waveform also exhibits the expected 6 dB increase in power spectral density resulting from the doubling of the photocurrent.

As described in Section 4.4.1, architectures to quantify photodiode-generated distortions are very important for microwave photonics applications. Shown in Figure 10.29 is an apparatus based on photonic AWG (McKinney et al. 2010) that simplifies some of the difficulties associated with the methods described in Section 4.4.1. A single laser is split into two paths feeding two phase modulators, which are driven by two different frequencies with amplitudes that push the phase modulators well into their large-signal regimes. This serves to generate two optical combs offset by the frequency difference of the driving signals. Because the two combs are derived from the same laser, unwanted frequency drift is minimized without the need for stabilized heterodyne techniques. The two combs are processed with a pulse shaper, and one line is selected from each to provide a signal to the photodiode under test. Finally, a second laser spaced sufficiently away from the selected combs lines such as to avoid in-band mixing terms is used to adjust the average photocurrent/modulation depth.

The setup in Figure 10.29 was employed to characterize a commercial photodiode, the results of which are shown in Figures 10.30 and 10.31 (McKinney et al. 2010). For these measurements, the optical filter was a double-pass Fourier transform pulse shaper with about 12-GHz frequency resolution (Weiner 2000). Shown in Figure 10.30 are measured spectra at the input and the output of the pulse shaper. The combs are offset by 1.24 GHz [Figure 10.30(a)], the desired stimulus frequency for the photodiode characterization. As can be seen in Figure 10.30(b), the output spectra exhibit at least 30-dB suppression (in optical power) of undesired frequency content. The photodiode characterization results are shown in Figure 10.31 for two bias voltages, 5 V and 8 V, both with $I_{dc} = 10\,\text{mA}$. The fundamental response at 1.24 GHz was identical for both bias voltages, but the second and third harmonics at 2.48 and 3.72 GHz, respectively, varied with photodiode bias. The data

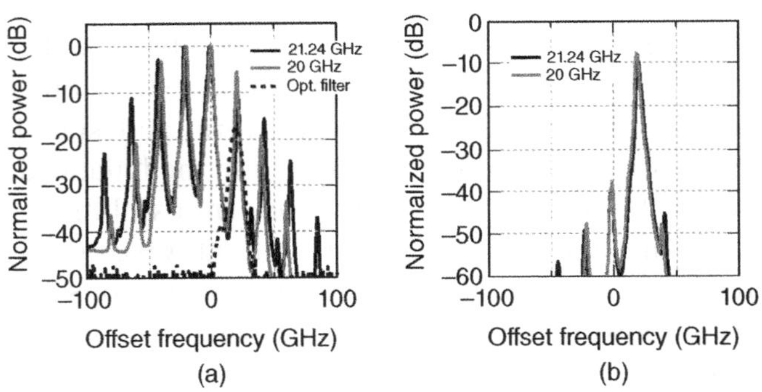

Figure 10.30. Measured optical spectra using the apparatus in Figure 10.29 showing (a) 20- and 21.24-GHz combs at the input to the pulse shaper with the pulse shaper response overlaid and (b) the output of the pulse shaper (McKinney et al. 2010). The frequency offset is relative to the unmodulated source laser.

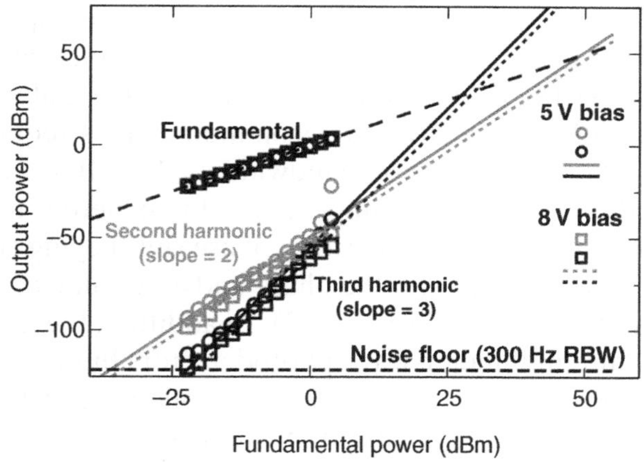

Figure 10.31. Measurement of photodiode second-harmonic (gray) and third-harmonic (black) distortion relative to a 1.24-GHz fundamental for a photodiode operating at 10-mA average current (McKinney et al. 2010). Data are shown for two photodiode bias voltages, 5 V (circles) and 8 V (squares).

shown in Figure 10.31 resulted in the following harmonic-referenced output intercept points: OIP2 = 49 dBm and OIP3 = 27 dBm at 5-V bias, and OIP2 = 53 dBm and OIP3 = 30 dBm at 8-V bias. These results were compared to those obtained with two phase-locked lasers, demonstrating excellent agreement (McKinney et al. 2010).

10.5 MILLIMETER-WAVE PHOTONICS

The precise frequency range associated with "millimeter-wave photonics" is not well defined. Some might designate it as the EHF band (30–300 GHz), whereas IEEE defines millimeter wave as 110–300 GHz (see Appendix II). While 30-GHz links are readily available, photonic architectures beyond 67 GHz are less prevalent. The biggest challenges for frequencies above 67 GHz are the electro-optic components, namely photodetectors and modulators. For reasons that will be discussed later in this section, the millimeter-wave modulator technology is presently lagging behind millimeter-wave photodetectors. Research in both areas produces ever-increasing capabilities driven by numerous important applications, which range from commercial wireless links to radio astronomy and military signal processing. This section will review these applications after discussing the photodetector and modulator components. The section concludes with a short discussion of graphene, a material that has emerged for high frequency photonics applications.

Both in terms of commercially available components and research-grade devices, the O/E conversion technology (photodetectors) surpasses the E/O conversion technology (modulators). This is largely due to a lack of a strong telecommunications requirement for links with single-channel analog bandwidths ≥100 GHz. For example, commercial 100-Gb/s links are implemented in coherent formats such as polarization-multiplexed quadrature phase shift keying (PM-QPSK) where 25 Gb/s is encoded onto the phase and amplitude of each of two polarizations (Xia and Wellbrock 2013). In addition, there are applications that require millimeter-wave photodetectors but not modulators. Two such applications are described in the following section: wireless links and radio astronomy. This trend leaves smaller-volume applications, such as military signal processing, lacking the much-needed modulator component technology. Commercial photodetectors with response into the W-Band (75–110 GHz) are available, whereas commercially packaged modulators beyond 67 GHz are just starting to emerge. Photodetectors operating across the full millimeter-wave band and into the terahertz regime have been demonstrated. For example, near-ballistic uni-traveling carrier (NBUTC) photodiodes have demonstrated DC compression currents of 37 mA at 110 GHz (Shi et al. 2010) and 17 mA at 170 GHz (Shi et al. 2012). In addition, a traveling-wave uni-traveling carrier photodetector has demonstrated a DC current of 13 mA, sourcing 10 μW at 612 GHz (Rouvalis et al. 2010). A recent survey (Seeds et al. 2013) provides additional information on photodetector performance in the millimeter wave. Electro-optic modulators are less prevalent at such operating frequencies. As described in earlier chapters, $LiNbO_3$ is widely employed for phase and MZM-based intensity modulators. Early work on millimeter-wave MZMs includes a 110-GHz Ti:$LiNbO_3$ demonstration (Noguchi et al. 1996). Phase modulators operating up to 300 GHz have also been implemented in $LiNbO_3$ (Macario et al. 2012). Other materials are receiving more attention for millimeter-wave photonics. For example, polymer-based MZMs have been shown to provide continuous electro-optic response to 110 GHz (Huang et al. 2012) and in excess of 150 GHz (Lee et al. 2002). Also being considered are compound semiconductors such as GaAs and InP (Dogru et al. 2013). As discussed in the following paragraphs, there are numerous applications of these high frequency components.

An architecture employing photonics for millimeter-wave wireless transmission is shown in Figure 10.32. Such links have been receiving increasing attention for both short-range indoor and long-range

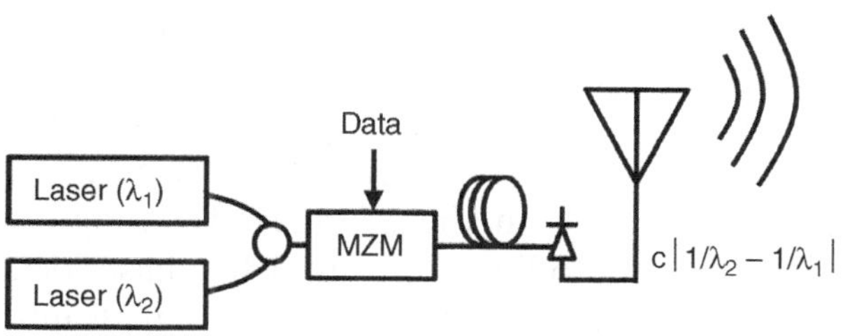

Figure 10.32. Basic block diagram of a wireless transmitter with a photonic front end, showing millimeter signal generation via optical heterodyning.

outdoor wireless applications, particularly at frequencies ≥60 GHz (Kukutsu et al. 2010, Chang and Liu 2013, Nagatsuma et al. 2013). As shown in Figure 10.32, a millimeter-wave signal can be generated optically without the need for a millimeter-wave modulator. The techniques used to this end typically employ some sort of optical heterodyne architecture (Yao 2009), such as mixing two lasers or individual lines from an optical comb source, to generate signals spanning 100 GHz–5 THz (Seeds et al. 2013). Following the photonic millimeter-wave generator is a modulator employed to encode data onto the signal; this modulator bandwidth needs to only support the data rate. The signal can then be remoted or co-located with a photodiode, which feeds an antenna for wireless transmission. The photodiode in this case must have enough bandwidth to generate the millimeter-wave signal from the optically modulated signal. While the photodiode is shown separately from the antenna in Figure 10.32, the two can be integrated as has been demonstrated in the microwave (Dorsey et al. 2010) and millimeter wave (Stohr et al. 2010). Although the heterodyne apparatus shown in Figure 10.32 is the simplest conceptually, advanced techniques described in the previous section can also be used for photonics-based millimeter-wave generation.

Millimeter-wave photonic systems have also proven useful in numerous radio-astronomy applications. The GBT link described in Section 10.1 eliminates the need for millimeter-wave modulators and photodiodes by downconverting to a microwave IF band. Large radio telescope arrays use the same concept but require that the LOs at each antenna in the array be phased locked. The Atacama large millimeter/submillimeter array (ALMA) is one such system employing photonics to accomplish this task. The ALMA was designed to produce high resolution images at received frequencies up to 942 GHz (Shillue et al. 2012), including relics of the 3 K blackbody radiation from the Big Bang (Wootten 2003). Photonics are used in the ALMA to generate and distribute LOs to each antenna site. The 27–122-GHz LOs are generated via optical heterodyning of a narrow-linewidth master laser and tunable semiconductor slave lasers (Shillue et al. 2012), which are distributed by a fiber optic network. The fiber optic links themselves carry LOs in the 27–122-GHz range, which, after demodulation with a photodiode, are used to phase-lock fixed electronic LOs to downconvert the millimeter-wave signals into 4–12-GHz IF bands (Shillue et al. 2012). The generation and/or distribution of these LO signals via an electronic approach would be difficult at best, but more likely inhibiting to the ALMA system performance. Similarly to the aforementioned wireless example, the ALMA millimeter-wave photonic system requires millimeter-wave photodiodes but not a millimeter-wave modulator, leveraging photonic generation of the LO signal.

As described in Section 10.3, the advantages of photonic-based signal processing over all-electronic techniques increase with frequency and bandwidth. Millimeter-wave signal processing, particularly for military applications, is therefore well suited for a photonic solution. However, as opposed to the previous two examples of millimeter-wave photonics, photonic millimeter-wave signal processing requires an electrical-to-optical conversion block (i.e., a millimeter-wave optical modulator). Thus, both the modulator and photodetector response must be considered for such applications. EW systems are particularly sensitive to gain ripple across their bandwidth, a detriment that is more difficult to mitigate in the millimeter wave. In addition to the novel wideband techniques described in Section 10.3, photonics offers continuous frequency coverage from below HF into the millimeter wave with relatively low ripple. Data demonstrating this point are plotted in Figure 10.33, a measured gain spectrum for a photonic link operating to 110 GHz. The roll off from low frequency to 110 GHz is about 14 dB, 11 dB attributed to the modulator (Macario et al. 2014) and 3 dB attributed to the photodiode.

Graphene is an emerging opto-electronic material for high frequency applications such as those described previously (Novoselov et al. 2004). There are numerous applications of graphene in optics, including sources, optical limiters, and optical frequency converters (Bonaccorso et al. 2010). Graphene-based modulators have been demonstrated at narrow bands near 600 GHz (Sensale-Rodriguez et al. 2012). Graphene photodetectors have already been demonstrated at 1.5 μm and exhibit promise for operation from 300 to 6000 nm (Mueller et al. 2010). In addition to a wide absorption band, graphene-based photodetectors

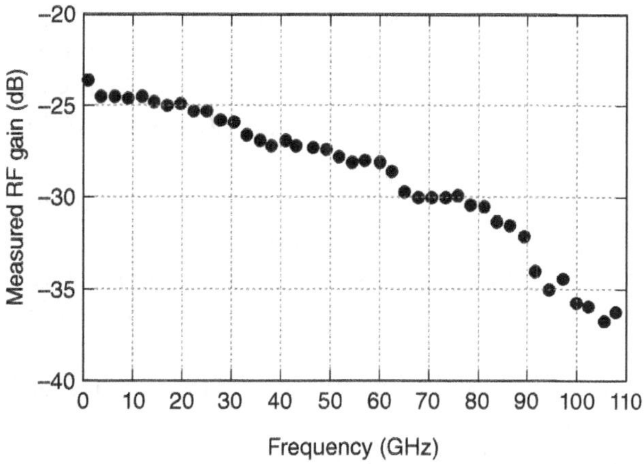

Figure 10.33. Measured response for a fiber optic link using a 110-GHz $LiNbO_3$ phase modulator (Phase Sensitive Innovations) and a 110-GHz photodiode (U2T) at I_{dc} = 4.5 mA. The symbols represent the peak response through an interferometer with a 375-ps path imbalance.

have the potential for high frequency operation. For example, the transit-time-limited bandwidth of a graphene photodetector was reported as 1.5 THz with an RC-limited bandwidth of 640 GHz (Xia et al. 2009). A big challenge of using graphene for photodetectors is their low quantum efficiency, typically resulting in responsivity being reported in milliamperes per Watt. Significant improvement in the quantum efficiency of graphene photodetectors is required for microwave photonics applications, an area that is a topic of present research.

10.6 INTEGRATED MICROWAVE PHOTONICS

Research on integrated photonics for telecommunications applications has been continually increasing over the past three decades. Just as microwave photonics leverages telecommunications-based bulk fiber optic components when possible, so it is for integrated microwave photonics. Many materials have been researched for large-scale photonic integration. Group III-V semiconductors have long been considered, particularly GaAs- and InP-based structures (Deri and Kapon 1991, Koch and Koren, 1991). Group IV materials such as Si and Ge also offer strong potential both in the standard telecommunications bands and in the mid-infrared (Soref 2010). Compound glasses offer low loss as has been demonstrated in numerous microring resonators architectures (Little and Chu 2000). Other compounds have also been studied such as ZnO and GaN for photonic nanowires (Yan et al. 2009). Given all the possibilities, two leading platforms for large-scale optical integration are Si- and InP-based photonic integrated circuits (PICs).

There are mixed opinions among experts as to which substrate, Si or InP, is most suitable for photonic integration. Indium phosphide is presently more attractive for lasers (Coldren et al. 2012) and active PICs (Coldren et al. 2011). The telecommunication industry presently employs more InP-based PICs than those on Si substrates. For example, Infinera Corporation deployed InP-based PICs as early as 2004 (Welch et al. 2007), which have stood the test of time (Nagarajan et al. 2010), and has demonstrated InP-based PICs capable of 1 Tb/s rates (Evans et al. 2011). The high speed photodiode manufacturer u2t Photonics also appears to consider InP-based PICs as the path ahead (Bottacchi et al. 2010). However, some experts believe that Si-based photonic integration is the more viable approach in the long run (Liang and Bowers 2009). Silicon-based photonics has distinct advantages over InP in several important areas such as low loss waveguides for routing and interconnects, an incredible infrastructure for processing (Orcutt et al. 2011), and compatibility with electronic integrated circuits. The remainder of this section concentrates on Si as a substrate for integrated microwave photonics, noting that hybrid integration may ultimately be the best solution.

Silicon photonics has been an area of active research since the mid-1980s (Reed et al. 2005) and a field receiving widespread attention since the turn of the century (Soref 2006). Recent textbooks (e.g., Deen and Basu 2012) and handbooks (e.g., Vivien and Pavesi 2013) are a testament to the maturity of silicon photonics as a research topic. Multifunctional silicon photonic chips are promising for a host of applications; driving application areas are large-scale industrial telecommunications (Jalali and Fatpour 2006), biomedicine (Fathpour and Jalali 2012), and optical chips for high performance computing (Taubenblatt 2012). Indeed, silicon photonics offers many promising features for microwave photonics applications as well (Capmany et al. 2013b, Marpaung et al. 2013). Silicon waveguides and active photonic devices have the potential to provide many of the advantages of bulk fiber optic systems for RF applications, including high bandwidth and efficiency. There are added benefits afforded by silicon as opposed to bulk fiber optics such as decreased size, better scalability in multichannel applications, environmental stability in optically coherent systems,

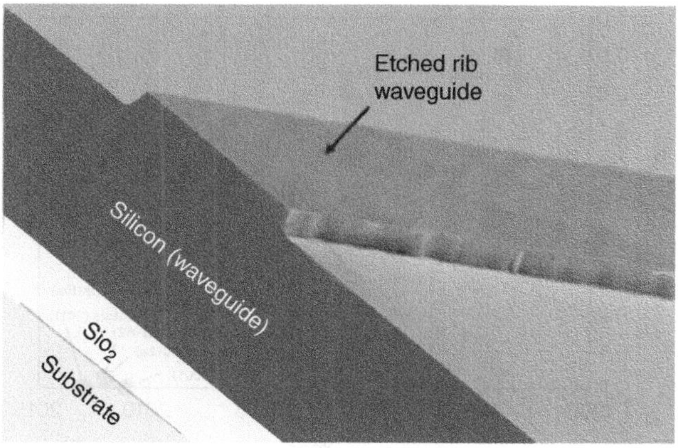

Figure 10.34. Scanning-electron micrograph (SEM) of a cleaved silicon waveguide (Pruessner et al. 2013). The rib waveguide is just under 5 μm across.

ease of manufacturability, and the potential for integration with other chip-level components. However, there remain issues to be addressed with silicon photonics, some of which are particular to microwave photonics applications.

A scanning-electron micrograph (SEM) of a cleaved silicon waveguide is shown in Figure 10.34 (Pruessner et al. 2013). Such structures have enormous integration potential, whether used as interconnects between multifunctional chips or as photonic delay lines. However, there are some issues that inhibit the immediate transition of silicon photonics technologies. Both the propagation loss and on/off chip coupling losses are issues that must be considered. High optical launch powers can offset large losses, but the limited linearity and power handling in single-mode silicon waveguides can inhibit link performance. The polarization sensitivity of silicon waveguides can also be a concern, particularly in nano-photonic waveguides but less so in lager rib waveguides (Figure 10.34). Multifunction monolithic integration is desirable, which requires all necessary components such as waveguides, couplers, switches, filters, sources, modulators, and detectors in a single material. The last three are critical but also particularly difficult to achieve in silicon all at the same optical wavelength. Ultimately, hybrid integration may provide a solution using silicon, which can incorporate discrete optical devices and/or other materials. In this case, the compatibility of silicon with the other platforms must be considered. The remainder of this section will review the progress in addressing the issues listed previously.

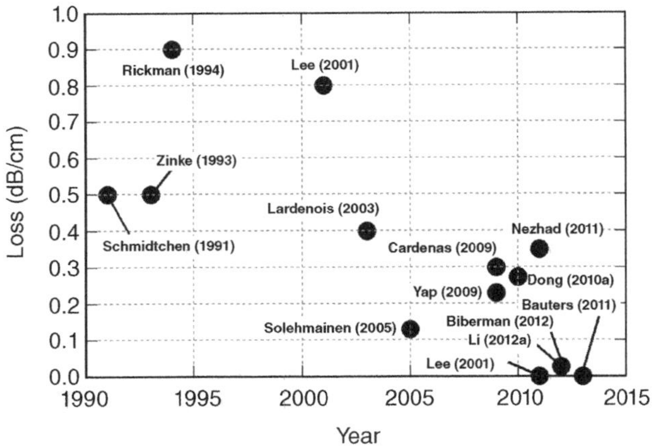

Figure 10.35. Reported attenuation coefficients for various silicon-based waveguides. (Bauters et al., 2011; Biberman et al., 2012; Cardenas et al., 2009; Dong et al., 2010a; Lardenois et al., 2003; Lee et al., 2012; Lee et al., 2001; Li et al., 2012; Nezhad et al., 2011; Rickman et al., 1994; Schmidtchen et al., 1991; Solehmainen et al., 2005; Yap et al., 2009; Zinke et al., 1993.)

While the optical attenuation coefficients in fiber are measured in decibels per kilometer, those for silicon waveguides are reported in decibels per centimeter and more recently in decibels per meter. For example, a typical propagation loss in single-mode fiber is 0.2 dB/km (0.04 dB/μs) as opposed to losses in silicon that are typically on the order of 0.2 dB/cm. The corresponding loss per unit delay is 2 dB/ns for silicon having a refractive index of 3.4 at room temperature. Therefore, while optical fiber is suitable for microsecond-class delay line signal processing, present silicon waveguide technology is restricted to nanosecond time scales. Shown in Figure 10.35 is a sampling of reported attenuation coefficients for various types of silicon-based waveguides. As noted by Vlasov and McNab (2004), it is difficult to construct a general comparative plot of this sort because the losses depend on many waveguide-specific parameters. Such parameters include waveguide core material (e.g., silicon, silicon nitride, or silicon dioxide), core thickness, and waveguide type (single- or multi-mode), to name a few. It must also be noted that the survey in Figure 10.35 includes many different waveguide technologies such as nano-photonic silicon waveguides, larger silicon rib waveguides, and low loss silica waveguides. Nonetheless, a trend of improving performance is evident with present reported values for silicon-based integrated waveguides down to fractions of 1 dB/m, as compared to early results of 15 dB/cm (Soref and Lorenzo 1986).

The three main sources of light attenuation in silicon waveguides are absorption, scattering, and radiation losses [see for example Chapter

10 in Deen and Basu (2012)]. Interband absorption occurs when photon energies are larger than the direct band gap, 4.2 eV for silicon. Therefore, longer wavelengths such as the telecommunication bands do not experience interband absorption in silicon. However, intraband free-carrier absorption (FCA) can cause loss at these wavelengths, which depends strongly on the free-carrier concentration (Soref and Bennett 1987). Scattering can occur due to impurities in the waveguide or roughness at the waveguide boundaries, requiring very pure and precise fabrication processes for low loss waveguides. Radiation losses are important in waveguide bends, such as for routing on chips or in spiral delay lines. Large input optical powers can obviously offset losses, but waveguide nonlinearities and a push for efficient circuits inhibit such a brute-force solution.

As in the case of analog fiber optics, it is desirable to maximize the received optical power in silicon photonics to achieve good RF link performance. However, the optical field in single-mode silicon waveguides is confined to a smaller area than that in optical fiber. In addition, the Kerr coefficient and Raman gain coefficient are about 200 and 1000 times larger in silicon, respectively, than silica glass (Boyraz et al. 2013). For example, $n_2 \approx 5 \times 10^{-18}\,\mathrm{m^2/W}$ for bulk silicon at 1550 nm (Bristow et al. 2007) as compared to the value cited for optical fiber in Chapter 5, $n_2 = 2.6 \times 10^{-20}\,\mathrm{m^2/W}$ (Antona et al. 2001). Two-photon absorption (TPA) also plays a role in the nonlinearity of silicon waveguides. The TPA process involves the simultaneous absorption of two photons whose combined energy exceeds the minimum indirect band gap, about 1.12 eV for silicon (Boyraz et al. 2013). In fact, TPA and TPA-induced FCA have been found to occur at lower powers than other nonlinear effects such as XPM and FWM. Thus, TPA and TPA-induced FCA can be primary causes of interchannel crosstalk in silicon waveguides (Okawachi et al. 2012). Although these effects are a detriment to signal-transport applications, the same nonlinear processes can potentially be harnessed for integrated microwave photonic signal processing (Marpaung and Eggleton 2013).

Although a monolithic chip set is the ideal scenario for PICs, off-chip coupling to optical fiber will most likely be needed in many applications of silicon photonics requiring long delays. For example, optical fiber is better suited for long-delay signal processing and high dynamic range multichannel systems because of the lower propagation loss and higher linearity as compared to silicon waveguides. Similarly, long-haul point-to-point links could benefit from PICs on the front- or back-end but would continue to use optical fiber for the transmission span. The primary problem with coupling from optical fiber to silicon waveguides is the disparity in the optical mode field diameters. A typical optical fiber core has an 8-μm diameter with a refractive index of 1.4, whereas silicon waveguides are considerably smaller (on the order of a few microns in diameter) and have a larger index (3.4). There are numerous approaches to making the transition from fiber to silicon with low optical loss, such as diffraction gratings, multimode interference, multipath interference, and resonant coupling (Bogaerts and Vermeulen 2013, Zaoui et al. 2014). Perhaps, the most straightforward technique is a mode conversion via an adiabatic taper where edge coupling can be used (Moerman et al. 1997). The main challenge with an adiabatic taper is coupling a symmetric fiber optic waveguide to a planar (asymmetric) silicon waveguide. However, fiber-to-chip coupling losses below 1 dB have been achieved using tapered couplers (Soganci et al. 2013). Intelligent and creative use of the available components in silicon can be used to minimize the number of on/off-chip transitions, thus controlling the associated losses and complexities.

Fully functional integrated optical circuits for microwave photonics require a considerable number of components or building blocks, many of which are presently available in silicon. One particularly important component is a polarization beam splitter (Dai et al. 2012) that can be used in a polarization diversity circuit (Fukuda et al. 2008) to mitigate PDL and PMD in silicon waveguides. Active components are also realizable in silicon, such as those employing the thermo-optic effect to achieve phase shifts. Polarization-diversity coherent receivers (Doerr et al. 2010) use a combination of these components. A phase shifter with two couplers, often multimode interference (MMI) couplers, can be arranged to construct an MZI in silicon. An MZI can be a very powerful tool, serving as a building block for optical phase demodulators (Voigt et al. 2008), optical attenuators, switches (Chen et al. 2012), and isolators (Ghosh et al. 2012). An MZI can be also be used as a modulator, with silicon ring-assisted MZIs (RAMZIs) capable of wideband operation (Gill et al. 2010).

Optical filters with RF-scale bandwidths are critical for microwave photonic signal processing. The possibility of chip-scale integration of RF filters can allow for a wide range of new applications. Low loss ring resonators make very narrow bandpass filters, especially when cascaded into high order resonators (Little et al. 1997). Such filters in silicon have been the topic of programs funded by the Defense Advanced Research Projects Agency (DARPA), achieving narrowband notch filters (Rasras et al. 2009), 1-GHz passband filters with 50 dB extinction (Dong et al. 2010b), and reconfigurable filters with sub-gigahertz bandwidth

(Alipour et al. 2011). Linear-cavity Fabry-Perot filters have also been constructed in silicon (Pruessner et al. 2009), with single-cavity devices demonstrating nearly 1 GHz bandwidths with large free spectral ranges approaching 100 GHz. Such filters can be used in unique ways, such as for add-drop filters (Pruessner et al. 2011) and cascaded microcavity filters capable of single-gigahertz bandwidths with terahertz-class FSRs (Pruessner et al. 2013). Given the availability of these components listed previously, it is important to consider monolithic integration in silicon. However, there are critical components that must be examined in this context, namely lasers, modulators, and photodiodes in silicon.

A coherent light source in silicon (for lasers and optical amplifiers) is a critical, but difficult to achieve, component. Stimulated emission is difficult to achieve in silicon owing to its indirect band structure. A tabulation of the history of silicon-based light sources is provided by Anopchenko et al. (2013). The recent concentrations have been categorized by both Liang and Bowers (2010) and Fang et al. (2013) into four areas: stimulated Raman scattering (SRS), rare-earth dopants in silicon, epitaxial growth or wafer bonding of other materials on silicon, and other hybrid approaches. The SRS process, described in Chapters 3 and 5, is attractive for light generation in silicon because of the tight mode confinement in silicon waveguides and the large Raman gain. For example, the Raman gain coefficient for a typical silicon waveguide is $g_R = 7.6 \times 10^{-10}\,\text{m/W}$ with a 1.4 μm pump having an associated frequency shift of 15.6 THz for 1.5 μm (Claps et al. 2002); the gain for standard optical fiber is much smaller at $g_R = 6.5 \times 10^{-14}\,\text{m/W}$ (see Chapter 5). The main issues for achieving efficient SRS in silicon are mitigating losses due to FCA and TPA. Nonetheless, SRS-based silicon lasers have been demonstrated, with the first CW operation reported by Rong et al. (2005). Given the success of rare-earth doping in optical fiber, particularly with erbium and ytterbium, rare-earth-doped silicon lasers and amplifiers are obvious light sources to consider. However, silicon is a poor host for erbium, but this remains an active research topic (Fang et al. 2013). There has been much consideration given to more attractive laser materials being epitaxially grown on silicon. Numerous III-V compound semiconductors on silicon have been researched, particularly GaAs and InP (Fang et al. 2013). Germanium is also a good candidate with the first germanium-on-silicon laser demonstrated by Liu et al. (2010). Finally, the hybrid approach to integrating lasers with silicon (Heck et al. 2013) involves bonding a variety of materials to silicon-on-insulator (SOI) substrates, as opposed to epitaxial growth of a laser medium on silicon.

Numerous physical effects can be employed to create an optical modulator, many of which have been considered for silicon (Marris-Morini et al. 2013). The Kerr effect, Franz–Keldysh effect, and Pockels effect are weak in silicon at 1.5 μm (Reed et al. 2010) and therefore not widely employed for silicon modulators. The thermo-optic effect, however, is strong in silicon (Cocorullo and Rendina 1992) but does not support high modulation frequencies. The main concentration for high speed modulators in silicon is based on the free carrier plasma dispersion (FCPD) effect, where changes in the carrier concentrations induce a change in the index of refraction or absorption. Typically, the FCPD effect is used to create a phase modulation, which can be converted to intensity modulation in an MZM (Dong et al. 2012, Gutierrez et al. 2013) or a micro-ring resonator (Xu et al. 2005). The former has been used to achieve silicon modulators with analog bandwidths up to 40 GHz (Thomson et al. 2012). As with lasers, hybrid approaches are also considered for modulators integrated with silicon. For example, GeSi electro-absorption modulators have been demonstrated leveraging an enhanced Franz-Keldysh effect in tensile-strained germanium (Liu et al. 2008). In addition, widely used $LiNbO_3$ has also been considered for integration with silicon for electro-optic modulators (Rabiei et al. 2013).

A 1.5-μm wavelength photodiode with silicon as the absorption medium is not easily obtained because the long-wavelength cutoff for silicon is 1.1 μm (1.12 eV). Migrating other components to 1 μm may not be a solution for silicon photonics because the waveguide losses would be large. There are some techniques that can be used to enhance the responsivity of silicon for 1.5-μm applications, including the use of mid-band-gap states and TPA (Geis et al. 2013). As in the case with lasers and modulators, heterogeneous integration is recognized as a feasible path to a silicon-compatible photodiode. Such hybrid devices have been demonstrated with 30 GHz of analog bandwidth. Representative technologies include Ge p–i–n structures coupled with SOI (Koester et al. 2006, Feng et al. 2009) and InP-based modified uni-traveling carrier photodiodes on SOI (Beling et al. 2013).

Integrated photonics will almost certainly continue to be a hot topic along with the subset of associated research in microwave photonics. Buzzwords have already emerged to describe work in this area, such as integrated microwave photonics (iMWP) and microwave photonic integrated circuits (MWPIC). Hybrid integration (Heck and Bowers 2013) approaches will most likely provide more possible solutions than monolithic efforts using only one material, with a heterogeneous silicon platform (Fish 2013) being a promising solution. Of particular interest is integration of silicon with III-V materials (Piels et al. 2014), such as InP (Coldren et al. 2011, Binetti et al. 2012). Integration of graphene (see the end of Section 10.5) with silicon is also an area of active research (Liu and Zhang 2013) with photodiodes coupled to SOI already demonstrated at 1.5 μm (Gan et al. 2013). Lattice mismatch and largely different thermal expansion coefficients present issues for hybrid integration. As mentioned previously, interfaces between on-chip waveguides and optical fiber can also be a concern. The level of effort presently being directed at these and other issues will most likely result in deployable integrated microwave photonic circuits in the next decade. Such circuits may replace or augment the present microwave photonics technologies and could enable new capabilities in many of the application areas covered in this book.

REFERENCES

Agarwal, A., Banwell, T., and Woodward, T. K., "Optically filtered microwave photonic links for RF signal processing applications," *Journal of Lightwave Technology*, 29(16), 2394–2401 (2011).

Alferness, R. C., "The evolution of configurable wavelength multiplexed optical networks—a historical perspective," *Proceedings of the IEEE*, 100(5), 1023–1034 (2012).

Alipour, P., Eftekhar, A. A., Atabaki, A. H., Li, Q., Yegnanarayanan, S., Madsen, C. K., and Adibi, A., "Fully reconfigurable compact RF photonic filters using a high-Q silicon microdisk resonators," *Optics Express*, 19(17), 15899–15907 (2011).

Anopchenko, A., Prokofiev, A., Yassievich, I. R., Ossicini, S., Tsybeskov, L., Lockwood, D. J., Saeed, S., Gregorkiewicz, T., Wojdak, M., Liu, J., and Meldrum, A., "Silicon-based light sources," in *Handbook of Silicon Photonics*, L. Vivien and L. Pavesi (editors), CRC Press (2013).

Antona, J. C., Bigo, S., and Kosmalski, S., "Nonlinear index measurements of various fibre types over C+L bands using four-wave mixing," in Proceedings of the 27th European Conf. on Optical Communications, 270–271 (2001).

Bauters, J. F., Heck, M. J. R., John, D. D., Barton, J. S., Blumenthal, D., Bowers, J. E., Bruinink, C. M., Leinse, A., and Heideman, R. G., "Ultra-low-loss (<0.1 dB/m) planar silica waveguide technology," in IEEE Photonics Conference Digest, 1–3 (2011).

Beling, A., Cross, A. S., Piels, M., Peters, J., Zhou, Q., Bowers, J. E., and Campbell, J. C., "InP-based waveguide photodiodes heterogeneously integrated on silicon-on-insulator for photonic microwave generation," *Optics Express*, 21(22), 25901–25906 (2013).

Biberman, A., Shaw, M. J., Timurdogan, E., Wright, J. B., and Watts, M. R., "Ultralow-loss silicon ring resonators," *Optics Letters*, 37(20), 4236–4238 (2012).

Binetti, P. R. A., Lu, M., Norberg, E. J., Guzzon, R. S., Parker, J. S., Sivananthan, A., Bhardwaj, A., Johansson, L. A., Rodwell, M. J., and Coldren, L. A., "Indium phosphide photonic integrated circuits for coherent optical links," *IEEE Journal of Quantum Electronics*, 48(2), 279–291 (2012).

Bogaerts, W. and Vermeulen, D., "Off-chip coupling," in *Handbook of Silicon Photonics*, L. Vivien and L. Pavesi (editors), pp. 97–138, CRC Press (2013).

Bonaccorso, F., Sun, Z., Hasan, T., and Ferrari, A. C., "Graphene photonics and optoelectronics," *Nature Photonics*, 4, 611–622 (2010).

Bottacchi, S., Beling, A., Matiss, A., Nielsen, M. L., Steffan, A. G., Unterborsch, G., and Umbach, A., "Advanced photoreceivers for high-speed optical fiber transmission systems," *IEEE Journal of Selected Topics in Quantum Electronics*, 16(5), 1099–1112 (2010).

Boyraz, O., Sang, X., Cazzanelli, M., and Huang, Y., "Nonlinear optics in silicon," in *Handbook of Silicon Photonics*, L. Vivien and L. Pavesi (editors), pp. 197–248, CRC Press (2013).

Bres, C. S., Zlatanovic, S., Wiberg, A. O. J., Adleman, J. R., Huynh, C. K., Jacobs, E. W., Kvavle, J. M., and Radic, S., "Parametric photonic channelized RF receiver," *IEEE Photonics Technology Letters*, 23(6), 344–346 (2011).

Bristow, A. D., Rotenberg, N., and van Driel, H. M., "Two-photon absorption and Kerr coefficients of silicon for 850–2200 nm," *Applied Physics Letters*, 90(191104), 1–3 (2007).

Capmany, J., Ortega, B., Pastor, D., and Sales, S., "Discrete-time optical processing of microwave signals," *Journal of Lightwave Technology*, 23(2), 702–723 (2005).

Capmany, J., Ortega, B., and Pastor, D., "A tutorial on microwave photonic filters," *Journal of Lightwave Technology*, 24(1), 201–229 (2006).

Capmany, J., Li, G., Lim, C., and Yao, J., "Microwave photonics: current challenges towards widespread application," *Optics Express*, 21(19), 22862–22867 (2013a).

Capmany, J., More, J., Gasulla, I., Sanco, J., Lloet, J., Sales, S., "Microwave photonic signal processing," *Journal of Lightwave Technology*, 31(4), 571–586 (2013b).

Cardenas, J., Poitras, C. B., Robinson, J. T., Preston, K., Chen, L., and Lipson, M., "Low loss etchless silicon photonic waveguides," in Conference on Lasers and Electro-Optics Digest, paper CThU6 (2009).

Chan, E. H. W., "Noise investigation of a large free spectral range high-resolution microwave photonic signal processor," *IEEE Photonics Journal*, 5(6), 5502109 (2013).

Chan, E. H. W., Zhang, W. W., and Minasian, R. A., "Photonic RF phase shifter based on optical carrier and RF modulation sidebands amplitude and phase control," *Journal of Lightwave Technology*, 30(23), 3672–3678 (2012).

Chang, G. K. and Liu, C., "1-100 GHz microwave photonic link technologies for next-generation WiFi and 5G wireless communications," in IEEE Topical Meeting on Microwave Photonics Digest, 5–8 (2013).

Chang, D. H., Fetterman, H. R., Erlig, H., Zhang, H., Oh, M. C., Zhang, C., and Steier, W. H., "39-GHz optoelectronic oscillator using broad-band polymer electrooptic modulator," *IEEE Photonics Technology Letters*, 14(2), 191–193 (2002).

Chen, W., Wang, W., Gong, Z., Zhou, Q., Jiang, X., Yang, J., and Zhou, H., "Mach-Zehnder-based four-port switching module on SOI," *IEEE Photonics Technology Letters*, 24(15), 1313–1315 (2012).

Chin, S., Thevenaz, L., Sancho, J., Sales, S., Capmany, J., Berger, P., Bourderionnet, J., and Dolfi, D., "Broadband true time delay for microwave signal processing, using slow light based on stimulated Brillouin scattering in optical fibers," *Optics Express*, 18(21), 22599–22613 (2010).

Claps, R., Diitropoulos, D., Han, Y., and Jalali, B., "Observation of Raman emission in silicon waveguides at 1.54 μm," *Optics Express*, 10(22), 1305–1313 (2002).

Clark, T. R. Jr., O'Connor, S. R., and Dennis, M. L., "A phase-modulation I/Q demodulation microwave-to-digital photonic link," *IEEE Transactions on Microwave Theory and Techniques*, 58(11), 3039–3058 (2010).

Clark, T. R. Jr., Callahan, P. T., and Dennis, M. L., "Wideband photonic compressive sampling analog-to-digital converter: architecture, device requirements and applications," *Proceedings of SPIE*, 8259 (2012).

Cocorullo, G. and Rendina, I. "Thermo-optical modulation at 1.5 μm in silicon etalon," *Electronics Letters*, 28(1), 83–85 (1992).

Coldren, L. A., Nicholes, S. C., Johansson, L., Ristic, S., Guzzon, R. S., Norberg, E. J., and Krishnamachari, U., "High performance InP-based photonic ICs—a tutorial," *Journal of Lightwave Technology*, 29(4), 554–570 (2011).

Coldren, L. A., Corzine, S. W., and Masanovic, M. L., *Diode Lasers and Photonic Integrated Circuits*, 2nd edition, Wiley (2012).

Coward, J. F., Yee, T. K., Chalfant, C. H., and Chang, P. H., "A photonic integrated-optic RF phase shifter for phased array antenna beam-forming applications," *Journal of Lightwave Technology*, 11(12), 2201–2205 (1993).

Cox, C. and Ackerman, E., "Bi-directional signal interface," U.S. Patent No. 7,555,219 (2009).

Cox, C. and Ackerman, E., "Bi-directional signal interface with enhanced isolation," U.S. Patent No. 8,433,163 (2013).

Dai, Y. and Yao, J., "Nonuniformly spaced photonic microwave delay-line filters and applications," *IEEE Transactions on Microwave Theory and Techniques*, 58(11), 3279–3289 (2010).

Dai, J., Xu, K., Sun, X., Niu, J., Lv, Q., Wu, J., Hong, X., Li, W., and Lin, J., "A simple photonic-assisted microwave frequency measurement system based on MZI with tunable measurement range and high resolution," *IEEE Photonics Technology Letters*, 22(15), 1162–1164 (2010).

Dai, D., Wang, Z., Peters, J., and Bowers, J. E., "Compact polarization beam splitter using an asymmetrical Mach-Zehnder interferometer based on silicon-on-insulator waveguides," *IEEE Photonics Technology Letters*, 24(8), 673–675 (2012).

Daryoush, A. S., "RF and microwave photonics in biomedical applications," in *Microwave Photonics Devices and Applications*, S. Iezekiel (editor), pp. 239–289, Wiley (2009).

Deen, M. J. and Basu, P. K., *Silicon Photonics*, Wiley (2012).

Deri, R. J. and Kapon, E., "Low-loss III-V semiconductor optical waveguides," *IEEE Journal of Quantum Electronics*, 27(3), 626–640 (1991).

Devgan, P., "A review of optoelectronic oscillators for high speed signal processing applications," *ISRN Electronics*, 401969, 1–16 (2013).

Devgan, P. S., Urick, V. J., McKinney, J. D., and Williams, K. J., "A low-jitter master–slave optoelectronic oscillator employing all-photonic gain," in IEEE Topical Meeting on Microwave Photonics Digest, 70–73 (2007).

Devgan, P. S., Urick, V. J., Diehl, J. D., and Williams, K. J., "Improvement in the phase noise of a 10 GHz optoelectronic oscillator using all-photonic gain," *Journal of Lightwave Technology*, 27(15), 3189–3193 (2009).

van Dijk, F., Enard, A., Buet, X., Lelarge, F., and Duan, G. H., "Phase noise reduction in a quantum dash mode-locked laser in a millimeter-wave coupled opto-electronic oscillator," *Journal of Lightwave Technology*, 26(15), 2789–2794 (2008).

Doerr, C. R., Winzer, P. J., Chen, Y. K., Chandrasekhar, S., Rasras, M. S., Chen. L., Liow, T. Y., Ang, K. W., and Lo, G. Q., "Monolithic polarization and phase diversity coherent receiver in silicon," *Journal of Lightwave Technology*, 28(4), 520–525 (2010).

Dogru, S., Shin, J. H., and Dagli, N., "Traveling wave electrodes for substrate removed electro-optic modulators with buried doped semiconductor electrodes," *IEEE Journal of Quantum Electronics*, 49(7), 599–606 (2013).

Dolfi, D., Michel-Gabriel, F., Bann, S., and Huignard, J. P., "Two-dimensional optical architecture for time-delay beam forming in a phased-array antenna," *Optics Letters*, 16(4), 255–257 (1991).

Dong, P., Qian, W., Liao, S., Liang, H., Kung, C., Feng, N., Shafiiha, R., Fong, J., Feng, D., Krishnamoorthy, A. V., and Asghari, M., "Low loss shallow-ridge silicon waveguides," *Optics Express*, 18(14), 14474–14479 (2010a).

Dong, P., Feng, N., Feng, D., Qian, W., Liang, H., Lee, D. C., Luff, B. J., Banwell, T., Agarwal, A., Toliver P., Menendez, R., Woodward, T. K., and Asghari, M., "GHz-bandwidth optical filters based on high-order silicon ring resonators," *Optics Express*, 18(23), 23784–23789 (2010b).

Dong, P., Chen, L., and Chen, Y., "High-speed low-voltage single-drive push-pull silicon Mach-Zehnder modulators," *Optics Express*, 20(6), 6163–6169 (2012).

Dorsey, W. M., McDermitt, C. S., Bucholtz, F., and Parent, M. G., "Design and performance of frequency selective surface with integrated photodiodes for photonic calibration of phased array antennas," *IEEE Transactions on Microwave Theory and Techniques*, 58(8), 2588–2593 (2010).

Draa, M. N., Hastings, A. S., and Williams, K. J., "Comparison of photodiode nonlinearity measurement systems," *Optics Express*, 19(13), 12635–12645 (2011).

Draa, M. N., Urick, V. J., and Williams, K. J., "Analysis of photonic phase-shifting technique employing amplitude-controlled fiber-optic delay lines," NRL Memorandum Report, NRL/MR/5650-12-9376 (2012).

Eliyahu, D., Seidel, D., and Maleki, L., "RF amplitude and phase-noise reduction of an optical link and an opto-electronic oscillator," *IEEE Transactions on Microwave Theory and Techniques*, 56(2), 449–456 (2008).

Esman, R. D., Monsma, M. J., Dexter, J. L., and Cooper, D. G., "Microwave true time-delay modulator using fibre-optic dispersion," *Electronics Letters*, 28(20), 1905–1907 (1992).

Esman, R. D., Frankel, M. Y., Dexter, J. L., Goldberg, L., Parent, M. G., Stilwell, D., and Cooper, D. G., "Fiber-optic prism true time-delay antenna feed," *IEEE Photonics Technology Letters*, 5(11), 1347–1349 (1993).

Evans, P., Fisher, M., Malendevich, R., James, A., Studenkov, P., Goldfarb, G., Vallaitis, T., Kato, M., Samra, P., Corzine, S., Strzelecka, E., Salvatore, R., Sedgwick, F., Kuntz, M., Lal, V., Lambert, D., Dentai, A., Pavinski, D., Zhang, J., Behnia, B., Bostak, J., Dominic, V., Nilsson, A., Taylor, B., Rahn, J., Sanders, S., Sun, H., Wu, K. T., Pleumeekers, J., Muthiah, R., Missey, M., Schneider, R., Stewart, J., Reffle, M., Butrie, T., Nagarajan, R., Joyner, C., Ziari, M., Kish, F., and Welch, D., "Multi-channel coherent PM-QPSK InP transmitter photonic integrated circuit (PIC) operating at 112 Gb/s per wavelength," in Optical Fiber Communication Digest, PDPC7 (2011).

Fandino, J. S. and Munoz, P., "Instantaneous frequency measurement system by means of a complementary optical filter monolithically integrated in InP," in IEEE International Manufacturing on Microwave Photonics Digest, 290–293 (2013).

Fang, Z., Chen, Q. Y., and Zhao, C. Z., "A review of recent progress in lasers on silicon," *Optics & Laser Technology*, 46, 103–110 (2013).

Fathpour, S. and Jalali, B. (editors), *Silicon Photonics for Telecommunications and Biomedicine*, CRC Press (2012).

Fathpour, S. and Riza, N. A., "Silicon-photonics-based wideband radar beamforming: basic design," *Optical Engineering*, 49(1), 018201 (2010).

Feng, D., Liao, S., Dong, P., Feng, N., Liang, H., Zheng, D., Kung, C., Fong, J., Shafiiha, R., Cunningham, J., Krishnamoorthy, A. V., and Asghari, M., "High-speed Ge photodetector monolithically integrated with large cross-section silicon-on-insulator waveguide," *Applied Physics Letters*, 95, 261105 (2009).

Fish, G., "Heterogeneous photonic integration for microwave photonic applications," in OFC/NFOCE Technical Digest, paper OW3D (2013).

Frankel, M. Y., Matthews, P. J., Esman, R. D., and Goldberg, L., "Practical optical beamforming networks," *Optical and Quantum Electronics*, 30(11–12), 1033–1050 (1998).

Frigyes, I. and Seeds, A. J., "Optically generated true-time delay in phased-array antennas," *IEEE Transactions on Microwave Theory and Techniques*, 43(9), 2378–2386 (1995).

Fukuda, H., Yamada, K., Tsuchizawa, T., Watanabe, T., Shinojima, H., and Itabashi, S., "Silicon photonic circuit with polarization diversity," *Optics Express*, 16(7), 4872–4880 (2008).

Gan, X., Shiue, R. J., Gao, Y., Meric, I., Heinz, T. F., Shepard, K., Hone, J., Assefa, S., and Englund, D., "Chip-integrated ultrafast graphene photodetector with high responsivity," *Nature Photonics*, 7(11), 883–887 (2013).

Gasulla, I. and Capmany, J., "Microwave photonic applications of multicore fibers," *IEEE Photonics Journal*, 4(3), 877–888 (2012).

Geis, M. W., Spector, S. J., Grein, M. E., Yoon, J. U., and Lyszczarz, T. M., "Near-infrared silicon photodetectors," in *Handbook of Silicon Photonics*, L. Vivien and L. Pavesi (editors), 518–535, CRC Press (2013).

Ghosh, S., Keyvaninia, S., Shoji, Y., Roy, W. V., Mizumoto, T., Roelkens, G., and Baets, R. G., "Compact Mach-Zehnder interferometer Ce:YIG/SOI optical isolators," *IEEE Photonics Technology Letters*, 24(18), 1653–1655 (2012).

Gill, D. M., Patel, S. S., Rasras, M., Tu, K. Y., White, A. E., Chen, Y. K., Pomerene, A., Carothers, D., Kamocsai, R. L., Hill, C. M., and Beattie, J., "CMOS-compatible Si-ring-assisted Mach-Zehnder interferometer with internal band equalization," *IEEE Journal of Selected Topics in Quantum Electronics*, 16(1), 45–52 (2010).

Gopalakrishnan, G. K., Burns, W. K., and Bulmer, C. H., "Microwave-optical mixing in $LiNbO_3$ modulators," *IEEE Transactions on Microwave Theory and Techniques*, 41(12), 2383–2391 (1993).

Goutzoulis, A. P. and Davies, D. K., "Switched fiber-optic delay architectures," in *Photonic Aspects of Modern Radar*, H. Zmuda and E. N. Toughlian (editors), Artech House, 351–380 (1994).

Goutzoulis, A. P., Davis, D. K., and Zomp, J. M., "Prototype binary fiber optic delay line," *Optical Engineering*, 28(11), 1193–1202 (1989).

Gutierrez, A. M., Brimont, A., Herrera, J., Aamer, M., Thomson, D. J., Gardes, F. Y., Reed, G. T., Fedeli, J. M., and Sanchis, P., "Analytical model for calculating the nonlinear distortion in silicon-based electro-optic Mach-Zehnder modulators," *Journal of Lightwave Technology*, 31(23), 3603–3613 (2013).

Hamidi, E., Leaird, D. E., and Weiner, A. M., "Tunable programmable microwave photonic filters based on an optical frequency comb," *IEEE Transactions on Microwave Theory and Techniques*, 58(11), 3269–3278 (2010).

Harmon, S. R., Urick, V. J., Diehl, J. F., Williams, K. J., "Tandem electrooptic modulation and interferometric detection: theory and application," *IEEE Photonics Journal*, 5(4), 55011211 (2013).

Heck, M. J. R. and Bowers, J. E., "Hybrid and heterogeneous photonic integration," in *Handbook of Silicon Photonics*, L. Vivien and L. Pavesi (editors), CRC Press (2013).

Heck, M. J. R., Bauters, J. F., Davenport, M. L., Doylend, J. K., Jain, S., Kurczveil, G., Srinivasan, S., Tang, Y., and Bowers, J. E., "Hybrid silicon photonic integrated technology," *IEEE Journal of Selected Topics in Quantum Electronics*, 19(4), (2013).

Helkey, R., Twichell, J. C., and Cox, C. III, "A down-conversion optical link with RF gain," *Journal of Lightwave Technology*, 15(6), 956–961 (1997).

Herczfeld, P. R., Daryoush, A. S., Kieli, M., Siegel, S., and Soref, R., "Wide-band true time delay phase shifter devices," in IEEE MTT-S Digest, 603–606 (1987).

Ho, K. P. and Kahn, J. M., "Linear propagation effects in mode-division multiplexing systems," *Journal of Lightwave Technology*, 32(4), 614–628 (2014).

Huang, H., Nuccio, S. R., Yue, Y., Yang, J. Y., Ren, Y., Wei, C., Yu, G., Dinu, R., Parekh, D., Chang-Hasnain, C. J., and Willner, A. E., "Broadband modulation performance of 100-GHz EO polymer MZMs," *Journal of Lightwave Technology*, 30(23), 3647–3652 (2012).

Hunter, D. B., Edvell, L. G., and Englund, M. A., "Wideband microwave photonic channelized receiver," in IEEE Topical Meeting on Microwave Photonics Digest, 249–252 (2005).

Jalali, B. and Fatpour, S., "Silicon photonics," *Journal of Lightwave Technology*, 24(12), 4600–4615 (2006).

Jalali, B., Fard, A. M., and Han, Y., "Tera sample-per-second time stretched analog-to-digital conversion," in *Microwave Photonics*, C. H. Lee (editor), CRC Press, 355–422 (2013).

Jewell, P. R., "The Green Bank Telescope," in Radio Telescopes, Proceedings of SPIE, 4015, 136–147 (2000).

Juodawlkis, P. W., Twichell, J. C., Betts, G. E., Hargreaves, J. J., Younger, R. D., Wasserman, J. L., O'Donnell, F. J., Ray, K. G., and Williamson, R. C., "Optically sampled analog-to-digital converters," *IEEE Transactions on Microwave Theory and Techniques*, 49(10), 1840–1853 (2001).

Karim, A. and Devenport, J., "High dynamic range microwave photonic links for RF signal transport and RF-IF conversion," *Journal of Lightwave Technology*, 26(15), 2718–2724 (2008).

Koch, T. L. and Koren, U., "Semiconductor photonic integrated circuits," *IEEE Journal of Quantum Electronics*, 27(3), 641–653 (1991).

Koester, S. J., Schaub, J. D., Dehlinger, G., and Chu, J. O., "Germanium-on-SOI infrared detectors for integrated photonic applications," *IEEE Journal of Selected Topics in Quantum Electronics*, 12(6), 1489–1502 (2006).

Kukutsu, N., Hirata, A., Yaita, M., Ajito, K., Takahashi, H., Kosugi, T., Song, H. J., Wakatsuki, A., Muramoto, Y., Nagatsuma, T., and Kado, Y., "Toward practical applications over 100 GHz," in IEEE International Microwave Symposium Digest, 1134–1137 (2010).

Lardenois, S., Pascal, D., Vivien, L., Cassan, E., Laval, S., Orobtchouk, R., Heitzmann, M., Bouzaida, N., and Mollard, L., "Low-loss submicrometer silicon-on-insulator rib waveguides and corner mirrors," *Optics Letters*, 28(13), 1150–1152 (2003).

Leaird, D. E. and Weiner, A. M., "Femtosecond direct space-to-time pulse shaping," *IEEE Journal of Quantum Electronics*, 37(4), 494–504 (2001).

Lee, K. K., Lim, D. R., Kimerling, L. C., Shin, J., and Cerrina, F., "Fabrication of ultralow-loss Si/SiO_2 waveguides by roughness reduction," *Optics Letters*, 26(23), 1888–1890 (2001).

Lee, M., Katz, H. E., Erben, C., Gill, D. M., Goplan, P., Heber, J. D., McGee, D. J., "Broadband modulation of light by using an electro-optic polymer," *Science*, 298, 1401–1403 (2002).

Lee, H., Chen, T., Painter, O., and Vahala, K. J., "Ultra-low-loss optical delay line on a silicon chip," *Nature Communications*, 3, 867 (2012).

Levy, E. C., Horowitz, M., and Menyuk, C. R., "Modeling optoelectronic oscillators," *Journal of Optical Society of America B*, 26(1), 148–159 (2009).

Levy, E. C., Okusaga, O., Horowitz, M., Menyuk, C. R., Zhou, W., and Carter, G. M., "Comprehensive computational model of single- and dual-loop optoelectronic oscillators with experimental verification," *Optics Express*, 18(20), 21461–21476 (2010).

Li, W. and Yao, J., "A wideband frequency tunable optoelectronic oscillator incorporating a tunable microwave photonic filter based on phase-modulation to intensity-modulation conversion using a phase-shifted fiber Bragg grating," *IEEE Transactions on Microwave Theory and Techniques*, 60(6), 1735–1742 (2012).

Li, G., Yao, J., Luo, Y., Thacker, H., Mekis, A., Zheng, X., Shubin, I., Lee, J. H., Raj, K., Cunningham, J. E., Krishnamoorthy, A. V., "Ultralow-loss, high-density SOI optical waveguide routing for macrochip interconnects," *Optics Express*, 20(11), 12035–12039 (2012).

Liang, D. and Bowers, J. E., "Photonic integration: Si or InP substrates," *Electronics Letters*, 45(12), 578–581 (2009).

Liang, D. and Bowers, J. E., "Recent progress in lasers on silicon," *Nature Photonics*, 4, 511–517 (2010).

Little, B. E. and Chu, S. T., "Toward very large-scale integrated photonics," *Optics & Photonics News*, 11(11), 24–29 (2000).

Little, B. E., Chu, S. T., Haus, H. A., Foresi, J., and Laine, J. P., "Microring resonator channel dropping filters," *Journal of Lightwave Technology*, 15(6), 998–1005 (1997).

Liu, M. and Zhang, X., "Graphene benefits," *Nature Photonics*, 7(11), 851–852 (2013).

Liu, J., Beals, M., Pomerene, A., Bernardis, S., Sun, R., Cheng, J., Kimerling, L. C., and Michel, J. "Waveguide-integrated, ultralow-energy GeSi electro-absorption modulators," *Nature Photonics*, 2, 433–437 (2008).

Liu, J., Sun, X., Aguilera, R. C., Kimerling, L. C., Michel, J., "Ge-on-Si laser operating at room temperature," *Optics Letters*, 35(5), 679–681 (2010).

Loayssa, A. and Lahoz, F. J., "Broad-band RF photonic phase shifter based on stimulated Brillouin scattering and single-sideband modulation," *IEEE Photonics Technology Letters*, 18(1), 208–210 (2006).

Loh, W., Yegnanarayanan, S., Klamkin, J., Duff, S. M., Plant, J. J., O'Donnel, F. J., and Juodawlkis, P. W., "Amplifier-free slab-coupled optical waveguide optoelectronic oscillator systems," *Optics Express*, 20(17), 19589–19598 (2012).

Macario, J., Yao, P., Shi, S., Zablocki, A., Harrity, C., Martin, R. D., Schuetz, C. A., and Prather, D. W., "Full spectrum millimeter-wave modulation," *Optics Express*, 20(21), 23623–23629 (2012).

Macario, J., Mercante, A., Yao, P., Zablocki, A., Shi, S., and Prather, D. W., "Ultra-broadband modulator packaging for millimeter-wave applications," accepted for publication in IEEE Transactions on Microwave Theory and Techniques (2014).

Mailloux, R. J., "Array grating lobes due to periodic phase, amplitude, and time delay quantization," *IEEE Transactions on Antennas and Propagation*, 32(12), 1364–1368 (1984).

Marpaung, D. "On-chip photonic-assisted instantaneous microwave frequency measurement system," *IEEE Photonics Technology Letters*, 25(9), 837–840 (2013).

Marpaung, D. and Eggleton, B. J., "Nonlinear integrated microwave photonics," in IEEE Topical Meeting on Microwave Photonics Digest, 84–87 (2013).

Marpaung, D., Roeloffzen, C., Heideman, R., Leinse, A., Sales, S., and Capmany, J., "Integrated microwave photonics," *Laser & Photonics Reviews*, 7(4), 506–538 (2013).

Marris-Morini, D., Soref, R. A., Thomson, D. J., Reed, G. T., Schaevitz, R. K., and Miller, D. A. B., "Optical modulation," in *Handbook of Silicon Photonics*, L. Vivien and L. Pavesi (editors), CRC Press (2013).

Matsko, A. B., Eliyahu, D., Koonath, P., Seidel, D., and Maleki, L., "Theory of coupled optoelectronic microwave oscillator I: expectation values," *Journal Optical Society of America B*, 26(5), 1023–1031 (2009).

McKenna, T. P., Sharp, M. D., Lucarelli, D. G., Nanzer, J. A., Dennis, M. L., and Clark, T. R. Jr., "Wideband photonic compressive sampling analog-to-digital converter for RF spectrum estimation," in OFC/NFOEC Technical Digest, paper OTh3D.1 (2013).

McKinney, J. D., "Background-free arbitrary waveform generation via polarization pulse shaping," *IEEE Photonics Technology Letters*, 22(16), 1193–1195 (2010).

McKinney, J. D. and Urick, V. J., "Radio-frequency down-conversion via sampled analog optical links," NRL Memorandum Report, NRL/MR/5650-10-9275 (2010).

McKinney, J. D. and Weiner, A. M., "Photonic synthesis of ultrabroadband arbitrary electromagnetic waveforms," in *Microwave Photonics*, C. H. Lee (editor), CRC Press, 243–306 (2013).

McKinney, J. D., Seo, D. S., and Weiner, A. M., "Direct space-to-time pulse shaping at 1.5 μm", *IEEE Journal of Quantum Electronics*, 39(12), 1635–1644 (2003).

McKinney, J. D., Leaird, D. E., Hastings, A. S., Weiner, A. M., and Williams, K. J., "Optical comb sources and high-resolution optical filtering for measurement of photodiode harmonic distortion," *Journal of Lightwave Technology*, 28(8), 1228–1235 (2010).

Merkel, K. D., Bekker, S. H., Stiffler, C. R., Traxinger, A., Woidtke, A., Sellin, P. B., Chase, M., and Babbit, W. R., "Continuous wideband spectrum analysis over 10 GHz bandwidth with 61 dBc spur-free dynamic range," in IEEE Topical Meeting on Microwave Photonics Digest, 297–300 (2013).

Minasian, R. A., "Photonic signal processing of microwave signals," *IEEE Transactions on Microwave Theory and Techniques*, 54(2), 832–846 (2006).

Minasian, R. A., Chan, E. H. W., and Yi, X., "Microwave photonic signal processing," *Optics Express*, 21(19), 22918–22936 (2013).

Moerman, I., Van Daele, P. P., and Demester, P. M., "A review on fabrication technologies for the monolithic integration of tapers with III-V semiconductor devices," *IEEE Journal of Selected Topics in Quantum Electronics*, 3(6), 1308–1320 (1997).

Mohan, R. K., Chang, T., Tian, M., Bekker, S., Olson, A., Ostrander, C., Kallaayoun, A., Dollinger, C., Babbit, W. R., Cole, Z., Reibel, R. R., Merkel, K. D., Sun, Y., Cone, R., Schlottau, F., and Wagner, K. H., "Ultra-wideband spectral analysis using S2 technology", *Journal of Luminescence*, 127, 116–128 (2007).

Montebugnoli, S., Boschi, M., Perini, F., Faccin, P., Brunori, G., and Pirazzini, E., "Large antenna array remoting using radio-over-fiber techniques for radio astronomical application," *Microwave and Optical Technology Letters*, 46(1), 48–54 (2005).

Mothe, N. and Bin, P. D., "Multichannel microwave photonics signals summation device," *IEEE Photonics Technology Letters*, 23(3), 140–142 (2011).

Mueller, T., Xia, F., and Avouris, P., "Graphene photodetectors for high-speed optical communications," *Nature Photonics*, 4, 297–301 (2010).

Nagarajan, R., Kato, M., Pleumeekers, J., Evans, P., Corzine, S., Hurtt, S., Dentai, A., Murthy, S., Missey, M., Muthiah, R., Salvatore, R. A., Joyner, C., Schneider, R., Ziari, M., Kish, F., and Welch, D., "InP photonic integrated circuits," *IEEE Journal of Selected Topics in Quantum Electronics*, 16(5), 1113–1125 (2010).

Nagatsuma, T., Horiguchi, S., Minamikata, Y., Yoshimitzu, Y., Hisatake, S., Kuwano, S., Yoshimoto, N., Terada, J., and Takahashi, H., "Terahertz wireless communications based on photonics technologies," *Optics Express*, 21(20), 23736–23747 (2013).

Nelson, C. W., Hati, A., Howe, D. A., Zhou, W., "Microwave optoelectronic oscillator with optical gain," in IEEE FCS Digest, 1014–1019 (2007).

Neyer, A. and Voges, E., "Nonlinear electrooptic oscillator using and integrated interferometer," *Optics Communications*, 37(3), 169–174 (1981).

Neyer, A. and Voges, E., "High-frequency electro-optic oscillator using an integrated interferometer," *Applied Physics Letters*, 40(1), 6–8 (1982).

Nezhad, M. P., Bondarenko, O., Khajavikhan, M., Simic, A., and Fainman, Y., "Etch-free low loss silicon waveguides using hydrogen silsesquioxane oxidation masks," *Optics Express*, 19(20), 18827–18832 (2011).

Ng, W., Walston, A. A., Tangonan, G. L., Lee, J. J., Newberg, I. L., and Bernstein, N., "The first demonstration of an optically steered microwave phased array antenna using true-time-delay," *Journal of Lightwave Technology*, 9(9), 1124–1131 (1991).

Nguyen, T. A., Chan, E. H. W., and Minasian, R. A., "Photonic radio frequency memory using frequency shifting recirculating delay line structure," *Journal of Lightwave Technology*, 32(1), 99–106 (2014).

Nichols, J. M. and Bucholtz, F., "Beating Nyquist with light: a compressively sampled photonic link," *Optics Express*, 19(8), 7339–7348 (2011).

Noguchi, K., Mitomi, O., and Miyazawa, H., "Low-voltage and broadband Ti:$LiNbO_3$ modulators operating in the millimeter wavelength region," in OFC Technical Digest, paper ThB2, 205–206 (1996).

Novak, D., Waterhouse, R., Nirmalathas, A., Lim, C., Bakaul, M., and Gamage, P., "Hybrid fiber radio concepts and prospects," in *Microwave Photonics*, C. H. Lee (editor), 175–199 (2013).

Novoselov, K. S., Geim, A. K., Morozov, S. V., Jiang, D., Zhang, Y., Dubonos, S. V., Grigorieva, I. V., and Firsov, A. A., "Electric field effect in atomically thin carbon films," *Science*, 306, 666–669 (2004).

Okawachi, Y., Kuzucu, O., Foster, M. A., Salem, R., Foster, A. C., Biberman, A., Ophir, N., Bergman, K., Lipson, M., and Gaeta, A. L., "Characterization of nonlinear optical crosstalk in silicon nanowaveguides," *IEEE Photonics Technology Letters*, 24(3), 185–187 (2012).

Okusaga, O., Adles, E. J., Levy, E. C., Zhou, W., Carter, G. M., Menyuk, C. R., and Horowitz, M., "Spurious mode rejection in dual injection-locked opto-electronic oscillators," *Optics Express*, 19(7), 5839–5854 (2011a).

Okusaga, O., Cahill, J., Zhou, W., Docherty, A., Carter, G. M., and Menyuk, C. R., "Optical scattering induced noise in RF-photonic systems," in IEEE FCS Digest, 1–6 (2011b).

Okusaga, O., Cahill, J. P., Docherty, A., Menyuk, C. R., Zhou, W., and Garter, G. M., "Suppression of Rayleigh-scattering-induced noise in OEOs," *Optics Express*, 21(19), 22255–22262 (2013).

Orcutt, J. S., Khilo, A., Holzwarth, C. W., Popovic, M. A., Li, H., Sun, J., Bonifield, H., Hollingsworth, R., Kartner, F. X., Smith, H. I., Stojanovic, V., and Ram, R. J., "Nanophotonic integration in state-of-the-art CMOS foundries," *Optics Express*, 19(3), 2335–2346 (2011).

Pan, S. and Yao, J., "Optical clock recovery using a polarization-modulator-based frequency-doubling optoelectronic oscillator," *Journal of Lightwave Technology*, 27(16), 3531–3539 (2009).

Pappert, S. A., Helkey, R., and Logan, R. T. Jr., "Photonic link techniques for microwave frequency conversion," in *RF Photonic Technology in Optical Fiber Links*, W. S. C. Chang (editor), Cambridge, 293–333 (2002).

Phillips, M. R. and Darcie, T. E., "Lightwave analog video transmission," in *Optical Fiber Telecommunications IIIA*, I. P. Kaminow and T. L. Koch (editors), Academic, 523–559 (1997).

Piels, M., Bauters, J. F., Davenport, M. L., Heck, M. J. R., and Bowers, J. E., "Low-loss silicon nitride AWG demultiplexer heterogeneously integrated with hybrid III-V/silicon photodetectors," *J. of Lightwave Technology*, 32(4), 817–823 (2014).

Pruessner, M. W., Stievater, T. H., Rabinovich, W. S., Devgan, P. S., and Urick, V. J., "Silicon-on-insulator integrated waveguide filters for photonic channelizer applications," in Conference on Lasers and Electro-Optics Digest, paper JWA39 (2009).

Pruessner, M. W., Khurgin, J. B., Stievater, T. H., Rabinovich, W. S., Bass, R., Boos, J. B., and Urick, V. J., "Demonstration of a mode-conversion cavity add-drop filter," *Optics Letters*, 36(12), 2230–2232 (2011).

Pruessner, M. W., Stievater, T. H., Goetz, P. G., Rabinovich, W. S., and Urick, V. J., "Cascaded integrated waveguide linear microcavity filters," *Applied Physics Letters*, 103, 011105 (2013).

Rabiei, P., Ma, J., Khan, S., Chiles, J., and Fathpour, S., "Heterogeneous lithium niobate photonics on silicon substrates," *Optics Express*, 21(21), 25573–25581 (2013).

Rasras, M. S., Tu, K. Y., Gill, D. M., Chen, Y. K., White, A. E., Patel, S. S., Pomerene, A., Carothers, D., Beattie, J., Beals, M., Michel, J., and Kimerling,

L. C., "Demonstration of a tunable microwave-photonic notch filter using low-loss silicon ring resonators," *Journal of Lightwave Technology*, 27(12), 2105–2110 (2009).

Reed, G. T., Headley, W. R., and Png, C. E. J., "Silicon photonics-the early years," in Proceedings of SPIE, 5730, 1–18 (2005).

Reed, G. T., Mashanovich, G., Gardes, F. Y., and Thomson, D. J., "Silicon optical modulators," *Nature Photonics*, 4, 518–526 (2010).

Ricchiuti, A. L., Suarez, I., Barrera, D., Rodriguez-Canto, P. J., Fernandez-Pousa, C. R., Abargues, R., Sales, S., Martinez-Pastor, J., and Capmany, J., "Colloidal quantum dots-PMMA waveguides as integrable microwave photonic phase shifters," *IEEE Photonics Technology Letters*, 26(4), 402–404 (2014).

Rickman, A. G., Reed, G. T., and Namavar, F., "Silicon-on-insulator optical rib waveguide loss and mode characteristics," *Journal of Lightwave Technology*, 12(10), 1771–1776 (1994).

Ridgway, R. W., "Microwave Photonics: Recent Programs at DARPA," presented at IEEE International Topical Mtg. on Microwave Photonics, paper Th2-1, 31 October (2013).

Rogge, M. S., Urick, V. J., Bucholtz, F., Williams, K. J., and Knapp, P., "Comparison of amplitude and phase modulation crosstalk in hyperfine WDM fiber optic links," in CLEO Technical Digest, paper CMH2 (2005).

Roman, J. E., Nichols, L. T., Williams, K. J., Esman, R. D., Tavik, G. C., Livingston, M., and Parent, M. G., "Fiber-optic remoting of an ultrahigh dynamic range radar," *IEEE Transactions on Microwave Theory and Techniques*, 46(12), 2317–2323 (1998).

Rong, H., Jones, R., Liu, A., Cohen, O., Hak, D., Fang, A., and Paniccia, M., "A continuous-wave Raman silicon laser," *Nature*, 433, 725–728 (2005).

Rouvalis, E., Renaud, C. C., Moodie, D. G., Robertson, M. J., and Seeds, A. J., "Traveling-wave uni-traveling carrier photodiodes for continuous wave THz generation," *Optics Express*, 18(11), 11105–11110 (2010).

Russell, P. S. J., "Photonic-crystal fibers," *Journal of Lightwave Technology*, 24(12), 4729–4749 (2006).

Sakamoto, T., Kawanishi, T., and Izutsu, M., "Optoelectronic oscillator using a LiNbO3 phase modulator for self-oscillating frequency comb generation," *Optics Letters*, 31(6), 811–813 (2006).

Salik, E., Yu, N., and Maleki, L., "An ultralow phase noise coupled optoelectronic oscillator," *IEEE Photonics Technology Letters*, 19(6), 444–446 (2007).

Savchenkov, A. A., Ilchenko, V. S., Byrd, J., Liang, W., Eliyahu, D., Matsko, A. B., Seidel, D., and Maleki, L., "Whispering-gallery mode based opto-electronic oscillators," in IEEE FCS Digest, 554–557 (2010).

Schermer, R. T., Bucholtz, F., and Villarruel, C. A., "Continuously-tunable microwave photonic true-time-delay based on a fiber-coupled beam deflector and diffraction grating," *Optics Express*, 19(6), 5371–5378 (2011).

Schermer, R. T., Villarruel, C. A., Bucholtz, F., and McLaughlin, C. V., "Reconfigurable liquid metal fiber-optic mirror for continuous, widely-tunable true-time-delay," *Optics Express*, 21(3), 2741–2747 (2013).

Schmidtchen, J., Splett, A., Schuppert, B., and Petermann, K., "Low loss singlemode optical waveguides with large cross-section in silicon-on-insulator," *Electronics Letters*, 27(16), 1486–1488 (1991).

Schneider, G. J., Murakowski, J. A., Schuetz, C. A., Shi, S., and Prather, D. W., "Radiofrequency signal-generation system with over seven octaves of continuous tuning," *Nature Photonics*, 7(2), 118–122 (2013).

Seeds, A. J., Fice, M. J., Balakier, K., Natrella, M., Mitrofanov, O., Lamponi, M., Chtioui, M., van Dijk, F., Pepper, M., Aeppli, G., Davies, A. G., Dean, P., Linfield, E., and Renaud, C. C., "Coherent terahertz photonics," *Optics Express*, 21(19), 22988–23000 (2013).

Sensale-Rodriguez, B., Yan, R., Kelly, M. M., Fang, T., Tahy, K., Hwang, W. S., Jena, D., Liu, L., and Xing, H. G., "Broadband graphene terahertz modulators enabled by intraband transitions," *Nature Communications*, 3, 1–7 (2012).

Shi, J. W., Kuo, F. M., Wu, C. J., Chang, C. L., Liu, C. Y., Chen, C. Y., and Chyi, J. I., "Extremely high saturation current-bandwidth product performance of a near-ballistic uni-traveling-carrier photodiode with a flip-chip bonding structure," *IEEE Journal of Quantum Electronics*, 46(1), 80–86 (2010).

Shi, J. W., Kuo, F. M., and Bowers, J. E., "Design and analysis of ultra-high-speed near-ballistic uni-traveling-carrier photodiodes under a 50-Ω load for high-power performance," *IEEE Photonics Technology Letters*, 24(7), 533–535 (2012).

Shillue, B., Grammer, W., Jacques, C., Brito, R., Meadows, J., Castro, J., Masui, Y., Treacy, R., Cliche, J. F., "The ALMA photonic local oscillator system," *Proceedings of SPIE*, 8452 (2012).

Singley, J. M., Diehl, J. F., McDermitt, C. S., Sunderman, C. E., and Urick, V. J., "Design and performance of a 560-microsecond Ku-Band binary fiber-optic delay line," NRL Memorandum Report (2014).

Soganci, I. M., Porta, A. L., and Offrein, B. J., "Flip-chip optical couplers with scalable I/O count for silicon photonics," *Optics Express*, 21(13), 16075–16085 (2013).

Solehmainen, K., Aalto, T., Dekker, J., Kapulainen, M., Harjanne, M., Kukli, K., Heimala, P., Kolari, K., and Leskela, M., "Dry-etched silicon-on-insulator waveguides with low propagation and fiber-coupling losses," *Journal of Lightwave Technology*, 23(11), 3875–3880 (2005).

Soref, R., "The past, present, and future of silicon photonics," *IEEE Journal of Selected Topics in Quantum Electronics*, 12(6), 1678–1687 (2006).

Soref, R., "Mid-infrared photonics in silicon and germanium," *Nature Photonics*, 4(9), 495–497 (2010).

Soref, R. A. and Bennett, B. R., "Electrooptical effects in silicon," *IEEE Journal of Quantum Elctronics*, 23(1), 123–129 (1987).

Soref, R. and Lorenzo, J. P., "All-silicon active and passive guided-wave components for $\lambda = 1.3$ and 1.6 μm," *IEEE Journal of Quantum Electronics*, 22(6), 873–879 (1986).

Spezio, A. E., "Electronic warfare systems," *IEEE Transactions on Microwave Theory and Techniques*, 50(3), 633–644 (2002).

Stohr, A., Babiel, S., Cannard, P. J., Charbonnier, B., van Dijk, F., Fedderwitz, S., Moodie, D., Pavlovic, L., Ponnampalam, L., Renaud, C. C., Rogers, D., Rymanov, V., Seeds, A., Steffan, A. G., Umbach, A., Weiss, M., "Millimeter-Wave Photonic Components for Broadband Wireless Systems," *IEEE Transactions on Microwave Theory and Techniques*, 58(11), 3071–3082 (2010).

Strekalov, D., Aveline, D., Yu, N., Thompson, R., Matsko, A. B., and Maleki, L., "Stabilizing and optoelectronic microwave oscillator with photonic filters," *Journal of Lightwave Technology*, 21(12), 3052–3061 (2003).

Sung, H. K., Zhao, X., Lau, E. K., Parekh, D., Chang-Hasnain, C. J., and Wu, M. C., "Optoelectronic oscillators using direct-modulated semiconductor lasers under strong optical injection," *IEEE Journal of Selected Topics in Quantum Electronics*, 15(3), 572–577 (2009).

Tang, Z., Pan, S., Zhu, D., Guo, R., Zhao, Y., Pan, M., Ben, D., and Yao, J., "Tunable optoelectronic oscillator based on a polarization modulator and a chirped FBG," *IEEE Photonics Technology Letters*, 24(17), 1487–1489 (2012).

Taubenblatt, M. A., "Optical interconnects for high-performance computing," *Journal of Lightwave Technology*, 30(4), 448–458 (2012).

Tavik, G. C., Hilterbrick, C. L., Evins, J. B., Alter, J. J., Crnkovich, J. G., de Graaf, J. W., Habicht, W., Hrin, G. P., Lessin, S. A., Wu, D. C., and Hagewood, S. M., "The advanced multifunction RF concept," *IEEE Transactions on Microwave Theory and Techniques*, 53(3), 1009–1020 (2005).

Thomson, D. J., Gardes, F. Y., Fedeli, J. M., Zlatanovic, S., Hu, Y., Kuo, B. P. P., Myslivets,, E., Alic, N., Radic, S., Mashanovich, G. Z., and Reed, G. T., "50-Gb/s silicon optical modulator," *IEEE Photonics Technology Letters*, 24(4), 234–236 (2012).

Urick, V. J. and Bucholtz, F., "Compensation of arbitrary chromatic dispersion in analogs links using a modulation-diversity receiver," *IEEE Photonics Technology Letters*, 17(4), 893–895 (2005).

Urick, V. J., Qui, J. X., and Bucholtz, F., "Wide-band QAM-over-fiber using phase modulation and interferometric demodulation", *IEEE Photonics Technology Letters*, 16(10), 2374–2376 (2004).

Urick, V. J., Rogge, M. S., Knapp, P. F., Swingen, L., and Bucholtz, F., "Wideband predistortion linearization for externally-modulated long-haul analog fiber optic-links," *IEEE Transactions on Microwave Theory and Techniques*, 54(4), 1458–1463 (2006).

Urick, V. J., Bucholtz, F., Devgan, P. S., McKinney, J. D., and Williams, K. J., "Phase modulation with interferometric detection as an alternative to intensity modulation with direct detection for analog-photonic links," *IEEE Transactions on Microwave Theory Technology*, 55(9), 1978–1985 (2007).

Urick, V. J., Diehl, J., Hastings, A., Sunderman, C., McKinney, J. D., Devgan, P. S., Dexter, J. L., and Williams, K. J., "Analysis of fiber-optic links for HF antenna remoting," NRL Memorandum Report, NRL/MR/5650-08-9101 (2008a).

Urick, V. J., Hastings, A., Sunderman, C., Diehl, J., Colladay, K., Dexter, J. L., Williams, K. J., "Field test on the feasibility of remoting HF antenna with fiber optics," NRL Memorandum Report, NRL/MR/5652-08-9137 (2008b).

Urick, V. J., Devagn, P. S., McKinney, J. D., Bucholtz, F., and Williams, K. J., "Channelisation of radio-frequency signals using optoelectronic oscillator," *Electronic Letters*, 45(24), 1242–1244 (2009).

Urick, V. J., McKinney, J. D., Diehl, J. F. and Singley, J. M., "Simultaneous optical phase and intensity modulation for analog signal processing," in IEEE AVFOP Conference Digest, paper TuD4 (2011).

Urick, V. J., Singley, J. M., Sunderman, C. S., Diehl, J. F., and Williams, K. J., "Design and performance of Ka-Band fiber-optic delay lines," NRL Memorandum Report, NRL/MR/5650-12-9456 (2012).

Urick, V. J., Bucholtz, B., and Funk, E. E., "High dynamic range, 100 km digital radio-over-fiber links," in *Microwave Photonics*, C. H. Lee (editor), CRC Press, 201–241 (2013).

Valley, G. C., "Photonic analog-to-digital converters," *Optics Express*, 15(5), 1955–1982 (2007).

Villarruel, C. A., McDermitt, C. A., and Bucholtz, F., "Microstructure fibre array for RF photonic signal processing applications," *Electronics Letters*, 42(5), 271–273 (2006).

Vivien, L. and Pavesi, L. (editors), *Handbook of Silicon Photonics*, CRC Press (2013).

Vlasov, Y. A. and McNab, S. J., "Losses in single-mode silicon-on-insulator strip waveguides and bends," *Optics Express*, 12(8), 1622–1631 (2004).

Voigt, K., Zimmermann, L., Winzer, G., Mitze, T., Bruns, J., Petermann, K., Huttl, B., and Schubert, C., "Performance of 40 Gb/s DPSK demodulator in SOI-technology," *IEEE Photonics Technology Letters*, 20(8), 616–616 (2008).

Wang, C. and Yao, J., "Fiber Bragg gratings for microwave photonics subsystems," *Optics Express*, 21(19), 22868–22884 (2013).

Weiner, A. M., "Femtosecond pulse shaping using spatial light modulators," *Review of Scientific Instruments*, 71(5), 1929–1960 (2000).

Weiner, A. M., *Ultrafast Optics*, John Wiley & Sons (2009).

Weiner, A. M., "Ultrafast optical pulse shaping: a tutorial review," *Optical Communications*, 284(15), 3669–3692 (2011).

Welch, D. F., Kish, F. A., Melle, S., Nagarajan, R., Kato, M., Joyner, C. H., Pleumeekers, J. L., Schneider, R. P., Back, J., Dentai, A. G., Dominic, V. G., Evans, P. W., Kauffman, M., Lambert, D. J. H., Hurtt, S. K., Mathur, A., Mitchell, M. L., Missey, M., Murthy, S., Nilsson, A. C., Salvatore, R. A., Van Leeuwen, M. F., Webjorn, J., Ziari, M., Grubb, S. G., Perkins, D., Reffle, M., and Mehuys, D. G., "Large-scale InP photonic integrated circuits: enabling efficient scaling of optical transport networks," *IEEE Journal of Selected Topics in Quantum Electronics*, 13(1), 22–31 (2007).

Williams, K. J. and Esman, R. D., "Optically amplified downconverting link with shot-noise-limited performance," *IEEE Photonics Technology Letters*, 8(1), 148–150 (1996).

Williams, K. J., Goldberg, L., Esman, R. D., Dagenais, M., and Weller, J. F., "6-34 GHz offset phase-locking of Nd:YAG 1319 nm nonplanar ring lasers," *Electronics Letters*, 25(18), 1242–1243 (1989).

Williamson, R. C. and Esman, R. D., "RF photonics," *Journal of Lightwave Technology*, 26(9), 1145–1153 (2008).

Willner, A. E., Khaleghi, S., Chitgarha, M. R., and Yilmaz, O. F., "All-optical signal processing," *Journal of Lightwave Technology*, 32(4), 660–680 (2014).

Wilson, B., Ghassemlooy, Z., and Darwazeh, I. (editors), *Analogue Optical Fibre Communications*, The Institution of Electrical Engineers (1995).

Wootten, A., "The Atacama Large Millimeter Array (ALMA)," *Proceedings of SPIE*, 4837, 110–118 (2003).

Xia, T. J. and Wellbrock, G. A., "Commercial 100-Gbit/s coherent transmission systems," in *Optical Fiber Telecommunications VIB*, I. P. Kaminow, T. Li, and A. E. Willner (editors), Academic (2013).

Xia, F., Mueller, T., Lin, Y., Valdes-Garcia, A., and Avouris, P., "Ultrafast graphene photodetector," *Nature Nanotechnology*, 4, 839–843 (2009).

Xu, Q., Schmidt, B., Pradhan, S., and Lipson, M., "Micrometre-scale silicon electro-optic modulator," *Nature*, 435, 325–327 (2005).

Xu, K., Dai, J., Duan, R., Dai, Y., Wu, J., and Lin, J., "Instantaneous microwave frequency measurement based on phase-modulated links with interferometric demodulation," *IEEE Photonics Technology Letters*, 23(18), 1328–1330 (2011).

Xue, W., Sales, S., Capmany, J., and Mork, J., "Wideband 360-degree microwave photonic phase shifter based on slow light in semiconductor optical amplifiers," *Optics Express*, 18(6), 6156–6163 (2010).

Yan, R., Gargas, D., and Yang, P., "Nanowire photonics," *Nature Photonics*, 3(10), 569–576 (2009).

Yang, A. C., Payne, R., and Soref, R. A., "Mode dependent, optical time delay system for electrical signals," United States Patent No. 4,714,314 (1987).

Yao, X. S., "Opto-electronic oscillators," in *RF Photonic Technology in Optical Fiber Links*, W. S. C. Chang (editor), Cambridge, 255–292 (2002).

Yao, J., "Microwave photonics," *Journal of Lightwave Technology*, 27(3), 314–335 (2009).

Yao, X. S. and Maleki, L., "High frequency optical subcarrier generator," *Electronics Letters*, 30(18), 1525–1526 (1994).

Yao, X. S. and Maleki, L., "Optoelectronic oscillator for photonic systems," *IEEE Journal of Quantum Electronics*, 32(7), 1141–1149 (1996a).

Yao, X. S. and Maleki, L., "Optoelectronic microwave oscillator," *Journal Optical Society of America B*, 13(8), 1725–1735 (1996b).

Yao, X. S. and Maleki, L., "Multiloop optoelectronic oscillator," *IEEE Journal of Quantum Electronics*, 36(1), 79–84 (2000).

Yao, X. S., Davis, L., and Maleki, L., "Coupled optoelectronic oscillators for generating both RF signal and optical pulses," *Journal of Lightwave Technology*, 18(1), 73–78 (2000).

Yao, X. S., Maleki, L., and Ilchenko, V., "Integrated opto-electronic oscillator having optical resonators," United States Patent No. US 6,873,631 B2 (2005).

Yap, K. P., Delage, A., Lapointe, J., Lamontagne, B., Schmid, J. H., Waldron, P., Syrett, B. A., and Janz, S., "Correlation of scattering loss, sidewall roughness and waveguide width in silicon-on-insulator (SOI) ridge waveguides," *Journal of Lightwave Technology*, 27(18), 3999–4008 (2009).

Yilmaz, H., Chakam, G. H., Breuel, A., Vogel, W., Berroth, M., and Freude, W., "Microstrip line fed patch antenna with liquid crystal phase shifter for optically generated RF-signals," in IEEE International Topical Meeting on Microwave Photonics Digest, 359–362 (2003).

Zaoui, W. S., Kunze, A., Vogel, W., Berroth, M., Butschke, J., Letzkus, F., and Burghartz, J., "Bridging the gap between optical fibers and silicon photonic integrated circuits," *Optics Express*, 22(2), 1277–1286 (2014).

Zhou, W. and Blasche, G. "Injection-locked dual opto-electronic oscillator with ultra-low phase noise and ultra-low spurious level," *IEEE Transactions on Microwave Theory and Techniques*, 53(3), 929–933 (2005).

Zhou, W., Okusaga, O., Nelson, C., Howe, D., and Carter, G., "10 GHz dual loop opto-electronic oscillator without RF-amplifiers," in Proceedings of SPIE, 6897 (2008).

Zhou, J., Fu, S., Aditya, S., Shum, P. P., and Lin, C., "Instantaneous microwave frequency measurement using photonic technique," *IEEE Photonics Technology Letters*, 21(15), 1069–1071 (2009).

Zhu, B., Fini, J. M., Yan, M. F., Liu, X., Chandrasekhar, S., Taunay, T. F., Fishteyn, M., Monberg, E. M., and Dimarcello, F. V., "High-capacity space-division-multiplexed DWDM transmissions using multicore fiber," *Journal of Lightwave Technology*, 30(4), 486–492 (2012).

Zinke, T. Fischer, U., Splett, A., Schuppert, B., and Petermann, K., "Comparison of optical waveguide losses in silicon-on-insulator," *Electronics Letters*, 29(23), 2031–2033 (1993).

Zmuda, H. and Toughlian, E. N. (editors), *Photonic Aspects of Modern Radar*, Artech (1994).

Zou, X., Chi, H., and Yao, J., "Microwave frequency measurement based on optical power monitoring using a complimentary optical filter pair," *IEEE Transactions on Microwave Theory and Techniques*, 57(2), 505–511 (2009a).

Zou, X., Pan, S., and Yao, J., "Instantaneous microwave frequency measurement with improved measurement range and resolution based on simultaneous phase modulation and intensity modulation," *Journal of Lightwave Technology*, 27(23), 5314–5320 (2009b).

Zou, X., Li, W., Pan, W., Yan, L., and Yao, J., "Photonic-assisted microwave channelizer with improved channel characteristics based on spectrum-controlled stimulated Brillouin scattering," *IEEE Transacions on Microwave Theory and Techniques*, 61(9), 3470–3478 (2013).

APPENDIX I

UNITS AND PHYSICAL CONSTANTS[1,2]

TABLE AI.1 SI Base Units

Physical quantity	Name	Symbol
Length	Meter	m
Mass	Kilogram	kg
Time	Second	s
Electric current	Ampere	A
Thermodynamic temperature	Kelvin	K
Amount of substance	Mole	mol
Luminous intensity	Candela	cd

[1]Haynes, W. M. (editor), *CRC Handbook of Chemistry and Physics*, 92nd edition, CRC Press, New York (2011).
[2]physics.nist.gov/constants.

Fundamentals of Microwave Photonics, First Edition.
Vincent J. Urick Jr, Jason D. McKinney, and Keith J. Williams.

TABLE AI.2 SI Prefixes

Factor	Prefix	Symbol	Factor	Prefix	Symbol
10^{24}	Yotta	Y	10^{-1}	Deci	d
10^{21}	Zetta	Z	10^{-2}	Centi	c
10^{18}	Exa	E	10^{-3}	Milli	m
10^{15}	Peta	P	10^{-6}	Micro	μ
10^{12}	Tera	T	10^{-9}	Nano	n
10^{9}	Giga	G	10^{-12}	Pico	p
10^{6}	Mega	M	10^{-15}	Femto	f
10^{3}	Kilo	k	10^{-18}	Atto	a
10^{2}	Hecto	h	10^{-21}	Zepto	z
10^{1}	Deka	da	10^{-24}	Yocto	y

TABLE AI.3 Quantities Derived from SI Base Units

Physical quantity	Unit
Wave number	1/m
Area	m^2
Volume	m^3
Velocity	m/s
Acceleration	m/s^2
Magnetic field strength	A/m
Current density	A/m^2
Luminance	cd/m^2

TABLE AI.4 Other SI-Derived Quantities

Physical quantity	Name	Symbol	In terms of other units
Frequency	Hertz	Hz	s^{-1}
Force	Newton	N	$m \cdot kg \cdot s^{-2}$
Pressure, stress	Pascal	Pa	$N/m^2 = m^{-1} \cdot kg \cdot s^{-2}$
Energy, work, heat	Joule	J	$N \cdot m = m^2 \cdot kg \cdot s^{-2}$
Power, radiant flux	Watt	W	$J/s = m^{-1} \cdot kg \cdot s^{-3}$
Electric charge	Coulomb	C	$s \cdot A$
Electric potential, emf	Volt	V	$J/C = m^2 \cdot kg \cdot s^{-3} \cdot A^{-1}$
Electric resistance	Ohm	Ω	$V/A = m^2 \cdot kg \cdot s^{-3} \cdot A^{-2}$
Electric conductance	Siemens	S	$1/\Omega = m^{-2} \cdot kg^{-1} \cdot s^3 \cdot A^2$
Electric capacitance	Farad	F	$C/V = m^{-2} \cdot kg^{-1} \cdot s^4 \cdot A^2$
Magnetic flux	Weber	Wb	$V \cdot s = m^2 \cdot kg \cdot s^{-2} \cdot A^{-1}$
Magnetic flux density	Tesla	T	$Wb/m^2 = kg \cdot s^{-2} \cdot A^{-1}$
Inductance	Henry	H	$Wb/A = m^2 \cdot kg \cdot s^{-2} \cdot A^{-2}$
Plane angle	Radian	rad	$m/m = 1$
Solid angle	Steradian	sr	$m^2/m^2 = 1$

(continued)

TABLE AI.4 (*Continued*)

Physical quantity	Name	Symbol	In terms of other units
Luminous flux	Lumen	lm	cd·sr = cd
Illuminance	Lux	lx	$lm/m^2 = m^{-2}\cdot cd$
Moment of force	–	–	$N\cdot m = m^2\cdot kg\cdot s^{-2}$
Surface tension	–	–	$N/m = kg\cdot s^{-2}$
Dynamic viscosity	–	–	$Pa\cdot s = m^{-1}\cdot kg\cdot s^{-1}$
Angular velocity	–	–	$rad/s = s^{-1}$
Angular acceleration	–	–	$rad/s^2 = s^{-2}$
Power spectral density	–	–	$W/Hz = m^2\cdot kg\cdot s^{-2}$
Energy density	–	–	$J/m^3 = m^{-1}\cdot kg\cdot s^{-2}$
Electric field strength	–	–	$V/m = m\cdot kg\cdot s^{-3}\cdot A^{-1}$
Electric flux	–	–	$C = s\cdot A$
Electric flux density	–	–	$C/m^2 = m^{-2}\cdot s\cdot A$
Electric charge density	–	–	$C/m^3 = m^{-3}\cdot s\cdot A$
Permittivity	–	–	$F/m = m^{-3}\cdot kg^{-1}\cdot s^4\cdot A^2$
Permeability	–	–	$H/m = m\cdot kg\cdot s^{-2}\cdot A^{-2}$
Radiant intensity	–	–	$W/sr = m^2\cdot kg\cdot s^{-3}$
Radiance	–	–	$W/(m^2\cdot sr) = kg\cdot s^{-3}$
Heat flux density, irradiance	–	–	$W/m^2 = kg\cdot s^{-3}$
Heat capacity, entropy	–	–	$J/K = m^2\cdot kg\cdot s^{-2}\cdot K^{-1}$
Specific heat capacity	–	–	$J/(kg\cdot K) = m^2\cdot s^{-2}\cdot K^{-1}$
Specific energy	–	–	$J/kg = m^2\cdot s^{-2}$
Thermal conductivity	–	–	$W/(m\cdot K) = m\cdot kg\cdot s^{-3}\cdot K^{-1}$

UNIT CONVERSIONS

Ångström = 1.000000×10^{-10} m.
Inch = 2.540000×10^{-2} m.
Foot = 3.048000×10^{-1} m.
Yard = 9.144000×10^{-1} m.
Mile = 1.609344×10^{3} m.
Nautical mile = 1.852000×10^{3} m.
Light year = 9.46073×10^{15} m.
Acre = 4.046873×10^{3} m^2.
Liter = 1.000000×10^{-3} m^3.
US fluid ounce = 2.957353×10^{-5} m^3.
US liquid gallon = 3.785412×10^{-3} m^3.
Avoirdupois pound = 4.535924×10^{-1} kg.
Minute = 6.000000×10^{1} s.
Hour = 3.600000×10^{3} s.
Day = 8.640000×10^{4} s.
Year = 3.153600×10^{7} s.
Miles/hour = 4.470400×10^{-1} m/s.
Temperature in K → temperature in Celsius: $T_C = T_K - 273.15$.
Temperature in K → temperature in Fahrenheit: $T_F = (1.8)T_K - 459.67$.
Temperature in K → temperature in Rankine: $T_R = (1.8)T_K$.
Dyne = 1.000000×10^{-5} N.
Standard atmosphere = 1.013250×10^{5} Pa.
Bar = 1.000000×10^{5} Pa.

TABLE AI.4 (*Continued*)

eV = 1.602177×10^{-19} J.
erg = 1.000000×10^{-7} J.
Foot-pound = 1.355818 J.
International table calorie = 4.186800 J.
Thermochemical calorie = 4.184000 J.
Horsepower (550 ft · lb/s) = 7.456999×10^{2} W.
Gauss = 1.000000×10^{-4} T.
Degree (angle) = 1.745329×10^{-2} rad.

QUANTITIES EXPRESSED IN A DECIBEL SCALE

Power, P_1, relative to power, P_2: [dB] = $10 \log\left(\frac{P_1}{P_2}\right)$.
Power, P, relative to carrier power, P_c: [dBc] = $10 \log\left(\frac{P}{P_c}\right)$.
Power, P, relative to 1 mW: [dBm] = $10 \log\left(\frac{P}{1\,\text{mW}}\right)$.
Power, P, relative to 1 W: [dBW] = $10 \log\left(\frac{P}{1\,\text{W}}\right)$.
Voltage, V, relative to 1 V: [dBV] = $10 \log\left(\frac{V}{1\,\text{V}}\right)$.
Antenna gain, g, relative to ideal isotropic antenna gain, g_i: [dBi] = $10 \log\left(\frac{g}{g_i}\right)$.
Radar cross section, σ, relative to 1 square meter: [dBsm] = $10 \log\left(\frac{\sigma}{1\,\text{m}^2}\right)$.

TABLE AI.5 Physical Constants

Constant	Symbol	Value
Speed of light in vacuum	c	2.99792458×10^{8} m/s
Elementary charge	q	$1.602176565(35) \times 10^{-19}$ C
Boltzmann's constant	k_B	$1.3806488(13) \times 10^{-23}$ J/K
Planck's constant	h	$6.62606957(29) \times 10^{-34}$ J · s
=	$\hbar$	$h/(2\pi) = 1.054571726(47) \times 10^{-34}$ J · s
Magnetic constant (permeability of vacuum)	μ_0	$4\pi \times 10^{-7}$ N/A^2 = $12.566370614 \ldots \times 10^{-7}$ N/A^2
Electric constant (permittivity of vacuum)	ε_0	$1/(\mu_0 c^2) = 8.854187817 \ldots \times 10^{-12}$ F/m
Characteristic impedance of vacuum	Z_0	$\mu_0 c = 376.730313461 \ldots \Omega$
Electron mass	m_e	$9.10938291(40) \times 10^{-31}$ kg
Proton mass	m_p	$1.672621777(74) \times 10^{-27}$ kg
Neutron mass	m_n	$1.674927351(74) \times 10^{-27}$ kg

Note: (·) is the standard uncertainty.

APPENDIX II

ELECTROMAGNETIC RADIATION

TABLE AII.1 Electromagnetic Spectrum Classifications[a]

Frequency	Wavelength	Notes
3–30 Hz	100–10 Mm	Extremely low frequency (ELF), ITU band 1
30–300 Hz	10–1 Mm	Super low frequency (SLF), ITU band 2
300 Hz–3 kHz	1 Mm–100 km	Ultra low frequency (ULF), ITU band 3
3–30 kHz	100–10 km	Very low frequency (VLF), ITU band 4
30–300 kHz	10–1 km	Low frequency (LF), ITU band 5
300 kHz–3 MHz	1 km–100 m	Medium frequency (MF), ITU band 6
3–30 MHz	100–10 m	High frequency (HF), ITU band 7
30–300 MHz	10–1 m	Very high frequency (VHF), ITU band 8
300 MHz–3 GHz	1 m–100 mm	Ultra high frequency (UHF), ITU band 9
3–30 GHz	100–10 mm	Super high frequency (SHF), ITU band 10
30–300 GHz	10–1 mm	Extremely high frequency (EHF), ITU band 11
300 GHz–3 THz	1 mm–100 μm	Tremendously high frequency (THF), ITU band 12

Fundamentals of Microwave Photonics, First Edition.
Vincent J. Urick Jr, Jason D. McKinney, and Keith J. Williams.

TABLE AII.1 (*Continued*)

Frequency	Wavelength	Notes
3–30 THz	100–10 μm	Far infrared
30–300 THz	10–1 μm	Infrared (0.1–1 eV)
300 THz–3 PHz	1 μm–100 nm	Near infrared, visible, near ultraviolet (1–10 eV)
3–30 PHz	100–10 nm	Ultraviolet (10–100 eV)
30–300 PHz	10–1 nm	Soft X-rays (100 eV–1 keV)
300 PHz–3 EHz	1nm–100 pm	Soft X-rays (1–10 keV)
3–30 EHz	100–10 pm	Hard X-rays, Soft γ-rays (10–100 keV)
30–300 EHz	10–1 pm	Soft γ-rays, Hard γ-rays (100 keV–1 MeV)
300–3000 EHz	1 pm–100 fm	Hard γ-rays, Cosmic rays (1–10 MeV)
3,000–30,000 EHz	100–10 fm	γ-rays produced by cosmic rays (10–100 MeV)

[a]Dolezalek, H., "Classification of electromagnetic radiation," in *CRC Handbook of Chemistry and Physics*, W. M. Haynes (editor), 92nd edition, p. 10–233, CRC Press, New York (2011).
ITU: International Telecommunication Union.

TABLE AII.2 IEEE Radar Frequency Bands[a]

Frequency range	Band designation
3–30 MHz	HF
30–300 MHz	VHF
300 MHz–1 GHz	UHF
1–2 GHz	L
2–4 GHz	S
4–8 GHz	C
8–12 GHz	X
12–18 GHz	Ku
18–27 GHz	K
27–40 GHz	Ka
40–75 GHz	V
75–110 GHz	W
110–300 GHz	mm

[a]*IEEE Standard Letter Designations for Radar-Frequency Bands*, IEEE Std 521™-2002, 8 January 2003.

TABLE AII.3 Radio Society of Great Britain Microwave Bands[a]

Frequency range(GHz)	Band designation
1–2	L
2–4	S
4–8	C
8–12	X
12–18	Ku
18–26.5	K
26.5–40	Ka
33–50	Q
40–60	U
50–75	V
60–90	E
75–110	W
90–140	F
110–170	D

[a]www.rsgb.org.

TABLE AII.4 NATO Band Designations

Frequency range	Band designation
0–250 MHz	A
250–500 MHz	B
0.5–1 GHz	C
1–2 GHz	D
2–3 GHz	E
3–4 GHz	F
4–6 GHz	G
6–8 GHz	H
8–10 GHz	I
10–20 GHz	J
20–40 GHz	K
40–60 GHz	L
60–100 GHz	M

APPENDIX III

POWER, VOLTAGE, AND CURRENT FOR A SINUSOID

TABLE AIII.1 Power, Voltage, and Current for a Sinusoid

Average power (W)	Average power (dBm)	RMS voltage (V)[a]	Peak voltage (V)[a]	RMS current (A)[a]	Peak current (A)[a]
0.00100	0.00	0.224	0.316	0.00447	0.00632
0.00200	3.01	0.316	0.447	0.00632	0.00894
0.00300	4.77	0.387	0.548	0.00775	0.0110
0.00400	6.02	0.447	0.632	0.00894	0.0126
0.00500	6.99	0.500	0.707	0.0100	0.0141
0.00600	7.78	0.548	0.775	0.0110	0.0155
0.00700	8.45	0.592	0.837	0.0118	0.0167
0.00800	9.03	0.632	0.894	0.0126	0.0179
0.00900	9.54	0.671	0.949	0.0134	0.0190
0.0100	10.0	0.707	1.00	0.0141	0.0200
0.0200	13.0	1.00	1.41	0.0200	0.0283
0.0300	14.8	1.22	1.73	0.0245	0.0346
0.0400	16.0	1.41	2.00	0.0283	0.0400
0.0500	17.0	1.58	2.24	0.0316	0.0447

(*continued*)

Fundamentals of Microwave Photonics, First Edition.
Vincent J. Urick Jr, Jason D. McKinney, and Keith J. Williams.

TABLE AIII.1 (*Continued*)

Average power (W)	Average power (dBm)	RMS voltage (V)[a]	Peak voltage (V)[a]	RMS current (A)[a]	Peak current (A)[a]
0.0600	17.8	1.73	2.45	0.0346	0.0490
0.0700	18.5	1.87	2.65	0.0374	0.0529
0.0800	19.0	2.00	2.83	0.0400	0.0566
0.0900	19.5	2.12	3.00	0.0424	0.0600
0.100	20.0	2.24	3.16	0.0447	0.0632
0.200	23.0	3.16	4.47	0.0632	0.0894
0.300	24.8	3.87	5.48	0.0775	0.110
0.400	26.0	4.47	6.32	0.0894	0.126
0.500	27.0	5.00	7.07	0.100	0.141
0.600	27.8	5.48	7.75	0.110	0.155
0.700	28.5	5.92	8.37	0.118	0.167
0.800	29.0	6.32	8.94	0.126	0.179
0.900	29.5	6.71	9.49	0.134	0.190
1.00	30.0	7.07	10.0	0.141	0.200

$P_{avg} = \langle V^2(t)\rangle/R = V_{rms}^2/R = V_{peak}^2/(2R)$.

$P_{avg} = \langle I^2(t)\rangle R = I_{rms}^2 R = I_{peak}^2 R/2$.

[a]For $R = 50\ \Omega$.

APPENDIX IV

TRIGONOMETRIC FUNCTIONS

AIV.1 TAYLOR SERIES EXPANSIONS

$$\sin x = \sum_{n=0}^{\infty} \frac{(-1)^n}{(2n+1)!} x^{2n+1} = x - \frac{x^3}{3!} + \frac{x^5}{5!} + \cdots$$

$$\cos x = \sum_{n=0}^{\infty} \frac{(-1)^n}{(2n)!} x^{2n} = 1 - \frac{x^2}{2!} + \frac{x^4}{4!} + \cdots.$$

AIV.2 TANGENT AND INVERSE FUNCTIONS

$$\tan x = \frac{\sin x}{\cos x} \quad \csc x = \frac{1}{\sin x}.$$

$$\cot x = \frac{1}{\tan x} \quad \sec x = \frac{1}{\cos x}.$$

Fundamentals of Microwave Photonics, First Edition.
Vincent J. Urick Jr, Jason D. McKinney, and Keith J. Williams.

AIV.3 PHASE SHIFT FORMULAE

$$\sin(-x) = -\sin x \qquad \cos(-x) = \cos x.$$

$$\sin x = \cos\left(\frac{\pi}{2} - x\right) \qquad \cos x = \sin\left(\frac{\pi}{2} - x\right).$$

AIV.4 SQUARE AND CUBIC RELATIONSHIPS

$$\sin^2 x = \frac{1}{2} - \frac{1}{2}\cos(2x) \qquad \cos^2 x = \frac{1}{2} + \frac{1}{2}\cos(2x).$$

$$\sin^3 x = \frac{3}{4}\sin x - \frac{1}{4}\sin(3x) \qquad \cos^3 x = \frac{3}{4}\cos x + \frac{1}{4}\cos(3x).$$

$$\sin^2 x + \cos^2 x = 1.$$

AIV.5 SUM ANGLE FORMULAE

$$\sin(x+y) = \sin x\cos y + \cos x\sin y \quad \cos(x+y) = \cos x\cos y - \sin x\sin y.$$

$$\sin(x-y) = \sin x\cos y - \cos x\sin y \quad \cos(x-y) = \cos x\cos y + \sin x\sin y.$$

AIV.6 DOUBLE ANGLE FORMULAE

$$\sin(2x) = 2\sin x\cos x \qquad \cos(2x) = \cos^2 x - \sin^2 x.$$

AIV.7 SUM AND DIFFERENCE FORMULAE

$$\sin x + \sin y = 2\sin\left(\frac{x+y}{2}\right)\cos\left(\frac{x-y}{2}\right)$$

$$\cos x + \cos y = 2\cos\left(\frac{x+y}{2}\right)\cos\left(\frac{x-y}{2}\right).$$

$$\sin x - \sin y = 2\cos\left(\frac{x+y}{2}\right)\sin\left(\frac{x-y}{2}\right)$$

$$\cos x - \cos y = -2\sin\left(\frac{x+y}{2}\right)\sin\left(\frac{x-y}{2}\right).$$

AIV.8 PRODUCT FORMULAE

$$\sin x \sin y = \tfrac{1}{2} \cos (x - y) - \tfrac{1}{2} \cos (x + y)$$

$$\sin x \cos y = \tfrac{1}{2} \sin (x - y) + \tfrac{1}{2} \sin (x + y).$$

$$\cos x \cos y = \tfrac{1}{2} \cos (x - y) + \tfrac{1}{2} \cos (x + y)$$

$$\cos x \sin y = -\tfrac{1}{2} \sin (x - y) + \tfrac{1}{2} \sin (x + y).$$

AIV.9 EULER'S FORMULA

$$e^{i\theta} = \cos \theta + i \sin \theta.$$

$$\sin \theta = \frac{1}{2i}(e^{i\theta} - e^{-i\theta}).$$

$$\cos \theta = \frac{1}{2}(e^{i\theta} + e^{-i\theta}).$$

(θ in radians).

APPENDIX V

FOURIER TRANSFORMS[1, 2]

The Fourier Transform (FT) of $f(t)$ is $F(\omega)$; $f(t)$ is the Inverse Fourier Transform (IFT) of $F(\omega)$, where ω is the angular frequency.

$$F(\omega) = \int_{-\infty}^{\infty} f(t)e^{-i\omega t}dt \quad \text{and} \quad f(t) = \frac{1}{2\pi}\int_{-\infty}^{\infty} F(\omega)e^{i\omega t}d\omega.$$

AV.1 PROPERTIES OF THE FOURIER TRANSFORM

Parity:	if $f(t)$ is even, then $F(\omega)$ is even; if $f(t)$ is odd, then $F(\omega)$ is odd.
Hermitian Symmetry:	if $f(t)$ is real, then $F(\omega)$ is Hermitian; if $f(t)$ is imaginary, then $F(\omega)$ is anti-Hermitian. (A Hermitian function is one whose real part is even and imaginary part is odd; an anti-Hermitian function is one whose real part is odd and imaginary part is even.)

[1]Vretblad, A., *Fourier Analysis and its Applications*, Springer (2003).
[2]Bracewell, R. N., *The Fourier Transform and its Applications*, McGraw Hill (2000).

Fundamentals of Microwave Photonics, First Edition.
Vincent J. Urick Jr, Jason D. McKinney, and Keith J. Williams.

Linearity: the FT of $x \cdot f(t) + y \cdot g(t) = x \cdot F(\omega) + y \cdot G(\omega)$, where $G(\omega)$ is the FT of $g(t)$.

Scaling: the FT of $f(x \cdot t) = F(\omega/x)/|x|$ for $x \neq 0$.

Negatives: the FT of $f(-t) = F(-\omega)$.

Time Shift: the FT of $f(t - \tau) = e^{-i\omega\tau}F(\omega)$.

Frequency Shift: the IFT of $F(\omega - \Omega) = e^{i\Omega t}f(t)$.

Time Derivative: the FT of $\frac{d^n f(t)}{dt^n} = (i\omega)^n F(\omega)$.

Frequency Derivative: the IFT of $\frac{d^n F(\omega)}{d\omega^n} = (-it)^n f(t)$.

Time Convolution: the IFT of $F(\omega) \cdot G(\omega) = \int_{-\infty}^{\infty} f(t - \tau)g(\tau)d\tau$.

Frequency Convolution: the FT of $f(t) \cdot g(t) = \frac{1}{2\pi}\int_{-\infty}^{\infty} F(\omega - \Omega)G(\Omega)d\Omega$.

Plancherel's Theorem: $\int_{-\infty}^{\infty} |f(t)|^2 dt = \frac{1}{2\pi}\int_{-\infty}^{\infty} |F(\omega)|^2 d\omega$.

APPENDIX VI

BESSEL FUNCTIONS[1]

Solutions to $x^2\frac{d^2y}{dx^2} + x\frac{dy}{dx} + (x^2 - n^2)y = 0$, where x is a real variable, are Bessel functions of order n.

AVI.1 PROPERTIES

$J_n(x) = \sum_{k=0}^{\infty} \frac{(-1)^k (x/2)^{n+2k}}{k!\Gamma(n+k+1)}$, where $\Gamma(n+k+1) = \int_0^{\infty} t^{n+k} e^{-t} dt$ for $(n+k) > -1$.

$J_{-n}(x) = (-1)^n J_n(x)$ for n an integer

$$J_n(x) = \frac{x^n}{2^n n!}\left[1 - \frac{x^2}{2^2 \cdot 1!\,(n+1)} + \frac{x^4}{2^4 \cdot 2!(n+1)(n+2)} - \frac{x^6}{2^6 \cdot 3!\,(n+1)\,(n+2)(n+3)} + \cdots\right].$$

$$J_n(x) \approx \frac{x^n}{2^n n!} \text{ for} x \ll 1.$$

[1]Watson, G. N., *A Treatise on the Theory of Bessel Functions*, Cambridge University Press (1995).

Fundamentals of Microwave Photonics, First Edition.
Vincent J. Urick Jr, Jason D. McKinney, and Keith J. Williams.

$$J_0(x) + 2\sum_{n=1}^{\infty} J_{2n}(x) = 1.$$

$$J_0^2(x) + 2\sum_{n=1}^{\infty} J_n^2(x) = 1.$$

AVI.2 RECURRENCE FORMULAE

$$J_{n+1}(x) = \frac{2n}{x} J_n(x) - J_{n-1}(x).$$

$$J_n'(x) = \frac{1}{2} J_{n-1}(x) - \frac{1}{2} J_{n+1}(x).$$

$$x \cdot J_n'(x) = x \cdot J_{n-1}(x) - n \cdot J_n(x) = n \cdot J_n(x) - x \cdot J_{n+1}(x).$$

AVI.3 GENERATING FUNCTION

$$\exp\left[\frac{x}{2}\left(t - \frac{1}{t}\right)\right] = \sum_{n=-\infty}^{\infty} t^n J_n(x) \quad \text{for } t \neq 0.$$

AVI.4 JACOBI'S SERIES

$$\cos(\phi \cos \Omega) = J_0(\phi) + 2\sum_{n=1}^{\infty} (-1)^n J_{2n}(\phi) \cos(2n\Omega).$$

$$\sin(\phi \cos \Omega) = 2\sum_{n=0}^{\infty} (-1)^n J_{2n+1}(\phi) \cos[(2n+1)\Omega].$$

AVI.5 ANGER'S SERIES

$$\cos(\phi \sin \Omega) = J_0(\phi) + 2\sum_{n=1}^{\infty} J_{2n}(\phi) \cos(2n\Omega).$$

$$\sin(\phi \sin \Omega) = 2\sum_{n=0}^{\infty} J_{2n+1}(\phi) \sin[(2n+1)\Omega].$$

AVI.6 JACOBI–ANGER EXPANSION

$$e^{i\phi\cos\Omega} = J_0(\phi) + 2\sum_{n=1}^{\infty}(i)^n J_n(\phi)\cos(n\Omega).$$

AVI.7 ROOTS

$J_0(a) = 0, J_1(b) = 0, J_2(c) = 0, J_3(d) = 0$

Root	a	b	c	d
1	2.4048	0	0	0
2	5.5201	3.8317	5.1356	6.3802
3	8.6537	7.0156	8.4172	9.7610
4	11.7915	10.1735	11.6198	13.0152

AVI.8 LOCAL MAXIMA AND MINIMA

$J_0(0) = 1$ $J_1(1.8412) = 0.5819.$
$J_0(3.8317) = -0.4028$ $J_1(5.3314) = -0.3461.$
$J_0(7.0156) = 0.3001$ $J_1(8.5363) = 0.2733.$
$J_0(10.1735) = -0.2497$ $J_1(11.7060) = -0.2333.$

$J_2(3.0542) = 0.4865$ $J_3(4.2012) = 0.4344.$
$J_2(6.7061) = -0.3135$ $J_3(8.0152) = -0.2912.$
$J_2(9.9695) = 0.2547$ $J_3(11.3459) = 0.2407.$
$J_2(13.1704) = -0.2209$ $J_3(14.5858) = -0.2110.$

INDEX

Fundamentals of Microwave Photonics, First Edition.
Vincent J. Urick Jr, Jason D. McKinney, and Keith J. Williams.

WILEY SERIES IN MICROWAVE AND OPTICAL ENGINEERING

KAI CHANG, Editor
Texas A&M University

FIBER-OPTIC COMMUNICATION SYSTEMS, Fourth Edition • *Govind P. Agrawal*

ASYMMETRIC PASSIVE COMPONENTS IN MICROWAVE INTEGRATED CIRCUITS • *Hee-Ran Ahn*

COHERENT OPTICAL COMMUNICATIONS SYSTEMS • *Silvello Betti, Giancarlo De Marchis, and Eugenio Iannone*

PHASED ARRAY ANTENNAS: FLOQUET ANALYSIS, SYNTHESIS, BFNs, AND ACTIVE ARRAY SYSTEMS • *Arun K. Bhattacharyya*

HIGH-FREQUENCY ELECTROMAGNETIC TECHNIQUES: RECENT ADVANCES AND APPLICATIONS • *Asoke K. Bhattacharyya*

RADIO PROPAGATION AND ADAPTIVE ANTENNAS FOR WIRELESS COMMUNICATION LINK NETWORKS: TERRESTRIAL, ATMOSPHERIC, AND IONOSPHERIC, Second Edition • *Nathan Blaunstein and Christos G. Christodoulou*

COMPUTATIONAL METHODS FOR ELECTROMAGNETICS AND MICROWAVES • *Richard C. Booton, Jr.*

ELECTROMAGNETIC SHIELDING • *Salvatore Celozzi, Rodolfo Araneo, and Giampiero Lovat*

MICROWAVE RING CIRCUITS AND ANTENNAS • *Kai Chang*

MICROWAVE SOLID-STATE CIRCUITS AND APPLICATIONS • *Kai Chang*

RF AND MICROWAVE WIRELESS SYSTEMS • *Kai Chang*

RF AND MICROWAVE CIRCUIT AND COMPONENT DESIGN FOR WIRELESS SYSTEMS • *Kai Chang, Inder Bahl, and Vijay Nair*

MICROWAVE RING CIRCUITS AND RELATED STRUCTURES, Second Edition • *Kai Chang and Lung-Hwa Hsieh*

MULTIRESOLUTION TIME DOMAIN SCHEME FOR ELECTROMAGNETIC ENGINEERING • *Yinchao Chen, Qunsheng Cao, and Raj Mittra*

DIODE LASERS AND PHOTONIC INTEGRATED CIRCUITS, Second Edition • *Larry Coldren, Scott Corzine, and Milan Masanovic*

EM DETECTION OF CONCEALED TARGETS • *David J. Daniels*

RADIO FREQUENCY CIRCUIT DESIGN • *W. Alan Davis and Krishna Agarwal*

RADIO FREQUENCY CIRCUIT DESIGN, Second Edition • *W. Alan Davis*

FUNDAMENTALS OF OPTICAL FIBER SENSORS • *Zujie Fang, Ken K. Chin, Ronghui Qu, and Haiwen Cai*

MULTICONDUCTOR TRANSMISSION-LINE STRUCTURES: MODAL ANALYSIS TECHNIQUES • *J. A. Brandão Faria*

PHASED ARRAY-BASED SYSTEMS AND APPLICATIONS • *Nick Fourikis*

SOLAR CELLS AND THEIR APPLICATIONS, Second Edition • *Lewis M. Fraas and Larry D. Partain*

FUNDAMENTALS OF MICROWAVE TRANSMISSION LINES • *Jon C. Freeman*

OPTICAL SEMICONDUCTOR DEVICES • *Mitsuo Fukuda*

MICROSTRIP CIRCUITS • *Fred Gardiol*

HIGH-SPEED VLSI INTERCONNECTIONS, Second Edition • *Ashok K. Goel*

FUNDAMENTALS OF WAVELETS: THEORY, ALGORITHMS, AND APPLICATIONS, Second Edition • *Jaideva C. Goswami and Andrew K. Chan*

HIGH-FREQUENCY ANALOG INTEGRATED CIRCUIT DESIGN • *Ravender Goyal (ed.)*

RF AND MICROWAVE TRANSMITTER DESIGN • *Andrei Grebennikov*

ANALYSIS AND DESIGN OF INTEGRATED CIRCUIT ANTENNA MODULES • *K. C. Gupta and Peter S. Hall*

PHASED ARRAY ANTENNAS, Second Edition • *R. C. Hansen*

STRIPLINE CIRCULATORS • *Joseph Helszajn*

THE STRIPLINE CIRCULATOR: THEORY AND PRACTICE • *Joseph Helszajn*

LOCALIZED WAVES • *Hugo E. Hernández-Figueroa, Michel Zamboni-Rached, and Erasmo Recami (eds.)*

MICROSTRIP FILTERS FOR RF/MICROWAVE APPLICATIONS, Second Edition • *Jia-Sheng Hong*

MICROWAVE APPROACH TO HIGHLY IRREGULAR FIBER OPTICS • *Huang Hung-Chia*

NONLINEAR OPTICAL COMMUNICATION NETWORKS • *Eugenio Iannone, Francesco Matera, Antonio Mecozzi, and Marina Settembre*

FINITE ELEMENT SOFTWARE FOR MICROWAVE ENGINEERING • *Tatsuo Itoh, Giuseppe Pelosi, and Peter P. Silvester (eds.)*

INFRARED TECHNOLOGY: APPLICATIONS TO ELECTROOPTICS, PHOTONIC DEVICES, AND SENSORS • *A. R. Jha*

SUPERCONDUCTOR TECHNOLOGY: APPLICATIONS TO MICROWAVE, ELECTRO-OPTICS, ELECTRICAL MACHINES, AND PROPULSION SYSTEMS • *A. R. Jha*

TIME AND FREQUENCY DOMAIN SOLUTIONS OF EM PROBLEMS USING INTEGTRAL EQUATIONS AND A HYBRID METHODOLOGY • *B. H. Jung, T. K. Sarkar, S. W. Ting, Y. Zhang, Z. Mei, Z. Ji, M. Yuan, A. De, M. Salazar-Palma, and S. M. Rao*

OPTICAL COMPUTING: AN INTRODUCTION • *M. A. Karim and A. S. S. Awwal*

INTRODUCTION TO ELECTROMAGNETIC AND MICROWAVE ENGINEERING • *Paul R. Karmel, Gabriel D. Colef, and Raymond L. Camisa*

MILLIMETER WAVE OPTICAL DIELECTRIC INTEGRATED GUIDES AND CIRCUITS • *Shiban K. Koul*

ADVANCED INTEGRATED COMMUNICATION MICROSYSTEMS • *Joy Laskar, Sudipto Chakraborty, Manos Tentzeris, Franklin Bien, and Anh-Vu Pham*

MICROWAVE DEVICES, CIRCUITS AND THEIR INTERACTION • *Charles A. Lee and G. Conrad Dalman*

ADVANCES IN MICROSTRIP AND PRINTED ANTENNAS • *Kai-Fong Lee and Wei Chen (eds.)*

SPHEROIDAL WAVE FUNCTIONS IN ELECTROMAGNETIC THEORY • *Le-Wei Li, Xiao-Kang Kang, and Mook-Seng Leong*

MICROWAVE NONCONTACT MOTION SENSING AND ANALYSIS • *Changzhi Li and Jenshan Lin*

COMPACT MULTIFUNCTIONAL ANTENNAS FOR WIRELESS SYSTEMS • *Eng Hock Lim and Kwok Wa Leung*

ARITHMETIC AND LOGIC IN COMPUTER SYSTEMS • *Mi Lu*

OPTICAL FILTER DESIGN AND ANALYSIS: A SIGNAL PROCESSING APPROACH • *Christi K. Madsen and Jian H. Zhao*

THEORY AND PRACTICE OF INFRARED TECHNOLOGY FOR NONDESTRUCTIVE TESTING • *Xavier P. V. Maldague*

METAMATERIALS WITH NEGATIVE PARAMETERS: THEORY, DESIGN, AND MICROWAVE APPLICATIONS • Ricardo Marqués, Ferran Martín, and Mario Sorolla

OPTOELECTRONIC PACKAGING • *A. R. Mickelson, N. R. Basavanhally, and Y. C. Lee (eds.)*

OPTICAL CHARACTER RECOGNITION • *Shunji Mori, Hirobumi Nishida, and Hiromitsu Yamada*

ANTENNAS FOR RADAR AND COMMUNICATIONS: A POLARIMETRIC APPROACH • *Harold Mott*

INTEGRATED ACTIVE ANTENNAS AND SPATIAL POWER COMBINING • *Julio A. Navarro and Kai Chang*

ANALYSIS METHODS FOR RF, MICROWAVE, AND MILLIMETER-WAVE PLANAR TRANSMISSION LINE STRUCTURES • *Cam Nguyen*

LASER DIODES AND THEIR APPLICATIONS TO COMMUNICATIONS AND INFORMATION PROCESSING • *Takahiro Numai*

FREQUENCY CONTROL OF SEMICONDUCTOR LASERS • *Motoichi Ohtsu (ed.)*

INVERSE SYNTHETIC APERTURE RADAR IMAGING WITH MATLAB ALGORITHMS • *Caner Özdemir*

SILICA OPTICAL FIBER TECHNOLOGY FOR DEVICE AND COMPONENTS: DESIGN, FABRICATION, AND INTERNATIONAL STANDARDS • *Un-Chul Paek and Kyunghwan Oh*

WAVELETS IN ELECTROMAGNETICS AND DEVICE MODELING • *George W. Pan*

OPTICAL SWITCHING • Georgios Papadimitriou, Chrisoula Papazoglou, and Andreas S. Pomportsis

MICROWAVE IMAGING • *Matteo Pastorino*

ANALYSIS OF MULTICONDUCTOR TRANSMISSION LINES • *Clayton R. Paul*

INTRODUCTION TO ELECTROMAGNETIC COMPATIBILITY, Second Edition • *Clayton R. Paul*

ADAPTIVE OPTICS FOR VISION SCIENCE: PRINCIPLES, PRACTICES, DESIGN AND APPLICATIONS • *Jason Porter, Hope Queener, Julianna Lin, Karen Thorn, and Abdul Awwal (eds.)*

ELECTROMAGNETIC OPTIMIZATION BY GENETIC ALGORITHMS • *Yahya Rahmat-Samii and Eric Michielssen (eds.)*

INTRODUCTION TO HIGH-SPEED ELECTRONICS AND OPTOELECTRONICS • *Leonard M. Riaziat*

NEW FRONTIERS IN MEDICAL DEVICE TECHNOLOGY • *Arye Rosen and Harel Rosen (eds.)*

ELECTROMAGNETIC PROPAGATION IN MULTI-MODE RANDOM MEDIA • *Harrison E. Rowe*

ELECTROMAGNETIC PROPAGATION IN ONE-DIMENSIONAL RANDOM MEDIA • *Harrison E. Rowe*

HISTORY OF WIRELESS • *Tapan K. Sarkar, Robert J. Mailloux, Arthur A. Oliner, Magdalena Salazar-Palma, and Dipak L. Sengupta*

PHYSICS OF MULTIANTENNA SYSTEMS AND BROADBAND PROCESSING • *Tapan K. Sarkar, Magdalena Salazar-Palma, and Eric L. Mokole*

SMART ANTENNAS • *Tapan K. Sarkar, Michael C. Wicks, Magdalena Salazar-Palma, and Robert J. Bonneau*

NONLINEAR OPTICS • *E. G. Sauter*

APPLIED ELECTROMAGNETICS AND ELECTROMAGNETIC COMPATIBILITY • *Dipak L. Sengupta and Valdis V. Liepa*

COPLANAR WAVEGUIDE CIRCUITS, COMPONENTS, AND SYSTEMS • *Rainee N. Simons*

ELECTROMAGNETIC FIELDS IN UNCONVENTIONAL MATERIALS AND STRUCTURES • *Onkar N. Singh and Akhlesh Lakhtakia (eds.)*

ANALYSIS AND DESIGN OF AUTONOMOUS MICROWAVE CIRCUITS • *Almudena Suárez*

ELECTRON BEAMS AND MICROWAVE VACUUM ELECTRONICS • *Shulim E. Tsimring*

FUNDAMENTALS OF GLOBAL POSITIONING SYSTEM RECEIVERS: A SOFTWARE APPROACH, Second Edition • *James Bao-yen Tsui*

SUBSURFACE SENSING • *Ahmet S. Turk, A. Koksal Hocaoglu, and Alexey A. Vertiy (eds.)*

RF/MICROWAVE INTERACTION WITH BIOLOGICAL TISSUES • *André Vander Vorst, Arye Rosen, and Youji Kotsuka*

InP-BASED MATERIALS AND DEVICES: PHYSICS AND TECHNOLOGY • *Osamu Wada and Hideki Hasegawa (eds.)*

COMPACT AND BROADBAND MICROSTRIP ANTENNAS • *Kin-Lu Wong*

DESIGN OF NONPLANAR MICROSTRIP ANTENNAS AND TRANSMISSION LINES • *Kin-Lu Wong*

PLANAR ANTENNAS FOR WIRELESS COMMUNICATIONS • *Kin-Lu Wong*

FREQUENCY SELECTIVE SURFACE AND GRID ARRAY • *T. K. Wu (ed.)*

PHOTONIC SENSING: PRINCIPLES AND APPLICATIONS FOR SAFETY AND SECURITY MONITORING • *Gaozhi Xiao and Wojtek J. Bock*

ACTIVE AND QUASI-OPTICAL ARRAYS FOR SOLID-STATE POWER COMBINING • *Robert A. York and Zoya B. Popović (eds.)*

OPTICAL SIGNAL PROCESSING, COMPUTING AND NEURAL NETWORKS • *Francis T. S. Yu and Suganda Jutamulia*

ELECTROMAGNETIC SIMULATION TECHNIQUES BASED ON THE FDTD METHOD • *Wenhua Yu, Xiaoling Yang, Yongjun Liu, and Raj Mittra*

SiGe, GaAs, AND InP HETEROJUNCTION BIPOLAR TRANSISTORS • *Jiann Yuan*

PARALLEL SOLUTION OF INTEGRAL EQUATION-BASED EM PROBLEMS • *Yu Zhang and Tapan K. Sarkar*

ELECTRODYNAMICS OF SOLIDS AND MICROWAVE SUPERCONDUCTIVITY • *Shu-Ang Zhou*

MICROWAVE BANDPASS FILTERS FOR WIDEBAND COMMUNICATIONS • *Lei Zhu, Sheng Sun, and Rui Li*

FUNDAMENTALS OF MICROWAVE PHOTONICS • *Vincent Jude Urick Jr., Jason Dwight McKinney, and Keith Jake Williams*

Printed and bound by CPI Group (UK) Ltd, Croydon, CR0 4YY
17/09/2026
14964286-0007